Molecularly Imprinted Polymers for Analytical Chemistry Applications

Polymer Chemistry Series

Titles in the series:

1: Renewable Resources for Functional Polymers and Biomaterials
2: Molecular Design and Applications of Photofunctional Polymers and Materials
3: Functional Polymers for Nanomedicine
4: Fundamentals of Controlled/Living Radical Polymerization
5: Healable Polymer Systems
6: Thiol-X Chemistries in Polymer and Materials Science
7: Natural Rubber Materials: Volume 1: Blends and IPNs
8: Natural Rubber Materials: Volume 2: Composites and Nanocomposites
9: Conjugated Polymers: A Practical Guide to Synthesis
10: Polymeric Materials with Antimicrobial Activity: From Synthesis to Applications
11: Phosphorus-Based Polymers: From Synthesis to Applications
12: Poly(lactic acid) Science and Technology: Processing, Properties, Additives and Applications
13: Cationic Polymers in Regenerative Medicine
14: Electrospinning: Principles, Practice and Possibilities
15: Glycopolymer Code: Synthesis of Glycopolymers and their Applications
16: Hyperbranched Polymers: Macromolecules in-between Deterministic Linear Chains and Dendrimer Structures
17: Polymer Photovoltaics: Materials, Physics, and Device Engineering
18: Electrical Memory Materials and Devices
19: Nitroxide Mediated Polymerization: From Fundamentals to Applications in Materials Science
20: Polymers for Personal Care Products and Cosmetics
21: Semiconducting Polymers: Controlled Synthesis and Microstructure
22: Bio-inspired Polymers

23: Fluorinated Polymers: Volume 1: Synthesis, Properties, Processing and Simulation
24: Fluorinated Polymers: Volume 2: Applications
25: Miktoarm Star Polymers: From Basics of Branched Architecture to Synthesis, Self-assembly and Applications
26: Mechanochemistry in Materials
27: Macromolecules Incorporating Transition Metals: Tackling Global Challenges
28: Molecularly Imprinted Polymers for Analytical Chemistry Applications

How to obtain future titles on publication:
A standing order plan is available for this series. A standing order will bring delivery of each new volume immediately on publication.

For further information please contact:
Book Sales Department, Royal Society of Chemistry, Thomas Graham House, Science Park, Milton Road, Cambridge, CB4 0WF, UK
Telephone: +44 (0)1223 420066, Fax: +44 (0)1223 420247
Email: booksales@rsc.org
Visit our website at www.rsc.org/books

Molecularly Imprinted Polymers for Analytical Chemistry Applications

Edited by

Wlodzimierz Kutner
Institute of Physical Chemistry, Polish Academy of Sciences, Warsaw, Poland and *Cardinal Stefan Wyszynski University in Warsaw, Poland*
Email: wkutner@ichf.edu.pl

and

Piyush Sindhu Sharma
Institute of Physical Chemistry, Polish Academy of Sciences, Warsaw, Poland
Email: psharma@ichf.edu.pl

ROYAL SOCIETY OF CHEMISTRY

Polymer Chemistry Series No. 28

Print ISBN: 978-1-78262-647-3
PDF ISBN: 978-1-78801-047-4
EPUB ISBN: 978-1-78801-427-4
ISSN: 2044-0790

A catalogue record for this book is available from the British Library

The Royal Society of Chemistry is a charity, registered in England and Wales, Number 207890, and a company incorporated in England by Royal Charter (Registered No. RC000524), registered office: Burlington House, Piccadilly, London W1J 0BA, UK, Telephone: +44 (0) 207 4378 6556.

For further information see our web site at www.rsc.org

Printed in the United Kingdom by CPI Group (UK) Ltd, Croydon, CR0 4YY, UK

Preface

The aim of the present book is to bring into focus an attractive research area and, particularly, recent advances of molecular imprinting in analytical chemistry applications. The use of molecularly imprinted polymers (MIPs) in supramolecular chemistry allows devising important materials for selective molecular recognition. For several materials, this selectivity is comparable to that of their biological counterparts. The present book provides several such examples of molecular imprinting covered by fourteen chapters, each devoted to different MIP analytical applications. Accordingly, Chapter 1 describes the most common synthetic approaches to manufacturing nanoMIPs and the use of these nanoMIPs in clinical diagnostics. Chapter 2 delivers comprehensive methods of syntheses of MIPs including designing of functional monomers, which provide different types of interactions with templates. Moreover, this chapter describes different strategies for devising MIPs for health monitoring. Chapter 3 reviews articles involving development of MIPs recognizing nucleosides, and their analogues. The second part of this chapter deals with the application of MIPs in separation and sensing of these compounds. Chapter 4 draws the reader's attention to integration of MIPs with nanomaterials, such as magnetic nanoparticles and conductive nanotubes. The derivatization methods used for this integration is described there in detail. This chapter critically discusses application of such materials. Chapter 5 focuses on commercially available MIP sorbents and patented MIPs, which are dedicated to determination of pharmaceuticals. Chapter 6 covers the advancement of techniques used in the last few decades for MIP micro-structuring and fabrication. Such nanostructured MIPs and nanocomposites allow improving selective target recognition. Theoretical and computational studies of molecular imprinting provided a valuable insight into the nature of the molecular-level imprinting events. Towards that, Chapter 7

Polymer Chemistry Series No. 28
Molecularly Imprinted Polymers for Analytical Chemistry Applications
Edited by Wlodzimierz Kutner and Piyush Sindhu Sharma

Published by the Royal Society of Chemistry, www.rsc.org

presents an overview of a thermodynamic treatment of factors governing the behavior of these MIPs. Moreover, this chapter summarizes the development and current status of computational strategies for studying different aspects of molecular imprinting. MIPs application in development of optical sensors is rapidly growing. Chapter 8 summarizes and critically evaluates main developments in this field over the past five years. Chapter 9 provides a comprehensive overview of the research accomplished towards devising MIP-based chemosensors for selective protein determination. Chapter 10 overviews what has recently been accomplished in devising water-compatible MIPs broadly covering the author's results in this field. Chapter 11 summarizes another exciting application of molecular imprinting, *i.e.*, in heterogeneous catalysis. This chapter enables comprehending of the state of the art of basics of artificial enzymes. Another important analytical application of MIPs that revealed a huge recent growth, is their use in sample preparation. Chapter 12 provides an overview of the MIP based sample preparation techniques. Other than acrylic based monomers, electroactive functional monomers have extensively been applied in molecular imprinting. Chapter 13 reports on typical electroactive functional monomers used in the electrosynthesis of MIPs for chemosensor applications. This chapter critically discusses efforts aiming at enhancement of the imprinting effect and sensing performance. Several literature examples are presented showing that, typically, a chemosensor based on individual MIPs reveal low selectivity and high cross-reactivity. Assembling multiple MIPs into a sensor array brings a convenient and effective solution to the problem. Chapter 14 discusses the key aspects of designing and fabricating MIP sensor arrays. This chapter provides representative literature examples highlighting applications of the MIP chemosensors arrays.

We would like to thank all authors that contributed and made publication of this book possible. They have accomplished a great job in arranging chapters in a reader friendly way. The chapters are summarized in a well readable form.

Interest in preparation and application of synthetic receptor based recognition units for chemical sensors is steadily growing. The book summarizes the latest developments and applications of molecular imprinting to both selective chemical sensing and separation. We strongly believe that it will guide scientists and graduate students interested in doing research in the field of molecular imprinting.

Piyush Sindhu Sharma
Wlodzimierz Kutner

Contents

Polymer Chemistry Series No. 28
Molecularly Imprinted Polymers for Analytical Chemistry Applications
Edited by Wlodzimierz Kutner and Piyush Sindhu Sharma

Published by the Royal Society of Chemistry, www.rsc.org

CHAPTER 1

Nano-sized Molecularly Imprinted Polymers as Artificial Antibodies

F. CANFAROTTA[a], A. CECCHINI*[b,c] AND S. PILETSKY[d]

[a]MIP Diagnostics Ltd, University Road, Leicester LE1 7RH, UK; [b]IMSaT, University of Dundee, Wilson House, 1 Wurzburg Loan, Dundee DD2 1FD, UK; [c]Bond Life Sciences Center, University of Missouri – Columbia, 1201 E. Rollins St., Columbia, MO 65211-7310, USA; [d]Department of Chemistry, University of Leicester, University Road, Leicester LE1 7RH, UK
*E-mail: cecchinia@missouri.edu

1.1 Molecularly Imprinted Polymers: Different Formats for Different Applications

Molecularly imprinted polymers (MIPs) emerged about 40 years ago, when Wulff and Sahan proposed the strategy of polymerisation performed in the presence of a target-template.[1] Since then, MIPs have been exploited for multiple applications thanks to their remarkable properties, such as high affinity and selectivity, and resistance to extremes of temperature, pressure and pH variations. In molecular imprinting, functional and cross-linking monomers are polymerised in an appropriate porogenic solvent in the presence of the compounds to be imprinted (called "templates"). After removal of the template, the polymer matrix retains recognition cavities that are complementary to this template in terms of size, shape and functionality.

Polymer Chemistry Series No. 28
Molecularly Imprinted Polymers for Analytical Chemistry Applications
Edited by Wlodzimierz Kutner and Piyush Sindhu Sharma

Published by the Royal Society of Chemistry, www.rsc.org

More precisely, the three-dimensional arrangement of cavity recognizing sites is driven by the structure of the template molecule. Therefore, after template removal the synthesized polymer is capable of binding the target, which can be the analyte itself. Compared with synthesis of monoclonal antibodies (mAbs), the synthesis of MIPs is simpler and cheaper, and can be performed without any preclinical development involving animals. In addition, MIPs show high stability and excellent mechanical properties. Moreover, they can be prepared for a wide variety of targets.[2,3] Template removal, however, is often difficult and incomplete, with the possible disadvantage of subsequent analyte leaching from the matrix, thus resulting in inaccurate performance in analytical applications.[4] Moreover, it is quite laborious to integrate them with signal transducers in sensors (unless electropolymerisation is involved, see Section 1.3.1.2, below) or to convert the template binding into an electric signal.[5] In general, MIPs can be manufactured in different ways (Table 1.1) and in several formats, for example as films or membranes, microparticles or nanoparticles. Compared to other formats, the nanoparticles present several advantages. In particular, this format allows the system to exhibit a much higher surface-to-volume ratio and larger total surface area per weight unit of polymer. The imprinted cavities are more easily accessible by the analytes, thus improving binding kinetics and template removal and, hence, enhancing their recognition capabilities.[6] Several authors have started developing nanoMIPs for diagnostic and therapeutic applications, for instance as drug delivery systems[7,8] and sensing elements in assays or sensors.[9,10] In virtue of their features, nanoMIPs represent an attractive option for a wide range of applications. One interesting characteristic of nanoMIPs is their property of remaining in solution, rendering them suitable for *in vitro* studies.[11] However, it is crucial to obtain particle batches with a very narrow size distribution and with high yield, especially for biomedical studies. Two other great advantages of MIPs, compared to natural ligands, are their relatively straightforward preparation and, particularly, their inexpensive fabrication. Indeed, the availability of cheap reagents for MIP syntheses has led researchers to explore novel polymerisation approaches for devising smaller and monodispersed MIPs, from the most intuitive techniques (*i.e.* precipitation polymerisation) to more sophisticated ones (*i.e.* the use of solid-phase polymerisation and automated reactors).

1.2 Advances in the Synthesis of NanoMIPs; Different Approaches to Preparation of MIPs as Nanoparticles

Molecular imprinting involves three main steps; *i.e.* (i) the formation of the monomer–template complexes, (ii) polymerisation, and (iii) removal of the template and collection of the MIPs. In general, MIPs can be fabricated by means of two main approaches: covalent and non-covalent

Table 1.1 List of synthetic procedures available for the synthesis of MIPs.[a]

Approach	Procedure	Advantages	Drawbacks	Ref.
Bulk	Performed using organic solvents. A block is obtained, and then crushed and sieved	Simple method	Wide particle size distribution and heterogeneity of active sites	37,38
Precipitation polymerisation	Polymer chains grow in solution, precipitating when their size makes them insoluble	Easy and fast with high yields. Low amount of reagents required	The low monomer concentration required might affect the interactions with the template	19,39
Emulsion polymerisation	Use of surfactants and high-shear homogenisation to emulsify the water phase with the organic one	Possible to obtain very small NPs (50 nm)	Surfactants might interfere with the imprinting process. Difficult removal of surfactants	19,40
Core-shell emulsion polymerisation	Deposition of an MIP layer on preformed nanoparticles (made of metals, silica, polymers)	Suitable for large-scale production. High yields	The presence of surfactants and the aqueous phase can decrease the imprinting effect	41,42
Core-shell grafting	Chemical linkage of MIP to preformed nanoparticles modified with double bonds or iniferter	Excellent control over shell thickness. Sequential shell polymerisation	Imprinted shell might be too thin for imprinting of bulky templates like proteins	43,44
Living radical polymerisation	Use of nitroxide species, metal-containing or dithiocarbonyl initiators. Polymer chains grow at similar rates	Excellent control over particle size and PD. Useful for thermolabile templates	Low yield. Removal of catalyst needed (in NMP and ATRP). Not suitable for photolabile templates	45,46
Solid-phase polymerisation	The template is immobilised on the surface of a solid support (typically micro-sized glass beads). High affinity nanoMIPs are collected by a temperature-based affinity separation step	High affinity and selectivity (nano- or picomolar). High purity with low template contamination. Fully automatable process	The template must have functional groups for immobilisation. Typically, one binding site per particle (low binding capacity)	9,32,33, 35

[a]PD: polydispersity. NMP: nitroxide-mediated polymerisation. ATRP: atom-transfer radical polymerisation. NP: nanoparticle.

imprinting. In the former, developed by Wulff,[12] reversible chemical bonds are created between the monomer and the template during the polymerisation, and the same bonds are then re-formed in the analyte binding step. The advantage is that only the monomer's functional groups interact with the template and cavities with more homogeneous recognizing sites are generated. However, not many compounds are suitable for this approach to be used and, therefore, they need preliminary derivatization with the monomer. Furthermore, the template removal is quite difficult and the analyte binding step is slower compared to that of other approaches.[5] In the non-covalent approach pioneered by Mosbach and co-workers,[13] hydrogen bonding as well as electrostatic and hydrophobic interactions are involved in the formation of the monomer–template pre-polymerisation complexes, as well as in the following analyte recognition. Since weak interactions are involved, an excess of monomer is usually employed to stabilise the monomer–template complex. This method is easier and more versatile than the covalent approach, although issues related to heterogeneity of the binding sites within cavities generated might arise.[14] Considering the advantages of the aforementioned two approaches, some authors have combined them, thus using a template covalently linked to the monomer and the following analyte binding step designed in a non-covalent way, thus introducing the concept of semi-covalent imprinting.[14] In this section, we will explore the main polymerisation modalities optimised so far for the synthesis of MIP nanoparticles (nanoMIPs), by focusing on (i) precipitation polymerisation, (ii) mini-/micropolymerisation and core–shell polymerisation (Scheme 1.1), (iii) atom transfer radical polymerisation (ATRP) and reversible addition–fragmentation chain

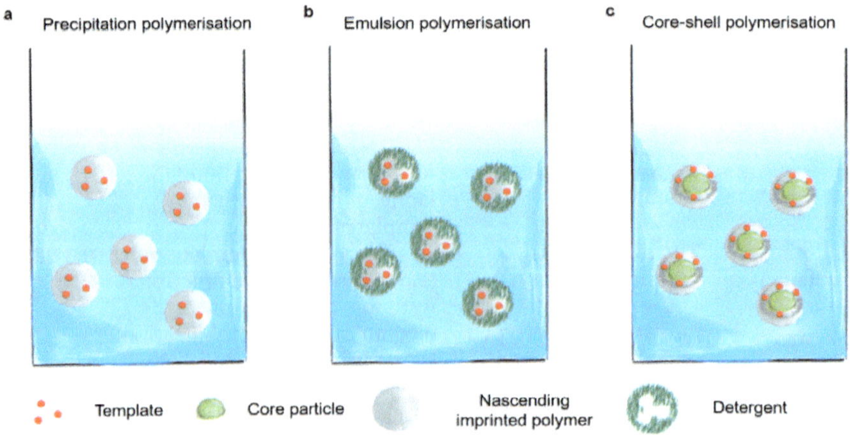

Scheme 1.1 Principal strategies to synthesise molecularly imprinted polymer particles: (a) precipitation, (b) emulsion, (c) core–shell. Adapted with permission from R. Schirhagl, *Anal. Chem.*, 2014, **85**, 250–261. Copyright 2016, American Chemical Society.

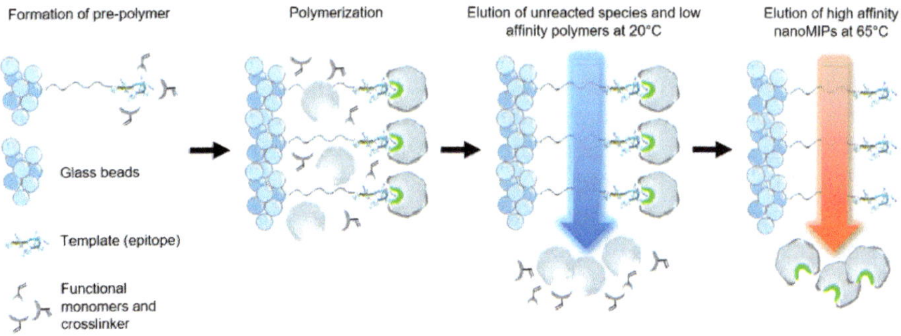

Scheme 1.2 The solid-phase synthesis strategy. The template molecule is covalently immobilised onto the surface of a bead, acting as the solid phase. After the polymerisation, the low-affinity MIPs and the unreacted species are easily removed through several washing steps at room temperature. Then, the high-affinity MIPs are eluted at 65 °C. Adapted with permission from Macmillan Publishers Ltd: *Nature Protocols*, (ref. 33), copyright 2016.

transfer polymerisation (RAFT), as well as (iv) solid-phase polymerisation without and with the use of automated reactors, with particular emphasis on pros and cons of each approach (Scheme 1.2).

1.2.1 Precipitation Polymerisation

Among all the techniques used to synthesise nanoMIPs, certainly precipitation polymerisation is the fastest and most straightforward strategy for obtaining monodispersed nanoparticles with high yield. Typically, this approach is based on a free-radical polymerisation occurring in highly diluted mixtures of monomers. The "growing" pre-polymer keeps enriching it with monomers, until it precipitates when its size renders the polymer insoluble. In this manner, spherical and uniform nanoparticles are rapidly produced with high yield and low consumption of reagents. However, the main drawback of this method lies in its strength point. In fact, the diluted solution of monomers may decrease the interaction between the active monomers and the template molecule. This decrease leads to a less efficient formation of the monomer-template pre-polymerisation complex and, therefore, it decreases the selectivity of the resulting nanoMIP. Moreover, the increase of the concentration of monomers leads to larger and less uniform nanoMIPs.[15] In the early 2010s, polymerisation in concentrated monomer solutions was performed.[16,17] The idea was to stop the polymerisation before the gelation point by abruptly diluting the solution of pre-polymers; subsequently, a poor solvent for the polymer was added for nucleation of the nanoMIPs.[6,7,18] Usually, precipitation polymerisation is performed using organic solvents. Nevertheless, it is also possible to precipitate nanoMIPs from aqueous solutions if a surfactant is added (at a low concentration) to the monomer mixture.

Obviously, after synthesis it is necessary to remove not only the template but also the surfactant. As discussed above, several parameters play a crucial role in this polymerisation, thus affecting quality of the resulting nanoMIPs. For instance, the ratio of different reagents, the type and concentration of the reagents (particularly the template) and the solvent can affect the resulting nanoMIP and reaction yield.[19] In conclusion, precipitation polymerisation is a cost-effective, time saving, and quick procedure for fabrication of nano- and microsized MIPs. This polymerisation allows the synthesis of uniform spherical nanoMIPs with high yield and the possibility to desirably tune their parameters.

1.2.2 Mini- and Micro-emulsion Polymerisation

As mentioned in Section 1.2.1 above, the polymerisation mixture can be optimised and enriched with different chemical species, *e.g.* surfactants. Notably, emulsion polymerisation involves organic solvent solutions of template-conjugated functional monomers and organic cross-linking monomers. The solution is then emulsified in an aqueous solution of surfactants and stabilisers (*e.g.* sodium dodecyl sulphate, SDS), and then the solution is usually stirred and sonicated. Generally, emulsion polymerisation can be classified as mini- and micro-emulsion polymerisation. The mini-emulsion approach exploits co-surfactant or surfactant monomers and, eventually, a stabiliser is added to produce nanoparticles with homogeneous size.[20] The peculiarity of mini-emulsion polymerisation lies in the synthesis of semi-covalently imprinted nanoparticles, *i.e.* the formation of the pre-polymerisation complex of monomers with the template engages covalent bonds while the target binding depends upon non-covalent interactions. NanoMIPs fabricated using this procedure show high affinity and selectivity for the target. On the other hand, however, micro-emulsion polymerisation generally occurs under more complicated conditions, namely, in the presence of water, oil, and one or more surfactants.[21] Moreover, spherical nanoMIPs are obtained using different procedures. A clever approach involves the addition of monomers or peptides conjugated to fatty acid chains to increase the confinement of the template to the surface of the growing nanoMIPs.[22] The main drawback of this technique is related to the presence of SDS and other chemicals. In fact, the presence of surfactant(s) and stabilisers in solution might affect monomer–target recognition and, therefore, it requires several washing steps, which may compromise the yield of the reaction as well as the efficacy and homogeneity of the nanoMIPs. Moreover, the application of these MIPs is limited to *in vitro* studies because it is impossible to completely remove those chemicals, a crucial step for any *in vivo* application. Therefore, nanoMIPs produced *via* emulsion polymerisation are not recommended for biological applications.[16] Nevertheless, mini- and microemulsions allow easy and rapid synthesizing of uniform nanoMIPs exhibiting remarkable recognition properties. Moreover, emulsion polymerisation, together with grafting polymerisation, can be exploited for assembling core–shell nanoMIPs. The most common "core materials" are polymeric structures, silica, or magnetic nanoparticles (*i.e.* Fe_3O_4 nanoparticles).[23] As discussed above, multiple advantages of emulsion polymerisation render this synthetic approach particularly suitable for devising core–shell nanoparticles. There are a few interesting strategies, which have been widely employed during the past years. In particular, the fabrication of core–shell nanoparticles is composed of two steps, *i.e.* (i) formation of the core and (ii) creation of the MIP shell by emulsion polymerisation. In all cases, from the use of latex cores to magnetic ones, the diameter of core–shell nanoMIPs was lower than 100 nm, thus opening up the gates to several different applications.[24,25] Generally, semi-covalent imprinting is the favourite approach, but also non-covalent imprinting has been exploited and was successful in terms of affinity and selectivity of the composite nanoparticles.[26] Despite the drawbacks listed above, the emulsion polymerisation still remains the most suitable approach for devising core–shell nanoMIPs, thanks to its straightforwardness and high yield.

1.2.3 Atom Transfer Radical Polymerisation (ATRP) and Reversible Addition-fragmentation Chain Transfer Polymerisation (RAFT)

Over the past decades, heat- and UV light-initiated polymerisations have been exploited for the synthesis of MIPs. Nevertheless, the desire to continuously optimise the polymerisation protocols has led to the development of new approaches, such as microwave-assisted synthesis initiation and controlled living radical polymerisation procedures (*i.e.* ATRP and RAFT).[27,28] In particular, these novel polymerisation strategies have been largely employed for either the fabrication of MIP films or core–shell nanoMIPs. Indeed, the big advantage of living radical polymerisation, compared to conventional free-radical polymerisation, lies in the control of the thickness of the MIP film, which is a crucial requirement for devising composite nanoparticles.[10,11] Moreover, the ATRP method exploits metal ion catalysts (usually copper ions) to coordinate the interactions operative in the formation of the monomer–template conjugate and the overall polymerisation reaction.[29] The ATRP mechanism is based on the equilibrium between radical and inactive species. Ions of metals, such as copper, iron, and molybdenum, mediate the entire catalytic process. When these metal ions are in their low oxidation state, they react intermittently with the inactive species. Therefore, "dormant" species act as both activators promoting the formation of the growing polymer and as deactivators, stopping the activity of the radicals, hence re-stabilising the equilibrium between the species in solution.[30] The equilibrium of the reaction and the metal ions employed are the strength point of this method. Moreover, they allow for control of the polymerisation itself. However, one drawback of ATRP is related to the presence of these metal ions. The metal ion catalyst might affect monomer–template recognition and it must be removed after the end of the polymerisation. Therefore, several and sophisticated purification steps are necessary (*e.g.* ion-exchange resins must be employed). With further optimisation of the polymerisation parameters, the amount of copper can be decreased, which is required for biomedical applications.

The RAFT method is a good candidate for controlled free-radical polymerisation, considering its versatility and simplicity. As with ATRP, the possibility to control the thickness of the polymer renders RAFT a suitable option for the fabrication of core–shell nanoMIPs, *via* the use of iniferter molecules or, more specially, RAFT agents (*e.g.* thiocarbonylthio compounds). With this method, uniformly distributed polymers with complex architectures can be synthesised (from linear block co-polymers to dendrimer-like polymers). While in the case of ATRP a crucial role is played by the metal ion catalyst, in RAFT polymerisation the RAFT agent is the principal effector. Once the reaction is initiated and the growing polymer is extending, this active species reacts with the RAFT agent to form the "RAFT adduct radical". The RAFT adduct radical is the most reactive species and, therefore, it is capable of triggering the formation of other growing polymers. In the case of core–shell nanoMIPs, the RAFT agent is immobilised onto the surface of the core particles (*e.g.* polystyrene, silica, Fe_3O_4).[31] Both grafting and precipitation polymerisation based on RAFT results in formation of uniform MIP shells with high affinity, selectivity, and with the possibility to tune the thickness of the grafted polymer. Although ATRP and RAFT are relatively new methods, they may be interesting starting points for further optimised polymerisation strategies thanks to the capability to control the thickness and architecture of the growing and final polymer.

1.2.4 Solid-phase Polymerisation

An option for conventional polymerisations in solution is solid-phase polymerisation. While other methods exploit free templates in solution, which requires several washing steps to remove the unreacted species (*e.g.* dialysis), solid-phase polymerisation overcomes this disadvantage by covalently immobilising the template molecule onto micrometre-size beads (Scheme 1.2). Typically, these beads are made of glass and, therefore, they need to be silanised with chemicals bearing suitable functional groups for further linking with the template. The template can be immobilised onto the glass beads using different strategies, depending on the characteristics of the template molecule itself (*i.e.* functional groups). For instance, generally, –SH group bearing templates are immobilised onto succinimidyl iodoacetate pre-functionalised beads, while $-NH_2$ and –COOH group bearing molecules can be immobilised *via* the EDC/NHS reaction. Once the solid phase is opportunely derivatised with the template of interest, the polymerisation proceeds as per other polymerisations in solution. Presently used imprinting techniques suffer from heterogeneous "polyclonal" distribution of binding sites, poor performance in water, low capacity, and slow mass transfer. A particular difficulty arises in the case of water-soluble targets.[4]

The use of a solid-phase polymerisation approach overcomes these difficulties, because the immobilised template enables an oriented immobilisation, thus decreasing the "polyclonality" of the imprinted sites. Furthermore, due to the effect of the solid phase, the binding sites are formed on the surface of the nanoparticles. This is in contrast to typical protocols with the template present in solution where most of the imprinted sites are formed within the polymer, thus leading to lower accessibility for the target molecule. The great advantage of solid-phase polymerisation lies in its independence from dialysis steps. Indeed, low affinity nanoMIPs and other components of the polymerisation mixture are first washed away at reaction temperature, whereas only high-affinity nanoMIPs are retained. The latter are subsequently eluted by increasing the temperature of the elution solvent. NanoMIPs produced by this solid-phase polymerisation approach have shown high affinity and selectivity for the target molecule.[32,33] The increasing number of applications, covering *in vitro* assays and *in vivo* studies as well as the interest in producing nanoMIPs in a large scale has led MIP experts to design automated systems. Thanks to their antibody mimicking features, nanoMIPs have attracted the attention of big companies and, therefore, the development of operator-independent procedures is now necessary. The great advantage of a polymerisation within an automated reactor lies in the total removal of human error and lifting operator fatigue, together with an increased yield of the reaction (and the possibility to scale up the protocol) and improvement of the batch-to-batch repeatability. The first automated reactor for nanoMIPs was proposed by Piletsky and co-workers.[34] They employed an automated reactor characterised by a column with a filter on the bottom with porosity of tens of micrometres. Template-functionalised micrometre-size glass beads were packed in the column placed inside the reactor. Both the polymerisation mixture and initiator were injected automatically into the column. After the polymerisation, several washing steps were performed to remove the non-reacted species and low affinity nanoMIPs, and then the most affine nanoparticles were eluted, all in an automated manner by using the reactor software.[35] Moreover, the use of iniferter in this automated reactor allowed for functionalisation of the nanoMIPs in a controlled fashion.[36] In this way, it is possible not only to obtain nanoMIPs with homogeneous distribution of binding site affinities (typically one recognition cavity per nanoparticle), but also functionalised nanoMIPs (*e.g.* with fluorophores or polyethylene glycol, PEG, layer to decrease the agglomeration tendency of the nanoparticles). In conclusion, nanoMIPs produced in an automated fashion by exploiting reactor polymerisation show high recognition capability, which, together with the high yield of the reaction, is making the use of this method increasingly common, especially for large scale applications.

1.3 NanoMIPs as Plastic Antibodies for Bioanalytical Applications

The use of Abs and enzymes in diagnostic assays is widely accepted and, to date, represents the gold standard in terms of sensitivity and affinity. Abs are routinely used in many diagnostic assays but, unfortunately, they suffer from short shelf-life, high costs of manufacturing and relatively low stability, especially in organic solvents and at extreme temperature and pH values. Furthermore, it is not easy to produce Abs against immunosuppressants or toxins because of their action on the immune response.[3] Abs also suffer from complex manufacturing procedures, lack of oral bioavailability, low cell membrane permeability and increased patient morbidity.[47] Indeed, even humanised types can elicit immunogenic reactions. Moreover, generating Abs against small molecules is not straightforward because chemical coupling to haptens is first required.[48] Finally, it is often difficult to immobilise Abs on the supports used in diagnostic assays.[49] In light of their attractive characteristics, nanoMIPs represent a viable alternative to Abs especially with regards to their stability and manufacturing costs. The fact that no animals are required for manufacturing these plastic antibodies is a step-change, which would enable companies to develop assay/sensors based on nanoMIPs at a fraction of the usual cost of Ab-based platforms. Furthermore, as discussed above, the possibility to generate nanoMIPs by solid-phase imprinting allows, for the first time, the production of surface-imprinted nanoparticles with antibody-like features (*e.g.* water dispersibility). Moreover, the automatic synthesis of nanoMIPs enables any party to develop their own plastic antibody, with minimal manual intervention and, therefore, training needed. At the same time, the low template contamination achievable with the solid-phase synthesis results in no need of dialysis, thus simplifying the overall MIP production. In contrast to their natural counterparts, nanoMIPs do not require any special storage conditions, being stable at room temperature for several months. This high stability makes them attractive for sensors or assays deployed in remote geographical areas, where cold chain supply might not be available. Interestingly, very recent studies demonstrated that nanoMIPs are biocompatible in several cell lines, not evoking any immunogenic response in macrophages.[50] In particular, cytokine release after exposure to nanoMIPs of different concentrations was monitored over 72 hours. The assay performed aims to flag the potential of a compound for generating an inflammatory response *in vivo*. Although the assay carried out is not predictive of *in vivo* inflammation, it can identify compounds that might lead to a potentially severe pro-inflammatory response (*i.e.* release of cytokines). This response might lead to the cytokine release syndrome (CRS) that causes serious systemic symptoms including fever, hypotension, and organ failure. Thanks to these properties, undoubtedly nanoMIPs hold great potential in diagnostics and in theranostics in particular. In the next paragraphs,

Table 1.2 Comparison of antibodies, nanoMIPs, and aptamers.

Characteristics	Antibodies	MIPs	Aptamers
Affinity	pico- to nanomolar	pico- to nanomolar	pico- to nanomolar
Size/nm	10–20	10–300 (hydrodynamic)	2–8
Immunogenicity	High	Unknown	Low
Thermal stability	Low	Very high	High
Organic solvent stability	Low	Very high	Low
pH/enzyme stability	Low	Very high	Low
Ease of functionalisation	Low	Very high	Very high
Availability of 'monomers'	Limited to the number of amino acids (21)	>4000 in the literature	Limited to the number of bases (4)
Manufacture	Mammalian cell systems	Chemical synthesis	Chemical synthesis
Cost for development of a new entity	$15000–25000	$4000–13000 (depending on template costs)	$6000–10000
Average lead time (development and production/validation)	>6–8 months	2–4 weeks (usually one day for production)	2–4 months (few days for production)
Batch-to-batch variability	Very high	Low (with automated reactor)	Low
Production costs	Very high	Low	Medium
Range of targets	Medium	Wide	Wide

we will review some of the most recent and innovative applications of plastic antibodies in sensors, assays, and for imaging purposes (Table 1.2).

1.3.1 NanoMIPs as Sensor Components

By definition, a sensor is a device capable of detecting events or changes in the surrounding environment, and able to process the information into a corresponding output (typically electrical or optical signals). Both chemical sensors and biosensors are now considered well-established tools in several fields of research, from analytical chemistry to clinical diagnostics as well as forensics and environmental sciences. A reliable chemo- and biosensor must exhibit a design and geometry to allow the most optimised interaction of its recognition part with the transducer, which is required to possess high sensitivity for the target. Biosensors bear the disadvantages related to the use of biomolecules, such as nucleic acids and proteins (*e.g.* Abs), as recognition elements. Exploiting macromolecules is advantageous for many reasons, among these their availability. On the other hand, their limited stability in non-physiological environments leads to difficulties in their storage and high costs for the production of more stable structures. Based on these considerations, researchers have moved their attention to more stable recognition tools, *i.e.* MIPs.[51] At the beginning of their development and optimisation, MIPs were mainly utilised as stationary phases for the purification of substances (*i.e.* chromatography). However, given their outstanding selectivity and stability, MIPs have been acquiring increasing relevance also in the field of chemosensors and bioassays. Particularly, MIP-based chemosensors,[52] which reveal the following integrated features, are needed: (i) a high-sensitivity transducer to monitor and process the binding event and (ii) a high-affinity and high-selectivity MIP capable of maintaining its recognition properties when implemented with the transducer. The versatility of MIP-based systems is worth highlighting. The binding of the molecule of interest can produce chemical and/or physical changes in the analyte itself or in the MIP, such as variations in fluorescence intensity or ionisation degree. Depending on the signal triggered upon binding and the variations in the properties of the analyte or the MIP, several transduction strategies can be employed, *e.g.* amperometry, potentiometry, conductometry (Scheme 1.3) as well as surface plasmon resonance (SPR) spectroscopy and surface enhanced Raman spectroscopy (SERS) (particularly for the detection of

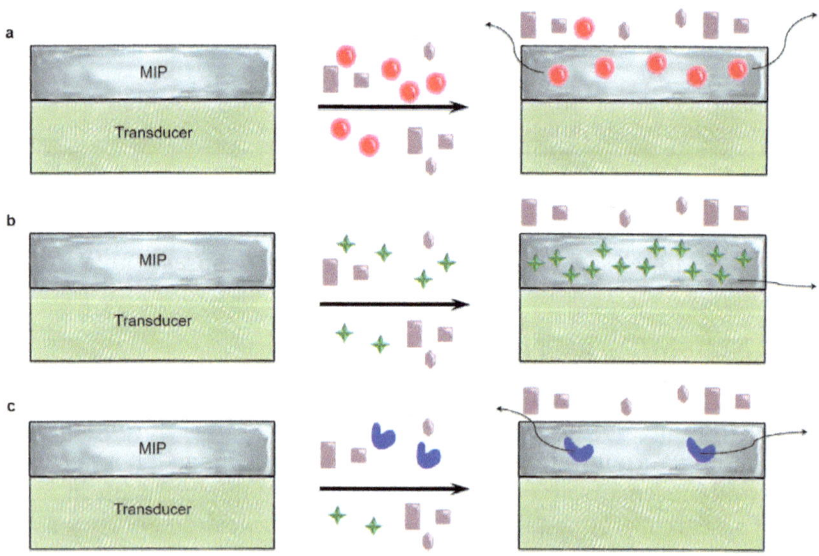

Scheme 1.3 Electrochemical sensors based on MIPs. (a) 'Affinity chemosensors' are those designed to produce a response upon the accumulation of electroactive target molecule; (b) 'receptor chemosensor' where interactions between the target and the MIP induce changes in polymer properties; (c) 'enzyme-mimicking chemosensor' detects changes in the environment induced by an MIP-catalysed reaction. Adapted from Piletsky and Turner, Electrochemical Sensors Based on Molecularly Imprinted Polymers, *Electroanalysis*,[53] John Wiley and Sons, © 2002 WILEY-VCH Verlag GmbH, Weinheim, Fed. Rep. of Germany.

proteins), and surface-acoustic wave (SAW) sensing. In this section, we will analyse and discuss possible applications of MIPs in chemosensors for the detection of biomolecules, focusing on electrochemical, optical, and other MIP-based chemosensors (*i.e.* calorimetric, SAW, and quartz crystal microbalance, QCM).[51,53–55]

1.3.1.1 *Electrical and Electrochemical Sensors*

The first electrochemical sensor (ECS) was developed in the 1950s to monitor the levels of oxygen. It was composed of an anode, a micro ammeter, a cathode and a source of electrolyte. ECSs are able to transform the electrochemical activity of the target molecule, or the changes in the electrochemical properties of the system, into electrical signals.[53] Each electrochemical sensor is rationally designed depending on its purpose. All ECSs are characterised by their size, shape, and components (*i.e.* recognition element for the detection of the molecule of interest and transducer). However, as previously stated, the most crucial step in devising a chemosensor is the integration of the recognition element within the transducer. Indeed, the ligand (*i.e.* the MIP) must be able to maintain its recognition properties once associated to the transducer. This can be achieved, for instance, by *in situ* polymerisation or by surface grafting by exploiting thermal initiation or UV irradiation.[55] Electropolymerisation is one of the most preferred strategies for fabrication of MIP chemosensors. Notably, electropolymerisation bears the big advantage of controlling the deposition of polymers onto the transducer in a precise and controlled manner, by measuring the time of applied potential or current. Nevertheless, the chemosensor is designed depending on the application and the properties of the systems itself as well as the physical and chemical characteristics of the target analyte. Planning the experiment and knowing in depth the principal features of the molecule of interest and the polymer lead to the most optimised design. If the analyte shows electrochemical activity or fluorescence, amperometry or ellipsometry can be exploited for the detection of the target. On the other hand, MIPs can be prepared by polymerisation starting from a mixture of monomers having different properties, such as electroconductive monomers. Additionally, solution for polymerisation could be enriched with metal ions or nanoparticles[19,56,57] able to improve sensitivity and enhance electron transfer or to coordinate the binding and, at the same time, inducing changes in the surrounding environment, *e.g.* pH changes. Moreover, MIPs have been coupled to carbon nanotubes or magnetic nanoparticles to further improve the sensitivity of the detection.[58,59] The most popular MIP-based chemosensors include (i) affinity chemosensors, (ii) receptor chemosensors, and (iii) catalytic chemosensors[53] (Scheme 1.3). Affinity chemosensors are similar to immune-sensors; the electroactive target accumulates at the surface of the transducer thanks to the presence of high-affinity MIPs. Receptor chemosensors are those based on changes related to the MIP; the binding with the analyte triggers variations in one or more features of the polymer, resulting in the generation of a signal subsequently processed by the transducer. Finally, catalytic or enzyme-like chemosensors are able to detect changes in the close surrounding environment due to the catalytic activity of the MIP itself.[53] Concerning electrochemical transducers, one can distinguish three main methods for the detection of target binding: (i) conductometry, (ii) amperometry, and (iii) potentiometry.[60] Conductometry is a technique that allows detecting conductivity alterations by exploiting the application of an electric field and, therefore, the migration of ions of opposite charges. However, it is impossible to discriminate between two or more ions in this case. Amperometry is based on the evaluation of the changes once a linear relationship between the concentration of the ionic species and their current at a fixed potential is established. For instance, Xue and co-workers developed an amperometric MIP-based electrochemical sensor.[61] To enhance the conductivity of the system, the MIP was doped with functional monomer-coated gold nanoparticles. The resulting chemosensor was successfully exploited for the detection of dopamine in human samples. The main disadvantage of this technique is the requirement of a porous membrane, which subsequently necessitates the removal of the target, eventually leading to false results.

On the other hand, potentiometry can solve the issue of "extraction". Indeed, potentiometry is based on generation of a potential difference with no need for the different charged species to diffuse through any membrane. In this case, neither size exclusion due to the porosity of the membrane occurs nor are extraction steps required. Anirudhan and Alexander[62] designed and fabricated a potentiometric sensor exploiting the recognition properties of MIPs and the physico-chemical features of multiwalled carbon nanotubes. This sensor showed several interesting features, starting from the possibility to tune the sensing of the target molecule at different pH, its reusability and very high detection capability of the target molecule (limit of detection, LOD, 10^{-10} M) and selectivity compared to other electrochemical sensors, either in water or in organic samples. It is also possible to measure a current at the time a potential sweep is applied. The generation of this current is due to oxidation or reduction of different species.[60] In the last 5 years, capacitance/impedance chemosensors have gained increasing importance by virtue of the possibility to investigate the binding of the analyte to the MIP by detecting variations in its thickness or electric permittivity. Notably, MIP-based chemosensors exploiting capacitance or impedance have been employed for the detection of biomarkers and organism-specific proteins.[55] Together with the development of new composite materials, these typologies of chemosensors are also leading to the preparation of even more innovative and challenging platforms for protein sensing. For instance, the impedance graphene-MIP-based chemosensor designed by Luo and colleagues[63] showed an excellent combination of molecular imprinting with electrochemical sensing. Notably, Luo *et al.* devised electrochemical sensors based on the detection of bovine haemoglobin protein *via* impedance measurement.

In conclusion, nowadays, electrochemical sensors are versatile, easily implemented and well-established tools for the detection of charged targets. The optimisation of MIP-functionalised electrochemical systems, using an imprinted polymer as the recognition element of the sensor, has allowed for the devising of high-sensitive sensors and also decreasing production costs and increasing robustness of the sensor itself.

1.3.1.2 Optical Sensors

As discussed above, electrical read-out is extremely useful when changes in MIP or analyte properties occur upon binding and can be transduced in electrical signals. However, there are several criteria to be satisfied.[64] Conversely, the development of optical sensors is constrained to only a few restrictions, mainly related to the polymer in the case of MIP-based sensors. Optical sensors convert electromagnetic radiation (typically in the UV–NIR range) into electric signals. With the rise of new polymerisation strategies, MIPs have been employed for devising optical sensors. Thanks to the progress in the covalent coupling of MIPs to optical transducers, innovative and robust MIP-based chemosensors have been exploited for the detection of target molecules with the LOD ranging from nM to fM.[19] Notably, there are several optical modalities that can be implemented to these systems: (i) fluorescence, fluorescence quenching and chemiluminescence, (ii) UV-vis and infrared spectroscopy; (iii) surface plasmon resonance (SPR) spectroscopy, and (iv) SERS[65] (Scheme 1.4). An MIP-based optical chemosensor can

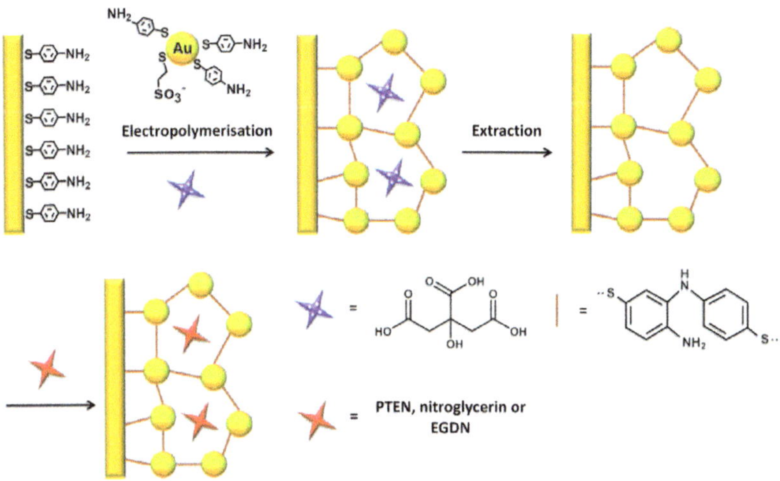

Scheme 1.4 The strategy exploited by Riskin *et al.*[78] for designing molecular imprinted gold composite chemosensors for the detection of small molecules *via* SPR spectroscopy and SERS. Reprinted with permission from Riskin *et al.*, *Anal. Chem.*, 2011, **83**, 3082–3088. Copyright 2016, American Chemical Society.

be designed in two main ways: (i) imprinting the polymer with fluorescent molecules, and (ii) rendering the MIP fluorescent by using fluorescent functional monomers or with the incorporation of a fluorescent agent.[19,64] In the case of the fluorescent target, the advantage lies in the straightforwardness of the strategy.[66,67] However, there is an intrinsic limit related to the availability of naturally fluorescent molecules.[68] Therefore, fluorescent MIPs appear to be more generic in nature and, therefore, suitable for this application.[69] Notably, since the 2000s several fluorescent monomers have been exploited to produce fluorescent MIPs for the detection of certain analytes in liquid samples by monitoring changes in the fluorescence intensity as well as quenching events upon binding.[69,70] Very recently, fluorescent nanoMIPs imprinted for tetracycline were used for quantification in real samples (bovine and pig serum).[71] An anthracene-derivative monomer was first synthesised, and then used as signalling functionality capable of being quenched upon addition of the drug, down to sub-micromolar concentrations. Additionally, MIP-based optical chemosensors have been recently exploited for the detection of protein biomarkers, with a detection limit of 10^{-8} M.[72] Quantum dots (QDs) appear to be the most suitable fluorescent agents for the development of fluorescent MIPs. Indeed, thanks to their small size together with their remarkable chemical and physical properties, QDs have been widely employed to develop molecularly imprinted optosensing materials (MIOMs), showing LODs down to 10^{-10} M.[73] Additionally, the incorporation of noble metals, particularly platinum, gold, and silver ions or their nanostructures may be a clever trick to enhance fluorescence variations. In a recent review, Wackerlig and Lieberzeit[19] explored the advantages and disadvantages of MIP-based sensors coupled to QDs and noble metal structures. Spectroscopy has been exploited for multiple applications, particularly in high-throughput assays as well as for studying the interaction with the analytes. Notably, (UV-vis spectroscopy)-coupled MIP-based chemosensors were used for high-throughput screening by using microtiter plates functionalised with MIPs.[74] On the other hand, infrared spectroscopy has been used to study in deep the interaction between the MIP recognition element and the target molecule upon binding. For instance, the thermal profile[75] or the changes in the frequencies of the vibration of NH bond stretching[76] before and after the polymerisation were employed.

Optical sensors based on either SPR spectroscopy or SERS appear to be even more appealing compared to other transducer modalities explored so far. They have gained increasing importance in the field of chemosensors, not only for bearing intrinsic elegance related to the measurement modality but also for the ability to detect the analyte at femtomolar concentrations.[77] SPR chemosensors typically couple a recognition element (*i.e.* MIPs) to either gold or silver; upon binding the target molecule, the chemosensor registers changes in the refractive index. Notably, variations in the stiffness after analyte binding leads to measurement of the amount of protein accumulated onto the chemosensor by detecting changes in the Bragg shift of the MIP.

The great advantage of these chemosensors lies in the very low biofouling observed after several measurements and, therefore, the reliability of the system.

The first SPR MIP-based chemosensors were devised in the 90s. During the past years, SPR chemosensors exploiting MIPs have undergone several improvements and now they have reached LODs in the order of tens of femtomolar for the detection of small molecules,[78] whereas macromolecules have been detected in the nM range.[79] In particular, nanoMIPs increased the sensitivity of the system compared to MIP films, likely due to the higher number of recognition cavities. While SPR MIP chemosensors have been optimised since the 90s, SERS systems are relatively novel.[80] The big advantage of SERS MIP-based chemosensors is their independence from variations of the size and stiffness of the polymer. In fact, these chemosensors allow observing changes in the vibrational spectrum upon target binding. For these reasons, the sensitivity of SERS chemosensors is expected to increase with further optimisation, especially in the case of nanoMIPs. Nevertheless, SERS based chemosensors still suffer from some issues, such as the long exposure time to the sample before starting to detect changes in Raman spectra as well as the robustness of the system itself, especially related to the integration of MIPs to the transducer,[65] which may be overcome by improving the design of the sensor. Indeed, Kamra and co-workers[81] designed a new approach for the devising an MIP optical chemosensor for label-free detection by exploiting SERS. The idea was to immobilise core–shell nanoMIPs onto the gold substrate, and subsequently couple the gold nanoparticles for SERS. They obtained and demonstrated the efficiency of this Au@MIP@AuNPs Raman system. Notably, the chemosensor showed reproducibility and very high recognition efficiency without the presence of a catalyst or an increase of temperature, ease of handling and most importantly robustness, that way opening the gates to further optimisation strategies.[81]

1.3.1.3 Other MIP Chemosensors

Other MIP-based systems worth mentioning are: (i) calorimetric, (ii) SAW, and (iii) QCM chemosensors. Calorimetric sensors are systems capable of detecting heat released upon binding of the analyte of interest,[51,82] whereas SAW chemosensors are acoustic systems. The latter are particularly interesting because they detect variations in the frequency of the acoustic waves upon analyte binding.[51] Very recently, Tretjakov and colleagues[83] deposited MIPs onto the surface of a SAW chip *via* electropolymerization. This innovative system demonstrated binding of the target protein with high affinity, *i.e.* three times higher than that of the corresponding NIP sensor, while also having the great advantage of being a label-free approach. However, the most popular among these systems are those of QCM chemosensors using sensing of the mass change. QCM-based chemosensors use a thin disc of a piezoelectric material, usually "sandwiched" between two metal film electrodes to complete the transducer, and integrated with the recognition element (*i.e.* an MIP film). The quartz crystal is a piezoelectric material capable to oscillate mechanically when a.c. voltage is applied to the electrodes. Therefore, these chemosensors are able to detect variations in the resonance frequency of the piezoelectric material. In particular, changes in this frequency are opposite to the mass change because of the analyte accumulation in the MIP film. There are three principal applications of piezoelectric MIP-based chemosensors: (i) diagnostics, (ii) environmental, and (iii) purification.[84] Since the technology for the production of MIPs was developed initially for purification of small-molecule compounds, it is not surprising to find in the literature a large number of these chemosensors for the detection of drugs and contaminants, some even in human serum.[85] It is worth mentioning the recent work published by Naklua and co-workers,[86] who devised a piezoelectric chemosensor coupling an MIP-QCM system to a dopaminergic receptor D1R for the detection of dopamine.[86] Additionally, MIP-based piezoelectric chemosensors have been employed for the detection of macromolecular compounds, achieving a sensitivity in the pM range.[87]

1.3.2 NanoMIPs in Assays

In the last decade, nanoMIPs have been successfully applied in diagnostic assays for analyte quantification. In one such example, nanoMIPs imprinted for vancomycin were used in the first ELISA-like assay, where the use of Abs was replaced by the said nanoparticles (Scheme 1.5).[88] In this assay, a horseradish peroxidase (HRP)-template conjugate was first prepared, and then employed in competitive binding experiments. That is, several solutions of free vancomycin (between 1 pM and 70 nM) were added to the wells together with the HRP conjugate. After incubation and washing, the substrate (3,3′,5,5′-tetramethylbenzidine, TMB) was added, leading to colour generation in solution due to reaction with HRP. The assay showed signal linearity from 1 pM to 70 nM and an LOD of 2.5 pM. Interestingly, this sensitivity is much higher than that of other ELISA tests reported in the literature, whose LOD was merely 0.1 nM.[89]

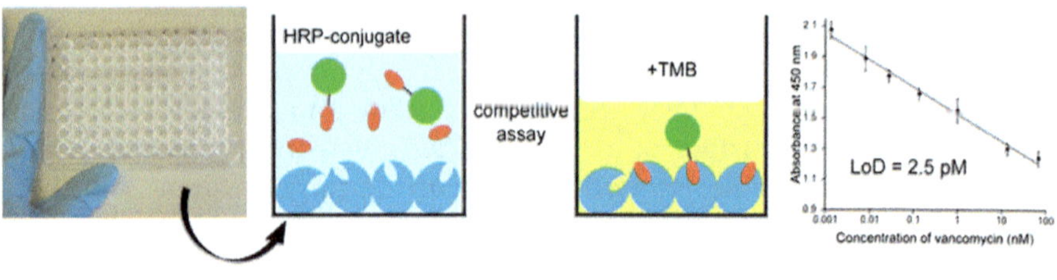

Scheme 1.5 Illustration of the ELISA protocol with vancomycin-imprinted nanoMIPs. Adapted with permission from Chianella *et al.*, *Anal. Chem.*, 2013, **85**, 8462–8468. Copyright 2016, American Chemical Society.

In a similar example, multifunctional nanoMIPs were recently used in a novel ELISA-like assay with no biomolecules involved.[90] Imprinted NPs, produced by solid-phase synthesis and embedding an iron oxide core with catalytic properties, act simultaneously as recognition and signalling elements. In light of the intrinsic peroxidase-like activity, iron oxide can be employed in different assays.[91] The iron oxide particles were first modified with carbon double bonds, and then used as "reactive seeds" for further polymerisation in the presence of a solid phase bearing vancomycin as the template. The assay was developed by initially conjugating vancomycin onto the well surface. Then, upon addition of free vancomycin and magnetic MIPs, competition occurred, in a similar manner as the aforementioned pseudo-ELISA. After addition of the TMB substrate, a blue colour was detected due to the catalytic activity of the core–shell MIPs, thus allowing detection of vancomycin in the nanomolar concentration range.

Very recently, a novel approach for synthesising nanoMIPs by using proteins as macro-functional monomers has been developed.[92] For proof-of-concept, HRP was chosen as a model and cross-linked using glutaraldehyde in the presence of a solid phase bearing immobilised templates (vancomycin and ampicillin). The cross-linking process in the presence of templates led to the formation of target-selective recognition cavities, without any significant adverse effects on the enzymatic activity. In contrast to complex protein engineering methods typically used to generate affinity proteins, this approach can be employed to prepare protein-based ligands in a short time. Since these affinity materials reveal both catalytic and molecular recognition properties, they are potentially useful in assays, for instance in an ELISA format where cross-linked HRP imprinted with a given template can replace traditional enzyme–antibody conjugates. However, this approach appears to work effectively only down to micromolar concentrations and, therefore, can be potentially a rapid alternative to raising Abs for targets that do not require high assay sensitivities.

1.3.3 NanoMIPs in Cells and *in vivo*

Very recently, several groups have started applying the molecular imprinting technology for the recognition of selected targets on the surface of cells. For instance, fluorescent nanoMIPs for molecular imaging of cells and tissues were devised for the first time by imprinting glucuronic acid, a monosaccharide present as the terminal unit on larger oligosaccharides.[93] The produced nanoMIPs were then employed to image the hyaluronan on human keratinocytes and on adult skin specimens (Figure 1.1a). Interestingly, molecules of other potentially interfering compounds, such as galactose, *N*-acetylglucosamine, *N*-acetylgalactosamine and glucose did not bind to the nanoMIPs, possibly because of the lack of the charged carboxyl group. Another target recently exploited by other groups for cellular targeting is sialic acid. Sellergren's group fabricated nanoMIPs targeting cell surface glycans, *via* sialic acid imprinting.[94] The overexpression of glycans terminating with sialic acid (SA) residues on the surface of cells

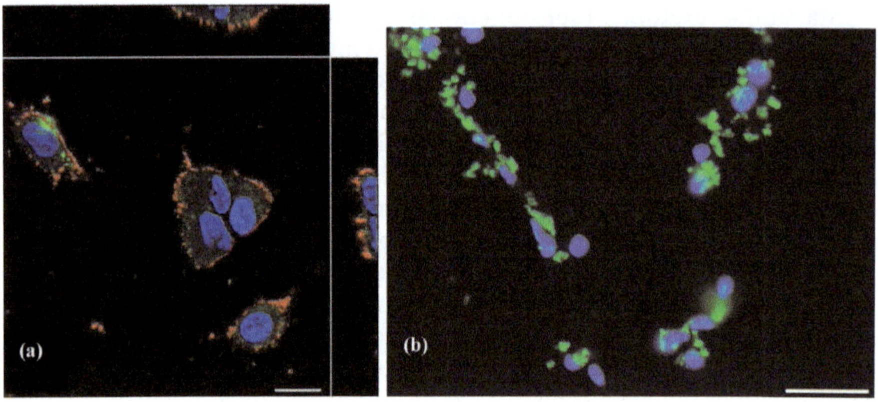

Figure 1.1 The confocal microscopy image of (a) fixated human keratinocytes (HaCaT) exposed to nanoMIPs. DAPI blue signal (cell nucleus), rhodamine red signal (MIPs), 3,3′-dioctadecyloxacarbocyanine perchlorate (DiO) green signal (cell membrane). Scale bar 20 μm. Reprinted with permission from Kunath *et al.*, Cell and Tissue Imaging with Molecularly Imprinted Polymers as Plastic Antibody Mimics, *Advanced Healthcare Materials*, John Wiley and Sons. © 2015 WILEY-VCH Verlag GmbH & Co. KGaA, Weinheim (b) DU145 cells incubated with SA-MIP (20 μg mL^{-1}). The scale bars correspond to 10 μm. Reprinted with permission from Shinde *et al.*, *J. Am. Chem. Soc.*, 2015, **137**, 13908–13912. Copyright 2016, American Chemical Society.

correlated with various diseases including cancer. Two-hundred-nanometre silica nanoparticles were used as cores for the subsequent grafting of the SA-imprinted layer. The authors speculate that a certain mixture of functional monomers is ideal for the nanoMIPs to bind strongly to the template. In particular, hydroxyl groups of SA would interact with a boronate-based monomer, while the carboxyl group would be targeted by a urea-based fluorescent monomer synthesised in-house. The produced fluorescent nanoMIPs were capable of selective staining different cell lines depending on the SA expression level (Figure 1.1b). Similarly, Yin *et al.* prepared SA-imprinted nanoMIPs for selective imaging of cancer cells.[95] In this case, the authors employed Raman-active nanotags (silver nanoparticles) as signaling cores. By surface imprinting based on silanes, an imprinted layer based on boronate, used as the functional monomer, was then created around the silver core. Raman spectroscopy provides significant advantages, such as high photostability and sensitivity, as well as multiplexing capacity. Healthy human hepatic cells and hepatocarcinoma cells (HepG-2) were used as a model to test the selectivity of the nanoMIPs. HepG-2 cells showed an evident SERS signals at 1435 cm^{-1}, much stronger than that for healthy cells, proving the selective binding of the nanoMIPs to SA on the cell surface.

An intriguing application of nanoMIPs involves their use as a therapeutic tool for selective protein/enzyme sequestration in cells. Once the nanoMIPs capture the protein/enzyme, the latter cannot carry out its physiological function, thus altering the cellular metabolism. Very recently, this principle was introduced by imprinting silica-coated iron oxide nanoparticles with DNAase I, a cytoplasmic enzyme involved in cell apoptosis.[96] The nanoparticles were also fluorescently tagged to visualize their distribution inside the cell. As done by several other groups, a silica layer was inserted between the magnetic core and the fluorescent reporter to minimise potential quenching issues. The nanoMIPs successfully inhibited the DNAase activity without affecting the short-term cell viability. These examples prove that nanoMIPs hold great potential as molecular recognition and imaging tools since, in contrast with Abs where multistep staining processes are required, multiple labelling can be easily achieved by using a panel of nanoMIPs, each one incorporating different dye for a given target. Furthermore, it is possible to link drugs or add magnetic functionalities to the nanoMIPs, thus allowing their use as drug delivery systems or in hyperthermia therapy. Therefore, it is expected that the number and complexity of these multifunctionalities in nanoMIPs will grow in the next years. Despite the aforementioned successful applications of nanoMIPs within sensors or assays, such nanoparticles have not been widely applied for *in vivo* diagnosis/therapy. The first *in vivo* application of imprinted nanoparticles was reported by Hoshino *et al.*,[97] who employed nanoMIPs imprinted with melittin (a peptide that is the principal component of bee venom) to remove the said molecule from the bloodstream of living mice (Scheme 1.6). The mice were intravenously injected with melittin and, afterwards, the nanoMIPs were administered *via* the tail vein. The MIPs successfully cleared melittin, improving the survival rate of mice over 24 h and decreasing the melittin toxic effects (*e.g.* weight loss and peritoneal phlogosis). This study demonstrates the potential of the nanoMIP for selective recognition of molecules *in vivo*.

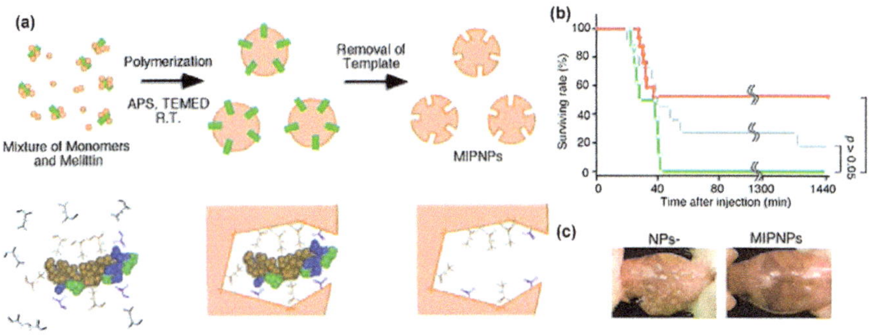

Scheme 1.6 (a) The melittin imprinting. (b) Neutralisation of the melittin toxicity by the nanoMIP. (c) Pathology of peritoneal inflammation in mice previously injected with melittin (4.0 mg kg^{-1}), with no treatment (left) or nanoMIP (30 mg kg^{-1}) injection (right). Adapted with permission from Hoshino *et al.*, *J. Am. Chem. Soc.*, 2010, **132**, 6644–6645. Copyright 2016, American Chemical Society.

The only other *in vivo* study has been recently performed by Wu and colleagues[98] who prepared amoxicillin loaded nanoMIPs imprinted with an epitope of Lpp20, a membrane lipoprotein expressed in *Helicobacter pylori*. *In vivo* imaging demonstrated a prolonged permanence of the nanoMIPs in stomachs of *H. pylori* infected mice. In particular, the nanoMIPs showed a more pronounced antibacterial effect than that of free amoxicillin, after intragastric administration.

1.4 Conclusion and Perspectives

Although currently Abs and enzymes are widely employed in diagnostics, they suffer from some drawbacks, such as high manufacturing costs, relatively poor stability (especially at extreme values of temperature and pH), short shelf life at room temperature, and long lead times. The use of nanoMIPs can overcome these problems. Compared with Abs, the synthesis of MIPs is simpler and more cost-effective, and it does not involve the use of animals. In addition, MIPs show high stability and excellent mechanical properties, and they can be prepared virtually for any target. No cold chain is required and, therefore, MIPs can be applied in chemosensors or assays to be deployed in remote geographical areas where storage at room temperature might be the only option available. NanoMIPs, sometimes called "plastic or synthetic antibodies", are nanostructured polymer particles capable of selective recognition of the target molecule. Thanks to their size, they represent a viable alternative to Abs, as demonstrated by their recent use in several diagnostic fields. The latest successful applications of nanoMIPs in cells and *in vivo* demonstrate their potential in diagnostics, and several other MIP-based composite nanosystems are expected to be fabricated in the coming years. However, improvements in the synthesis of nanoMIPs are required, in particular in relation with their size distribution and large scale production. In this regard, the herein described solid-phase approach – fully automatable – can become a viable option for large scale manufacturing of nanoMIPs to be applied for chemosensor devising and assay developing. Molecular imprinting of polymers by solid-phase synthesis represents the most generic, versatile and cost-effective approach to formation of synthetic molecular receptors/binders to date with antibody-like features. Plastic antibodies represent an entirely new compound class, which can be deployed to address both extracellular protein targets and, thanks to their capability to cross the cell membrane, potentially to currently intractable intracellular proteins. In fact, both Abs and aptamers are not cell-permeable. Furthermore, depending on the imprinted template, the polymers can be used repeatedly without loss of their "memory effect".[32] Moreover, MIPs are not proteins and, therefore, are not susceptible to proteolysis or bacterial degradation. Undoubtedly, nanoMIPs will play a crucial role in the development of novel robust assays/sensors, possibly in a home environment, where expensive and sensitive components like Abs would require extra care by the end user. Due to the robust nature of MIPs and their possibility to be reused, MIP-based systems able to work in continuous and/or under harsh conditions will soon be commercialised. Moreover, lateral flow tests and microfluidics based on nanoMIPs are likely to be deployed in the next years, potentially allowing multiplexed analysis due to the possibility to embed different types of coloured/fluorescent cores within a single nanoMIP.

List of Abbreviations

Ab	Antibody
ATRP	Atom transfer radical polymerisation
CRS	Cytokine release syndrome
D1R	Dopaminergic receptor 1
DAPI	4′,6-Diamidino-2-phenylindole
DiO	3,3′-Dioctadecyloxacarbocyanine perchlorate
DU145 cells	Human prostatic carcinoma cells
ECS	Electrochemical sensor
EDC	*N*-(3-Dimethylaminopropyl)-*N*′-ethylcarbodiimide hydrochloride
ELISA	Enzyme-linked immunosorbent assay
HaCaT	Human keratinocytes
HepG-2 cells	Human liver hepatocellular carcinoma cells
HRP	Horseradish peroxidase
LOD	Limit of detection
Lpp20 membrane	Lipoprotein expressed in *Helicobacter pylori*
mAb	Monoclonal antibody
MIOM	Molecularly imprinted optosensing material
MIP	Molecularly imprinted polymer
nanoMIP	MIP nanoparticle
NHS	*N*-Hydroxysuccinimide
NIR	Near-infrared
NMP	Nitroxide-mediated polymerisation
PD	Polydispersity
PEG	Polyethylene glycol
QCM	Quartz crystal microbalance
QD	Quantum dot
RAFT	Reversible addition–fragmentation chain transfer polymerisation
SA	Sialic acid
SAW	Surface-acoustic wave
SDS	Sodium dodecyl sulphate
SERS	Spectroscopy and surface enhanced Raman spectroscopy
SPR	Surface plasmon resonance
TBM	3,3′,5,5′-Tetramethylbenzidine
UV	Ultra violet

References

1. G. Wulff and A. Sarhan, *Angew. Chem., Int. Ed.*, 1972, **11**, 341.
2. E. V. Piletska, A. Guerreiro, M. J. Whitcombe and S. A. Piletsky, *Macromolecules*, 2009, **42**, 4921.
3. A. Poma, A. P. F. Turner and S. A. Piletsky, *Trends Biotechnol.*, 2010, **28**, 629.
4. R. A. Lorenzo, A. M. Carro, C. Alvarez-Lorenzo and A. Concheiro, *Int. J. Mol. Sci.*, 2011, **12**, 4327.
5. A. G. Mayes and M. J. Whitcombe, *Adv. Drug Delivery Rev.*, 2005, **57**, 1742.
6. S. Tokonami, H. Shiigi and T. Nagaoka, *Anal. Chim. Acta*, 2009, **641**, 7.
7. E. Asadi, S. Azodi-Deilami, M. Abdouss and S. Khaghani, *Appl. Biochem. Biotechnol.*, 2012, **167**, 2076.
8. D. Turkmen, N. Bereli, M. E. Corman, H. Shaikh, S. Akgol and A. Denizli, *Artif. Cells, Nanomed., Biotechnol.*, 2014, **42**, 316.
9. I. Basozabal, A. Guerreiro, A. Gomez-Caballero, M. Aranzazu Goicolea and R. J. Barrio, *Biosens. Bioelectron.*, 2014, **58**, 138.
10. W. J. Cheong, S. H. Yang and F. Ali, *J. Sep. Sci.*, 2013, **36**, 609.
11. Y. Ge and A. P. F. Turner, *Chem.–Eur. J.*, 2009, **15**, 8100.
12. G. Wulff, *Pure Appl. Chem.*, 1982, **54**, 2093.
13. B. Sellergren, M. Lepisto and K. Mosbach, *J. Am. Chem. Soc.*, 1988, **1**, 5853.
14. Y. Hu, J. Pan, K. Zhang, H. Lian and G. Li, *Trends Anal. Chem.*, 2013, **43**, 37.
15. P. Li, F. Rong and C. Yuan, *Polym. Int.*, 2003, **52**, 1799.
16. R. Schirhagl, P. A. Lieberzeit and F. L. Dickert, *Adv. Mater.*, 2010, **22**, 2078.
17. R. Schirhagl, U. Latif, D. Podlipna, H. Blumenstock and F. L. Dickert, *Anal. Chem.*, 2012, **84**, 3908–3913.
18. R. Schirhagl, E. W. Hall, I. Fuereder and R. N. Zare, *Analyst*, 2012, **137**, 1495.
19. J. Wackerlig and P. A. Lieberzeit, *Sens. Actuators, B*, 2015, **207**, 144.
20. D. Vaihinger, K. Landfester, I. Kräuter, H. Brunner and G. E. M. Tovar, *Macromol. Chem. Phys.*, 2002, **203**, 1965.
21. R. S. Slomkowski, J. V. Alemán, R. G. Gilbert, M. Hess, K. Horie, R. G. Jones, P. Kubisa, I. Meisel, W. Mormann, S. Penczek and R. F. T. Stepto, *Pure Appl. Chem.*, 2011, **83**, 2229.
22. Z. Zeng, Y. Hoshino, A. Rodriguez, H. Yoo and K. J. Shea, *ACS Nano*, 2010, **4**, 199.
23. C. J. Tan and Y. W. Tong, *Anal. Bioanal. Chem.*, 2007, **389**, 369.
24. N. Pérez, M. J. Whitcombe and E. N. Vulfson, *J. Appl. Polym. Sci.*, 2000, **77**, 1851.
25. N. Pérez, M. J. Whitcombe and E. N. Vulfson, *Macromolecules*, 2001, **34**, 830.
26. N. Pérez-Moral and A. G. Mayes, *Langmuir*, 2004, **20**, 3775.
27. M. M. Titirici and B. Sellergren, *Chem. Mater.*, 2006, **18**, 1773.
28. X. Wei, X. Li and S. M. Husson, *Biomacromolocules*, 2005, **6**, 1113.

29. K. Matyjaszewski, *Macromolecules*, 2012, **45**, 4015.
30. F. di Lena and K. Matyjaszewski, *Prog. Polym. Sci.*, 2010, **35**, 959.
31. S. Xu, J. Li and L. Chen, *J. Mater. Chem.*, 2011, **21**, 4346.
32. A. Poma, A. Guerreiro, M. J. Whitcombe, E. V. Piletska, A. P. F. Turner and S. A. Piletsky, *Adv. Funct. Mater.*, 2013, **23**, 2821.
33. F. Canfarotta, A. Poma, A. Guerreiro and S. A. Piletsky, *Nat. Protoc.*, 2016, **11**, 443.
34. S. A. Piletsky, E. V. Piletska, A. Guerreiro, M. J. Whitcombe and A. Poma, Pat US 20140228472 A1, 2011.
35. A. Poma, A. Guerreiro, S. Caygill, E. Moczko and S. A. Piletsky, *RSC Adv.*, 2014, **4**, 4203.
36. E. Moczko, A. Guerreiro, E. V. Piletska and S. A. Piletsky, *Langmuir*, 2013, **29**, 9891.
37. A. Pardo, L. Mespouille, P. Dubois, B. Blankert and P. Duez, *Chemistry*, 2014, **20**, 3500.
38. Y. Hu, S. Feng, F. Gao, E. C. Y. Li-Chan, E. Grant and X. Lu, *Food Chem.*, 2015, **176**, 123.
39. M. K. Bojdi, M. H. Mashhadizadeh, M. Behbahani, A. Farahani, S. S. H. Davarani and A. Bagheri, *Electrochim. Acta*, 2014, **136**, 59.
40. M. Pourfarzib, M. Shekarchi, H. Rastegar, B. Akbari-Adergani, A. Mehramizi and R. Dinarvand, *J. Chromatogr. B: Anal. Technol. Biomed. Life Sci.*, 2015, **974**, 1.
41. Z. Zhang, L. X. Chen, F. F. Yang and J. H. Li, *RSC Adv.*, 2014, **4**, 31507.
42. X. Gu, R. Xu, G. Yuan, H. Lu, G. Hui, B. Gu and H. Xie, *Anal. Chim. Acta*, 2010, **675**, 64.
43. L. Zhao, F. Zhao and B. Zeng, *Biosens. Bioelectron.*, 2014, **62**, 19.
44. L. Tan, W. Li, H. Li and Y. Tang, *J. Chromatogr. A*, 2014, **1336**, 59.
45. S. E. Shim, Y. Shin, J. W. Jun, K. Lee, H. Jung and S. Choe, *Macromolecules*, 2003, **36**, 7994.
46. V. D. Salian and M. E. Byrne, *Macromol. Mater. Eng.*, 2013, **298**, 379.
47. T. Funakoshi, M. Suzuki and K. Tamura, *Cancer Treat. Rev.*, 2014, **40**, 1221.
48. N. Lavignac, C. J. Allender and K. R. Brain, *Anal. Chim. Acta*, 2004, **510**, 139.
49. L. Ye and K. Haupt, *Anal. Bioanal. Chem.*, 2004, **378**, 1887.
50. F. Canfarotta, A. Waters, R. Sadler, P. McGill, A. Guerreiro, D. Papkovsky, K. Haupt and S. A. Piletsky, *Nano Res.*, 2016, **9**, 3463.
51. Y. Fuchs, O. Soppera and K. Haupt, *Anal. Chim. Acta*, 2012, **717**, 7.
52. S. Suriyanarayanan, P. J. Cywinski, A. J. Moro, G. J. Mohr and W. Kutner, *Top. Curr. Chem.*, 2012, **325**, 165.
53. S. A. Piletsky and A. P. F. Turner, *Electroanalysis*, 2002, **14**, 317.
54. K. Haupt and K. Mosbach, *Chem. Rev.*, 2000, **100**, 2495.
55. M. J. Whitcombe, I. Chianella, L. Larcombe, S. A. Piletsky, J. Noble and R. Porter, *Chem. Soc. Rev.*, 2011, **40**, 1547.
56. D. Yu, Y. Zeng, Y. Qi, T. Zhou and G. Shi, *Biosens. Bioelectron.*, 2012, **38**, 270.

57. T. Alizadeh and L. A. Fard, *Anal. Chim. Acta*, 2013, **769**, 100.
58. A. Afkhami, H. Ghaedi, T. Madrakian, M. Ahmadi and H. Mahmood-Kashani, *Biosens. Bioelectron.*, 2013, **44**, 34.
59. L. Zhu, Y. Cao and G. Cao, *Biosens. Bioelectron.*, 2014, **54**, 258.
60. M. C. Blanco-López, M. J. Lobo-Castañón, A. J. Miranda-Ordieres and P. Tuñón-Blanco, *Trends Anal. Chem.*, 2004, **23**, 36.
61. C. Xue, Q. Han, Y. Wang, J. Wu, T. Wen, R. Wang, J. Hong, Z. Zhou and H. Jiang, *Biosens. Bioelectron.*, 2013, **49**, 199.
62. T. S. Anirudhan and S. Alexander, *Biosens. Bioelectron.*, 2015, **64**, 586.
63. J. Luo, S. Jiang and X. Liu, *Sens. Actuators, B*, 2014, **203**, 782.
64. R. Schirhagl, *Anal. Chem.*, 2014, **85**, 250.
65. O. Y. F. Henry, D. C. Cullen and S. A. Piletsky, *Anal. Bioanal. Chem.*, 2005, **382**, 947.
66. X. A. Ton, V. Acha, K. Haupt and B. Tse Sum Bui, *Biosens. Bioelectron.*, 2012, **36**, 22.
67. A. Valero-Navarro, A. L. Medina-Castillo, J. F. Fernandez-Sanchez and A. Fernández-Gutiérrez, *Biosens. Bioelectron.*, 2011, **26**, 4520.
68. S. Al-Kindy, R. Badía, J. L. Suárez-Rodríguez and M. E. Díaz-García, *Crit. Rev. Anal. Chem.*, 2000, **30**, 291.
69. P. Turkewitsch, B. Wandelt, G. D. Darling and W. S. Powell, *Anal. Chem.*, 1998, **70**, 2025.
70. A. Tong, H. Dong and L. Li, *Anal. Chim. Acta*, 2002, **466**, 31.
71. H. Niu, Y. Yang and H. Zhang, *Biosens. Bioelectron.*, 2015, **74**, 440.
72. Y. Lv, Y. Qin, F. Svec and T. Tan, *Biosens. Bioelectron.*, 2016, **80**, 433.
73. G. Fang, C. Fan, H. Liu, M. Pan, H. Zhu and S. Wang, *RSC Adv.*, 2014, **4**, 2764.
74. S. A. Piletsky, E. V. Piletska, B. Chen, K. Karim, D. Weston, G. Barrett, P. Lowe and A. P. F. Turner, *Anal. Chem.*, 2000, **72**, 4381.
75. S. L. Gong, Z. J. Yu, L. Z. Meng, L. Hu and Y. B. He, *J. Appl. Polym. Sci.*, 2004, **93**, 637.
76. D. J. Duffy, K. Das, S. L. Hsu, J. Penelle, V. M. Rotello and H. D. Stidham, *J. Am. Chem. Soc.*, 2002, **124**, 8290.
77. L. Chang, Y. Ding and X. Li, *Biosens. Bioelectron.*, 2013, **50**, 106.
78. M. Riskin, Y. Ben-Amram, R. Tel-Vered, V. Chegel, J. Almog and I. Willner, *Anal. Chem.*, 2011, **83**, 3082.
79. G. Sener, L. Uzun, R. Say and A. Denizli, *Sens. Actuators, B*, 2011, **160**, 791.
80. S. Kostrewa, M. Emgenbroich, D. Klockow and G. Wulff, *Macromol. Chem. Phys.*, 2003, **204**, 481.
81. T. Kamra, S. Chaudhary, C. Xu, L. Montelius, J. Schnadt and L. Ye, *J. Colloid Interface Sci.*, 2016, **461**, 1.
82. H. N. Kim, Z. Guo, W. Zhu, J. Yoon and H. Tian, *Chem. Soc. Rev.*, 2011, **40**, 79.
83. A. Tretjakov, V. Syritski, J. Reut, R. Boroznjak and A. Öpik, *Anal. Chim. Acta*, 2016, **902**, 182.
84. Y. Uludağ, S. A. Piletsky, A. P. F. Turner and M. A. Cooper, *FEBS J.*, 2007, **274**, 5471.

85. Y. Tan, J. Yin, C. Liang, H. Peng, L. Nie and S. Yao, *Bioelectrochemistry*, 2001, **53**, 141.
86. W. Naklua, R. Suedee and P. A. Lieberzeit, *Biosens. Bioelectron.*, 2016, **81**, 117.
87. D. F. Tai, M. H. Jhang, G. Y. Chen, S. C. Wang, K. H. Lu, Y. D. Lee and H. T. Liu, *Anal. Chem.*, 2010, **82**, 2290.
88. I. Chianella, A. Guerreiro, E. Moczko, J. S. Caygill, E. V. Piletska, I. M. P. De Vargas Sansalvador, M. J. Whitcombe and S. A. Piletsky, *Anal. Chem.*, 2013, **85**, 8462.
89. K. Fujiwara, Y. Yoshizaki, M. Shin, T. Miyazaki, T. Saita and S. Nagata, *Antimicrob. Agents Chemother.*, 2012, **56**, 5883.
90. R. V. Shutov, A. Guerreiro, E. Moczko, I. M. P. De Vargas-Sansalvador, I. Chianella, M. J. Whitcombe and S. A. Piletsky, *Small*, 2014, **10**, 1086.
91. L. Gao, J. Zhuang, L. Nie, J. Zhang, Y. Zhang, N. Gu, T. Wang, J. Feng, D. Yang, S. Perrett and X. Yan, *Nat. Nanotechnol.*, 2007, **2**, 577.
92. J. Czulak, K. Metran, F. Canfarotta, A. Goddard, R. H. Cowan, A. W. Trochimczuk and S. A. Piletsky, *Nanoscale*, 2016, **8**, 11060.
93. S. Kunath, M. Panagiotopoulou, J. Maximilien, N. Marchyk, J. Sänger and K. Haupt, *Adv. Healthcare Mater.*, 2015, **4**, 1322.
94. S. Shinde, Z. El-Schich, A. Malakpour, W. Wan, N. Dizeyi, R. Mohammadi, K. Rurack, A. Gjörloff Wingren and B. Sellergren, *J. Am. Chem. Soc.*, 2015, **137**, 13908.
95. D. Yin, S. Wang, Y. He, J. Liu, M. Zhou, J. Ouyang, B. Liu, H. Y. Chen and Z. Liu, *Chem. Commun. (Camb.)*, 2015, **51**, 17696.
96. Y. Liu, S. Wang, C. Zhang, X. Su, S. Huang and M. Zhao, *Anal. Chem.*, 2013, **85**, 4853.
97. Y. Hoshino, H. Koide, T. Urakami, H. Kanazawa, T. Kodama, N. Oku and K. J. Shea, *J. Am. Chem. Soc.*, 2010, **132**, 6644.
98. Z. Wu, J. Hou, Y. Wang, M. Chai, Y. Xiong, W. Lu and J. Pan, *Int. J. Pharm.*, 2015, **496**, 1006.

CHAPTER 2

Synthetic Chemistry for Molecular Imprinting

TAN-PHAT HUYNH*[a,b] AND TRUNG-ANH LE[a]

[a]Laboratory of Physical Chemistry, Faculty of Science and Engineering, Åbo Akademi University, Porthaninkatu 3-5, 20500 Turku, Finland; [b]Center of Functional Materials, Åbo Akademi University, 20500 Turku, Finland
*E-mail: tan.huynh@abo.fi

2.1 Introduction

It has been 60 years since Dickey had established the "specific adsorption" concept, in which his developed artificial material (silica gel prepared in the presence of methyl orange) recognized the target analyte (methyl orange) with selectivity much higher than that to other interfering compounds.[1] The artificial material was later named "molecularly imprinted polymer" or MIP because it is fabricated using a molecular imprinting process.[2,3] MIPs are chemical analogs of biological enzymes because they not only are able of recognizing molecules selectively but also their functions in catalysis are very similar to those of enzymes.[4] Shortly after development of early MIPs, researchers realized the huge potential of these materials in chemical sensing. Chemical sensors based on MIPs have emerged as competitive candidates in the sensor field because of their high selectivity (nearly equaling that of biosensors) and long-term stability in many applications.[5–8] The present chapter aims to sketch out the synthetic approaches for developing MIPs as well as their analytical applications in

Polymer Chemistry Series No. 28
Molecularly Imprinted Polymers for Analytical Chemistry Applications
Edited by Wlodzimierz Kutner and Piyush Sindhu Sharma

Published by the Royal Society of Chemistry, www.rsc.org

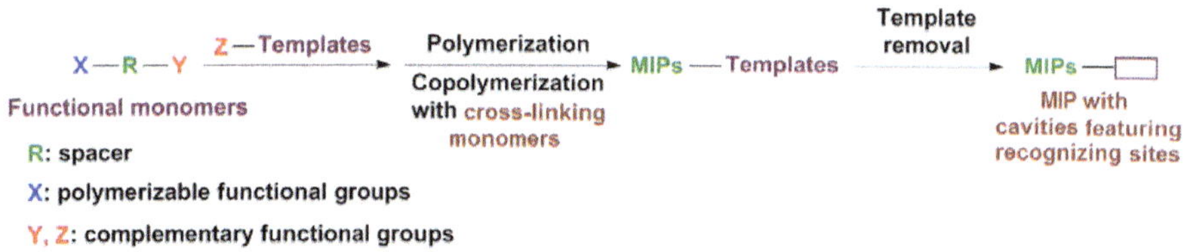

Scheme 2.1 A simplified procedure for MIP preparation.

an interdisciplinary research. It can be considered as an introductory chapter prior to more advanced topics of MIPs for different applications in analytical chemistry introduced in subsequent chapters.

2.2 Strategic Syntheses of MIPs

Fabrication of polymers that can selectively interact with different target analytes is of paramount importance. From a synthetic standpoint, the preparation of MIPs in Scheme 2.1 starts by choosing suitable functional monomers for analytes of interest. Then, polymerization of the functional monomers, or the copolymerization of the functional monomers with other cross-linking monomers, in the presence of the analytes acting as templates follows. The removal of the templates is then carried out to achieve the final MIPs with cavities comprising numerous recognizing sites. From the chemical component perspective, MIPs can be made of organic compounds,[9,10] inorganic compounds,[11,12] or their combination to devise hybrid materials.[13–15] Moreover, depending on the properties and applications of MIPs of interest, different fabrication methods and techniques have been reported over the years.[16]

2.2.1 Interactions Between Templates and Functional Monomers

Potentially, a variety of different templates can be studied and imprinted. Imprinting can easily be adapted for analytes of different structural complexity and morphology, such as small-molecule compounds, macromolecular compounds, biological cells, viruses, and even whole bacteria.[17–20] On the one hand, functional monomers bear recognizing functional sites complementary to binding sites of imprinted templates and polymerizable moieties, and on the other they allow formation of a polymer matrix in the presence of cross-linking monomers (Scheme 2.2).

Clearly, understanding the interaction between the templates of interest and the chosen functional monomers plays a crucial role in the fabrication of MIPs. In general, there are five main types of interaction corresponding to five major molecular imprintings, which include (i) weak non-covalent, (ii) electrostatic/ionic, (iii) covalent, (iv) semi-covalent, and (v) metal centre coordinative approaches (Scheme 2.2).[21,22] These interactions not only determine

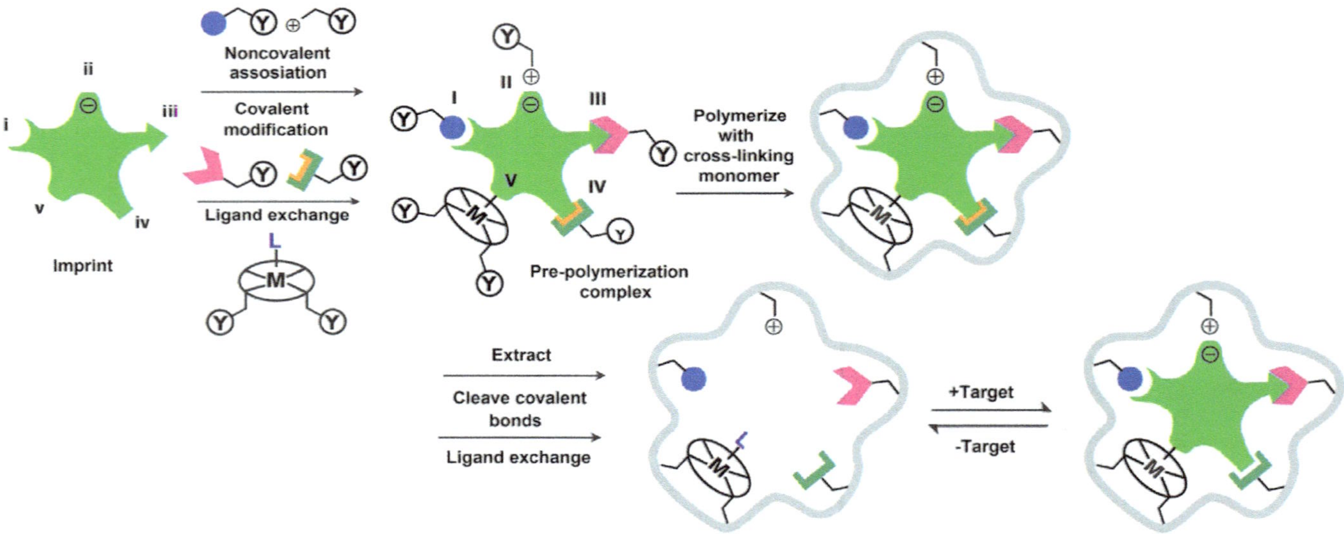

Scheme 2.2 Five main types of interactions used in molecular imprinting: (I) non-covalent, (II) electrostatic or ionic, (III) covalent, (IV) semi-covalent, and (V) metal-centre coordination. An imprint/template is recognized by one or many appropriately chosen functional monomers using the above interactions through complementary functionalities. A pre-polymerization complex of the imprint and functional monomer molecules is formed when the functional monomers bind to the imprint molecule. The functional monomer contains a polymerizable moiety Y, which can undergo a cross-linking reaction with an appropriate cross-linking monomer. After polymerization of the complex with a cross-linking monomer to form a solid polymer matrix (grey), the template–polymer is intact. The template is then removed through extracting, cleavage of chemical bonds, or ligand exchange, and leaves behind in the MIP an imprinted cavity with recognizing sites. Subsequent uptake of a target molecule is achieved by replicating interactions from (I) to (V) between the MIP cavity and the analyte. Reprinted from *J. Mol. Recognit.*, **19**, C. Alexander, H. S. Andersson, L. I. Andersson, R. J. Ansell, N. Kirsch, I. A. Nicholls, J. O'Mahony and M. J. Whitcombe, Molecular imprinting science and technology: a survey of the literature for the years up to and including 2003, 106, Copyright 2006, with permission from Elsevier.

the spatial arrangement of molecules of functional monomers around the analyte molecules, which gives MIPs the memory and recognition of the chosen analytes, but also the bonding strength of MIPs to particular templates leading to different sorption kinetics and selectivity of MIPs. We will discuss thoroughly possible interactions between templates and functional monomers because they are essential for fabricating MIPs and enhancing their selectivity towards different analytes.

2.2.1.1 Covalent and Semi-covalent Imprinting

Different types of covalent bonds between templates and functional monomers leading to boronic esters,[23–25] Schiff bases,[26,27] ketals or acetals,[28] disulfides,[29,30] *etc.*, have been utilized to synthesize MIPs (Scheme 2.3). Typically, these covalent bonds must be reversible to allow removing of the template and then binding of the analytes. The covalent approach offers several advantages, namely, high MIP selectivity because of dedicated covalent bonds and homogeneous recognizing sites. This homogeneity is due to the 1 : 1 stoichiometry of templates and functional monomers. However, reversible covalent bonds are scarce. Therefore, in case of sophisticated templates, much effort is required for preparing pre-polymerization complex mixtures. Hence, alternatives employing other interactions between the template and functional monomers have been used. *Semi-covalent imprinting* consists of covalent binding of functional monomers to a template during MIP synthesis and reversible non-covalent binding of the MIP cavity to the analyte molecule.[31,32] Therefore, recognizing sites that covalently interact with the template are always designed with a spacer (Scheme 2.2), so that the subsequent non-covalent analyte binding is feasible.

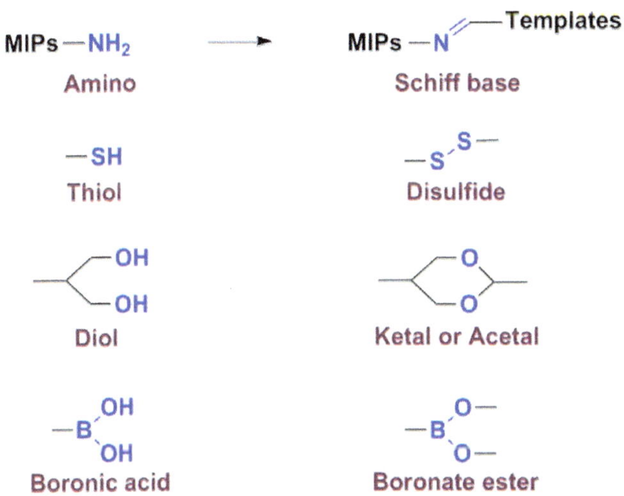

Scheme 2.3 Examples of covalent bonds used for MIP preparation.

2.2.1.2 Weak Non-covalent Interactions

Non-covalent imprinting is so far the most common approach to MIP synthesis because of its simplicity[33] and efficiency even for complex asymmetric analytes.[34] In fact, these kinds of interactions cover a wide range of different interactions. Herein, we wish to classify several weak forces including Lewis acid-base, van der Waals (vdW), π–π stacking, and hydrophobic interactions which share the same coulombic nature with other interactions in chemistry, for instance, much stronger electrostatic forces between ions.

2.2.1.2.1 Lewis Acid–Base Interactions. Similar to Lewis acids and bases, which behave as electron acceptors and donors, respectively, Lewis acid–base interactions involve electrostatic attractions between two oppositely charged parts of molecules. One of the distinct properties of these interactions includes an electron density shift, *i.e.*, a partial charge transfer, from the Lewis base to the Lewis acid. The concept of the Lewis acid–base interactions was suggested to describe various chemical bonds, including hydrogen, halogen, chalcogen bonds, *etc.*[35–37]

According to IUPAC, the hydrogen bond can be understood as "an attractive interaction between a hydrogen atom from a molecule or a molecular fragment X–H in which X is more electronegative than H, and an atom or a group of atoms in the same or a different molecule, in which there is evidence of bond formation"[38] while the halogen bonds "occurs when there is evidence of a net attractive interaction between an electrophilic region associated with a halogen atom in a molecular entity and a nucleophilic region in another, or the same, molecular entity".[39] These two bonds have been used in MIP preparation.[40–44]

2.2.1.2.2 van der Waals Interactions. Sharing the same electrostatic nature, vdW interactions between molecules are normally divided into three types of forces, namely, Keesom forces (between two permanent dipoles), Debye forces (between a permanent dipole and an induced dipole), and London dispersion forces (between two instantaneously induced dipoles).[45] Even though they are considered as weak interactions, the importance and application of vdW forces have been found in various fields, ranging from new structures and devices in materials science,[46–49] protein–protein recognition,[50] peptide self-assembly,[51] interactions between different components of cellular surfaces and intercellular media,[40] in complex biological systems, in catalytic processes,[52,53] and in molecular recognition.[54] Moreover, vdW forces engaging together with other types of interactions have been considered for MIPs over the years.[55–57]

2.2.1.2.3 Hydrophobic Interactions. Hydrophobic interactions appear between non-polar molecules in polar solvents. Polar medium presence is the critical condition for the formation of hydrophobic forces because solute molecules surrounded by a solvent cage tend to interact each other and result in the entropy increase.[58] The most plausible example for this interaction used in the MIP study is the interaction between cyclodextrin and non-polar binding sites of a template.[59,60] However, non-selective binding is the common drawback of this approach.

2.2.1.3 Electrostatic Interactions

Electrostatic interactions occur between charged ions, manifested as either attractions or repulsions. The attractive electrostatic interactions between the charged groups on the side chain of polymers with amino acids of templates have been often utilized in MIP designing.[61,62] This approach leads to more selective and stronger MIP-template interactions. However, charged residues can also cause non-specific binding of the template, thus decreasing the imprinting effect.

2.2.1.4 Metal–Ligand Coordination

The study of different ligands and their selective complexation of metal ions have drawn great attention over the years, especially for fabricating ion-imprinted polymers (IIPs). While the Pearson hard and soft acid and base principle is normally regarded as the starting point for selecting donor atoms of the appropriate ligands, Hancock and coworkers[63–65] have suggested more factors for further consideration including solvent effects, linear free-energy relationships, ligand field theory, electronic and steric effects, chelate and macrocyclic effects, and the coordinating capacity of different donor groups (neutral and negative oxygen, neutral saturated and unsaturated nitrogen, heavier atoms, such as S, Se, P, and As).

2.2.2 MIP Synthesis

After selecting appropriate types of interactions and functional monomers, the next study involves engineering MIPs for target analytes. This study is divided into three consecutive steps including (i) pre-polymerization complex formation, (ii) polymerization of this complex, and (iii) template removal from the resulting MIP. Success in this study will help to generate MIP cavities perfectly matching analyte molecules.

2.2.2.1 Imprinting Strategies

As mentioned above, the imprinted compounds can vary from low-molecular simple species, such as metal cations, to highly complex biological macromolecular compounds. In order to develop chemical insights at atomic and molecular levels into different imprinting strategies, we now discuss several ways to incorporate ligands into IIPs for metal cation analytes[66,67] as the simplest model for other MIPs summarized in Scheme 2.4. Therefore, the strategies are applicable to imprint more complicated analytes.

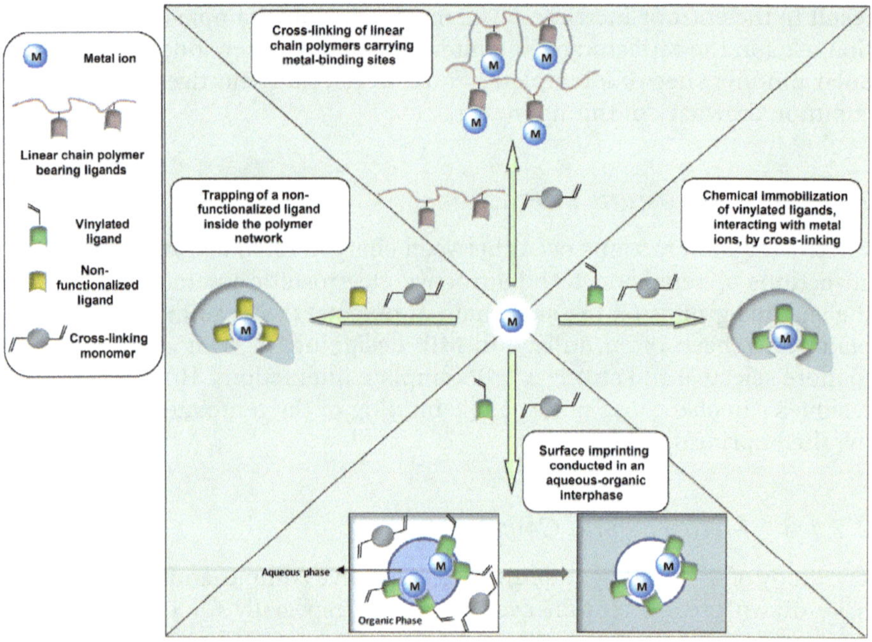

Scheme 2.4 Imprinting approaches for IIP preparation. Reprinted from *React. Funct. Polym.*, 73, C. Branger, W. Meouche and A. Margaillan, Recent advances on ion-imprinted polymers, 859–875,[67] Copyright 2013 with permission from Elsevier.

2.2.2.1.1 Linear Chain Polymer Cross-linking. Synthetic or natural polymer chains that have coordinating functional groups are cross-linked *via* copolymerization to form IIPs. The polymer matrix is mainly made of coordinating functional monomers and, therefore, all of the physical, chemical, and biological properties of IIPs are derived from those of functional monomers. In many cases, the properties of IIPs cannot simultaneously fulfill numerous criteria for their applications in different media. To overcome this major obstacle, other strategies were suggested, in which the coordinating functional monomers were incorporated into a polymer matrix instead (Scheme 2.4, top). As a result, IIPs inherited recognition capabilities from functional monomers while other properties, such as electrical conductivity, magnetism, mechanical durability, and bio-compatibility, are attributed to the polymer matrix.

2.2.2.1.2 Ligand Chemical Immobilization. Multifunctional ligands, including a polymerizable vinyl moiety and coordinating groups to selectively bind to the metal ions, are chosen (Scheme 2.4, right). The vinyl functional group is of great interest because it offers polymerization capability without affecting the coordination chemistry of the ligands. Since the ligands are chemically incorporated into the polymer substrate, highly stable IIPs with special properties can easily be obtained. These vinylated ligands,[68] however, are not widely commercialized. Therefore, sophisticated syntheses are usually required on the one hand or simple vinylated ligands are employed on the other for the fabrication of low-selectivity IIPs.[67]

2.2.2.1.3 Ligand Physical Entrapment. In order to overcome disadvantages of chemical immobilization, the physical entrapment of non-vinylated ligands inside polymer matrices was suggested as a promising alternative (Scheme 2.4, left).[67] IIPs are fabricated by copolymerization of a mixture of metal complexes, functional monomers, and cross-linking monomers. Since the introduction of polymerizable functional groups, *e.g.*, vinyl moieties, onto ligands of interest is not required, this procedure clearly simplifies the synthesis of IIPs. Moreover, this fabrication approach leads to highly porous and stable IIPs. However, a sufficiently high number of recognition sites remains as the major challenge of this approach because of slow leaching of physically bound templates.[67]

2.2.2.1.4 Surface Imprinting. As the bulk IIPs or MIPs synthesized by the three above approaches share the same possible problems of incomplete template removal and slow mass transfer because of low accessibility of cavities and, even more importantly, recognition sites in these cavities, surface immobilization of templates might become an alternative (Scheme 2.4, bottom).[69] This approach is useful for detecting macromolecules, cells, bacteria, *etc.*, which partially bind to the MIP surfaces.[70] In general, the polymer core can be of different chemical compositions, structures, and morphologies. More importantly, these polymer substrates must be surface-modified to allow further functionalization towards introduction of imprinted layers onto the surface. With the advances in developing various sorbent materials and heterocatalysts, the employment of different material substrates to fabricate surface-imprinted polymers (SIPs) has become more feasible than ever.[70]

2.2.2.2 Polymerization

2.2.2.2.1 Polymerization Reactions. Traditionally, polymerization reactions are classified as condensation or addition processes depending on the composition and structure of monomers; and step-growth or chain-growth polymerization based on the reaction mechanisms.

In the step-growth polymerization, bifunctional monomers are required for the fabrication of step-growth polymers. In the presence of multifunctional monomers, cross-linking between polymer chains might occur to give 3-D polymer matrices of different morphology and properties. Scheme 2.5 presents several routes for preparing MIPs based on, *e.g.*, polyamide,[71,72] polyurea,[73,74] polyurethane,[75] and polysiloxane[76,77] (Scheme 2.5a), as well as several commonly-used cross-linking monomers, such as citric acid,[78] methyltrichlrosilane,[79] phloroglucinol,[80] and *p,o,p'*-diisocyanato diphenylmethane[81] (Scheme 2.5b).

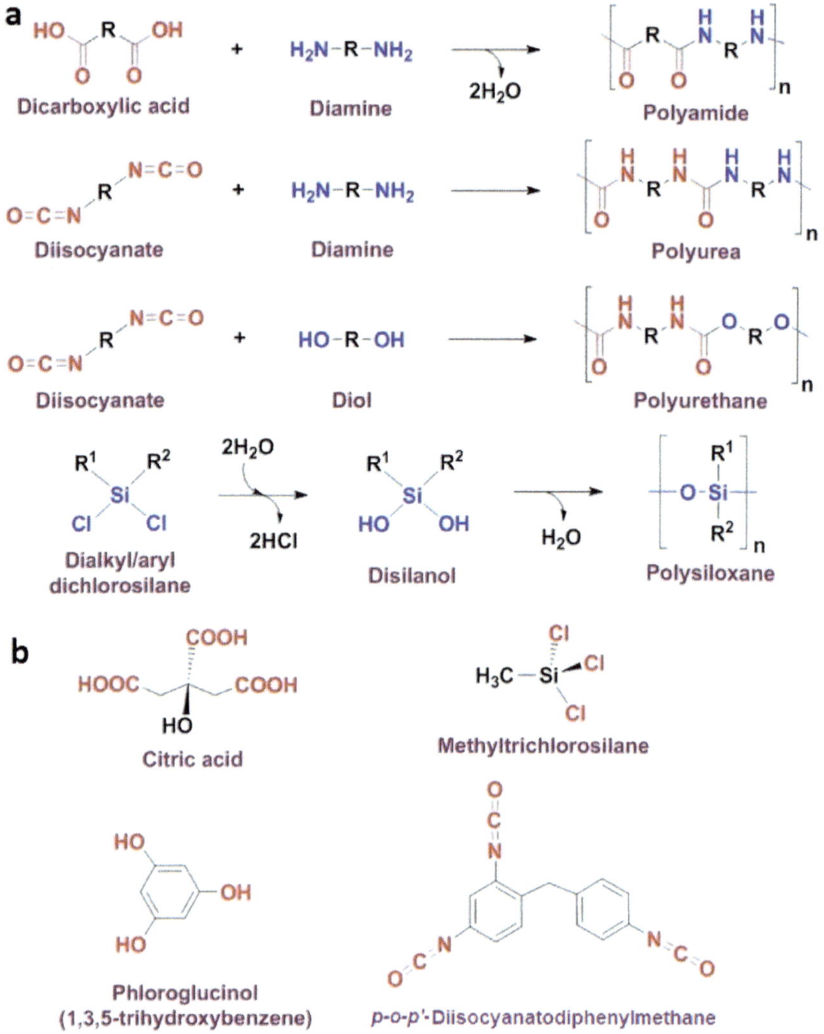

Scheme 2.5 (a) Synthetic chemistry of common polyamide, polyurea, polyurethane, and polysiloxane and (b) some common cross-linking monomers used in step-growth polymerization for preparing MIPs.

Besides, chain-growth polymerization involves the growth of a polymer chain that proceeds exclusively by reactions between monomers and reactive sites on the polymer chain with regeneration of the reactive sites at the end of each growth step.[82] One of most important active centers in radical polymerization for MIP preparation is a free radical. Free radicals can be generated by chemical initiators, light, applied electrical potential, or enzymes, thus allowing realization of different polymerization methods, such as chemical polymerization, photopolymerization, electropolymerization, or enzyme-initiated polymerization, respectively.[83–85] Free-radical polymerization can be

Table 2.1 Examples of vinyl-, acrylate-, and methacrylate-based cross-linking monomers used for MIP preparation.

	Bifunctional ($x = 2$)	Trifunctional ($x = 3$)
Vinyl-based 	Divinyl benzene[90]	
Acrylate-based 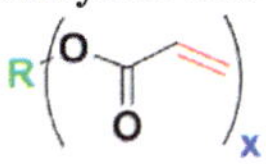	6,6′-Diacryloyl-trehalose[91] *N*,*N*′-Methylene bis(acrylamide)[86]	Ethylene glycol maleic rosinate acrylate[87]
Methacrylate-based 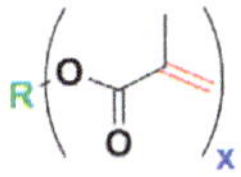	Ethylene glycol dimethacrylate (EGDMA)[88] Poly(ethylene glycol)-dimethacrylate[92]	Trimethylolpropane trimethylacrylate[89]

performed under mild reaction conditions (*e.g.*, ambient temperature and atmospheric pressure) in bulk or in solution. Moreover, it is very tolerant with respect to functional groups in the monomers and impurities in the system (*e.g.*, water).[83] Indeed, many low-cost vinyl monomers were used for preparing MIPs, notably acrylate-based polymers.[83] Similarly to polymers fabricated *via* the step-growth polymerization route, multifunctional cross-linking monomers also play a crucial role in synthesizing 3-D polymers by chain-growth polymerization reactions and, therefore, several typical examples of cross-linking monomers[86–92] are listed in Table 2.1.

2.2.2.2.2 Polymerization Techniques. Based on the free-radical polymerization mechanism mentioned above, bulk and solution polymerization reactions have been mostly used in MIP preparation.[93–95] This was because of both a high yield of these reactions and high product purity, as well as simplicity compared to other reactions. The polymerization started with a functional monomer, cross-linking monomer, initiator in the solvent presence at high (solution polymerization) or low (bulk polymerization) volume.[96] Between them, solution polymerization has some advantages over bulk polymerization, such as the final product with the shape and size control and thus no need of further processing like crushing, grounding, and sieving.[97] Solution polymerization is either homogeneous if the polymer remains soluble, or heterogeneous if the polymer is insoluble, thus leading to precipitation polymerization.

Similar to that of the bulk and solution polymerization, suspension polymerization produces suspended polymer beads with a spherical form (0.001 to 1 cm in diameter) and rough surface because of the pores, which are formed during the polymerization.[96] Advantages of this procedure include low viscosity and heat as well as stable dispersion to be directly used in MIP preparation. However, the main disadvantage is the necessity to separate and purify the resulting polymer beads, or to accept a contaminated product.

In the sol–gel process, tetraalkoxysilane precursors, such as tetramethoxysilane and tetraethoxysilane, are firstly hydrolyzed to form colloids (sols), and then polycondensated to form highly cross-linked silica materials (gels).[98] The sol–gel process has distinct advantages including room-temperature fabrication, thus avoiding thermal or chemical decomposition and the use of non-toxic solvents, such as water and ethanol.

2.2.2.3 Template Removal

The last but not least of the MIP preparation steps is template removal. It strongly affects sensitivity of the MIP to the target analyte. If template removal is successful, MIP cavities are formed with the exact size, shape as well as position and space orientation of recognizing sites. The following methods of extraction[99,100] are currently being used for template removal.

- Solvent extraction, where aqueous or organic solvents, in which the template is highly soluble, are used to elute the template. This method is simple but features several drawbacks, such as time consumption and waste of solvents. Therefore, it often requires the assistance of temperature and convection (to accelerate the solubility of the template),[101] acids or bases (to protonate or deprotonate the template towards weakening complexation of the template by recognizing MIP cavity sites),[102] or enzymes (to decompose a biomolecular template),[103] *etc.*
- Extraction using a supercritical fluid.

Nevertheless, template removal is never exhaustive because of two main reasons including (i) shrinking, swelling, or damaging cavities after extraction, which mostly happens to soft and low cross-linked MIPs and (ii) incomplete template extraction. Therefore, it is important to develop chemical techniques to characterize the extent of template removal and the faultlessness of the MIP cavity.

There are two common ways of checking the extent of completion of template removal, *i.e.*, either direct template detection or its indirect detection through physical or chemical changes of MIP properties. Of course, the first way is preferable because removed templates are directly detected in extracting solutions using different techniques, *e.g.*, UV-vis, fluorescence, or NMR spectroscopy, electroanalytical techniques, *etc.*[104–108] Nevertheless, when the direct confirmation of template removal does not work, the indirect evaluation through changes in, for instance, electrochemical current of a redox probe in solution or fingerprints of elements or functional groups from X-ray photoelectron spectroscopy (XPS) or IR spectra, respectively, is used.[105,109]

Completion of template removal is just one step towards MIP readiness for use. The remaining structure of the MIP cavity after template extraction is believed to be even more important and has to be characterized. However, to our best knowledge, there is no direct technique to confirm the cavity structure yet, except for micrometre-size analytes, like yeast cells.[110] This is because of a limit in technology for imaging a nanoscale cavity. Typically, the template-extracted cavity is evaluated indirectly through examining selectivity of the resulting MIPs, which is discussed in the next section.

2.3 Analytical Applications of MIPs

Due to their enzyme mimicking structures that are highly selective to target (imprinted) analytes, MIPs are involved in many applications from applied chemistry, such as purification and separation,[111] through catalysis,[4] *etc.*, to drug development.[112,113] The application of MIPs in analytical chemistry, *i.e.*, their use as recognition units in chemosensors,[114–117] is also very important, confirmed by the hundreds of papers that have been published annually and that are still increasing in number. Therefore, the fabrication of a chemosensor based on MIPs using different transducers will be firstly described in Section 2.3.1, and then their potential uses in different areas of analytical applications will be highlighted.

2.3.1 Fabrication of MIP-based Chemosensors

Basically, the construction of the MIP-based chemosensor is similar to that of an ordinary chemo- or biosensor, *i.e.*, it consists of a sensing part and a transducer (Scheme 2.6).[118] Both are integrated with each other so that chemical signals of analyte recognition generated by the MIP–analyte interactions can be converted into physical signals. Based on the integration level of the transducers and the MIPs, these sensors fall into two categories, *i.e.*, the

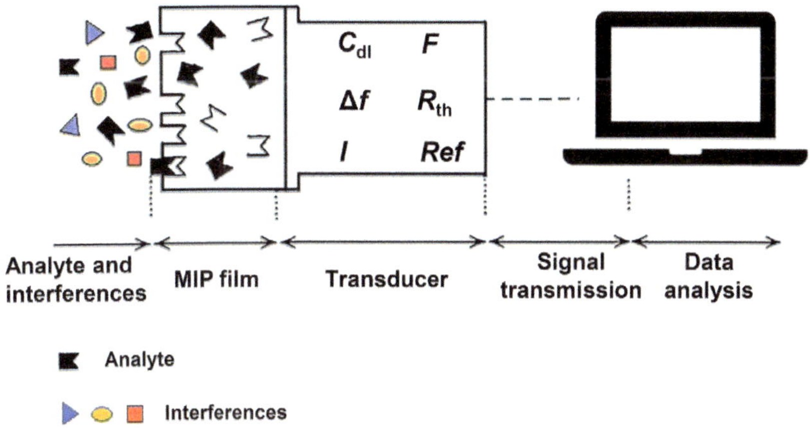

Scheme 2.6 Comprehensive process for analyte sensing using an MIP-based chemosensor. The transducer converts the molecular recognition signal to different physical signals, for example, C_{dl} – double-layer capacitance, Δf – resonant frequency change, I – electronic/electrochemical current, F – fluorescence, R_{th} –heat transfer resistance, R_{ef} – optical reflectance.

internal (Section 3.1.1) and external (Section 3.1.2) parts of the MIPs. This classification will give readers a brief overview on MIP-based chemosensors before elucidating their use in analytical applications.

2.3.1.1 MIPs with Internal Transducers

The internal transducers, herein the inseparable moieties or fillers, are fused with MIPs at the stage of either the synthesis of functional monomers or the preparation of MIPs.[119–121] These sensors behave as indicators. That is, they generate physical (mostly spectroscopic, such as absorbance, reflectance, fluorescence, and luminescence) signals when triggered by external stimulants, herein analytes occupying imprinted cavities. They often appear as nanoparticles either dispersed in media or coated on substrates including quantum dots, conducting polymers, or fluorescent particles (Scheme 2.7a).[122–127] Therefore, an external optical detector and light source are always combined with these sensors. Because of the nano size of these MIP sensors, analytes efficiently influence, for example, the fluorescence response of the transducers; as a result, fluorescence sensors are usually more sensitive than other sensors in many analytical applications.[126]

Because of tight integration of MIPs and transducers, this architecture exhibits several disadvantages including (i) disposal of MIP-based chemosensors directly into media because they are rarely collected after use, (ii) instability because of photoactive (narrow band-gap) moieties or fillers, (iii) detection under batch conditions in most cases, and (iv) high toxicity.[128]

2.3.1.2 MIPs with External Transducers

The external transducers in the second type of MIP-based chemosensors can be used as independent sensors themselves, *e.g.*, a metal electrode of an electrochemical sensor. However, they are non-selective and, indeed, they must be integrated with MIPs. This sensing architecture is more applicable than that of MIPs with internal transducers because MIP with external transducers are (i) reusable (removal and deposition of an MIP layer in one cycle), (ii) operable under both batch and flow solution conditions, (iii) potentially integrated with battery or other power supplies leading to fabrication of wearable or portable sensors. These sensors include electronic, electrochemical, mass, optical, and HTM-based sensors (Scheme 2.7b–e).[129–132] However, these sensors require dedication in design, miniaturization, and compatibility between the MIP film and the transducer surface.

2.3.2 MIP-based Chemosensors for Analytical Study

This section will address how MIP-based chemosensors can be used in analytical applications, *i.e.*, for detecting different types of analytes including metal ions, small-molecule compounds, such as toxins,[133] drugs,[134]

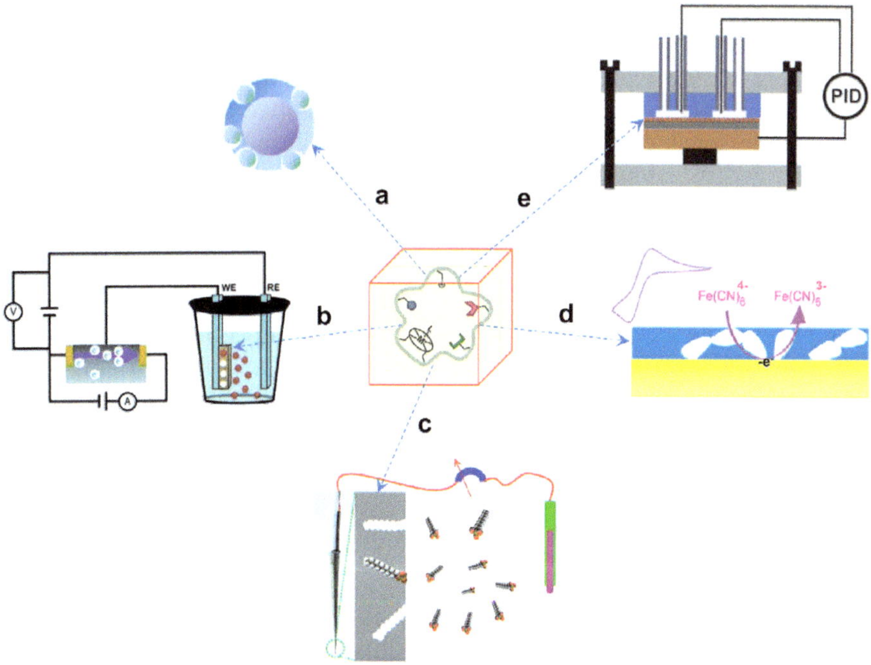

Scheme 2.7 Illustration of several types of MIP chemosensors: (a) nanoparticale-based sensor, (b) chemical field-effect transistor, Chem-FET, (c) potentiometric sensor, (d) electrochemical sensor using a redox proble, and (e) heat-transfer method (HTM) based chemosensor. PID stands for a proportional-integral-derivative controller. (a) Reproduced from ref. 122 with permission from the Royal Society of Chemistry. (b) reprinted with permission from Z. Iskierko, P. S. Sharma, D. Prochowicz, K. Fronc, F. D'Souza, D. Toczydłowska, F. Stefaniak, and K. Noworyta, *ACS Appl. Mater. Interfaces*, 2016, **8**, 19860, Copyright 2016 American Chemical Society. (c) Reprinted from *Environ. Technol. Innovation*, **5**, C. Fang, Z. Chen, M. Megharaj and R. Naidu, Potentiometric detection of AFFFs based on MIP, 52–59, Copyright 2016, with permission from Elsevier. (d) Reprinted from *Biosens. Bioelectron.*, **50**, N. Arimian, M. Vagin, M. H. A. Zavar, M. Chamsaz, A. P. F. Turner and A. Tiwari, An ultrasensitive molecularly-imprinted human cardiac troponinsensor, 492–498, Copyright 2013, with permission from Elsevier. (e) Reproduced from G. Wackers, T. Vandenryt, P. Cornelis, E. Kellens, R. Thoelen, W. D. Ceuninck, P. Losada-Pérez, B. v. Grinsven, M. Peeters and P. Wagner, *Sensors*, 2014, **14**, 11016, http://dx.doi.org/10.3390/s140611016, © 2014 The Authors. Published under the terms of the CC BY 3.0 licence, https://creativecommons.org/licenses/by/3.0/legalcode.

biomarkers,[135,136] and even macromolecular compounds, such as proteins,[61] peptides,[137] nucleic acids,[138,139] *etc*. This interesting versatility of MIPs in analyte recognition is due to their high selectivity. Therefore, MIP-based chemosensors dominate as *selective* sensors, *i.e.*, the sensors capable of recognizing only a certain type of analytes and, hence, being selective with respect to interferences present in the same media. The activity of selective MIP-based chemosensors is similar to that of the enzymes. There are only a few examples where MIPs are prepared in sensor arrays and their global selectivity is derived from not only chemical selectivity mentioned above but also from algorithmic selectivity, *i.e.*, through pattern recognition or any multivariate analysis.[140,141] However, this array-type sensor is out of the scope of this chapter.

For evaluating analytical performance of a selective chemosensor, several parameters are determined, as follows.

- *Linear dynamic concentration range* is the range of concentrations where a sensor response is directly proportional to the concentration of the analyte in the sample.
- *Sensitivity* is the change in the sensors response divided by a corresponding change in the concentration or, simply, the gradient of the calibration plot.
- *Signal-to-noise (S/N) ratio* is the ratio of the sensor response to the analyte and to the blank solution (without the analyte).
- *Limit of detection (LOD)* is the smallest amount or concentration of analyte in the sample that can reliably be distinguished from zero at $S/N = 3$.
- *Limit of quantitation (LOQ)* is the smallest amount or concentration of analyte in the sample that can reliably be distinguished from zero at $S/N = 10$
- *Selectivity* is the sensitivity to the analyte divided by the sensitivity to the interference.

Several more analytical parameters of sensors can be found elsewhere.[142]

The understanding of the above-mentioned important analytical terms regarding MIP-based chemosensors allows us to proceed to the next section on MIP applications in analyte sensing for (i) health monitoring and (ii) environment and food quality control (Table 2.2). A list of various analytes is discussed herein, with the exception of nucleotides, proteins, and pharmaceutical analytes, which shall be addressed in more detail in other chapters of the book.

2.3.2.1 *Health Monitoring*

For health monitoring, MIP-based chemosensors are used as tools for early-stage detection of diseases through selective determination of whole bacteria/cells/virus or molecular biomarkers released into body fluids (urine, blood, saliva, *etc.*).[61,117,143,144] Mostly, molecules of these analytes are macro-sized. Therefore, both bulk and surface imprinting strategies are used (Table 2.2). Some micrometre-size and bulky analytes including bacteria, cells, and virus were detected only by surface imprinting.[145] The MIP surface recognized target cells by generating hydrogen bonds between carboxyl, amide, or catechol groups of MAA, PU, or PDA, respectively, with protein

Table 2.2 Typical applications of MIP-based chemosensors in analyte sensing.[a]

Analyte	MIP	Transduction	LOD/nM	Ref.
Health monitoring				
Cell, bacteria and virus				
Escherichia coli	PU$^{(SIP)}$	HTM	10^4 cfu mL^{-1}	80
	Histidine-functionalized acrylate polymer	Capacitance	70 cfu mL^{-1}	187
	Electropolymerized PDA$^{(SIP)}$	Electrochemi-luminescence	8 cfu mL^{-1}	146
	PU$^{(SIP)}$	PM at QCM	1.4×10^7 cfu mL^{-1}	188
Staphylococcus epidermidis	Poly(3-aminophenyl boronic acid)$^{(SIP)}$	EIS	10^3 cfu mL^{-1}	189
Yeast cells	Poly(DA)$^{(SIP)}$	Potentiometry	50 cfu mL^{-1}	190
Cancer cells	PU$^{(SIP)}$	HTM	2.9×10^4 cells mL^{-1}	191
Japanese encephalitis virus (JEV)	NH_2-silica microsphere$^{(SIP)}$	Fluorescence spectroscopy	9.6×10^{-3}	192
Hepatitis A virus	DOPA-silica microsphere$^{(SIP)}$	Fluorescence spectroscopy	8.6×10^{-3}	193
HIV-p24	Acrylamine polymer	DPV	8.6×10^{-6}	194
Biomarkers				
Prostate specific antigen (PSA)	Poly(MAA-EGDMA) $^{(SIP)}$	SPR	2.0×10^{-4}	195
Cholesterol	Poly(aminothiophenol)	DPV	3.3×10^{-6}	153
	Poly(dansyl-modified β-cyclodextrin)	Fluorescence spectroscopy	100	196
3SLac and 6SLac	Poly(vinylphenylboronic acid-EGDMA)	MIP-gate FET	10^4	150
Calcitonin	Surface-functionalized ZnO nanostructure	DPSV	0.9×10^{-3}	197
Neutrophil gelatinase-associated lipocalin (NGAL)	Siloxane-based polymer	LSPR	13	198
8-OHdG	Polyphenol	EIS	3.5×10^{-3}	136
Neopterin	Boronic & cytosine functionalized polythiophene	Potentiometry	22×10^3	199
NNAL	Poly(MAA-TRIM)	LC-MS	8.1	151
Environment and food control				
Metal ions and aromatic contaminants				
Cr^{6+}	Poly(4-vinylpyridine)	Photoelectro-chemistry	0.11	200
Cd^{2+}	Poly(ethyleneimine-MAA)	Colorimetry	3.6	201
Cu^{2+}	Glycine	DPV	0.04	161
Cu^{2+} and Hg^{2+}	3-Aminopropyl triethoxysilane	Fluorescence spectroscopy	0.55 and 0.28	202
TNT	Poly(MAA)	SWV	0.5	170
	NH_2 functionalized polythiophene	PM at QCM	70×10^3	105

(continued)

Table 2.2 (*continued*)

Analyte	MIP	Transduction	LOD/nM	Ref.
TNP	SiO_2	Fluorescence spectroscopy	43	171
	NH_2 functionalized polythiophene	Fluorescence spectroscopy	0.9×10^{-3}	108
Nitrotriazolone	Chitosan	Chemiresistor	77×10^3	172
Triclosan	Poly(AA-MMA)	Potentiometry	1.9	174
Leucomalachite green	Poly(MAA-EGDMA)	Fluorescence spectroscopy	30	173
Pesticides				
Fenvalerate	Poly(allyl fluorescein-TRIM)	Fluorescence spectroscopy	0.07	164
Carbendazim	Poly(MAA) and magnetite	Fluorescence spectroscopy	12	165
λ-Cyhalothrin	Poly(MAA-EGDMA)	Fluorescence spectroscopy	4×10^3	166
Methidathion	Acrylic based monomer	EIS	17	177
Estrogens				
Diethylstilbestrol	Mesoporous silica	Fluorescence spectroscopy	0.19	167
	Poly(DA)	Fluorescence spectroscopy	0.04	168
β-Estradiol	Poly(NPEDMA)	CV	0.7×10^{-3}	203
	Hollow mesoporous silica nanoparticle(SIP)	HPLC	0.08	169
Food toxins				
PG	Poly(phenylenediamine)	DPV	0.2	178
Ethyl carbamate	Poly(*o*-aminophenol)	DPV	37	179
Citrinin	Polypyrrole	DPV	0.2×10^{-3}	180
Aspartame	Acetic acid functionalized-terthiophene	PM at QCM	0.04	181
Melamine	Poly(MAA-EGDMA)	SWV	1.8×10^{-3}	175
	18-Crown-6 functionalized polythiophene	PM at QCM	5	204
Histamine	Poly(MAA-EGDMA)	Passive radio frequency	50	205
	18-Crown-6 and boronic acid functionalized polythiophene	PM at QCM	5	206
α-Tocopherol	Poly(MAA-EGDMA)	SERS	23	207
D-glutamic acid	Poly(*o*-phenylenediamine)	DPV	0.2	182

[a]SIP, surface imprinted polymer; cfu, colony forming unit.

expression and the presence of carbohydrate groups on the outer surface of the membrane of the cells. For instance, an MIP film prepared by PDA electropolymerization, imprinted with *E. coli*, which was specifically surface modified with polyclonal antibody (pAb) labeled nitrogen-doped graphene quantum dots (N-GQD), changed its electrochemiluminescence in the *E. coli* presence in a test solution.[146] This determination technique appeared to be very sensitive yielding an LOD as low as 8 cfu mL^{-1}, *viz.* ten, a thousand, and a million times lower than those of the HTM, capacitive, and quartz crystal microbalance (QCM) based sensors (Table 2.2). The MIP chemosensor for *E. coli* was highly selective to *Salmonella* (a bacterial interference of a similar size) because of not only selective binding of PDA MIP but also specific immunoreaction of *E. coli* O157 : H7 with pAb-(N-GQDs).

The piezoelectric microgravimetry (PM) technique using QCM is not as sensitive (mass resolution up to ~10 ng cm^{-2}) as other techniques in most applications (Table 2.2) because of the frequency limit (up to 30 MHz) of the quartz resonator caused by the technological limit of making resonators thinner than 55 μm.[147,148] Moreover, PM is a mass-dependent technique. Hence, it operates well for high-molecular weight compounds and sensitivity decreases in viscous liquid media compared to that in air. Moreover, quartz resonators are fragile and expensive, which is disadvantageous for applications.

When the size of the analyte molecules is lower than that of cells, they can be recognized by either surface or bulk imprinting (Table 2.2).[70,149] Diverse types of transducers have been used to detect a broad range of analytes including cholesterol, oligosaccharides [3′-sialyllactose (3SLac) and 6′-sialyllactose (6SLac)] as well as biomarkers, such as 8-hydroxy-2′-deoxyguanosine (8-OHdG), neopterin, and 4-(methylnitrosamino)-1-(3-pyridyl)-1-butanol (NNAL).[136,150,151] MIP-based electrochemical sensors dominate in this application because electrochemistry provides abundant techniques to detect analytes in solution,[152] for instance, chronoamperometry, differential pulse voltammetry (DPV), potentiometry, and EIS, as well as devices, such as an extended-gate field-effect transistor (EG-FET) or electrochemical quartz crystal microbalance (EQCM) (Table 2.2). When a DPV transduction was applied to an electrode of a large specific surface area of Au nanoflowers causing signal enhancement, the resulting chemosensor exhibited an extremely low (in the femtomolar range) LOD, for cholesterol determination.[153] Cholesterol or, particularly, saturated fats or trans fats, are known to cause many fatal diseases including heart disease, stroke, and diabetes.[154] Aminothiophenol was herein used as the functional monomer to synthesize an MIP film by amine templating, and then recognized cholesterol through hydrogen bonds on the one hand and S anchoring to the Au micro flowers on the other. This led to the synergistic effect of PDA, Au micro flowers, DGr, and MIP that was proven to lower the LOD of the sensor; in other words, if even one of these conditions were not met, this sub-femtomolar LOD would not be reached. The sensor was highly selective to as many as nine different interferences, among which vitamin D3 and progesterone are structurally similar to cholesterol.

2.3.2.2 *Environment and Food Control*

Controlling the quality of our living environment as well as everyday food is now more demanding than ever because contaminants in the environment and food directly affect the future of organisms. These threats include heavy metals, polyaromatic hydrocarbons, organophosphorus compounds, radioactive substances, *etc.*, originating from many different sources. Their pollution can be widespread contaminating waters, soil, food, drinks, *etc.*[155–157] Therefore, this part of the chapter will address the issue of how MIP-based chemosensors has dealt with these problems. Because molecules of most of these analytes are usually much smaller than those of compounds appearing in health monitoring (see Section 3.2.1, above), only bulk imprinting was used for the former (Table 2.2).

Heavy metal contaminants, emerged from industrial processes and laboratories, can enter plant, animal, and human tissues *via* inhalation, diet, and manual handling. Then, they bind to and interfere with the functioning of vital cellular components. Some example elements, *i.e.*, Cr, Cu, Hg, Cd (Table 2.2) regarded as toxic heavy metals, are essential, in small quantities for human health and, fortunately, their presence in waters can be detected using IIPs.[158] Spectroscopic (fluorescent, colorimetric, or photoelectrochemical) based chemosensors were mainly applied because of metal–ligand coordination chemistry leading to the change in spectroscopic behavior of the MIPs.[159,160] This phenomenon allowed determining heavy metals in waters with selectivity higher than that to other ions and low LOD, *i.e.*, in the picomolar range, that is barely achievable by other techniques. However, it is quite difficult to apply MIP chemosensors with internal transducers for real samples. Interestingly, the glycine-based IIP coordinated to Cu^{2+} could reach down to an LOD of 40 pM (Table 2.2) when the imprinting signal was amplified through electrocatalytic effect by using horse radish peroxidase immobilized on the IIP particles. However, the disadvantage of this method comprises involvement of toxic components (H_2O_2, hydroquinone, *etc.*).[161]

The fluorescence spectroscopy technique also exhibits high sensitivity to different types of analytes including fenvalerate and carbendazim pesticides, which cause different toxicity effects, like infertility, demonstrated through damage to the testicles of laboratory animals.[162,163] The LOD ranged from tens of nanomoles up to a few picomoles.[164,165] However, fluorescence spectroscopy is inapplicable for determination of λ-cyhalothrin,[166] most likely because of dense MIP particle formation during synthesis by Pickering emulsion polymerization. This preparation procedure results in difficulty for an analyte to diffuse into the particles to cause a fluorescence change. However, the technique succeeded in detection, at a sub-nanomole concentration, of diethylstilbestrol, an estrogen occurring as a cattle feed supplement and a medical drug for treatment of breast and prostate cancers. This determination was performed using both selective SiO_2 or PDA-based MIPs and surface enhancement of mesoporous silica nanoparticles or core–shell Fe_3O_4@SiO_2 and Cd-QDs, respectively.[167,168] Moreover, the mesoporous silica nanoparticles were used as an HPLC column support for detecting β-estradiol at a picomole level.[169]

Compared to metal ions, determination of organic contaminants in water and food is even more challenging because of the irregularity of their structures and charges requiring dedicated design of the MIPs. We now start with the most straightforward organic toxins including nitroaromatic explosives, namely, 2,3,6-trinitrotoluene (TNT), 2,3,6-trinitrophenol (TNP), and nitrotriazolone,[105,108,170–172] a leucomalachite green organic dye,[173] and a triclosan antibacterial.[174] Herein, the MIPs were recognizing target analytes through formation of hydrogen bonds and π–π stacking. Combination of MAA-based MIP recognition with square wave voltammetry (SWV) transduction led to highly sensitive (in a pM range) chemosensor for TNT and melamine.[170,175] Not only the designed magnetic MIP but also the developed detection procedure, *i.e.*, pre-reducing TNT and then oxidizing its reduced state, allowed selection of TNT among other ions and organic interferences in water. That was because the SWV background current is very small. Moreover, the square wave net current is insensitive to currents arising from convective mass transport, thus making SWV more sensitive than DPV in most applications.[176] Another advantage of SWV is the possibility of applying a relatively fast potential scan. Replacing SWV by potentiometry or EIS led to the nanomole concentration range of the LOD in determination of triclosan[174] and methidathion[177] (an organophosphorous insecticide), respectively.

Some analytes are derivatives of carboxylic acids, *e.g.*, propyl gallate (PG), ethyl carbamate, citrinin, D-glutamic acid or aspartame, that are used as additives in food. They may cause toxic effects. MIPs, dedicated for determination of these additives, are designed to form hydrogen bonds between binding sites of these analytes with recognizing sites of different functional monomers including phenylenediamine, *o*-aminophenol, polyoxometalate, poly(*o*-phenylenediamine), and acetic acid functionalized-terthiophene.[178–182] Propyl gallate (PG), ethyl carbamate, and citrinin were detected at nanomole and sub-picomole concentrations by using DPV for the MIP-coated glassy carbon electrodes. However, the very low LOD is not useful in some practical applications, in which detection of a citrinin mycotoxin in food is an example.[183] Citrinin is mainly found in stored grains. However, sometimes it is encountered in fruits and other plant products at a micro- to nanomolar concentration.[184] This makes the MIP-based chemosensor described above inapplicable.

Detection of aromatic amines, *e.g.*, histamine and melamine, are crucial in control of food quality. For instance, histamine is an indicator of food spoilage while melamine has well been known as an adulterant for feedstock and milk in China for several years now. That was possible because melamine can make diluted or low-quality food material appear to be of higher protein content by elevating the total nitrogen content, confirmed by a simple protein test. Although combination of the acrylic-based MIP and the SWV transducer could synergistically determine melamine in the femtomolar concentration range, it is worth mentioning that recognizing primary amine of histamine or melamine using a polymer of crown-6-ether functionalized thiophene is also a great approach because of strong inclusion chemistry revealed by computational calculations.[106] In effect, the PM chemosensors based on these MIP recognition films exhibited high selectivity to most interferences at the LOD of a few nanomoles.

Similar to histamine, α-tocopherol is a significant indicator of quality of food lipids and related formulations,[185] *e.g.*, vegetable oils. Herein, Au-coated dendrites are used as an effective SERS substrate because they are demonstrated to be better SERS enhancers than uncoated dendrites and also to present higher spectral resolution. Moreover, Au-coated Ag dendrite substrates are stable for SERS analysis, although exhibiting slow degradation.[186] Therefore, the LOD of MAA-based chemosensor reached down to 10 ppb of α-tocopherol.

2.4 Conclusion

In an effort to cover a very broad and important topic of synthetic chemistry for molecular imprinting, we discussed in detail the strategic approach including (i) selection of functional monomers for the target analyte using appropriate interactions and (ii) fabrication of an MIP through abundant imprinting methods involving pre-polymerization complex formation, and then this complex polymerization followed by template extraction (Section 2.2). Hopefully, the present chapter is useful as an effective tool for solving the molecular-imprinting "jigsaw" in analytical application where target analytes are sophisticated, *i.e.*, varying in size, in binding sites, and even in charges. Not only are highly selective MIPs demanded but also sensitive transducers must be used towards analyte sensing. Therefore, Section 2.3 summarizes most recent and efficient MIP-based chemosensors being used in health monitoring as well as in the food and environment quality control.

List of Abbreviations

3Slac	3′-Sialyllactose
6SLac	6′-Sialyllactose
8-OHdG	8-Hydroxy-2′-deoxyguanosine
cfu	Colony forming unit
DGr	Dopamine graphene
DPV	Differential pulse voltammetry
DPSV	Differential pulse stripping voltammetry
EGDMA	Ethylene glycol dimethacrylate
EG-FET	Extended-gate field-effect transistor
EIS	Electrochemical impedance spectroscopy
EQCM	Electrochemical quartz crystal microbalance
FET	Field-effect transistor
HPLC	High-performance liquid chromatography
HTM	Heat-transfer method
IIP	Ion-imprinted polymer
LOD	Limit of detection
LOQ	Limit of quantitation

LSPR	Local surface plasmon resonance
MAA	Methacrylic acid
MMA	Methyl methacrylate
MIP	Molecularly imprinted polymer
NGAL	Neutrophil gelatinase-associated lipocalin
N-GQDs	Nitrogen-doped graphene QDs
NNAL	4-(Methylnitrosamino)-1-(3-pyridyl)-1-butanol
NPEDMA	*N*-Phenylethylene diamine methacrylamide
pAb	Polyclonal antibody
PDA	Polydopamine
PID	Proportional-integral-derivative
PG	Propyl gallate
PSA	Prostate specific antigen
PM	Piezoelectric microgravimetry
PU	Polyurethane
QCM	Quartz crystal microbalance
QD	Quantum dot
S/N	Signal-to-noise (ratio)
SERS	Surface-enhanced Raman spectroscopy
SIP	Surface-imprinted polymers
SWV	Square wave voltammetry
SPR	Surface plasmon resonance (spectroscopy)
TRIM	Trimethylpropane trimethacrylate
TNP	2,3,6-Trinitrophenol
TNT	2,3,6-Trinitrotoluene
vdW	van der Waals
XPS	X-Ray photoelectron spectroscopy

Acknowledgements

TPH thanks the starting fund from AAU's Research Profiling (Grant No. 301843) for financial support.

References

1. F. H. Dickey, Specific adsorption, *J. Phys. Chem.*, 1955, **59**, 695.
2. G. Wulff and A. Sarhan, Über die Anwendung von enzymanalog gebauten Polymeren zur Racemattrennung, *Angew. Chem.*, 1972, **84**, 364.
3. G. Wulff, A. Sarhan and K. Zabrocki, Enzyme-analogue built polymers and their use for the resolution of racemates, *Tetrahedron Lett.*, 1973, **14**, 4329.
4. G. Wulff, Enzyme-like catalysis by molecularly imprinted polymers, *Chem. Rev.*, 2002, **102**, 1.
5. E. Hedborg, F. Winquist, I. Lundström, L. I. Andersson and K. Mosbach, Some studies of molecularly-imprinted polymer membranes in combination with field-effect devices, *Sens. Actuators, A*, 1993, **37–38**, 796.

6. N. F. Starodub, S. A. Piletsky, N. V. Lavryk and A. V. El'skaya, Template sensors for low weight organic molecules based on SiO_2 surfaces, *Sens. Actuators, B*, 1993, **13–14**, 708.
7. S. A. Piletsky, Y. P. Parhometz, N. V. Lavryk, T. L. Panasyuk and A. V. El'skaya, Sensors for low-weight organic molecules based on molecular imprinting technique, *Sens. Actuators, B*, 1994, **18–19**, 629.
8. D. Kriz, O. Ramstrum, A. Svensron and K. Mosbach, Introducing biomimetic sensors based on molecularly imprinted polymers as recognition elements, *Anal. Chem.*, 1995, **67**, 2142.
9. J. Matsui, M. Higashi and T. Takeuchi, Molecularly imprinted polymer as 9-ethyladenine receptor having a porphyrin-based recognition center, *J. Am. Chem. Soc.*, 2000, **122**, 5218.
10. R. Xing, S. Wang, Z. Bie, H. He and Z. Liu, Preparation of molecularly imprinted polymers specific to glycoproteins, glycans and monosaccharides *via* boronate affinity controllable–oriented surface imprinting, *Nat. Protoc.*, 2017, **12**, 964.
11. X. Luo, F. Deng, L. Min, S. Luo, B. Guo, G. Zeng and C. Au, Facile one-step synthesis of inorganic-framework molecularly imprinted TiO2/WO3 nanocomposite and its molecular recognitive photocatalytic degradation of target contaminant, *Environ. Sci. Technol.*, 2013, **47**, 7404.
12. F. Deng, Y. Xubiao, L. Shaolin, W. Sheng, L. Luo, C. Au and R. Qi, Sol-hydrothermal synthesis of inorganic-framework molecularly imprinted TiO2/SiO2 nanocomposite and its preferential photocatalytic degradation towards target contaminant, *J. Hazard. Mater.*, 2014, **278**, 108.
13. X. Kan, Y. Zhao, Z. W. Zhirong Geng and J.-J. Zhu, Composites of multiwalled carbon nanotubes and molecularly imprinted polymers for dopamine recognition, *J. Phys. Chem. C*, 2008, **112**, 4849.
14. Y.-K. Lv, L.-M. Wang, L. Yang, C.-X. Zhao and H.-W. Sun, Synthesis and application of molecularly imprinted poly(methacrylic acid)–silica hybrid composite material for selective solid-phase extraction and high-performance liquid chromatography determination of oxytetracycline residues in milk, *J. Chromatogr. A*, 2012, **1227**, 48.
15. T. Shahar, T. Sicron and D. Mandler, Nanosphere molecularly imprinted polymers doped with gold nanoparticles for high selectivity molecular sensors, *Nano Res.*, 2017, **10**, 1056.
16. M. J. Whitcombe, N. Kirsch and I. A. Nicholls, Molecular imprinting science and technology: a survey of the literature for the years 2004–2011, *J. Mol. Recognit.*, 2014, **27**, 297.
17. H. Diliën, M. Peeters, J. Royakkers, J. Harings, P. Cornelis, P. Wagner, E. S. Redeker, C. E. Banks, K. Eersels, B. V. Grinsven and T. J. Cleij, Label-free detection of small organic molecules by molecularly imprinted polymer functionalized thermocouples: toward *in vivo* applications, *ACS Sens.*, 2017, **2**, 583.
18. S. Wagner, J. Bell, M. Biyikal, K. Gawlitza and K. Rurack, Integrating fluorescent molecularly imprinted polymer (MIP) sensor particles with

a modular microfluidic platform for nanomolar small-molecule detection directly in aqueous samples, *Biosens. Bioelectron.*, 2018, **99**, 244.

19. Z. Iskierko, P. SindhuSharma, K. Bartold, A. Pietrzyk-Le, K. Noworyta and W. Kutner, Molecularly imprinted polymers for separating and sensing of macromolecular compounds and microorganisms, *Biotechnol. Adv.*, 2016, **34**, 30.
20. X. Shen, J. S. Bonde, T. Kamra, L. Bülow, J. C. Leo, D. Linke and L. Ye, Bacterial Imprinting at Pickering Emulsion Interfaces, *Angew. Chem., Int. Ed.*, 2014, **53**, 10687.
21. J. E. Lofgreen and G. A. Ozin, Controlling morphology and porosity to improve performance of molecularly imprinted sol–gel silica, *Chem. Soc. Rev.*, 2014, **43**, 911.
22. C. Alexander, H. S. Andersson, L. I. Andersson, R. J. Ansell, N. Kirsch, I. A. Nicholls, J. O'Mahony and M. J. Whitcombe, Molecular imprinting science and technology: a survey of the literature for the years up to and including 2003, *J. Mol. Recognit.*, 2006, **19**, 106.
23. A. R. Martin, J.-J. Vasseur and M. Smietana, Boron and nucleic acid chemistries: merging the best of both worlds, *Chem. Soc. Rev.*, 2013, **42**, 5684.
24. D. Li, Y. Chen and Z. Liu, Boronate affinity materials for separation and molecular recognition: structure, properties and applications, *Chem. Soc. Rev.*, 2015, **44**, 8097.
25. R. W. Gunasekara and Y. Zhao, A general method for selective recognition of monosaccharides and oligosaccharides in water, *J. Am. Chem. Soc.*, 2017, **139**, 829.
26. H. Sunayama, T. Ohta, A. Kuwahara and T. Takeuchi, Fluorescence signaling molecularly imprinted polymers for antibiotics prepared *via* site-directed post-imprinting introduction of plural fluorescent reporters within the recognition cavity, *J. Mater. Chem. B*, 2016, **4**, 7138.
27. M. Hussain, K. Kotova and P. A. Lieberzeit, Molecularly imprinted polymer nanoparticles for formaldehyde sensing with QCM, *Sensors*, 2016, **16**, 1011.
28. K. J. Shea and D. Y. Sasaki, An analysis of small-molecule binding to functionalized synthetic polymers by 13C CP/MAS NMR and FT-IR spectroscopy, *J. Am. Chem. Soc.*, 1991, **113**, 4109.
29. T. Takeuchi, T. Mori, A. Kuwahara, T. Ohta, A. Oshita, H. Sunayama, Y. Kitayama and T. Ooya, Conjugated-protein mimics with molecularly imprinted reconstructible and transformable regions that are assembled using space-filling prosthetic groups, *Angew. Chem., Int. Ed.*, 2014, **53**, 12765.
30. H. V. R. Burri and D. Yu, Covalent imprinting and covalent rebinding of benzyl mercaptan: towards a facile detection of proteins, *Anal. Lett.*, 2017, **50**, 866.
31. M. Cieplak, K. Szwabinska, M. Sosnowska, B. K. C. Chandra, P. Borowicz, K. Noworyta, F. D'Souza and W. Kutner, Selective electrochemical sensing of human serum albumin by semi-covalent molecular imprinting, *Biosens. Bioelectron.*, 2015, **74**, 960.

32. M. Dabrowski, M. Cieplak, P. S. Sharma, P. Borowicz, K. Noworyta, W. Lisowski, F. D'Souza, A. Kuhn and W. Kutner, Hierarchical templating in deposition of semi-covalently imprinted inverse opal polythiophene film for femtomolar determination of human serum albumin, *Biosens. Bioelectron.*, 2017, **94**, 155.
33. D. A. Spivak, Optimization, evaluation, and characterization of molecularly imprinted polymers, *Adv. Drug Delivery Rev.*, 2005, **57**, 1779.
34. L. I. Andersson and K. Mosbach, Enantiomeric resolution on molecularly imprinted polymers prepared with only non-covalent and non-ionic interactions, *J. Chromatogr. A*, 1990, **316**, 313.
35. S. J. Grabowski, in *Practical Aspects of Computational Chemistry IV*, ed. S. J. Leszczynski, Springer, Boston, MA, 2016, p. 245.
36. P. Politzer, J. S. Murray and P. Lane, σ-Hole bonding and hydrogen bonding: Competitive interactions, *Int. J. Quantum Chem.*, 2007, **107**, 3046.
37. A. Bauzá, T. J. Mooibroek and A. Frontera, The bright future of unconventional σ/π-hole interactions, *ChemPhysChem*, 2015, **16**, 2496.
38. E. Arunan, G. R. Desiraju, R. A. Klein, J. Sadlej, S. Scheiner, I. Alkorta, D. C. Clary, R. H. Crabtree, J. J. Dannenberg, P. Hobza, H. G. Kjaergaard, A. C. Legon and B. Mennucci, Definition of the hydrogen bond (IUPAC Recommendations 2011), *Pure Appl. Chem.*, 2011, **83**, 1637.
39. G. R. Desiraju, P. S. Ho, L. Kloo, A. C. Legon, R. Marquardt, P. Metrangolo, P. Politzer, G. Resnati and K. Rissanen, Definition of the halogen bond (IUPAC Recommendations 2013), *Pure Appl. Chem.*, 2013, **85**, 1711.
40. I. A. Nicholls, O. Ramström and K. Mosbach, Insights into the role of the hydrogen bond and hydrophobic effect on recognition in molecularly imprinted polymer synthetic peptide receptor mimics, *J. Chromatogr. A*, 1995, **691**, 349.
41. T. Zhang, F. Liu, W. Chen, J. Wang and K. Li, Influence of intramolecular hydrogen bond of templates on molecular recognition of molecularly imprinted polymers, *Anal. Chim. Acta*, 2001, **450**, 53.
42. Y. Zhou, Z.-b. Qu, Y. Zeng, T. Zhou and G. Shi, A novel composite of graphene quantum dots and molecularly imprinted polymer for fluorescent detection of paranitrophenol, *Biosens. Bioelectron.*, 2014, **52**, 317.
43. T. Takeuchi, Y. Minato, M. Takase and H. Shinmori, Molecularly imprinted polymers with halogen bonding-based molecular recognition sites, *Tetrahedron Lett.*, 2005, **46**, 9025.
44. H. Shi, J. Zhao, Y. Wang and G. Zhao, A highly selective and picomolar level photoelectrochemical sensor for PCB 101 detection in environmental water samples, *Biosens. Bioelectron.*, 2016, **81**, 503.
45. K. Berland, V. R. Cooper, K. Lee, E. Schröder, T. Thonhauser, P. Hyldgaard and B. I. Lundqvist, van der Waals forces in density functional theory: a review of the vdW-DF method, *Rep. Prog. Phys.*, 2015, **78**, 066501.
46. O. Lopez-Sanchez, E. A. Llado, V. Koman, A. F. I. Morral, A. Radenovic and A. Kis, Light generation and harvesting in a van der Waals heterostructure, *ACS Nano*, 2014, **8**, 3042.

47. A. Tkatchenko, Current understanding of van der waals effects in realistic materials, *Adv. Funct. Mater.*, 2014, **25**, 2054.
48. Y. Liu, N. O. Weiss, X. Duan, H.-C. Cheng, Y. Huang and X. Duan, Van der Waals heterostructures and devices, *Nat. Rev. Mater.*, 2016, **1**, 16042.
49. K. S. Novoselov, A. Mishchenko, A. Carvalho and A. H. C. Neto, 2D materials and van der Waals heterostructures, *Science*, 2016, **353**, aac9439.
50. C. Chothia and J. Janin, Principles of protein–protein recognition, *Nature*, 1975, **256**, 705.
51. J. Wang, K. Liu, R. Xing and X. Yan, Peptide self-assembly: thermodynamics and kinetics, *Chem. Soc. Rev.*, 2016, **45**, 5589.
52. A. D. Cort, L. Mandolini and L. Schiaffino, The role of attractive van der waals forces in the catalysis of michael addition by a phenyl decorated uranyl-salophen complex, *J. Org. Chem.*, 2008, **73**, 9439.
53. J. C. F. Rodriguez-Reyes, C. G. F. Siler, W. Liu, A. Tkatchenko, C. M. Friend and R. J. Madix, van der Waals interactions determine selectivity in catalysis by metallic gold, *J. Am. Chem. Soc.*, 2014, **136**, 13333.
54. V. v. A. Castelli, A. D. Cort, L. Mandolini, V. Pinto, D. N. Reinhoudt, F. Ribaudo, C. Sanna, L. Schiaffino and B. H. M. Snellink-Ruël, Molecular recognition of carbonyl compounds by uranyl-salophen based neutral receptors driven by van der Waals forces, *Supramol. Chem.*, 2002, **14**, 211.
55. L. Schwarz, M. C. Bowyer, C. I. Holdsworth and A. McCluskey, Synthesis and evaluation of a molecularly imprinted polymer selective to 2,4,6-trichloroanisole, *Aust. J. Chem.*, 2006, **59**, 129.
56. Y. Li, H.-H. Yang, Q.-H. You, Z.-X. Zhuang and X.-R. Wang, Protein recognition *via* surface molecularly imprinted polymer nanowires, *Anal. Chem.*, 2006, **78**, 317.
57. I. Gavrilović, K. Mitchell, A. D. Brailsford, D. A. Cowan, A. T. Kicman and R. J. Ansell, A molecularly imprinted receptor for separation of testosterone and epitestosterone, based on a steroidal cross-linker, *Steroids*, 2011, **76**, 478.
58. P. Atkins and J. d. Paula, *ATKINS' Physical Chemistry*, Oxford University Press, Great Britain, 2006, p. 1053.
59. G. Z. Kyzas, N. K. Lazaridis and D. N. Bikiaris, Optimization of chitosan and B-cyclodextrin molecularly imprinted polymer synthesis for dye adsorption, *Carbohydr. Polym.*, 2013, **91**, 198.
60. S. A. Piletsky, H. S. Andersson and I. A. Nicholls, Combined hydrophobic and electrostatic interaction-based recognition in molecularly imprinted polymers, *Macromolecules*, 1999, **32**, 633.
61. M. J. Whitcombe, I. Chianella, L. Larcombe, S. A. Piletsky, J. Noble, R. Porter and A. Horgan, The rational development of molecularly imprinted polymer-based sensors for protein detection, *Chem. Soc. Rev.*, 2011, **40**, 1547.
62. S. H. Ou, M. C. Wu, T. C. Chou and C. C. Liu, Polyacrylamide gels with electrostatic functional groups for the molecular imprinting of lysozyme, *Anal. Chim. Acta*, 2004, **504**, 163.

63. R. D. Hancock and A. E. Martell, Ligand design for selective complexation of metal ions in aqueous solution, *Chem. Rev.*, 1989, **89**, 1875.
64. R. D. Hancock, Chelate ring size and metal ion selection. The basis of selectivity for metal ions in open-chain ligands and macrocycles, *J. Chem. Educ.*, 1992, **69**, 615.
65. A. E. Martell and R. D. Hancock, *Metal Complexes in Aqueous Solutions*, Springer, NY, 1st edn, 1996.
66. T. P. Rao, R. Kala and S. Daniel, Metal ion-imprinted polymers—novel materials for selective recognition of inorganics, *Anal. Chim. Acta*, 2006, **578**, 105.
67. C. Branger, W. Meouche and A. Margaillan, Recent advances on ion-imprinted polymers, *React. Funct. Polym.*, 2013, **73**, 859.
68. Y.-K. Tsoi, Y.-M. Ho and K. S.-Y. Leung, Selective recognition of arsenic by tailoring ion-imprinted polymer for ICP-MS quantification, *Talanta*, 2012, **89**, 162.
69. O. Hayden, C. Haderspock, S. Krassnig, X. Chen and F. L. Dickert, Surface imprinting strategies for the detection of trypsin, *Analyst*, 2006, **131**, 1044.
70. K. Eersels, P. Lieberzeit and P. Wagner, A review on synthetic receptors for bioparticle detection created by surface-imprinting techniques - from principles to applications, *ACS Sens.*, 2016, **1**, 1171.
71. Y. Kondo, M. Yoshikawa and H. Okushita, Molecularly imprinted polyamide membranes for chiral recognition, *Polym. Bull.*, 2000, **44**, 517.
72. K. Lettau, A. Warsinke, A. Laschewsky, K. Mosbach, E. Yilmaz and F. W. Scheller, An esterolytic imprinted polymer prepared *via* a silica-supported transition state analogue, *Chem. Mater.*, 2004, **16**, 2745.
73. F. Locatelli, P. Gamez and M. Lemaire, Molecular imprinting of polymerised catalytic complexes in asymmetric catalysis, *J. Mol. Catal. A: Chem.*, 1998, **135**, 89.
74. M. Tada and Y. Iwasawa, Design of molecular-imprinting metal-complex catalysts, *J. Mol. Catal. A: Chem.*, 2003, **199**, 115.
75. F. L. Dickert, P. Forth, P. Lieberzeit and M. Tortschanoff, Molecular imprinting in chemical sensing – Detection of aromatic and halogenated hydrocarbons as well as polar solvent vapors, *Fresenius. J. Anal. Chem.*, 1998, **360**, 759.
76. M. Glad, O. Norrlöw, B. Sellergren, N. Siegbahn and K. Mosbach, Use of silane monomers for molecular imprinting and enzyme entrapment in polysiloxane-coated porous silica, *J. Chromatogr. A*, 1985, **347**, 11.
77. K. Wang, X. Guan, S. Chai, Q. Zou, X. Zhang and J. Zhang, A novel, molecularly imprinted polymer sensor made using an oligomeric methyl silsesquioxane–TiO2 composite sol on a glassy carbon electrode for the detection of procainamide hydrochloride, *Biosens. Bioelectron.*, 2015, **64**, 94.
78. N. Blanchemain, S. Haulon, B. Martel, M. Traisnel, M. Morcellet and H. F. Hildebrand, Vascular PET prostheses surface modification with cyclodextrin coating: development of a new drug delivery system, *Eur. J. Vasc. Endovasc. Surg.*, 2005, **29**, 628.

79. R. C. P. Cubbon, *Polymeric Precursors for Ceramic Materials*, iSmithers Rapra Publishing, Shropshire, UK, 1994, vol. 7, p. 86.
80. B. v. Grinsven, K. Eersels, O. Akkermans, S. Ellermann, A. Kordek, M. Peeters, O. Deschaume, C. Bartic, H. Diliën, E. S. Redeker, P. Wagner and T. J. Cleij, Label-Free detection of *Escherichia coli* based on thermal transport through surface imprinted polymers, *ACS Sens.*, 2016, **1**, 1140.
81. F. L. Dickert, M. Tortschanoff, W. E. Bulst and G. Fischerauer, Molecularly imprinted sensor layers for the detection of polycyclic aromatic hydrocarbons in water, *Anal. Chem.*, 1999, **71**, 4559.
82. A. D. Jenkins, P. Kratochvíl, R. F. T. Stepto and U. W. Suter, Glossary of basic terms in polymer science (IUPAC Recommendations 1996), *Pure Appl. Chem.*, 1996, **68**, 2287.
83. P. A. G. Cormack and A. Z. Elorza, Molecularly imprinted polymers: synthesis and characterisatio, *J. Chromatogr. A*, 2004, **804**, 173.
84. T.-P. Huynh, P. S. Sharma, M. Sosnowska, F. D'Souza and W. Kutner, Functionalized polythiophenes: Recognition materials for chemosensors and biosensors of superior sensitivity, selectivity, and detectability, *Prog. Polym. Sci.*, 2015, **47**, 1.
85. M. D. Attieh, Y. Zhao, A. Elkak, A. Falcimaigne-Cordin and K. Haupt, Enzyme-initiated free-radical polymerization of molecularly imprinted polymer nanogels on a solid phase with an immobilized radical source, *Angew. Chem., Int. Ed.*, 2017, **56**, 3339.
86. S. A. Piletsky, H. Matuschewski, U. Schedler, A. Wilpert, E. V. Piletska, T. A. Thiele and M. Ulbricht, Surface functionalization of porous polypropylene membranes with molecularly imprinted polymers by photograft copolymerization in water, *Macromolecules*, 2000, **33**, 3092.
87. P. Li, T. Wang, F. Lei, P. Tang, X. Tan, Z. Liu and L. Shen, Rosin-based molecularly imprinted polymers as the stationary phase in high-performance liquid chromatography for selective separation of berberine hydrochloride, *Polym. Int.*, 2014, **63**, 1699.
88. H. J. Naghasha, O. Okay and Y. Yaĝci, Gel formation by chain-crosslinking photopolymerization of methyl methacrylate and ethylene glycol dimethacrylate, *Polymer*, 1997, **38**, 1187.
89. S. Zeng, Y. She, B. Jiao, G. Liu, J. Wang, X. Su, X. Ma, M. Jin, F. Jin and S. Wang, Molecularly imprinted polymer for selective extraction and simultaneous determination of four tropane alkaloids from Przewalskia tangutica Maxim. fruit extracts using LC-MS/MS, *RSC Adv.*, 2015, **5**, 94997.
90. V. M. Biju, J. M. Gladis and T. P. Rao, Ion imprinted polymer particles: synthesis, characterization and dysprosium ion uptake properties suitable for analytical applications, *Anal. Chim. Acta*, 2003, **478**, 43.
91. H. Taguchi, H. Sunayama, E. Takano, Y. Kitayama and T. Takeuchi, Preparation of molecularly imprinted polymers for the recognition of proteins *via* the generation of peptide-fragment binding sites by semi-covalent imprinting and enzymatic digestion, *Analyst*, 2015, **140**, 1448.

92. S. He, M. J. Yaszemski, A. W. Yasko, P. S. Engel and A. G. Mikos, Injectable biodegradable polymer composites based on poly(propylene fumarate) crosslinked with poly(ethylene glycol)-dimethacrylate, *Biomaterials*, 2000, **21**, 2389.
93. S. Rimmer, Synthesis of molecular imprinted polymer networks, *Chromatographia*, 1998, **47**, 470.
94. A. G. Strikovsky, D. Kasper, M. Grün, B. S. Green, J. Hradil and G. Wulff, Catalytic molecularly imprinted polymers using conventional bulk polymerization or suspension polymerization: selective hydrolysis of diphenyl carbonate and diphenyl carbamate, *J. Am. Chem. Soc.*, 2000, **122**, 6295.
95. S. A. Mohajeri, G. Karimi, J. Aghamohammadian and M. R. Khansari, Clozapine recognition *via* molecularly imprinted polymers; bulk polymerization *versus* precipitation method, *J. Appl. Polym. Sci.*, 2011, **121**, 3590.
96. G. Odian, *Principles of Polymerization*, John Wiley & Sons, New Jersey, 4th edn, 2004, p. 832.
97. L. D. Mafu, T. A. M. Msagati and B. B. Mamba, Ion-imprinted polymers for environmental monitoring of inorganic pollutants: synthesis, characterization, and applications, *Environ. Sci. Pollut. Res.*, 2013, **20**, 790.
98. L. Chen, X. Wang, W. Lu, X. Wu and J. Li, Molecular imprinting: perspectives and applications, *Chem. Soc. Rev.*, 2016, **45**, 2137.
99. A. Ellwanger, C. Berggren, S. Bayoudh, C. Crecenzi, L. Karlsson, P. K. Owens, K. Ensing, P. Cormack, D. Sherrington and B. Sellergren, Evaluation of methods aimed at complete removal of template from molecularly imprinted polymers, *Analyst*, 2001, **126**, 784.
100. P. Zahedi, M. Ziaee, M. Abdouss, A. Farazincan and B. Mizaikoff, Biomacromolecule template-basedmolecularly imprinted polymers with anemphasis on their synthesis strategies: areview, *Polym. Adv. Technol.*, 2016, **27**, 1124.
101. C. Wierzbicka, S. B. Torsetnes, O. N. Jensen, S. Shinde and B. Sellergren, Hierarchically templated beads with tailored pore structure for phosphopeptide capture and phosphoproteomics, *RSC Adv.*, 2017, **7**, 17154.
102. H. R. Culver, S. D. Steichen and N. A. Peppas, A closer look at the impact of molecular imprinting on adsorption capacity and selectivity for protein templates, *Biomacromolecules*, 2016, **17**, 4045.
103. J. Erdossy, E. Kassa, A. Farkas and V. Horvath, Enzymatic digestion as a tool for removing proteinaceous templates from molecularly imprinted polymers, *Anal. Methods*, 2017, **9**, 4496.
104. T.-P. Huynh, P. Pieta, F. D'Souza and W. Kutner, Molecularly imprinted polymer for recognition of 5-fluorouracil by the RNA-type nucleobase pairing, *Anal. Chem.*, 2013, **85**, 8304.
105. T.-P. Huynh, M. I. Sosnowska, J. W. Sobczak, C. B. KC, V. N. Nesterov, F. D'Souza and W. Kutner, Simultaneous chronoamperometry and piezoelectric microgravimetry determination of nitroaromatic explosives

using molecularly imprinted thiophene polymers, *Anal. Chem.*, 2013, **85**, 8361.

106. T.-P. Huynh, C. B. KC, W. Lisowski, F. D'Souza and W. Kutner, Molecularly imprinted polymer of bis(2,2′-bithienyl)methanes for selective determination of adrenaline, *Bioelectrochemistry*, 2013, **93**, 37.
107. P. S. Sharma, M. Dabrowski, K. Noworyta, T.-P. Huynh, C. B. KC, J. W. Sobczak, P. Pieta, F. D'Souza and W. Kutner, Fullerene derived molecularly imprinted polymer for chemosensing of adenosine-5′-triphosphate (ATP), *Anal. Chim. Acta*, 2014, **844**, 61.
108. T.-P. Huynh, A. Wojnarowicz, A. Kelm, P. Woznicki, P. Borowicz, A. Majka, F. D'Souza and W. Kutner, Chemosensor for selective determination of 2,4,6-trinitrophenol using a custom designed imprinted polymer recognition unit cross-linked to a fluorophore transducer, *ACS Sens.*, 2016, **1**, 636.
109. A. Wojnarowicz, P. S. Sharma, M. Sosnowska, W. Lisowski, T.-P. Huynh, M. Pszona, P. Borowicz, F. D'Souza and W. Kutner, An electropolymerized molecularly imprinted polymer for selective carnosine sensing with impedimetric capacity, *J. Mater. Chem. B*, 2016, **4**, 1156.
110. U. Latif, S. Can, O. Hayden, P. Grillberger and F. L. Dickert, Sauerbrey and anti-Sauerbrey behavioral studies in QCM sensors—Detection of bioanalytes, *Sens. Actuators, B*, 2013, **176**, 825.
111. W. J. Cheong, S. H. Yang and F. Ali, Molecular imprinted polymers for separation science: a review of reviews, *J. Sep. Sci.*, 2012, **36**, 609.
112. R. Langer and N. A. Peppas, Advances in biomaterials, drug delivery, and bionanotechnology, *AIChE J.*, 2003, **49**, 2990.
113. B. Sellergren and C. J. Allender, Molecularly imprinted polymers: A bridge to advanced drug delivery, *Adv. Drug Delivery Rev.*, 2005, **57**, 1733.
114. D. Kriz, O. Ramström and K. Mosbach, Molecular imprinting new possibilities for sensor technology, *Anal. Chem.*, 1997, **69**, 345A.
115. A. P. F. Turner, Biosensors: sense and sensibility, *Chem. Soc. Rev.*, 2013, **42**, 3184.
116. A. A. Volkert and A. J. Haes, Advancements in nanosensors using plastic antibodies, *Analyst*, 2014, **139**, 21.
117. P. S. Sharma, Z. Iskierko, A. Pietrzyk-Le, F. D'Souza and W. Kutner, Bioinspired intelligent molecularly imprinted polymers for chemosensing: A mini review, *Electrochem. Commun.*, 2015, **50**, 81.
118. J. Janata, *Principles of Chemical Sensors*, Plenum Press, New York, 2nd edn, 2009, p. 373.
119. J. Matsui, K. Akamatsu, S. Nishiguchi, D. Miyoshi, H. Nawafune, K. Tamaki and N. Sugimoto, Composite of Au nanoparticles and molecularly imprinted polymer as a sensing material, *Anal. Chem.*, 2004, **76**, 1310.
120. P. Turkewitsch, B. Wandelt, G. D. Darling and W. S. Powell, Fluorescent Functional recognition sites through molecular imprinting. a polymer-based fluorescent chemosensor for aqueous cAMP, *Anal. Chem.*, 1998, **70**, 2025.

121. R. C. Stringer, S. Gangopadhyay and S. A. Grant, Detection of nitroaromatic explosives using a fluorescent-labeled imprinted polymer, *Anal. Chem.*, 2010, **82**, 4015.
122. N. Murase, S.-i. Taniguchi, E. Takano, Y. Kitayama and T. Takeuchi, A molecularly imprinted nanocavity-based fluorescence polarization assay platform for cortisol sensing, *J. Mater. Chem. B*, 2016, 1770.
123. L. Ye and K. Mosbach, Polymers recognizing biomolecules based on a combination of molecular imprinting and proximity scintillation: a new sensor concept, *J. Am. Chem. Soc.*, 2001, **123**, 2901.
124. O. Y. F. Henry, D. C. Cullen and S. A. Piletsky, Optical interrogation of molecularly imprinted polymers and development of MIP sensors: a review, *Anal. Bioanal. Chem.*, 2005, **382**, 947.
125. L. Basabe-Desmonts, D. N. Reinhoudt and M. Crego-Calama, Design of fluorescent materials for chemical sensing, *Chem. Soc. Rev.*, 2007, **36**, 993.
126. H. N. Kim, Z. Guo, W. Zhu, J. Yoon and H. Tian, Recent progress on polymer-based fluorescent and colorimetric chemosensors, *Chem. Soc. Rev.*, 2011, **40**, 79.
127. W. Wan, S. Wagner and K. Rurack, Fluorescent monomers: "bricks" that make a molecularly imprinted polymer "bright", *Anal. Bioanal. Chem.*, 2016, **408**, 1753.
128. F. Canfarotta, M. J. Whitcombe and S. A. Piletsky, Polymeric nanoparticles for optical sensing, *Biotechnol. Adv.*, 2013, **31**, 1585.
129. C. Fang, Z. Chen, M. Megharaj and R. Naidu, Potentiometric detection of AFFFs based on MIP, *Environ. Technol. Innovation*, 2016, **5**, 52.
130. N. Karimian, M. Vagin, M. H. A. Zavar, M. Chamsaz, A. P. F. Turner and A. Tiwari, An ultrasensitive molecularly-imprinted human cardiac troponinsensor, *Biosens. Bioelectron.*, 2013, **50**, 492.
131. Z. Iskierko, P. S. Sharma, D. Prochowicz, K. Fronc, F. D'Souza, D. Toczydłowska, F. Stefaniak and K. Noworyta, Molecularly Imprinted Polymer (MIP) film with improved surface area developed by using Metal–Organic Framework (MOF) for sensitive lipocalin (NGAL) determination, *ACS Appl. Mater. Interfaces*, 2016, **8**, 19860.
132. G. Wackers, T. Vandenryt, P. Cornelis, E. Kellens, R. Thoelen, W. D. Ceuninck, P. Losada-Pérez, B. v. Grinsven, M. Peeters and P. Wagner, Array formatting of the Heat-Transfer Method (HTM) for the detection of small organic molecules by molecularly imprinted polymers, *Sensors*, 2014, **14**, 11016.
133. P. S. Sharma, F. D'Souza and W. Kutner, in *Portable Chemical Sensors: Weapons against Bioterrorism, NATO Science for Peace and Security Series a: Chemistry and Biology*, ed. D. P. Nikolelis, Springer, Dordrecht, Netherland, 2012, p. 63.
134. P. S. Sharma, F. D'Souza and W. Kutner, Molecular imprinting for selective chemosensing of environmental hazards and drugs of abuse, *Trends Anal. Chem.*, 2012, **34**, 59.

135. Y. Wang, Z. Zhang, V. Jain, J. Yi, S. Mueller, J. Sokolov, Z. Liu, K. Levon, B. Rigas and M. H. Rafailovich, Potentiometric sensors based on surface molecular imprinting: Detection of cancer biomarkers and viruses, *Sens. Actuators, B*, 2010, **146**, 381.
136. G. V. Martins, A. C. Marques, E. Fortunato, M. Goreti and F. Sales, 8-hydroxy-2′-deoxyguanosine (8-OHdG) biomarker detection down to picoMolar level on a plastic antibody film, *Biosens. Bioelectron.*, 2016, **86**, 225.
137. Y. Hoshino, T. Kodama, Y. Okahata and K. J. Shea, peptide imprinted polymer nanoparticles: a plastic antibody, *J. Am. Chem. Soc.*, 2008, **130**, 15242.
138. M. Muti, M. Soysal, F. M. Nacak, K. Gençdağ and A. E. Karagözler, A novel DNA probe based on molecularly imprinted polymer modified electrode for the electrochemical monitoring of DNA, *Electroanalysis*, 2015, **27**, 1368.
139. K. Bartold, A. Pietrzyk-Le, T.-P. Huynh, Z. Iskierko, M. Sosnowska, K. Noworyta, W. Lisowski, F. Sannicolò, S. Cauteruccio, E. Licandro, F. D'Souza and W. Kutner, Programmed transfer of sequence information into a molecularly imprinted polymer for hexakis(2,2′-bithien-5-yl) DNA analogue formation toward single-nucleotide-polymorphism detection, *ACS Appl. Mater. Interfaces*, 2017, **9**, 3948.
140. K. D. Shimizu and C. J. Stephenson, Molecularly imprinted polymer sensor arrays, *Curr. Opin. Chem. Biol.*, 2005, **14**, 743.
141. T.-P. Huynh and W. Kutner, Molecularly imprinted polymers as recognition materials for electronic tongues, *Biosens. Bioelectron.*, 2015, **74**, 856.
142. D. C. Harris, *Quantitative Chemical Analysis*, W. H. Freeman, New York, USA, 7th edn, 2006, p. 1008.
143. D.-F. Tai, C.-Y. Lin, T.-Z. Wu, J.-H. Huang and P.-Y. Shu, Artificial receptors in serologic tests for the early diagnosis of dengue virus infection, *Clin. Chem.*, 2006, **52**, 1486.
144. R. D. Kryscio and A. N. Peppas, Critical review and perspective of macromolecularly imprinted polymers, *Acta Biomater.*, 2012, **8**, 461.
145. N. W. Turner, C. W. Jeans, K. R. Brain, C. J. Allender, V. Hlady and D. W. Britt, From 3D to 2D: a review of the molecular imprinting of proteins, *Biotechnol. Prog.*, 2006, **22**, 1474.
146. S. Chen, X. Chen, L. Zhang, J. Gao and Q. Ma, Electrochemiluminescence detection of *Escherichia coli* O157:H7 based on a novel polydopamine surface imprinted polymer biosensor, *ACS Appl. Mater. Interfaces*, 2017, **9**, 5430.
147. Z. Lin, C. M. Yip, I. S. Joseph and M. D. Ward, Operation of an ultrasensitive 30-mhz quartz crystal microbalance in liquids, *Anal. Chem.*, 1993, **65**, 1546.
148. H. Furusawa, T. Ozeki, M. Morita and Y. Okahata, Added mass effect on immobilizations of proteins on a 27 MHz Quartz crystal microbalance in aqueous solution, *Anal. Chem.*, 2009, **81**, 2268.

149. Y. Ge and A. P. F. Turner, Too large to fit? Recent developments in macromolecular imprinting, *Trends Biotechnol.*, 2008, **26**, 218.
150. S. Nishitani, T. Kajisa and T. Sakata, Development of molecularly imprinted polymer-based field effect transistor for sugar chain sensing, *Jpn. J. Appl. Phys.*, 2017, **56**, 04CM02.
151. Y. Xia, J. E. McGuffey, S. Bhattacharyya, B. Sellergren, E. Yilmaz, L. Wang and J. T. Bernert, Analysis of the tobacco-specific nitrosamine 4-(methylnitrosamino)-1-(3-pyridyl)-1-butanol in urine by extraction on a molecularly imprinted polymer column and liquid chromatography/atmospheric pressure ionization tandem mass spectrometry, *Anal. Chem.*, 2005, **77**, 7639.
152. A. J. Bard and L. R. Faulkner, Electrochemical Methods: Fundamentals and Applications, John Wiley & Sons, Inc., New York, 2nd edn, 2001, p. 864.
153. H. Yang, L. Li, Y. Ding, D. Ye, Y. Wang, S. Cui and L. Liao, Molecularly imprinted electrochemical sensor based on bioinspired Au microflowers for ultra-trace cholesterol assay, *Biosens. Bioelectron.*, 2017, **92**, 748.
154. *Comparative Quantification of Health Risks*, ed. M. Ezzati, A. D. Lopez, A. Rodgers and C. J. L. Murray, World Health Organization, Switzerland, 2004, vol. 1.
155. T. Shibamoto and L. F. Bjeldanes, *Introduction to Food Toxicology*, Elsevier, Burlington, MA, 2nd edn, 2009, p. 320.
156. S. Lesage and R. E. Jackson, *Groundwater Contamination and Analysis at Hazardous Waste Sites*, Marcel Dekker, Inc., New York, USA, 1992, p. 552.
157. P. Patnaik, *A Comprehensive Guide to the Hazardous Properties of Chemical Substances*, John Wiley & Sons, New Jersey, 3rd edn, 2007, p. 691.
158. P. B. Tchounwou, C. G. Yedjou, A. K. Patlolla and D. J. Sutton, in *Molecular, Clinical and Environmental Toxicology*, ed. A. Luch, Springer, Basel, 2012, vol. 101, p. 133.
159. S. Al-Kindy, R. Badfa, J. L. Suarez-Rodrfguez and M. E. Diez-Garcia, Molecularly imprinted polymers and optical sensing applications, *Crit. Rev. Anal. Chem.*, 2000, **30**, 291.
160. S. M. Ng and R. Narayanaswamy, Fluorescence sensor using a molecularly imprinted polymer as a recognition receptor for the detection of aluminium ions in aqueous media, *Anal. Bioanal. Chem.*, 2006, **386**, 1235.
161. J. Li, L. Zhang, G. Wei, Y. Zhang and Y. Zeng, Highly sensitive and doubly orientated selective molecularly imprinted electrochemical sensor for Cu^{2+}, *Biosens. Bioelectron.*, 2015, **69**, 316.
162. J. Zhou, K. Xiong, Y. Yang, X. Ye, J. Liu and F. Li, Deleterious effects of benomyl and carbendazim on human placental trophoblast cells, *Reprod. Toxicol.*, 2015, **51**, 64.
163. P. Sengupta and R. Banerjee, Environmental toxins: Alarming impacts of pesticides on male fertility, *Hum. Exp. Toxicol.*, 2014, **33**, 1017.

164. W. Han, L. Gao, X. Li, L. Wang, Y. Yan, G. Che, B. Hu, X. Lin and M. Song, A fluorescent molecularly imprinted polymer sensor synthesized by atom transfer radical precipitation polymerization for determination of ultra trace fenvalerate in the environment, *RSC Adv.*, 2016, **6**, 81346.
165. R. İlktaç, N. Aksuner and E. Henden, Selective and sensitive fluorimetric determination of carbendazim in apple and orange after preconcentration with magnetite-molecularly imprinted polymer, *Spectrochim. Acta, Part A*, 2017, **174**, 86.
166. C. Liu, Z. Song, J. Pan, X. Wei, L. Gao, Y. Yan, L. Li, J. Wang, R. Chen, J. Dai and P. Yu, Molecular imprinting in fluorescent particle stabilized pickering emulsion for selective and sensitive optosensing of λ-cyhalothrin, *J. Phys. Chem. C*, 2013, **117**, 10445.
167. Y. Kim, K. M. Lee and J. Y. Chang, Highly luminescent tetra(biphenyl-4-yl) ethene-grafted molecularly imprinted mesoporous silica nanoparticles for fluorescent sensing of diethylstilbestrol, *Sens. Actuators, B*, 2017, **242**, 1296.
168. Q. Jiang, D. Zhang, Y. Cao and N. Gan, An antibody-free and signal-on type electrochemiluminescence sensor for diethylstilbestrol detection based on magnetic molecularly imprinted polymers-quantum dots labeled aptamer conjugated probes, *J. Electroanal. Chem.*, 2017, **789**, 1.
169. H. Lu and S. Xu, Hollow mesoporous structured molecularly imprinted polymers for highly sensitive and selective detection of estrogens from food samples, *J. Chromatogr. A*, 2017, **1501**, 10.
170. T. Alizadeh, Preparation of magnetic TNT-imprinted polymer nanoparticles and their accumulation onto magnetic carbon paste electrode for TNT determination, *Biosens. Bioelectron.*, 2014, **61**, 532.
171. M. Li, H. Liu and X. Ren, Ratiometric fluorescence and mesoporous structured imprinting nanoparticles for rapid and sensitive detection 2,4,6-trinitrophenol, *Biosens. Bioelectron.*, 2017, **89**, 899.
172. S. Avaz, R. B. Roy, V. R. S. S. Mokkapati, A. Bozkurt, S. Pandit, I. Mijakovic and Y. Z. Menceloglu, Graphene based nanosensor for aqueous phase detection of nitroaromatics, *RSC Adv.*, 2017, **7**, 25519.
173. J. Yang, Z.-Z. Lin, H.-P. Zhong, X.-M. Chen and Z.-Y. Huang, Determination of leucomalachite green in fish using a novel MIP-coated QDs probe based on synchronous fluorescence quenching effect, *Sens. Actuators, B*, 2017, **252**, 561.
174. R. Liang, L. Kou, Z. Chen and W. Qin, Molecularly imprinted nanoparticles based potentiometric sensorwith a nanomolar detection limit, *Sens. Actuators, B*, 2013, **188**, 972.
175. I. Bakas, Z. Salmi, M. Jouini, F. Geneste, I. Mazerie, D. Floner, B. Carbonnier, Y. Yagci and M. M. Chehimi, Picomolar detection of melamine using molecularly imprinted polymer-based electrochemical sensors prepared by uv-graft photopolymerization, *Electroanalysis*, 2015, **27**, 429.

176. J. G. Osteryoung and R. A. Osteryoung, Square wave voltammetry, *Anal. Chem.*, 1985, **57**, 101A.
177. I. Bakas, A. Hayat, S. Piletsky, E. Piletska, M. M. Chehimi, T. Noguer and R. Rouillo, Electrochemical impedimetric sensor based on molecularly imprinted polymers/sol–gel chemistry for methidathion organophosphorous insecticide recognition, *Talanta*, 2014, **130**, 294.
178. Y. Dai, X. Li, L. Fan, X. Lu and X. Kan, "Sign-on/off" sensing interface design and fabrication for propyl gallate recognition and sensitive detection, *Biosens. Bioelectron.*, 2016, **86**, 741.
179. X. Zhao, J. Zuo, S. Qiu, W. Hu, Y. Wang and J. Zhang, Reduced Graphene Oxide-Modified Screen-Printed Carbon (rGO-SPCE)-based disposable electrochemical sensor for sensitive and selective determination of ethyl carbamate, *Food Anal. Methods*, 2016, **160**, 113.
180. N. Atar, M. L. Yola and T. Eren, Sensitive determination of citrinin based on molecular imprinted electrochemical sensor, *Appl. Surf. Sci.*, 2016, **362**, 315.
181. B. D. B. Tiu, R. B. Pernites, S. B. Tiu and R. C. Advincula, Detection of aspartame *via* microsphere-patterned and molecularly imprinted polymer arrays, *Colloids Surf., A*, 2016, **495**, 149.
182. L. Ge, S. Wang, J. Yu, N. Li, S. Ge and M. Yan, Molecularly imprinted polymer grafted porous Au-paper electrode for an microfluidic electro-analytical origami device, *Adv. Funct. Mater.*, 2013, **23**, 3115.
183. K. K. Sinha and G. Prasad, Effects of citrinin on pigment, protein and nucleic acid contents in maize seeds, *Biol. Plant.*, 1996, **38**, 317.
184. J. Alexander, D. Benford, A. Boobis, S. Ceccatelli, B. Cottrill, J.-P. Cravedi, A. D. Domenico, D. Doerge, E. Dogliotti, L. Edler, P. Farmer, M. Filipič, J. F. Gremmels, P. Fürst, T. Guérin, H. K. Knutsen, M. Machala, A. Mutti, M. Rose, J. Schlatter and R. v. Leeuwen, Scientific Opinion on the risks for public and animal health related to the presence of citrinin in food and feed, *EFSA J.*, 2012, **10**, 2605.
185. M. T. Satue, S.-W. Huang and E. N. Frankel, Effect of natural antioxidants in virgin olive oil on oxidative stability of refined, bleached, and deodorized olive oil, *J. Am. Oil Chem. Soc.*, 1995, **72**, 1131.
186. A. Gutés, R. Maboudian and C. Carraro, Gold-coated silver dendrites as SERS substrates with an improved lifetime, *Langmuir*, 2012, **28**, 17846.
187. N. Idil, M. Hedström, A. Denizli and B. Mattiasson, Whole cell based microcontact imprinted capacitive biosensor for the detection of Escherichia coli, *Biosens. Bioelectron.*, 2017, **87**, 807.
188. A.-M. Poller, E. Spieker, P. A. Lieberzeit and C. Preininger, Surface imprints: advantageous application of ready2use materials for bacterial quartz-crystal microbalance sensors, *ACS Appl. Mater. Interfaces*, 2017, **9**, 1129.
189. M. Golabi, F. Kuralay, E. W. H. Jager, V. Beni and A. P. F. Turner, Electrochemical bacterial detection using poly(3-aminophenylboronic acid)-based imprinted polymer, *Biosens. Bioelectron.*, 2017, **93**, 87.

190. R. Liang, J. Ding, S. Gao and W. Qin, Mussel-inspired surface-imprinted sensors for potentiometric label-free detection of biological species, *Angew. Chem., Int. Ed.*, 2017, **56**, 6833.
191. K. Eersels, B. v. Grinsveny, T. Vandenryt, K. L. Jiménez-Monroy, M. Peeters, V. Somers, C. Püttmann, C. Stein, S. Barth, G. M. J. Bos, W. T. V. Germeraad, H. Diliën, T. J. Cleij, R. Thoelen, W. D. Ceuninck and P. Wagner, Improving the sensitivity of the heat transfer method (HTM) for cancer cell detection with optimized sensor chips, *Phys. Status Solidi A*, 2015, **212**, 1320.
192. C. Liang, H. Wang, K. He, C. Chen, X. Chen, H. Gong and C. Cai, A virus-MIPs fluorescent sensor based on FRET for highly sensitive detection of JEV, *Talanta*, 2016, **160**, 360.
193. B. Yang, H. Gong, C. Chen, X. Chen and C. Cai, A virus resonance light scattering sensor based on mussel-inspired molecularly imprinted polymers for high sensitive and high selective detection of Hepatitis A Virus, *Biosens. Bioelectron.*, 2017, **87**, 679.
194. Y. Ma, X.-L. Shen, Q. Zeng, H.-S. Wang and L.-S. Wang, A multi-walled carbon nanotubes based molecularly imprinted polymers electrochemical sensor for the sensitive determination of HIV-p24, *Talanta*, 2017, **164**, 121.
195. G. Ertürk, H. Özen, M. A. Tümer, B. Mattiasson and A. Denizli, Microcontact imprinting based surface plasmon resonance (SPR) biosensor for real-time and ultrasensitive detection of prostate specific antigen (PSA) from clinical samples, *Sens. Actuators, B*, 2016, **224**, 823.
196. Y. Cheng, P. Jiang, S. Lin, Y. Li and X. Dong, An imprinted fluorescent chemosensor prepared using dansyl-modified β-cyclodextrin as the functional monomer forsensing of cholesterol with tailor-made selectivity, *Sens. Actuators, B*, 2014, **193**, 838.
197. S. Patra, E. Roy, R. Madhuri and P. K. Sharma, Imprinted ZnO nanostructure-based electrochemical sensing of calcitonin: A clinical marker for medullary thyroid carcinoma, *Anal. Chim. Acta*, 2015, **853**, 271.
198. A. Abbas, L. Tian, J. J. Morrissey, E. D. Kharasch and S. Singamaneni, Hot spot-localized artifi cial antibodies for label-free plasmonic biosensing, *Adv. Funct. Mater.*, 2013, **23**, 1789.
199. P. S. Sharma, A. Wojnarowicz, M. Sosnowska, T. Benincori, K. Noworyta, F. D'Souza and W. Kutner, Potentiometric chemosensor for neopterin, a cancer biomarker, using an electrochemically synthesized molecularly imprinted polymer as the recognition unit, *Biosens. Bioelectron.*, 2016, **77**, 565.
200. T. Fang, X. Yang, L. Zhang and J. Gong, Ultrasensitive photoelectrochemical determination of chromium(VI) in water samples by ion-imprinted/formate anion-incorporated graphitic carbon nitride nanostructured hybrid, *J. Hazard. Mater.*, 2016, **312**, 106.
201. K. Huang, Y. Chen, F. Zhou, X. Zhao, J. Liu, S. Mei, Y. Zhou and T. Jing, Integrated ion imprinted polymers-paper composites for selective and sensitive detection of Cd(II) ions, *J. Hazard. Mater.*, 2017, **333**, 137.

202. J. Qi, B. Li, X. Wang, Z. Zhang, Z. Wang, J. Han and L. Chen, Three-dimensional paper-based microfluidic chip device formultiplexed fluorescence detection of Cu2+and Hg2+ions based on ionimprinting technology, *Sens. Actuators, B*, 2017, **251**, 224.
203. S. D. Azevedo, D. Lakshmi, I. Chianella, M. J. Whitcombe, K. Karim, P. K. Ivanova-Mitseva, S. Subrahmanyam and S. A. Piletsky, Molecularly imprinted polymer-hybrid electrochemical sensor for the detection of β-estradiol, *Ind. Eng. Chem. Res.*, 2013, **23**, 13917.
204. A. Pietrzyk, W. Kutner, R. Chitta, M. E. Zandler, F. D'Souza, F. Sannicolo and P. R. Mussini, Melamine acoustic chemosensor based on molecularly imprinted polymer film, *Anal. Chem.*, 2009, **81**, 10061.
205. D. Croux, T. Vangerven, J. Broeders, J. Boutsen, M. Peeters, S. Duchateau, T. Cleij, W. Deferme, P. Wagner, R. Thoelen and W. D. Ceuninck, Molecular imprinted polymer films on RFID tags: a first step towards disposable packaging sensors, *Phys. Status Solidi A*, 2013, **210**, 938.
206. A. Pietrzyk, S. Suriyanarayanan, W. Kutner, R. Chitta and F. D'Souza, Selective histamine piezoelectric chemosensor using a recognition film of the molecularly imprinted polymer of bis(bithiophene) derivatives, *Anal. Chem.*, 2009, **81**, 2633.
207. S. Feng, F. Gao, Z. Chen, E. Grant, D. D. Kitts, S. Wang and X. Lu, Determination of α-tocopherol in vegetable oils using a molecularly imprinted polymers–surface-enhanced raman spectroscopic biosensor, *J. Agric. Food Chem.*, 2013, **61**, 10467.

CHAPTER 3

Molecularly Imprinted Polymers-based Separation and Sensing of Nucleobases, Nucleosides, Nucleotides and Oligonucleotides

P. FAVETTA, M. G. AYARI AND L. A. AGROFOGLIO*

University of Orléans, CNRS, ICOA, UMR 7311, F-45067 Orléans, France
*E-mail: luigi.agrofoglio@univ-orleans.fr

3.1 Introduction

Nucleosides are structural subunits of nucleic acids and the biochemical precursors of nucleotides, which play an important role in cell metabolism. Their analogues are largely used to combat cancer and viral infections. For over four decades now, nucleoside mimetics have attracted considerable attention in synthetic organic chemistry owing to their antiviral activities against human immunodeficiency virus (HIV), hepatitis B virus (HBV), herpes simplex virus (HSV), and emerging viruses.[1] Actually, with over 40 approved compounds and 30 nucleosides in the "pipeline", antiviral chemotherapy has a huge panel of powerful nucleoside analogues. Nucleosides, the RNA metabolites, are also well known as potential biomarkers in various types of cancer. They are mainly detected in urine wherein their

Polymer Chemistry Series No. 28
Molecularly Imprinted Polymers for Analytical Chemistry Applications
Edited by Wlodzimierz Kutner and Piyush Sindhu Sharma

Published by the Royal Society of Chemistry, www.rsc.org

concentrations have been found to be elevated in cancer patients. Thus, because of the central role of nucleosides and nucleotides in the cell metabolism and the importance of their analogs in treatment of antiviral diseases and clinical diagnosis of cancers, rapid and sensitive analysis of these compounds is needed.

Over the past three decades, research on the development of molecularly imprinted polymers (MIPs) that recognize nucleobases and nucleotide analogues has grown tremendously. The initial studies focused on designing MIPs capable of showing high selectivity towards these nucleobases, nucleosides, and nucleotides. Research on the pre-polymerisation formulation, the optimization of this formulation, the use of covalent, semi-covalent or non-covalent approaches and examination of interactions involved was initially carried out by a number of teams from different backgrounds including polymer chemists, analytical chemists, and biochemists. Then, these results were used to transfer theory into practice by developing analytical techniques where the possible use of MIPs improved the method, for example as an adsorbent phase in the sample pre-treatment techniques (SPE, SPME, and SMPD), as a stationary phase in separation techniques (HPLC, CEC, and GE), or even as a selective permeability membrane. Moreover, the imprinted materials have been developed for direct determination of the analyte of interest in complex matrices, such as environmental, biological and food matrices, by integrating the MIP recognition unit with, *e.g.*, electrochemical, optical, and microgravimetric transducers.

For instance, Ersoz *et al.*[2] have developed a quartz crystal microbalance (QCM) MIP chemosensor in order to determine a human marker of the oxidative stress, 8-hydroxy-2′-deoxyguanosine, 8-OHdG (**1** in Scheme 3.1). This marker is one of the most abundant oxidative DNA lesions, resulting from reactive oxygen species (ROS), in blood serum from a breast cancer patient. The 8-OHdG imprint uses a methacryloyl based metal-chelate polymer [MAAP-Fe(III)]. With the MAAP-Fe based QCM chemosensor, the 8-OHdG concentration was determined as 0.297 μM with the lowest limit of detection (LOD) among other literature 8-OHdG detection methods. Gemcitabine (**2** in Scheme 3.1), an anticancer approved drug, was detected in serum samples with an LOD of 3 fM by a highly sensitive MIP film formed *in situ* on gold electrodes *via* electropolymerization of pressure assisted thermal processing, PATP-functionalized gold nanoparticles (AuNPs).[3] Adenosine-3′,5′-cyclic monophosphate, cAMP (**3** in Scheme 3.1), a second

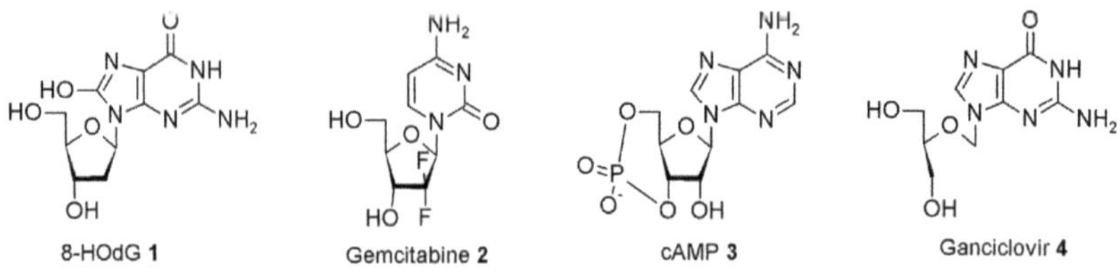

Scheme 3.1 Structural formulas of nucleosides cited for MIP chemosensors.

messenger, is an important intracellular regulator involved in a cascade of events that transduce the signal into the change of many cells. A biomimetic sensor for cAMP was fabricated in combination with an ion-sensitive field-effect transistor (ISFET) as the transducer and a cAMP-imprinted polymer as the molecular recognition material.[4] Binding ability and selectivity of cAMP to a cAMP-imprinted polymer were high in aqueous media. The antiviral ganciclovir (**4** in Scheme 3.1) was quantified in human serum plasma by a selective and sensitive voltammetric chemosensor with an MIP prepared by electropolymerization with AuNPs and deposited onto multi-walled carbon nanotubes (MWCNTs)/glassy carbon electrode (GCE).[5] Beside electrochemical and QCM sensors, fluorescent MIP-based chemosensors have also been devised.[6]

This topic has not been reviewed for several years now. Therefore, it has been decided to provide highlights of the fascinating world of nucleoside analogues along with the MIPs. We will initially review the articles that are interested in the development of MIPs for the recognition of nucleoside analogues, and then the second part will describe the work published on the use of MIPs in separation techniques and pre-treatment of samples. The final part will summarize the research where a new MIP technology can improve analytical selectivity of the sensors of all types. The reader is also referred to the earlier and excellent reviews of the literature about MIP synthesis,[7] as well as "basic" pyrimidines functionalization,[8] and to the recent contributions in "Modified Nucleosides" edited by Piet Herdewijn.[9]

3.2 Various Approaches to Synthesize MIPs for Nucleic Acids

We present herein different approaches used to prepare MIPs for recognition of nucleobases, nucleos(t)ides and their analogues, defined by their purine and pyrimidine substructure, such as those described in the general remarks on nucleosides and their analogues part above.

3.2.1 Nucleoside Structures and Conformation

Nucleosides are *N*-glycosides of ribose and deoxyribose, which consist of a ribofuranosyl sugar linked through its anomeric position to a base. This base is either a heterocycle containing nitrogen purine or pyrimidine. Adenine **5** and guanine **6** are purines while cytosine **7**, uracil **8**, and thymine **9** are pyrimidines. Along with a sugar, they form adenosine **10**, guanosine **11**, cytidine **12**, uridine **13**, and thymidine **14** nucleosides, respectively (Scheme 3.2).

Natural nucleosides have the D-configuration at the 4′-position; the base substituent is designated β because it is oriented on the same face of the sugar moiety as the 5′-hydroxymethyl group. A D-nucleoside, in which the

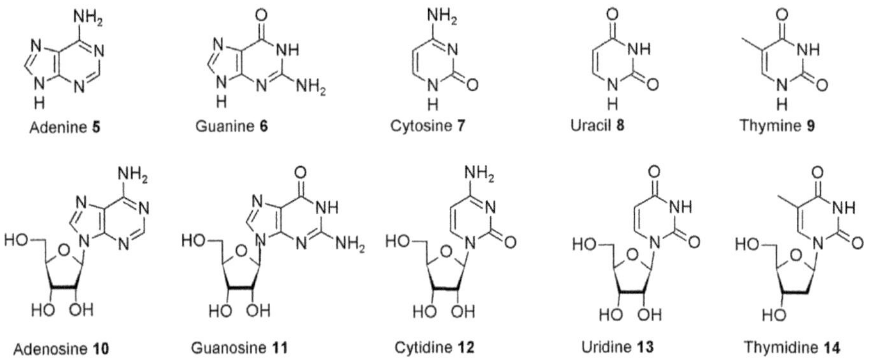

Scheme 3.2 Structural formulas of natural nucleobases and nucleosides.

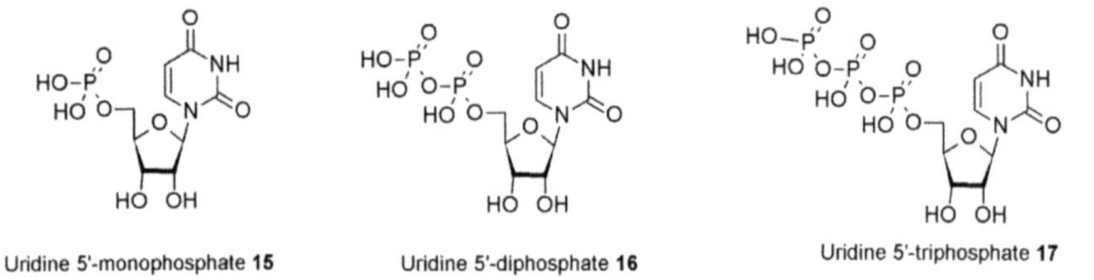

Scheme 3.3 Structural formulas of phosphorylated nucleosides, applied for uridine.

base and sugar are oriented towards opposite faces, would be designated as α-D-. If the 4′-position has the L-configuration, it is to be regarded as the enantiomer of the corresponding D-isomer and may, likewise, have the base in either the α- or β-orientation. Several conformational variations are possible, because of rotation around the base-to-sugar bond, and puckering the sugar ring.

Nucleotides are phosphate esters of nucleosides; most commonly, the phosphoryl group is attached to the oxygen atom of the 5′-hydroxyl group; thus, nucleotides are typically assumed to be 5′- unless otherwise stated (Scheme 3.3). Monophosphates can be further phosphorylated to produce di- and tri-phosphates, as illustrated in Scheme 3.3 for uridine. Noticeably, the phosphates are ionized at the physiological pH value.

3.2.2 Molecularly Imprinted Polymers for Recognition of Purines

The first highly cross-linked imprinted material made for nucleobases was prepared by Shea *et al.*,[10] in 1993, as a mix of methacrylic acid (MAA), *N*,*N*′-1,3-phenylenebis(2-methy1-2-propenamide), ethyleneglycol dimethacrylate (EGDMA) and 9-ethyladenine (**18** in Scheme 3.4) as the dummy

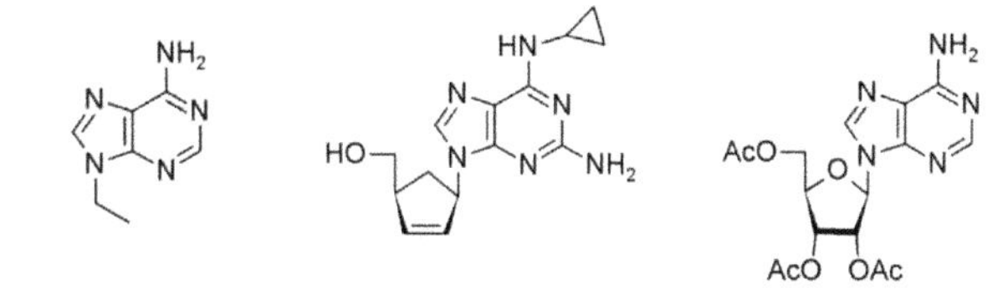

9-Ethyladenine **18** Abacavir **19** tri-OAc-Adenosine **20**

Scheme 3.4 Structural formulas of nucleosides cited in Section 3.2.1.

template, in chloroform. The determined complex stability constant was $7.6 \times 10^4\ M^{-1}$ for most selective sites of adenine moiety and the polymer selectivity was high towards adenine analogues even in aqueous media. Following their work of 1993, Shea *et al.* published three articles in 1997,[11] 1998[12] and 2001,[13] where the group explored interactions and selectivity of the 9-ethyladenine templated MIP to nucleotide bases. Primarily, the researchers showed that the choice of the porogen and polymerization type was crucial for the MIP selectivity. For MAA as the functional monomer and EGDMA as the cross-linking monomer, the acetonitrile was a better porogen than chloroform. The polymerization photoinitiated at low temperature was more successful than the thermopolymerization at 80 °C. Moreover, it seemed that the polymer imprinted with 2-aminopyridine showed the highest binding affinity for 9-ethyladenine.[11] Furthermore, the binding and selectivity to nucleosides were investigated using triacetylated derivatives, soluble in an organic solvent. The results showed that the binding affinity was, by decreasing force, in the order of adenosine, guanosine, cytidine, while thymidine and uridine were not retained. Apparently, the pattern 2-aminopyridine had a greater role in MIP binding of the purine and pyrimidine bases as well as the corresponding nucleosides, and so in their selectivity towards carboxyl groups. It was demonstrated that the MAA monomer was not very suitable for preparation of an MIP imprinted with the thymine and uracil derivatives.[12] In the 2001 study, the team sought to optimize the formulation to devise an MIP selective to oligonucleotides. The research on the adenine dimer showed that the formulation used with 9-ethyladenine led to the best results, even using chloroform as the solvent, provided that a surfactant was added to the solution to solubilize the oligonucleotide. The results were dependent on the ionization state of the phosphate group. For the best results, it should be in its non-dissociated acid form.[13]

To continue the work, in 2000 Takeuchi *et al.* decided to add another functional monomer along with the MAA, namely the 5,10,15-*tris*(4-isopropylphenyl)-20-(4-methacryloyloxyphenyl) porphyrin zinc(II) complex, to increase the affinity of the 9-ethyladenine model nucleobase template and an MIP.[14] A higher stability for this MIP, compared to those for other MIPs polymerized only with the MAA or the porphyrin derivative, was found: 7.5×10^5 M^{-1} in dichloromethane. This constant was much higher than that previously determined by Shea *et al.* with only MAA as the functional monomer.[12]

In 1994, Buchardt *et al.*[15] have added an additional functional monomer, *viz.*, 4(5)-vinylimidazole, to MAA to synthetize an MIP capable of recognizing adenine in water. Moreover, due to the presence of the imidazole group mimicking the histidine enzymatic amino acid, the MIP allowed for hydrolyzation of adenosine-5′-triphosphate (ATP) to adenosine-5′-diphosphate (ADP) in aqueous media. The possibility of operating at low pH led to protonation of the imino nitrogen atom of imidazole and activated the MIP in enhancing the hydrolysis rate. The results showed that the MIP could be used as a catalyst for hydrolysis of nucleotides.[15] Working on the recognition of the nucleotide by "footprint", Shinkai *et al.*[16] in 1999 adopted another strategy. They used a mixture of oppositely charged polymers, *viz.*, poly(diallyldimethylammonium) and polyboronate, without cross-linking them. The molar formulation ratio of the AMP/(polycation monomer unit)/(polyanion monomer unit) mixture was 1:2:1 in 0.1 M carbonate buffer solution (pH = 10.3). Then, a dedicated cavity was left vacated when adenosine-5′-monophosphate (AMP) was removed. Capturing, and then releasing of AMP was correlated to the precipitation and swelling of the polyion complex. This mixture, which self-organized, may find certain analytical applications, for instance, as a way to fabrication of chemosensors.[16] In the same way, Robertson *et al.*[17] showed that using a commercial polymer, here a polyamide–imide polymer named Torlon 4000T, deposited as a thin film on a pre-treated gold coated glass slide by spin casting, allowed for the distinguishing of adenosine from guanosine selectively. The stability constant of the polymer-adenosine complex in a 0.02% NaN_3 aqueous buffer reached 7.86×10^4 M^{-1} and the polymer was more selective than four other polymers. All data were obtained by surface plasmon resonance (SPR) spectroscopy.[17]

To continue the idea that simple non-cross-linked polymers are useful as MIPs, in three papers between 2004 and 2007 Sreenivasan[18–20] demonstrated that conjugated polyurethane and polyaniline were promising as materials to recognize nucleotides or nucleosides. The first work showed that a polyurethane film derivatized with 3-aminophenylboronic acid in the presence of AMP was able to capture AMP in an amount larger than that of ADP, by structural recognition.[18] In the second and third paper, the author devised a polyaniline MIP. This MIP was synthetized in the presence of either nucleotides showing affinity to ATP stronger than that to ADP, which was stronger than that to AMP,[19] or that in the presence of nucleobases showing a quick sample recovery. This affinity increased in the order of adenine < cytosine < uracil < thymine.[20]

Moreover, Pyshnyi *et al.*[21] managed to build a highly selective to nucleotide MIP by dissolving a commercial nylon-6 in a mixture of dimethylformamide (DMF) and trifluoroethanol (TFE), in the presence of the ATP template. In effect, polyamide, with its carboxylate and ammonium end groups and amide functions, showed molecular recognition through a well-defined spatial arrangement. The selectivity coefficient was estimated as 3.5.[21] Other teams rather attempted to examine different functional monomers in order to increase the selectivity of a polymer imprinted with nucleosides, nucleotides or nucleobases. Piletsky *et al.*[22] have demonstrated that the itaconic acid, a natural compound, had a very high affinity for abacavir (**19** in Scheme 3.4) and that the specific capacity of the synthetized MIP was 68 mg g^{-1} in 50 mM sodium acetate buffer solution (pH = 4.0).[22]

As another type of monomer, Vasapollo *et al.*[23] have proposed phthalocyanine derivatives to complex the template, tri-*O*-acetyladenosine (**20** in Scheme 3.4). Eventually, the Zn(II) tetra(4′-methacryloyloxyphenoxy) phthalocyanine showed the highest affinity for adenosine derivatives with the complex stability constant of 1.35×10^4 M^{-1}.[23] Then, the authors managed to increase binding ability for tri-*O*-acetyladenosine by adding MAA along with their phthalocyanine monomer, thus attaining the stability constant of 2.96×10^4 M^{-1}. Moreover, there was a very high selectivity with respect to uridine derivatives.[24] Finally, in 2009, the same authors showed[25] that the zinc metal centre, complexed in the monomer, played a decisive role in nucleoside binding, as demonstrated by comparing this binding to that of the non-metallated phthalocyanine. The adenosine interacted with the polymer through multi-point interactions involving coordination and hydrogen bonding with zinc-complexed phthalocyanine and MAA, respectively.[25] This approach opened doors to the recognition of nucleic bases by metal cation complexation. To bind nucleotide analytes by the MIP and reach selectivity to AMP and guanosine-5′-monophosphate (GMP), Agrofoglio *et al.*[26] tried to find the proper combination of functional monomers in order to generate an MIP highly selective to the AMP. Their first study demonstrated that the simultaneous use of non-covalent and covalent imprinting, by forming a vinylphenylboronate diester of adenosine and acrylamide, respectively, generating an MIP recognizing 1,2-*cis*-diol ribose and nucleotide base parts selectively with respect to other nucleotides. The use of a surfactant, tetrabutylammonium hydroxide, was necessary for nucleotide binding to the MIP.[26] Therefore, the team has enhanced the selectivity of the MIP AMP by using a dual approach of ionic and hydrogen bonding. That is, 2-(dimethylamino)ethyl methacrylate was selected to target the anionic phosphate moiety and acrylamide in order to link the adenine and sugar units. The monomer selection and the interaction type were guided by the design of the enzyme adenylate kinase in order to design an adsorbent mimicking the active site of the target enzyme to screen among the synthetic analogues of an AMP-like drug.[27] For the same purpose, Tsakalof *et al.*[28] have developed an MIP capable of selective recognizing triacetyl-*O*-adenosine (**20** in Scheme 3.4) in order to generate recognition sites for ATP in their synthetic enzyme. They used a monomer that approximated the MAA, namely, 3-vinylbenzoic acid. The imprinting factor (IF) equaled 11 and the complex stability constant was 8.3×10^5 M^{-1}.[28]

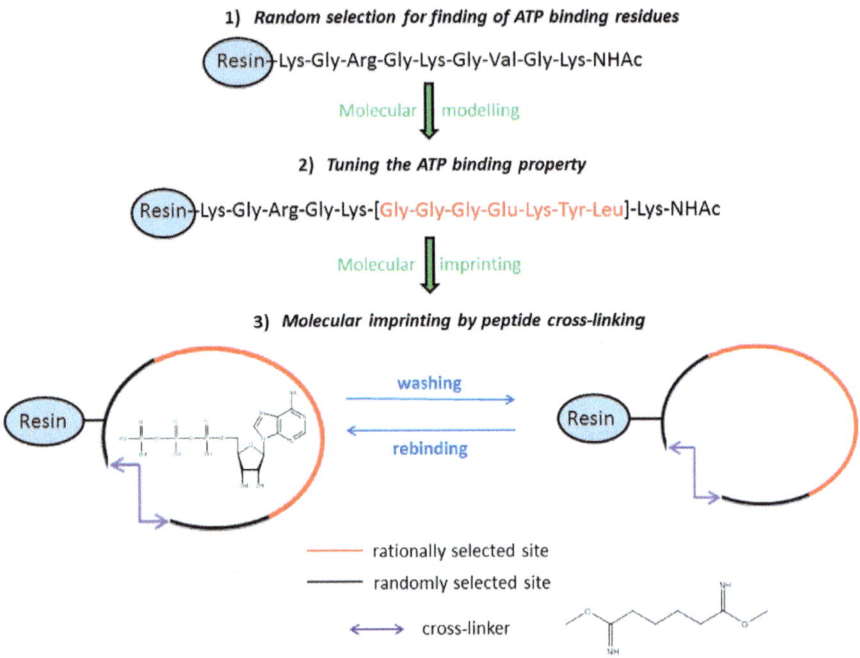

Scheme 3.5 The flow chart of molecular imprinting with the use of a peptide, which was (1) randomly selected, (2) rationally designed, and (3) cross-linked in the presence of the ATP template in the peptide-MIP. Adapted from ref. 30.

For Matsui *et al.*,[29] the functional monomer used to recognize ATP selectively was a rationally designed peptide, based on a peptide motif present in the active site of biotin carboxylase. The bead resin grafted with a peptide of the sequence Lys-Gly-Arg-Gly-Lys-Gly-Gly-Gly-Glu-Lys-Tyr-Leu-Lys-NHAc was mixed with ATP and dimethyl adipimidate as the cross-linking monomer. This rational peptide selection allowed the complex stability constant ratios to reach 19 and 10, for ATP *versus* ADP and guanosine-5′-triphosphate (GTP), respectively. Selectivity was demonstrated to nucleotides including ADP, AMP, GDP, GTP, cytidine-5′-triphosphate (CTP) and uridine-5′-triphosphate (UTP).

In another research,[30] a peptide monomer was functionalized with a vinyl group at the N-terminal group. This monomer was copolymerized with *N*-isopropylacrylamide (NIPAM) and cross-linked with *N*,*N*′-methylenebisacrylamide in the presence of the ATP template to form a molecularly imprinted hydrogel (Scheme 3.5). The complex stability constant was 5.5×10^4 M^{-1} for ATP on the ATP-imprinted on-bead-peptide hydrogel composite.[30] For both MIPs, selectivity to ATP and other nucleotides, such as ADP, AMP, GDP, GTP, CTP and UTP was demonstrated.

In the study of Gong *et al.*,[31] NIPAM was used to generate a temperature-sensitive MIP capable of recognizing adenine in a certain temperature range.

The quantity of NIPAM added to formulation containing adenine, MAA, and EGDMA was very large, thus imparting temperature phase behavior to the MIP. At temperatures below the transition temperature, the polymer swelled and was highly selective to adenine. However, above the transition temperature, the polymer shrank and lost the ability for analyte recognition. Thus, the feasibility of module recognition by temperature-sensitive MIP was demonstrated for adenine.

For a number of other teams, the MIP improvement consisted of changing the type of polymerization and the use of an inorganic or hybrid material. In the former case, Byrne *et al.*[32,33] showed that the use of controlled/living radical polymerization incorporating an iniferter, here the tetraethylthiuram disulfide, allowed for an increase in the number of binding sites in an MIP templated with the ethyladenine-9-acetate nucleobase analogue. In fact, the polymer networks were more homogenous, the population of imprinted cavities was larger and more uniform than those found in the classical free-radical polymerization, while keeping equivalent selectivity and affinity for the nucleoside template. Byrne *et al.* concluded that the double-bond conversion was the principal parameter to control in order to improve capacity or affinity of the MIP.[32,33] Moreover, the type of polymerization was in the center of focus of Kobayashi *et al.*[34] The inosine-templated MIP was synthetized by emulsion polymerization with laurylbenzenesulfonic acid as the surfactant. The formulation contained MAA and divinylbenzene (DVB) as the functional and cross-linking monomer, respectively. The size of polymer particles ranged between 0.1 and 0.4 μm. The complex stability constant was 7.2×10^4 M^{-1} for inosine. Because of such a high stability constant, not surprisingly the selectivity with respect to hypoxanthine and inosinic acid was also high. However, the MIP synthetized by emulsion polymerization has been imprinted also with the surfactant and, therefore, showed a cooperative effect between the surfactant and the template.[34] In the latter case, Nowakowska *et al.*[35] generated a hybrid material using the layer-by-layer technique. For that, they used submicrometre-sized silica gel particles coated with polyanionic and/or polycationic organic polymers. The polyionic polymers contained a co-monomer with thymine moiety capable to recognize adenine and ATP selectively by complementary binding, which involved Watson–Crick type interactions. Because of the ability to form a dimer under UV light illumination, the thymine served as the cross-linking monomer. The adenine imprinted polymer was highly selective to ATP and other nucleobases or nucleotides while the adenosine imprinted polyanionic polymer preferably bound adenosine rather than other nucleosides.[35–37]

Finally, Mujahid *et al.*[38] used a material based on a metal oxide polymer to synthesize an inorganic MIP selective to guanine. The study on titania NPs showed that a nucleic base interacted with titania NPs by hydrogen bonding and that the inorganic polymer had selective recognition cavities for guanine.[38]

3.2.3 Molecularly Imprinted Polymer for Recognition of Pyrimidines

The number of publications devoted to fabrication of MIPs for pyrimidines is fewer than those relating to purines. Although the biological significance of the pyrimidine family is high, the need to have an adenine derivative as the first model molecule used in the development of nucleic acid MIPs directed a number of theoretical studies to purines and their analogues. Nevertheless, there has been rich scientific development conducted on MIPs sensitive to pyrimidine bases.

The first reference in the domain of pyrimidine-templated MIPs was the study by Sellergren *et al.*[39] in 2002. Here, the most suitable monomer to imprint uracil was sought among three, including *N*,*N′*-1,3-phenylenebis-(2-methy1-2-propenamide) previously tested by Shea *et al.*[10] The templates were cyclohexyl- or benzyluracil used to increase the solubility of the derivatives in an apolar porogenic solvent solution. For the uracil family, the highest stability constant of the pre-polymerization complex was reached with the 1,5-bis(acrylamide)pyridine functional monomer. This was because the use of the monomer having the amido functionality was better than the amino one for hydrogen bonding with the carbonyl group of uracil. The IF of the MIP with *N*-1-benzyluracil was 4.64.[39] This study was confirmed by the same team in 2005, where 2,6-bis(acrylamido)pyrimidine functional monomer, substituted in position 6, was used to bind a uracil analogue. The ^{1}H-NMR spectroscopy titration and fluorescence quenching results showed that the uracil template was strongly bound in an appreciable amount in the MIP presenting a monomer with a donor–acceptor–donor of hydrogen bond array capable of strongly recognizing the amide function of uracil.[40] In 2006, Ding *et al.*[41] used the same functional monomer, 2,6-bis(acrylamido)pyridine (BAAPy), to recognize selectively alkyluracil and alkylthymine by postfunctionalization of a diblock copolymer, poly[(tert-butylmethacrylate)-*block*-(2-hydroxylethyl methacrylate)], in the presence of the template. The polymer was cross-linked with EGDMA and the final diblock copolymer formed NPs by micelle formation in a mixed solvent solution of chloroform and cyclohexane. As described above, the triple hydrogen functionality present in the BAAPy allowed binding of the uracil or thymine base of the template. The nanomicelles showed a template binding capacity 2.4-fold higher than that of particles prepared by grinding a bulk polymer.[41]

Piletsky *et al.*,[42] when working on the MIP for an anti-HIV drug, 3′-azido-3′-deoxythymidine (AZT) (**21** in Scheme 3.6), demonstrated that the choice of a cross-linking monomer was very important for the non-specific binding and selectivity parameters. The AZT-ester dummy template was mixed and polymerized with MAA and trimethylolpropane trimethacrylate (TRIM), EGDMA or DVB. The IF was the highest for AZT with DVB in acetonitrile. This cross-linked MIP and NIP were highly selective to AZT-ester and anti-HIV 3′-thiacytidine (3-TC) (**22** in Scheme 3.6).[42]

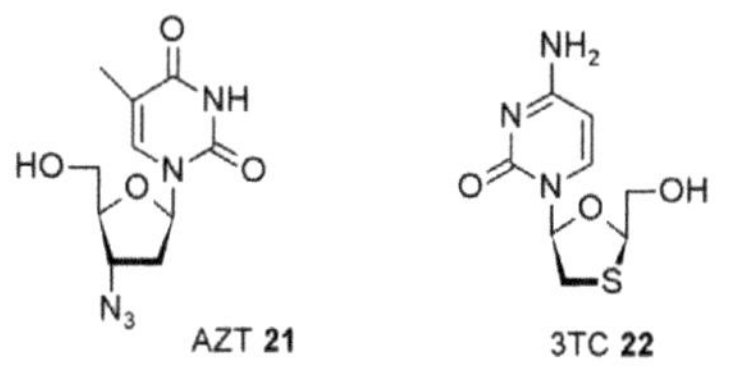

Scheme 3.6 Structural formulas of anti-HIV drugs, 3′-azido-3′-deoxythymidine (AZT) **21**, and 3′-thiacytidine (3TC) **22**.

It seems that also additives, such as MWCNTs, can be used to increase the MIP surface area, as Mathew *et al.*[43] demonstrated. The vinyl functionalized MWCNTs were mixed with thymine, MAA, and EGDMA, and then thermopolymerized to form an MIP selective to thymine. Selectivity factors were calculated to be 4.56 and 5.05 to uracil and 5-fluorouracil, respectively. Selectivity and capacity for MWCNT-MIP were higher than those for a conventional MIP for thymine.[43]

More recently, the team of H.-Y. Lin[44] prepared NPs of MIP by precipitation of 75% deacetylated chitosan mixed with magnetite nanoparticles (MNPs) and thymine. MNPs were used to simplify particle collection. The binding behavior of this MIP was higher for thymine than that for adenine or cytosine. As the chitosan NPs are known to penetrate the cell nucleus, these MNPs were used to deliver highly defined telomeric gene sequences. These sequences contained thymine capable to express the tumor suppressor p53 gene in oncology.[44]

Another way of improving selectivity of a pyrimidine-templated MIP was to change the method of polymerization by, for example, synthetizing core-shell particles by distillation–precipitation polymerization. This was used by Luo *et al.*[45] who fabricated microparticles of the poly(methyl methacrylate-*co*-ethyleneglycol dimethacrylate) imprinted shell embedded with the poly[acrylamide-*co*-*N*,*N*′-methylenebis(acrylamide)] core to incur hydrophilicity to the adsorbent particles. The selectivity study showed a high difference between uridine, used as the template, and its close analogue, thymidine or 2′-deoxy-5-methyluridine.[45]

The choice of a dummy template was of primary importance in the synthesis of MIPs selective to such a highly polar analyte as pyrimidine. In order to develop hydrogen bonding by a non-covalent method, aprotic polar or apolar solvents were used as the porogens for this analyte imprinting. Solubility of pyrimidines in these solvents was very low. So, in 2014, Agrofoglio's and Hall's teams[46] proposed to study the effect of length of the chain of tri-*O*-acyluridine derivatives on the capacity and selectivity of MIPs synthetized with 2,6-bis(acrylamido)pyridine as the functional, and EGDMA as the cross-linking monomer in chloroform. The longer the carbon chain linked to three hydroxyl groups of ribose, the lower was the IF for uridine analogues. The highest complex stability constant was $1.7 \times 10^4\ M^{-1}$ for the MIP, imprinted with 2′,3′,5′-*O*-triacetyluridine, toward its template. Finally, the team concluded that the molecular size of the dummy template affected the recognition of the target analyte. Indeed, compared to the ethyl and propyl ester derivatives, the less bulky dummy template, the methyl ester, was the best choice to selectively recognize uridine.[46]

3.3 MIPs for Extraction and Separation of Nucleic Acid Analogues

Molecular imprinting has received much attention in chemical analysis because of its high selectivity leading to enhanced sensitivity of the method used.

In this part, we will see different applications using MIPs for detection and separation of nucleotides or nucleosides. This part contains four sections composed according to the technique used for these applications including solid phase extraction (SPE), solid phase microextraction (SPME), molecularly imprinted solid-phase dispersion (MSPD) as well as stationary phases and membranes, and formulations (Table 3.1).

3.3.1 Molecularly Imprinted Solid-phase Extraction (MISPE)

SPE is one of several sample preparation techniques that has been applied for 30 years in analytical chemistry fields, such as food, toxicology, environmental and other analyses. It is based on the principles of adsorption, partition or ion chromatography. The adsorbent types are practically as numerous as stationary phases used in column liquid chromatography. The SPE technique allows extracting, concentrating, distributing, and/or adsorbing one or more compounds from a liquid sample. In the SPE field, MIPs have been used as very selective adsorbents for detection or pre-concentration of different nucleic acid analogues.

Agrofoglio *et al.*[47] in 2008 first used MISPE in the domain of analysis of human nucleoside biomarkers. The synthesis of an MIP capable of retaining selectively 5-methyluridine (**23** in Scheme 3.7) from a urine sample was developed using the combination of a covalent bond, between the ribose moiety and a vinyl-phenylboronate ester derivative, and a non-covalent bond, between acrylamide and the base moiety. The number of high-affinity sites in cavities of the MIP and the complex stability constant for 5-methyluridine were extracted from the Scatchard plot equaling 18 μmol g^{-1} and 2.17×10^4 M^{-1}, respectively. MISPE was highly selective to pyrimidine nucleosides and selective to purines or N^4-acetylcytidine. Dog urine samples were analyzed by MISPE showing a clear LC-UV chromatogram.[47] Moreover, Scorrano *et al.*[48] devised, one year later, an MIP selective to 1-methyladenosine (**24** in Scheme 3.7), another urinary cancer biomarker. However, they used a non-covalent approach with MAA, which is the main functional monomer used with purines, in an acetonitrile–water (4 : 1, *v* : *v*) solvent mixture as the porogen, to dissolve the polar template. The complex stability

Table 3.1 Formulations and applications of MIPs described in Section 3.4 according to target compounds (number in parentheses is the molar ratio).[a]

Target	Medium	Template	Monomer	Cross-linking monomer	Porogen	Initiator	Polymerization		Ref.
Solid-phase extraction (SPE)									
1-Methyl-adenosine	Spiked human urine	1-MA (1)	MAA (14)	EGDMA (75)	MeCN : water (4 : 1, v : v)	AIBN	Batch	Thermo	48
5-Methyluridine	Spiked dog urine	5-Methyluridine-2′,3′-(4-vinyl-benzyl)boronate (1)	AA (4)	PETRA (20)	MeCN	AIBN	Batch	Photo	47
7-Methyl-guanosine	Synthetic urine	Triacetylated guanosine (1)	MAA (8)	EGDMA (20)	MeCN : MeOH (70 : 30, v : v)	ABDV	Batch	Thermo	49
8-Hydroxy-2′-deoxy-guanosine	Urine sample	Guanosine (1)	AA (13) + 4-VP (9)	MBAA (7)	DMSO : dodecanol (1 : 1, v : v)	AIBN	Monolith	Thermo	56–59
Acyclovir	Urine samples	Theophylline (1)	APTEOS-MAA (3.3)	TEOS (4.5) and EGDMA (25)	Chloroform	AIBN	Batch	Thermo	52
Acyclovir	Spiked human serum	Acyclovir (1)	allyl-cytosine (1)	EGDMA (5)	DMSO and MeCN	AIBN	Core-shell on vinylated silica micro-spheres	Thermo	51
Guanine	Beer	Guanine	MAPDIA	Triethanol-amine tri-methacrylate	DMF	AIBN	Bulk	Thermo	60
Imiquimod	Aqueous samples, urine and blood serum	9-Isobutyl-adenine (1)	MAA (4)	EGDMA (20)	MeCN	ACCN	Bulk	Photo	53
Lamivudine	Spiked human serum and urine	Lamivudine (1)	MAA (8)	EGDMA (40)	MeCN : THF (4 : 1, v : v)	AIBN	Bulk	Photo	55
Tegafur	Spiked urine sample	Tegafur (1)	BAAPy (1)	EGDMA (20)	Chloroform	ABDV	Bulk	Thermo	54

(*continued*)

Table 3.1 (*continued*)

Target	Medium	Template	Monomer	Cross-linking monomer	Porogen	Initiator	Polymerization		Ref.
Zidovudine and stavudine	Spiked human serum	Zidovudine (1)	MAA (4)	EGDMA (20)	MeCN	AIBN	Bulk	Photo	50
Solid-phase microextraction (SPME)									
5-Fluorouracil and uracil	Diluted plasma	Uracil (1) or 5-FU (1)	Melamine (1) chloranil (1)		DMF	PMMA pre-coating	Uniform mono-layer	Thermo	61
Abacavir	Environmental and urine samples	Abacavir (1)	Acrylic acid (10)	EGDMA (58)	DMF	AIBN	Glass tubes in solution	Thermo	62
Matrix solid-phase dispersion (MSPD)									
Acyclovir	Pork, chicken, and beef tissues	Theophylline (1)	MAA (9)	EGDMA (20)	Chloroform	AIBN	Bulk	Thermo	63
Adenine	Healthy human urine	AMP	APTEOS and APBA	TEOS	Water	GPTES	Nanoparti-cles	Sol–gel	65
Hypoxanthine and inosine	Fish muscle	Theophylline (1)	Acrylamide (4)	EGDMA (20)	Chloroform	AIBN	Precipitation	Thermo	64
Molecularly imprinted polymer stationary phase									
Adenine	HPLC	9-Ethyladenine	*N*,*N*′-1,3-phenylenebis-(2-methyl-2-propenamide and MAA	EGDMA	Chloroform	AIBN		Photo	10
Nucleobases	CEC	9-Ethyladenine (1)	MAA (12)	EGDMA (18)	MeCN	AIBN		Thermo	69
Nucleotides	HPLC	2′-Deoxyadenos-ine monophos-phate (1)	Zn^{2+}-(4-vinyl) benzylcyclen (1)	EGDMA	MeCN	ABDV		Thermo	67
Nucleotides	HPLC	Diphenyl phos-phate (1)	4-VP (2)	Glycerol dimethacry-late (25)	1-Hexanol	ADVN		Thermo	68

Nucleotides prodrugs	HPLC	Nucleotide prodrug of d4T (1)	4-VP (6)	EGDMA (30)	Chloroform	AIBN		Photo	66
dsDNA	GE	λDNA 564bp	2-Vinyl-4,6-diamino-1,3,5-triazine	AA and MBAA	Water	APS and TEMED		Thermo	70 and 71
Molecularly imprinted polymer membranes									
2-Deoxy-adenosine	Spiked human urine	2-Deoxy-adenosine (1)	MAA (14)	EGDMA (75)	MeCN : water (4 : 1, *v* : *v*)	AIBN	PVDF	Thermo	86
6-Benzyladenine	0.1 M acetate buffer (pH = 3.8)	6-Benzyladenine (1)	MAA (4)	EGDMA (10)	DMF	AIBN	Cellulose acetate	Photo	85
Adenosine	MeOH : Chloroform (6 : 94, *v* : *v*)	9-Ethyladenine	MAA (5)	EGDMA (22)	DMF	AIBN		Thermo	74
Adenosine	EtOH : water (1 : 1, *v* : *v*)	9-Ethyladenine	Tetrapeptide derivative		THF		PS resin		78
Adenosine	H_2O : EtOH, (5 : 95, *v* : *v*)	9-Ethyladenine		Proteins	TFE		PTFE	Thermo	79
AMP		AMP	Dimethyl-amino ethyl methacrylate	EGDMA	DMF	AIBN			72 and 73
cAMP	Water (pH = 4.2)	cAMP (1)	Dimethylamino ethyl methacrylate (17)	TRIM (67)	Ethanol : Water (70 : 30, *v* : *v*)	Benzoin ethyl ether	PVDF	Photo	75 and 76
cAMP	Water (pH = 4.2)	cAMP (1)	Dimethylamino ethyl methacrylate (8)	TRIM (33)	Ethanol : Water (70 : 30, *v* : *v*)	Benzoin ethyl ether	PVDF	Photo	77
Thymine	MeOH	Thymine (1)	9-Vinyladenine (4)	DEGDMA (87.5)	DMF	AIBN	Porous cellulosic membrane	Photo	84
Uracil	Water	Uracil (2%, *w* : *w*)	MAA	Poly (AN-*co*-MAA) (5%, *w* : *w*)	DMSO and liquid CO_2		Phase inversion imprinting		82

(*continued*)

Table 3.1 (*continued*)

Target	Medium	Template	Monomer	Cross-linking monomer	Porogen	Initiator	Polymerization	Ref.
Uracil	Water	Uracil (2%, *w*:*w*)		poly(AN-*co*-MAA) without or with poly(AN-*co*-SMA) (10%, *w*:*w*)	DMSO and water		Phase inversion imprinting	80 and 81
Uracil	Water	Uracil (2%, *w*:*w*)		Poly(styrene-*co*-maleic acid (15%, *w*:*w*)	DMSO and supercritical CO_2		Phase inversion imprinting	83

[a] 4-VP, 4-vinylpyridine; 5-FU, 5-fluorouracil; AA, acrylamide; ABDV, azobis(4-methoxy-2,4-dimethylvaleronitrile); ACCN, 1,10-azobis(cyclohexanecarbonitrile); ADVN, 2,2′-azobis(2,4-dimethyl valeronitrile); AIBN, azobisisobutyronitrile; AMP, adenosine-5′-monophosphate; APBA, 3-aminophenylboronic acid; APS, ammonium persulfate; APTEOS, 3-aminopropyltriethoxysilane; BAAPy, 2,6-bis(acrylamido)pyridine; cAMP, adenosine 3′,5′-cyclic monophosphate; CEC, capillary electrochromatography; DEGDMA, diethyleneglycol dimethacrylate; DMF, *N*,*N*-dimethylformamide; DMSO, dimethylsulfoxide; EGDMA, ethyleneglycol dimethacrylate; EtOH, ethanol; GPTES, 3-glycidyloxypropyl triethoxysilane; MAA methacrylic acid, MAPDIA, 5-{[4-(methacryloyloxy)phenyl]diazenyl}isophthalic acid; MBAA, *N*,*N*′-methylenebisacrylamide; MeCN, acetonitrile; MeOH, methanol; PETRA, pentaerythritol tetra-acrylate; poly(AN-*co*-MAA), poly(acrylonitrile-*co*-methacrylic acid; poly(AN-*co*-SMA), poly(acrylonitrile-*co*-vinylbenzyl-stearyldimethylamine chloride); PS,(polystyrene; PTFE, polytetrafluoroethylene; PVDF, polyvinylidene-difluoride; TEMED, *N*,*N*,*N*′,*N*′-tetramethylethylenediamine; TEOS, tetraethoxysilane; TFE, 2,2,2-Trifluoroethanol; THF, tetrahydrofuran; TRIM, trimethylopropane trimethacrylate.

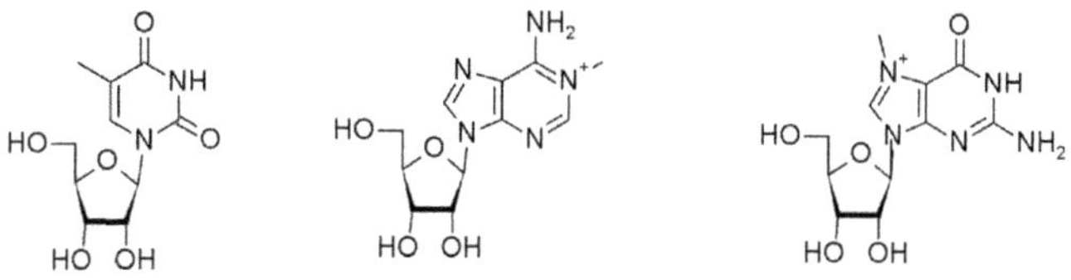

Scheme 3.7 Structural formulas of various methylated nucleosides cited.

constant and the maximum number of high affinity sites on the MIP were also calculated from the Scatchard plot equaling 3.2×10^4 M^{-1} and 20 μmol g^{-1}, respectively. High selectivity to adenosine, 2′-deoxyadenosine, cytidine, and inosine was demonstrated and selectivity of the MIP was high compared to that of the NIP or C18 adsorbent. With the MISPE, the 1-methyladenosine biomarker was extracted from spiked human urine with 95% recovery.[48] In the same domain of endogenous nucleoside extraction by MISPE, Manesiotis and Agrofoglio *et al.*[49] have more recently synthesized an MIP selective to 7-methylguanosine (**25** in Scheme 3.7).

Here, the problem of template solubility in the porogen was solved using 2′,3′,5′-tri-*O*-acetylguanosine. The best functional monomer, *viz.*, MAA, was found after ^{1}H-NMR spectroscopy titration. The average recognizing site complex stability constant and the highest concentration of recognizing sites were 2.3×10^4 M^{-1} and 5.27 μmol g^{-1}, respectively, for the selected MIP. This MIP was highly selective to the methylated guanosine derivatives. Therefore, it was applied as the sorbent in the extraction of cancer biomarkers of nucleoside derivatives from synthetic urine samples. For 7-methylguanosine, the sample clean-up was significant and recovery was up to 90% (Figure 3.1).[49]

The other type of analysis of nucleosides in biological fluids involved determination of exogenous antiviral and anti-cancerous drugs. For that purpose, since 2009, Pichon *et al.*[50] have developed MISPE to extract two inhibitors of nucleoside reverse transcriptase, *viz.*, AZT (**21** in Scheme 3.6) and stavudine (**26** in Scheme 3.8), in human serum. MAA, EGDMA and AZT were selected as the template, and the functional and cross-linking monomer, respectively. The MIP selectivity was 0.2 μmol g^{-1} for AZT, and the recovery of AZT and stavudine from 10^{-6} M spiked human serum was 80 and 85%, respectively. With the LC-MS technique, the LOD and limit of quantification (LOQ) for AZT was 5×10^{-7} and 1×10^{-7} M, respectively.[50] To determine acyclovir (**27** in Scheme 3.8), an antiviral guanosine derivative in human serum, Tang *et al.*[51] had an idea to use a functional monomer bearing a cytidine moiety, the nucleobase complementary to guanosine, to develop a three-hydrogen bond pattern of recognition. The resulting adsorbent was polymerized on silica microbeads grafted with 3-methacryloyloxypropyltrimethoxysilane using (allyl-cytosine)-acyclovir pre-polymerization complex and EGDMA in dimethylsulfoxide (DMSO)-acetonitrile (1 : 4, *v* : *v*). The MISPE was employed

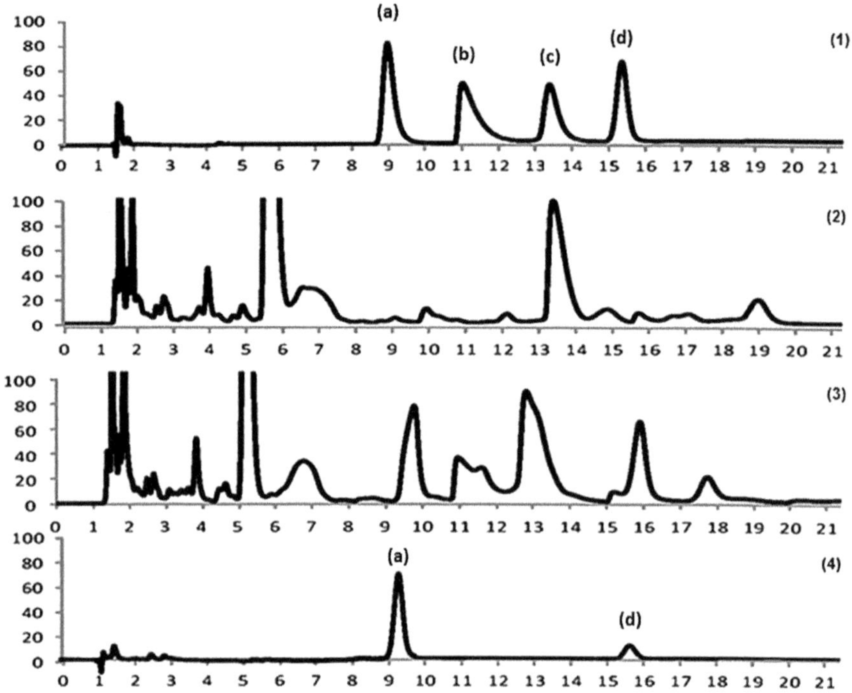

Figure 3.1 LC chromatograms obtained during urine sample extraction with the 7-methylguanosine-templated MIP. (1) Injection of standards in synthetic urine, (2) non-spiked healthy patient urine, (3) healthy patient urine spiked with the four guanosine analogues, (4) healthy patient urine after MISPE treatment. Peak assignment: (a) 7-methylguanosine, (b) guanosine, (c) 1-methylguanosine, and (d) 8-hydroxy-2′-deoxyguanosine (Agrofoglio *et al.* unpublished data).

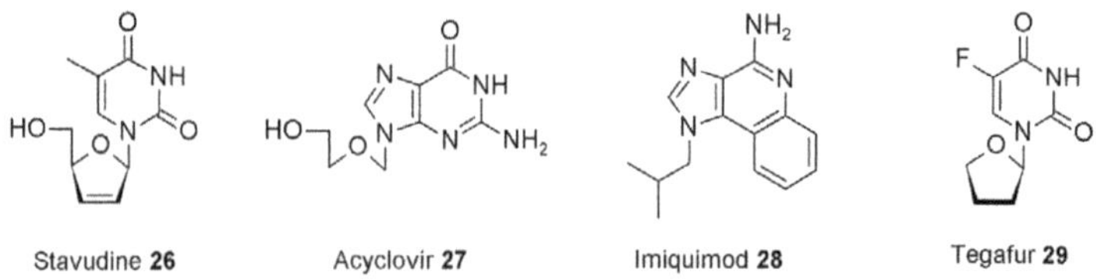

Scheme 3.8 Structural formulas of the stavudine **26**, acyclovir **27**, imiquimod **28**, and tegafur **29** purine anticancer drugs.

to extract acyclovir serum samples spiked with 10 to 400 ng mL^{-1} acyclovir, with a mean recovery of 95.6%. After the MISPE step followed by that of LC-UV separation, the LOD for acyclovir was 1.8 ng mL^{-1}.[51] Another team had the same goal, *i.e.*, the MISPE determination of acyclovir in human biofluids. That is, Row *et al.*[52] used a 1.5 mL tapered plastic centrifuge tube filled with a hybrid inorganic-organic MIP prepared from a mixture of

3-aminopropyltriethoxysilane (3-APTEOS)-MAA, EGDMA and tetraethoxysilane (TEOS), chloroform, and theophylline, as the functional monomer, cross-linking monomer, porogen and template, respectively. Recovery of this determination was high, reaching 93.2% of acyclovir compared to MISPE determination using other cartridges of sorbents, such as C18, HLB, and SCX. The recovery of the nucleoside analogue in healthy human urine and that of a human having absorbed a tablet of a drug was in the range of 91.6–103.3%, and the LOD was ~8.1 ng mL^{-1}.[52]

In the same field, Manesiotis *et al.*[53] synthesized MIPs selective to anticancer drugs; one was selective to imiquimod (**28** in Scheme 3.8), an *N*-alkylated fused purine, a drug used in superficial basal cell carcinoma, and the other was selective to tegafur (**29** in Scheme 3.8), a chemotherapeutic fluorouracil prodrug used in the treatment of cancers, and the MIPs allowed for extraction of them from the blood and urine samples. On the one hand, MAA was chosen as the functional monomer for the MIP for imiquimod, with 9-isobutyladenine as the template, because of its higher solubility in the porogen solvent, while maintaining structural similarity to the molecule of the target drug and eliminating template bleeding. The results showed a high IF of 17 for imiquimod determination with the filled column technique. For the sample spiked with imiquimod urine and blood serum samples, the MISPE technique allowed recovery up to 90% and 80% of 10 mg mL^{-1} imiquimod, respectively.[53] On the other hand, however, the MIP selective to tegafur was prepared using a dedicated functional monomer for recognition of the uracil moiety, namely, the BAAPy. Tegafur was used directly as the template because it was soluble in the chloroform porogen. Tegafur analyte binding tests showed high MIP selectivity toward 5-tegafur, but not to fluorouracil and its analogues, and the IF of 25 for the tegafur template. The MIP templated with tegafur was applied to the MISPE technique and was capable of recovering 92%, on average, of the prodrug in urine samples spiked with 200 μg mL^{-1} tegafur, 5-fluorouracil, caffeine and theophylline.[54] Another group, Dinarvand *et al.*,[55] has fabricated a lamivudine-MIP (**22** in Scheme 3.6) with MAA as the functional monomer, EGDMA as the cross-linking monomer, a mixture of acetonitrile and tetrahydrofuran as the porogen and a drug as the template. The resulting MIP was used as the MISPE adsorbent, showing sensitivity to the tegafur analyte higher than that of zidovudine and acyclovir. The assay of lamivudine in human serum and urine spiked with tegafur between 60 and 700 ng mL^{-1} resulted in recovery of 84.2–93.5% and 82.5–90.8%, respectively. After MISPE of the urine and serum samples, the LOQ for lamivudine, by HPLC-UV, was 24.05 μg L^{-1} and 58.6 μg L^{-1}, respectively.[55]

To measure a very low concentration of an endogenous nucleoside biomarker of oxidative stress, 8-OHdG, (**1** in Scheme 3.1), Zhang *et al.*[56–59] employed a miniaturized MISPE technique. This technique resembles that of solid-phase dynamic extraction but the needle internal wall is not coated with the extraction phase. All of the development and applications were described in four papers.[56–59] The first one[56] showed the synthesis of the in-tube monolith MISPE with a mixture of guanosine as the dummy template, 4-vinylpyridine and acrylamide as the functional monomers, *N*,*N*′-methylene bisacrylamide as the cross-linking monomer and DMSO and dodecanol as the porogens. The enrichment factor was 100 and 76 for guanosine and 8-OHdG, respectively. When MISPE was coupled to HPLC-UV, the LOD and LOQ for 8-OHdG was 3.2 and 11 nM, respectively. For urine samples, the recovery was 81–86%, and the method demonstrated that the urinary 8-OHdG level in cancer patients was significantly higher than that in healthy humans.[56] The second paper[57] showed a modification of the method developed in the previous paper in a monolith MIP filled capillary as the on-line SPE element of the HPLC-UV technique. The MIP was the same and its incorporation as the on-line extraction bed resulted in an LOD of 2.04 nM, which was lower than that for the previously described off-line MISPE. One hundred and eleven urinary samples were collected from three groups (healthy, exposed to chemical stress, and diverse cancer type patients), and then analyzed by the validated method. The urinary 8-OHdG level in patients with the lung, breast, liver, and cervical cancer was higher than that in healthy individuals. The concentration of 8-OHdG is assayed routinely in human urine with this technique.[57] In the third paper,[58] the technique of separation and determination was replaced by capillary electrophoresis with electrochemical detection (CE-ECD) and the monolith MIP was used in-tube micro-SPE to determine 8-OHdG in human urine. This paper demonstrates the feasibility of the use of MISPE coupled to electrophoresis. The LOD reached was the lowest compared to other reported conventional SPEs coupled to CE-ECD.[58] In the last paper,[59] a panel of urine of patients, including taxi drivers, traffic policemen, coking plant workers, and healthy volunteers, was analyzed and an enrichment factor of 73 for 8-OHdG was determined, the LOD and LOQ being 2.61 nM and 8.63 nM, respectively. An elevated excretion of 8-OHdG was found in subjects exposed to environmental pollutants, physical labor, and a day–night shift work. The new approach of the MIP monolith MISPE hyphenated to CE-ECD demonstrated sensitivity sufficient for determination of low levels of human urinary 8-OHdG.[59]

To complete this part on the use of MIPs in SPE for extraction of nucleoside analogues, Tang *et al.*[60] have more recently opened a new way of the MISPE technique by developing a photocontrolled SPE of guanine. The photoresponsive imprinted adsorbent was prepared using guanine as the template, 5-{[4-(methacryloyloxy)phenyl]diazenyl}isophthalic acid as the functional monomer and triethanolamine trimethacrylate as the cross-linking monomer. A high population of binding sites led to a complex stability constant of 3.70×10^4 M^{-1} for guanine in the methanol-(phosphate buffer) solution. The MIP bound guanine more strongly than adenine and uric acid. The photocontrolled MISPE allowed for extraction of guanine from food and beer samples by loading and washing the adsorbent at 440 nm and eluting at 365 nm, with a mean recovery of 98% at a mean guanine concentration of 7.9 μM.

The contribution of this photoregulated release and uptake of the guanine target analyte in the SPE sample allowed for more control over the selectivity of this important step.[60]

3.3.2 Molecularly Imprinted Solid-phase Microextraction (MISPME)

Solid-phase microextraction (SPME) is an SPE technique, proposed in the early 1990s by Janusz Pawliszyn at the University of Waterloo. The difference betwen SPME and SPE is that in the first technique the sorbent is coated or grafted onto the outside of fused silica fibers or on the internal surface of a capillary tube. The SPME technique allows extracting and accumulating compounds that are present in trace quantities in a liquid or a gas. Advantages of SPME include a relatively simple procedure of sorption and desorption, a high compatibility with separative analytical systems, an easy automation, and a high preconcentration ability. However, as in all extraction techniques, the selectivity has to be high because of matrix effects, which alter the distribution constant. So, in SPME, modification of the adsorbent by incorporating in it selective cavities can be one of solutions for increasing selectivity.

That was why Prasad *et al.*[61] in 2008 introduced the MIP-SPME technique for uracil and 5-fluorouracil (5-FU) determination in human blood plasma. The MIP was based on equimolar DMF solutions of melamine, chloranil, and uracil or 5-FU, brush grafted on a sol–gel film linked to an activated poly(methyl methacrylate) PMMA optical fiber. Thickness of the monolayer film reached 22 or 25 μm, which is low compared to that of conventional SPME films. However, the procedure showed an anti-clogging effect against proteins and a more facile mass transfer, which improved sorption ability. The selectivity with respect to interferences, such as barbituric acid, hypoxanthine, adenine, caffeine, cytosine, uric acid, guanine, and ascorbic acid was demonstrated. The IF of the MIP-coated SPME fibers for uracil and 5-FU was 7.5 and 6.4, respectively, and the enrichment factor in aqueous environment was ~10 and ~8, respectively. The LOD for uracil and 5-FU in human blood plasma was 0.0245 ng mL^{-1} and 0.0484 ng mL^{-1}, respectively, when MIP-SPME was coupled to an MIP-based chemosensor. The developed method permitted determining uracil and 5-FU at ultra-trace levels in biological samples.[61]

More recently, an MISPE fiber was used to extract abacavir (**19** in Scheme 3.4), a nucleoside inhibitor of human immunodeficiency virus reverse transcriptase, in environmental and biological samples.[62] As abacavir is a purine analogue, acrylic acid was chosen as the functional monomer, and EGDMA and DMF were retained as the cross-linking monomer and the porogenic solvent, respectively. Glass tubes were coated with MIP films, and then these tubes were mounted on a steel adapter. The effect of the process parameters on the adsorption behavior was studied and the equilibrium data were fitted with different models. For abacavir, the maximum adsorption capacity was

149 mg g^{-1}, reached by the Langmuir–Freundlich model. The SPME analyte enrichment was off-line with LC-MS. The samples of river water, sea water, lake water, influent and effluent wastewaters as well as urine samples spiked with abacavir were analyzed and recovery was 88–99% for spiked levels of 100 to 1000 ng L^{-1}. To illustrate the sensitivity of the method, for example, in urine, the LOD and LOQ was 12.7 and 41.9 ng mL^{-1}, respectively. This new MIP-SPME technique selectively enriched abacavir in complex aqueous matrices.[62]

3.3.3 Molecularly Imprinted Matrix Solid-phase Dispersion

The dispersion of solid adsorbent in the matrix, so called matrix solid-phase dispersion, is dedicated to extraction of compounds from the solid, semi-solid, or viscous samples. It consists of introducing a solid phase, conventionally employed in solid-phase extraction, in the sample and to mix the resulting suspension to extract analytes from complex matrices. An SPE cartridge is then filled with the solid mixture or suspension and a sequence of solvents percolation is programmed, similarly as in the SPE technique, to wash out interferences and to elute analytes. Here, too, the selectivity of the technique is an important parameter, and solid supports of an MIP have been used to meet this requirement.

The first example was the work of Han *et al.*[63] to extract acyclovir (**27** in Scheme 3.8) from edible animal tissues and its determination by LC. This work aimed at developing the molecularly imprinted matrix solid-phase dispersion (MI-MSPD) technique to improve greatly selectivity of the commonly used dispersants. Theophylline was used as the dummy template and MAA as the functional monomer. The MIP, in a powder form, was blended with pork, beef, or chicken tissues in a mortar, and then the homogenized mixture was introduced into an SPE cartridge between two frits. After a washing step, the acyclovir was eluted with a methanol (MeOH)–(acetic acid) mixture. For the acyclovir antiviral medication, the IF was 5.74, the linear dynamic concentration range extended from 0.10 to 50.0 μg g^{-1}, and the recovery at three spiked levels was in the range of 85.5–108.1%. For this medication, the LOQ was 0.09 mg g^{-1}. The MI-MSPD technique combined the advantages of MIP and MSPD and, therefore, could successively be applied for determination of antivirals in animal meat.[63] Other work describing the use of MI-MSPD was the study of González Rodríguez *et al.*[64] that monitored fish freshness by analyzing the ATP degradation products. Theophylline was used as the dummy template of hypoxanthine and inosine, acrylamide as the functional monomer, EGDMA as the cross-linking monomer, and chloroform as the porogenic solvent. For hypoxanthine and inosine, the maximum IF value was 6.82 and 9.51 and the MIP binding capacity value was 368.3 mg g^{-1} and 508.8 mg g^{-1}, respectively, in the water–EtOH (9:1, *v*:*v*) solvent. Selectivity to inosine, hypoxanthine, and other ATP degradation products, such as xanthine and uric acid, was evaluated. The validated MI-MSPD method, where fish samples were blended with MIP particles, and then transferred to SPE cartridges, washed, eluted, evaporated, dissolved in water, and then injected to the UHPLC-DAD device, reached a linearity of 10 to 40 μg mL^{-1} and a recovery ranging from 103.1% to 113.4%. The proposed MI-MSPD based method was simple, selective, and reliable in extracting and purifying hypoxanthine and inosine in complex food matrices.[64]

In the same year, a new approach to MI-MSPD was proposed by Liu *et al.*,[65] to obtain a highly efficient extractive adsorbent for urinary ribonucleoside biomarkers of cancer, and was named "dual-template docking oriented molecular imprinting", DTD-OMI. AMP, used as the template, was bound to a cationic surfactant by ionic interactions at a high concentration. This mixture formed rod-shape micelles of a lamellar phase with adenosine moieties extending outside of the micelles. Then, mesoporous silica NPs were synthetized around these micelles, thus incorporating the nucleoside moieties in the sol–gel matrix. After removing the surfactant and the AMP, the NPs presented the MCM-41 form with the adenosine cavities imprinted on inside walls of mesopores. The silane functional monomers selectively interacted with the ribose and adenine moieties. The mesopore diameter was 2.3 nm while the BET determined surface area was 833 m^2 g^{-1}. For adenosine, the IF and the binding capacity was 43.6 and 191.7 μmol g^{-1}, respectively. Selectivity with respect to deoxyadenosine, guanosine, cytidine, uridine, AMP, GMP, CMP, deoxyadenosine monophosphate, and UMP was evaluated. Urine samples donated by healthy individuals were mixed with the MIP NPs, and then these NPs were collected by centrifugation, and subsequently rinsed several times. Finally, the MIP NPs were extracted in acidic media, and afterwards the supernatant was analyzed by micellar electrokinetic chromatography with UV detection. The resulting electropherograms showed a significantly enriched adenosine peak and a greatly improved separation performance. This MI-MSPD technique seemed to be favorable for further identification and quantification of adenosine in biological samples.[65]

3.3.4 Molecularly Imprinted Polymers as Stationary Phases

Generally, after extraction, a test solution is analyzed mainly by a separative technique coupled to spectrometric detection. The use of chromatography can be helpful to separate a reaction mixture where expected analogues of molecules structurally close to those of analytes are present. The separation power of the LC used depends upon column efficiency, resolution, and selectivity. So, it was natural that molecular imprinting combined the stationary phase into the separative part of this analytical technique.

In 1993, Shea *et al.* described the first study relating the use of MIP as the stationary phase to separate nucleic acid analogues.[10] In this study, the MIP column was used to determine physico–chemical parameters of the MIP rather than to separate nucleoside analogues from the mixture. However, the way of using and managing the chromatographic part for calculating the intrinsic behavior of MIP permitted numerous other teams to devise MIPs as the selective liquid chromatography, capillary electrochromatography, or gel electrophoresis stationary phases in the separation of (oligo)nucleos(t)ides. Allender *et al.*[66] were the first to use an MIP stationary phase to separate diastereoisomers of nucleoside monophosphates prodrugs from synthetic mixtures. Because of the considerable efficiency of phosphoramidate derivatized nucleotides in inhibiting activity of nucleoside reverse transcriptase, the synthesis of this type of nucleotide was investigated in detail. However, in the final reaction media, diastereoisomers of the expected pro-drug were present. It was challenging to separate both forms because one was more active than the other. Thanks to the selectivity of the imprinting technology, the team synthesized an MIP with a pure nucleotide prodrug of stavudine (**26** in Scheme 3.8) as the template, MAA as the functional monomer, and EGDMA as the cross-linking monomer in chloroform as the porogen. The grounded MIP particles, with a mean size of 45 μm, were packed into a 15 cm × 4.6 mm stainless-steel HPLC column. The mobile phase was a mixture of acetonitrile and chloroform. The ability to resolve diastereomeric mixtures of several anti-HIV prodrugs was examined by proper adjustment of chromatographic parameters. The separation factor of each couple of compounds injected was determined and it appeared to exceed 1.7 for 31 mixtures and was equal to 1.0 just for only one mixture. So, the separation achieved practically in all cases was sufficient for preparative separation. To conclude, the nucleotide prodrug diastereomers were purified using MIP columns, as exemplified with the dinucleotide family synthetized.[66] Also for HPLC, Aoki *et al.*[67] devised a new MIP capable of selective binding of the phosphomonoester dianions for separation of monophosphate nucleotides just in one run. Here, a Zn^{2+}-(4-vinyl)benzylcyclen complex was used as the functional monomer to develop a metal–ligand coordination bond between the phosphate group and the metal cation, 2′-deoxyadenosine monophosphate as the template, and EGDMA as the cross-linking monomer. The MIP powder was suspended in acetonitrile, and then packed into a 10 cm × 4.6 mm stainless-steel HPLC column. The nucleotide and nucleoside mixture was eluted for 25 min with the gradient of the diammonium phosphate salt in a solution of MeOH and the HEPES buffer solution. The retention order was cytidine < guanosine < adenosine and deoxyadenosine < 3′,5′-cAMP and 5′-CMP < thymidine < 5′-GMP < 3′-AMP < 5′-AMP < 5′-dTMP < 5′-dAMP. The mononucleotides were retained because of the Zn^{2+}-(R-OPO_3^{2-}) interaction and the selective retention of the thymidine analogue was attributed to the Zn^{2+}-dT^- interaction, at the mobile phase pH employed (pH = 7.0). So, fabrication of MIPs capable of discriminating the phosphate group could be a solution in complex nucleotide separations encountered in different fields.[67]

The team of Haginaka[68] has worked on the synthesis of a new MIP-HPLC stationary phase for highly selective separation of adenosine phosphates. The MIP was prepared by a multi-step swelling using polystyrene seed particles and diphenyl phosphate as the template, 4-vinylpyridine as the functional monomer, glycerol dimethacrylate as the cross-linking monomer, and 1-hexanol as the porogen. A mean size of monodisperse microbeads was

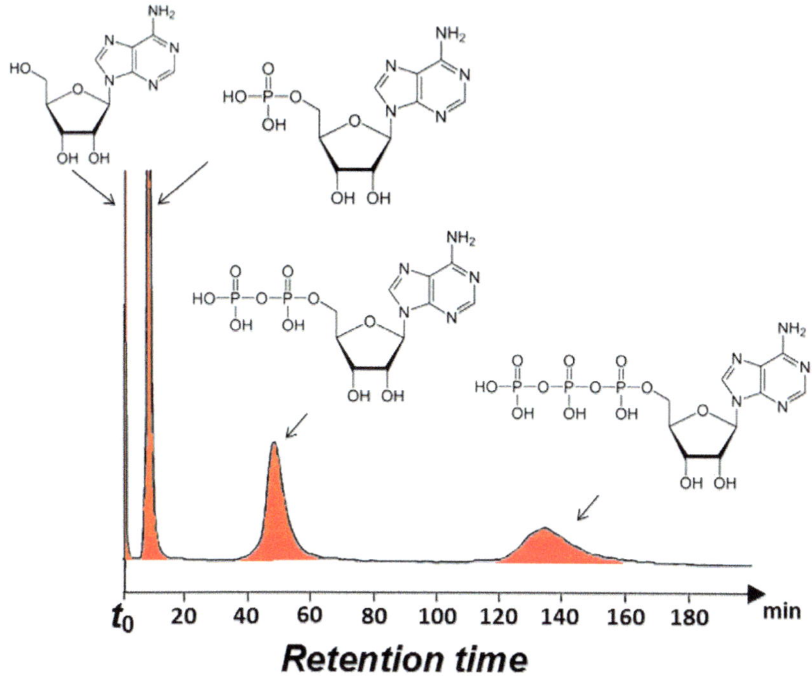

Figure 3.2 The HPLC chromatogram of separation of adenosine, AMP, ADP, and ATP on the diphenylphosphate-templated MIP stationary phase. HPLC conditions: column size, 50 mm × 4.6 mm I.D.; eluent, 10 mM sodium dihydrogenphosphate–acetonitrile (15:85, *v*:*v*); column temperature, 25 °C; flow rate, 1.0 mL min^{-1}; UV detection wavelength, 210 nm; injected sample mass, 1000 ng each. Adapted from ref. 68.

3.8 μm with a specific surface area of 167 m^2 g^{-1}. The retention and molecular recognition properties of this MIP for organophosphorus compounds were investigated using the 50 mm × 4.6 mm HPLC column packed with microbeads of this MIP and a mixture of (10 mM sodium dihydrogenphosphate)–acetonitrile (15:85, *v*:*v*) as the mobile phase. At high acetonitrile content, retention was governed by hydrophilic interactions and the retention of adenosine phosphates, AMP, ADP, and ATP on the MIP column was higher the higher was the number of phosphate groups. Moreover, the selectivity factor and resolution were then elevated (Figure 3.2).[68]

In 2004, Liu *et al.*[69] described the only study showing the feasibility of using an MIP as the stationary phase in capillary electrochromatography (CEC) to separate nucleotides.[69] The MIP was *in situ* prepared inside a 75 μm × 70 cm fused silica capillary with the internal surface functionalized with 3-trimethoxysilylpropyl methacrylate and 1,1-diphenyl-2-picrylhydrazyl. The MIP formulation involved 9-ethyladenine as the dummy template, MAA as the functional monomer, EGDMA as the cross-linking monomer, and acetonitrile as the porogenic solvent. The optimized analysis conditions included 25 mM phosphate buffer solution (pH = 8.0)–MeOH (20:80, *v*:*v*) and 15 kV as the applied voltage. The migration time for cytosine, thymine, guanine and adenine was 27.5, 31.3, 34.8, and 36.5 min, respectively. The efficiency, calculated as the number of theoretical plates per metre, was 75 300 for cytosine, 50 200 for thymine, 14 800 for guanine and only 2600 for adenine. This was because of a broad peak for adenine as a result of stronger interaction with the MIP and its selective cavities generated by the adenine analogue template. In conclusion, molecular imprinting is an interesting approach to produce stationary phases for CEC with high selectivity to a given class of compounds.[69]

After development of a hybrid CEC technique where chromatography under an electric field condition was operated, it was natural that research groups have become interested in introducing MIPs into the electrophoresis technique, and gel electrophoresis in particular. Two papers related to this field, written in 2006 by Minoura *et al.*,[70,71] presented studies on the use of molecular imprinting for recognition of a double-stranded DNA (dsDNA) target by MIP gel electrophoresis. In the first paper,[70] a λDNA 564-bp fragment, isolated from a λDNA HindIII marker, was used as the template. It was mixed with 2-vinyl-4,6-diamino-1,3,5-triazine, used as the functional monomer, acrylamide as the classical functional monomer used in polyacrylamide gel electrophoresis, methylenebisacrylamide as the cross-linking monomer, and ammonium persulfate in a HEPES buffer. The mixture was polymerized in a glass tube. Results of migration showed that the relative size of the dsDNA 564-bp fragment was well calculated because it was placed between the oligonucleotide standards of 500 and 600-bp in MIP-PAGE (polyacrylamide gel electrophoresis), while migration of the same dsDNA fragment was the same as that of the standard of 500-bp in non-imprinted PAGE. So, electrophoresis using the imprinted gel allowed differentiation of ds-DNA fragments according to their sizes more accurately than the non-imprinted one. The results were confirmed with the separation of verotoxin dsDNA and its base-pair counterpart on the MIP gel electrophoresis (MIP-GE) for recognizing verotoxin dsDNA. The MIP-GE system could contribute to separating target DNA in general and, particularly, in forensic medicine.[70] In the second paper,[71] by working with the same MIP-GE, the team analyzed a mixed-DNA sample. Results showed that the method was able to detect the analyte if it was present in a mixed-DNA sample. So, the target dsDNA could be detected with this method in an unknown dsDNA sample. The selectivity of the MIP-GE was confirmed with the separation of c-Ha-ras codon 61 from its point mutants, because the functional monomer was able to interact with the A–T base pair and to distinguish the mutations from the A–T to G–C and from G–C to A–T base pairs. So, this detection method, based on the MIP-PAGE, could be used for point mutation analysis.[71]

3.3.5 Molecularly Imprinted Polymer as Membranes

The development of artificial membranes and synthetic biological membranes has been important in bio-organic and environmental chemistry because of their high selectivity. Receptor–ligand interactions are responsible for the effectiveness of biological membranes. There are many synthetic membranes capable of selective transport of solutes. Among these, membranes are formed of an active thin layer deposited on a porous support material and their selectivity is due to the formation of a reversible complex between the active film while the permeability is due to the porosity of a solid porous filter.

Moreover, the membranes can be used in catalysis, to separate the target compound from complex mixtures as well as to form a receptor for sensor systems. So, to incur selectivity to this technology, an MIP can be incorporated in the membrane in the form of particles or deposited as a film. Alternately, an MIP can be the membrane itself. Several techniques have been used by different teams for the past twenty years to make MIP membranes.

The first, a 1990 study showing fabrication of a nucleoside selective membrane based on an imprinted polymeric material was that of Piletsky *et al.*[72] The membrane was synthesized with AMP as the template, diethylaminoethyl methacrylate as the functional monomer and EGDMA as the cross-linking monomer in DMF as the porogen. The total pore number of the AMP polymer, measured by sorption, was 0.68 $cm^3 g^{-1}$. With the electrodialysis in phosphate buffer solution measurements, it was found that the analogues' separation could be effective because of their different transport rates through the membrane, as a result of different binding in imprinted cavities. There was a limiting transport rate of AMP, which further did not increase with the increase of the electrolyte concentration. The permeability of the membrane was correlated with the structure of the nucleotide analogues, AMP and GMP. That is, the templated membranes demonstrated a considerably higher selectivity than the usual dialysis membranes and, therefore, they could be used in sensor technology.[72,73] Following the work of Piletsky *et al.*,[73] and also their work in the field of MIP recognition of nucleobases, Shea *et al.*[74] published, in 1996, a study on the development of a selective membrane for adenosine. Molecularly imprinted films were synthesized using 9-ethyladenine as the template and MAA as the functional monomer, in DMF as the porogen, and simultaneously deposited on silanized glass slides. The membrane selectivity was determined by the transport rates of adenine, thymine, guanine, and cytosine. The selectivity factor, *i.e.*, the ratio of the flux of the target compound to that of the other substrate, in the MeOH–chloroform (6 : 94, *v* : *v*) solution, was 3.4 for adenosine against guanosine. Affinity of the MIP templated with 9-ethyladenine was low for other purine and pyrimidine bases, namely, adenine and adenosine, because there was no pathway available for transport of these analogues through the membrane. Thus, the imprinted membranes, because of their stability, mechanical strength, and selectivity, could be used for continuous separation processing, chiral separation, or protein transport.[74]

A large study has been accomplished by Kochkodan *et al.*, presented in three papers[75–77] between 2002 and 2005, on the development of imprinted membranes selective to a derivative of ATP, *viz.*, cAMP, known to play an important role in intracellular signal transduction in many different organisms. First,[75] this group described the formation of a thin polymer film on a polyvinylidene difluoride (PVDF) membrane where the target molecule was recognized by ionic interaction and shape recognition using dimethylaminoethyl acrylate as the functional monomer and TRIM as the cross-linking monomer. The pre-polymerization complex mixture was deposited in the PVDF membrane, and then photopolymerized with benzoin ethyl ether in MeOH. The behavior of the membrane was studied by filtration of aqueous solution of cAMP. The degree of modification of the membrane was followed as the function of concentration of the initiator, cross-linking monomer, and functional monomer. The optimization of membrane capacity was thus described. Moreover, this latter factor depended upon the medium pH; the highest capacity was at pH = 4.2. The film thickness was 10 nm. The full capacity was reached quickly but the high value of water flux authorized by this membrane allowed using it for SPE.[75] In the second paper,[76] the team found that the results obtained using hydrophobic polyvinylidene fluoride were better than those using hydrophilic PVDF. The atomic force microscopy was used to examine surface morphology of MIP membranes and the selectivity of cAMP *versus* cGMP was demonstrated. The increase of surface roughness caused the increase of cAMP sorption and, hence, the effect of the salt concentration on the MIP capacity in the template solution was evaluated. Recognition ability of the MIP membrane toward cAMP was high because of ionic binding and proper positioning functional groups involved in binding in the MIP.[76] In the third work,[77] the same team prepared a PVDF coated MIP membrane, keeping the pre-polymerization complex solution composition same as that described previously. The goal here was to compare binding properties of membranes prepared using three different functional monomers, *viz.*, dimethylaminoethyl acrylate, 2-hydroxyethyl methacrylate, or MAA. Concentrations of functional and cross-linking monomers were optimized to maximize the capacity of the membrane. The best performing monomer seemed to be dimethylaminoethyl acrylate. Finally, the capacity increased when the degree of modification and the amount of MIP deposited on the surface of the PVDF membrane increased. For example, the MIP membrane with the degree of modification of 920 mg cm^{-2} retained 67% of cAMP and only 30% of cGMP, in 5×10^{-5} M aqueous solution. The binding capacity of the cAMP-templated MIP membranes was governed by the degree of the modification.[77]

Yoshikawa *et al.*,[78,79] since 2000, have attempted synthetizing a membrane selective to adenosine. Toward this goal, they tried to generate selective cavities in different polymers, including cellulose acetate with acetyl content of 40%, carboxylated polysulfone with the degree of substitution of 0.88, and a polystyrene resin grafted by a tetrapeptide derivative. The polymer was dissolved, together with the 9-ethyladenine template, in tetrahydrofuran, and then poured into a flat laboratory dish, and subsequently the support was dried. A mean thickness of different membranes was 145(±5) μm. Adsorption and permeation was measured with an aqueous EtOH solution of a mixture of 1 mM adenosine and 1 mM guanosine. The molecular cavities were well imprinted with the adenosine template during membrane preparation. The higher affinity and the higher selective permeation toward adenosine and the adenosine–guanosine mixture, respectively, were found for the polystyrene–tetrapeptide derivative polymer. That is, it was demonstrated that, in the field of membranes, the molecularly recognizing cavity could be generated during membrane casting without radical polymerization.[78] The second paper from this group,[79] published in 2005, demonstrated that the use of proteins, here water-soluble proteins from a microorganism, could be converted into a material molecularly recognizing adenosine. For that, they mixed 9-ethyl-adenosine as the template in a 2,2,2-trifluoroethanol solution of the *G. thermodenitrificans* DSM465 protein fraction, used as an environment-friendly 'green' polymer. This mixture was directly poured into a polytetrafluoroethylene dish. Results showed an adsorption selectivity of 4.11 for adenosine *versus* guanosine when the mixture of both was used. From the isotherm of adsorption of adenosine both a complex stability constant of 1.75×10^5 M^{-1} and the active sites concentration of 1.02×10^{-2} M for the MIP membrane of microorganism proteins were determined. This biomacromolecule based recognition material could be used as an alternative molecular imprinting method.[79]

Moreover, Kobayashi's group was very active in the field of imprinted polymer membranes for pyrimidine by publishing four papers[80–83] between 2004 and 2007. In 2004, the team used the immersion-precipitation phase-inversion method to fabricate a membrane selective to uracil with poly(acrylonitrile-*co*-methacrylic acid) or poly(AN-*co*-MAA).[80] A mixture containing uracil as the template and the polymer was dissolved in DMSO, and then submerged in a coagulation bath containing water. Because of the exchange of the DMSO solvent by the water non-solvent, a solid precipitated. Uracil interacted with the polymer by formation of hydrogen bonds, demonstrated by results of the FT-IR and ^{1}H NMR spectroscopy measurements. The imprinted material surface roughness was higher than that of the non-imprinted membrane. The membrane selectively bound uracil compared to dimethyluracil and caffeine. The uracil binding at saturation was 6.9 mol g^{-1}. Apparently, the uracil-imprinted poly(AN-*co*-MAA) membrane was highly selective to uracil in permeation studies. Therefore, the phase-inversion imprinting method easily allowed synthesizing a selective MIP membrane.[80] In their second paper,[81] the team modified the previously described formulation by incorporating a second polymer, namely, poly(acrylonitrile-*co*-vinylbenzyl-stearyldimethylamine chloride) or poly(AN-*co*-SMA). A copolymer membrane was cross-linked with ionic interactions between the positively charged ammonium moiety of poly(AN-*co*-SMA) and negatively charged carboxyl group of poly(AN-*co*-MAA), as demonstrated by the FT-IR spectroscopy analysis. In this new membrane, uracil simultaneously interacted with the stearyldimethylamine and carboxyl group of the copolymer. Finally, the imprinted membrane, prepared from a mixture of the poly(AN-*co*-SMA)-poly(AN-*co*-MAA) in a molar ratio of 1:1, showed the highest binding capacity at 0.38 μmol g^{-1}. The separation factor for uracil against dimethyluracil and caffeine was 6.3 for both. Therefore, the cross-linked polymer based on the ion complex of two copolymers showed a dense morphology, which played a role in the recognition behavior of the imprinted membrane.[81] In the third study,[82] with the copolymer and the template the same as that described in the first paper,[80] the team investigated the use of liquid carbon dioxide as the non-solvent, instead of water, to prepare the uracil-imprinted membrane. With the immersion method, the separation factor was determined as 3.0, 3.8, and 2.5 for the uracil *versus* cytosine, dimethyluracil, and thymine, respectively. From the adsorption isotherm derived for uracil, the complex stability constant and the active site number were determined as 1.67×10^4 M^{-1} and 12.8 μmol g^{-1}, respectively, for the MIP membrane phase-inversed in the compressed CO_2 liquid. Advantageously, these values were higher than those for the membrane prepared in water. Apparently, liquid carbon dioxide was a promising solvent to form an imprinted membrane by the phase-inversion method.[82] Finally, to conclude the study on the uracil selective MIP membrane, the group studied carbon dioxide in the supercritical fluid state as the non-solvent in the inversion-phase method.[83] Here, the copolymer used was poly(styrene-*co*-maleic acid) and uracil was the template. As previously, the determined separation factor was 3.33, 4.1, and 5.2, for uracil *versus* cytosine, dimethyluracil, and thymine, respectively, and the adsorbed quantity of uracil was 12.6 μmol g^{-1} for the imprinted membrane prepared at 50 °C, with *N*-methyl-2-pyrrolidone as the casting solvent. So, it seemed that carbon dioxide can be used as the supercritical fluid to freeze the shape of cavities imprinted in the MIP membrane formed by the phase-inversion method.[83]

Another team that has worked on preparation of a pyrimidine MIP membrane, is the Zhou group.[84] The pre-polymerization mixture containing thymine as the template, 9-vinyladenine as the functional monomer and diethyleneglycol dimethacrylate (DEGDMA) as the cross-linking monomer was impregnated on a porous cellulose membrane in DMF, and then photopolymerized. The use of the complementary base of thymine, *viz.*, adenine, to generate Watson–Crick type interactions, allowed reaching a stability constant of 6.49×10^3 M^{-1} of the complex of the template with the functional monomer. A competitive transport experiment involving the MIP membrane showed that in MeOH the rate of diffusion of thymine and its uracil analogue was higher than that of adenine, guanine, and cytosine. This rate could depend upon the shape and size of the recognizing thymine and its uracil close analogue, and the arrangement of functional groups in the membrane channels. This imprinted membrane could be used in biological matrixes to recognize thymine and uracil in DNA and RNA hydrolysates, respectively.[84] The second work of Zhou *et al.*[85] was related to the synthesis of an MIP membrane, selective to cytokinin, based on the previous paper.[84] 6-Benzyladenine served as the dummy template, MAA was the functional monomer, EGDMA was the cross-linking monomer, and DMF was the porogenic solvent. A cellulose acetate membrane filter was immersed in the solution of a pre-polymerization complex, which was then photopolymerized. The permeation and adsorption study involving the membrane was followed by electrochemical measurements using 0.1 M acetate buffer solution (pH = 3.8). The apparent stability constant and stoichiometry of the pre-polymerization complex of the template with the functional monomer was $2.04 \times 10^2\ M^{-1}$ and 1 : 1, respectively. The membrane transport was studied with 6-benzyladenine and an interference analogue of cytokinin, the 6-(furfurylamino) purine. The flux was a very important factor playing a significant role on the selectivity of the membrane, *i.e.*, the higher the flow rate, the lower was the selectivity between these two test compounds. Finally, this imprinted membrane revealed a remarkable selectivity for the target cytokinin compared to that of the 6-(furfurylamino)purine. This selectivity was the same during six months of membrane operation. Application of this MIP membrane to detection of 6-benzyladenine in plant extracts could be envisioned.[85] Most recently Scorrano *et al.*[86] demonstrated the use of this MIP membrane for determination of nucleosides. The goal was to extract 2′-deoxy-adenosine, an important tumor biomarker in humans, from urine in order to determine it at low concentrations by removing interferences. For that, a PVDF hydrophilic filter was immersed in the solution of a pre-polymerization complex of the 2-deoxyadenosine template with the MAA functional monomer in the presence of the EGDMA cross-linking monomer, in an acetonitrile–MeOH mixture used as the porogen. The polymer was synthetized by thermopolymerization with azobisisobutyronitrile (AIBN). The MIP membrane showed a modification degree of 2.4 mg cm^{-2} and a mean pore diameter of 0.15 μm. The adsorption kinetics allowed measuring, in a mixture, adsorption of 2-deoxyadenosine, which was merely 0.5 μmol cm^{-2} in 30 min, while adenosine and 1-methyladenosine were not adsorbed at all. Apparently, the selectivity was very high. Finally, the MIP membrane was immersed in 0.5 mM 2-deoxyadenosine spiked human urine, and then 30 min adsorption allowed. The analysis of the extract by HPLC-UV enabled calculation of an 85% recovery in the spiked samples. This work demonstrated the use of that membrane type for biological sample pretreatment to monitor cancer biomarkers in humans.[86]

3.4 MIPs as Nucleos(t)ide and Analogue Recognition Units in Chemosensors

Mimetic enzymes provide an effective and practical method for chemical sensors. However, selectivity of these artificial enzymes is compromised by the inconsistent recognition of the binding sites of the enzyme towards the substrates. To improve stability of mimetic enzymes, several researchers incorporated catalytic groups in these enzymes or coupled them with other functional materials, *e.g.*, hairpin DNA, different ligands, and metal–organic framework (MOF) materials. These mimetic enzymes not only maintain the original properties, but also reveal prominent properties of hybrid functional materials, so that the stability of these mimetic enzymes can largely be increased. Historically, the first study was published in 1987 by a Japanese team.[87] It involved grafting a molecular recognition featuring monolayer on an electrode. The covalently grafted monolayer of octadecylsilane (ODS), in the presence of *n*-hexadecane, on an SnO_2 substrate, was capable of strong adsorption of molecules with a thin hydrophobic tail, *e.g.*, vitamin K1, yielding a pronounced electrochemical response. In the same year, the same team demonstrated that the above sites were present in the active SnO_2–ODS monolayer enabling selective binding of molecules with the size recognition effect and shape.[87] This 'chemoselective layer' was used as the recognizing and binding unit of an electrochemical sensor. The Mosbach team[88] relied on these two studies to evaluate the same strategy for fabrication of an ellipsometric chemosensor. In this chemosensor, the film was bonded to a silicon wafer. The complex stability constant of vitamin K1 on the ODS-silicon layer was calculated as $2.8 \times 10^5\ M^{-1}$, thus opening the field of the use of the guest-selective surfaces in optical sensors.[88]

The MIPs were introduced into the field of electrochemical sensors only after these initial studies by Mosbach *et al.*[89] with the development of a sensing layer in field-effect capacitors and Piletsky *et al.*[90] with the work on a highly substrate-selective polymer membrane in an electrochemical cell with platinum electrodes. The development and use of the MIP technology in the field of sensors have not stopped ever since. A schematic representation of an MIP chemosensor is presented in Scheme 3.9.

The part developed here, based on determination of nucleic bases, nucleosides, and nucleotides using sensors with MIPs as full recognition elements, will be presented in three sections, for more clarity, depending upon the sensor type, *i.e.*, electrochemical, mechanical, and optical, as Scheme 3.9 shows, where the MIP film is the recognition unit.

3.4.1 Electrochemical Sensors with MIPs

The number of publications addressing the use of MIPs as elements for selective recognition of the analytes to be determined electrochemically is the highest compared to other sensors. Partly, this is explained by electrochemical activity of nucleobases and, therefore, nucleosides, nucleotides and their analogues (Table 3.2). The other explanation is that the electrochemical analyzers are fairly common and show the highest sensitivity among different analyzers, except, perhaps, the radiometric type analyzers.

To return to the beginning of the combination of electrochemical sensors with MIP technology for the determination of nucleic-like compounds, we must go back to 1989. At that time, a French research center, working on implementation of conducting polymers as electronic sensors coated a platinum electrode with a polypyrrole film by electropolymerization of pyrrole, in an aqueous solution of a biological anion, to result in a chemosensor selective to AMP.[96] Then, it was not until 1996 that another team was interested in voltammetric determination of the adenosine nucleoside

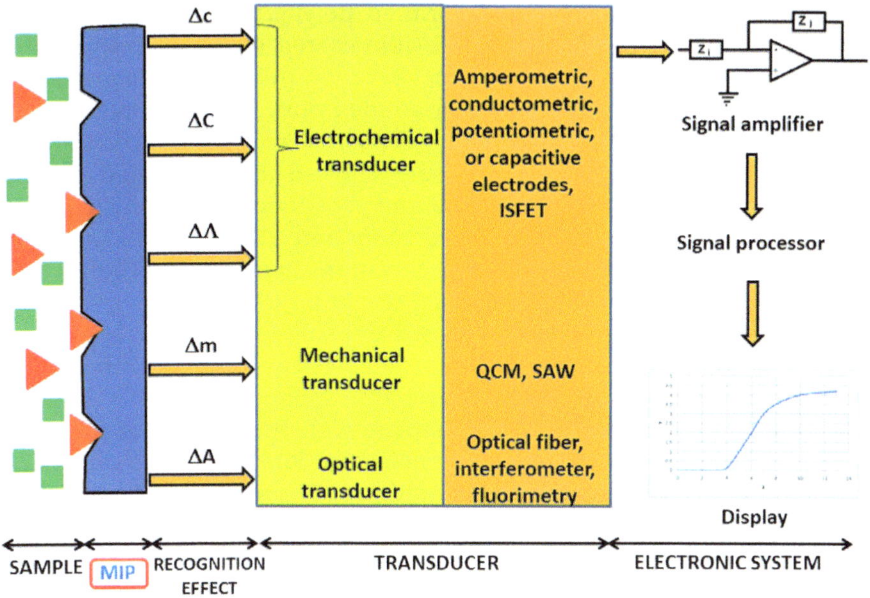

Scheme 3.9 Representation of the operating principle of a biosensor (Δ*c*, Δ*C* and ΔΛ stand for the change in concentration, capacity, and conductivity, respectively. Δ*m* is the change in mass and Δ*A* is the change in absorbance).

Table 3.2 One-electron redox potentials for different nucleic bases, nucleosides, and nucleotides.[a]

Redox species	E_{ox} (pH = 13)/V *vs.* NHE[91]	E_{ox}/V *vs.* NHE[92]	E_{red}/V *vs.* NHE[93]	E_{ox}/V *vs.* Ag/AgCl[93]	E_{red}/V *vs.* SHE[94]	E_{ox}/V *vs.* NHE in ACN[95]	E_{red}/V *vs.* NHE in DMF[95]
Adenine	0.75	1.75	−2.86	0.96	—	1.96	−2.52
Adenosine	0.81	—	—	—	−2.1	—	—
AMP	—	—	—	1.19	−2.0	—	—
Cytidine	—	—	—	—	−2.4	—	—
Cytosine	0.81	2.18	−2.41	1.31	—	2.14	−2.35
CMP	—	—	—	1.46	−1.9	—	—
Guanine	0.63	1.40	−3.31	0.70	—	1.49	−2.76
Guanosine	0.71	—	—	—	−1.6	—	—
GMP	—	—	—	0.89	−1.5	—	—
Thymine	0.79	2.00	−2.32	1.16	—	2.11	−2.18
TMP	—	—	—	1.41	—	—	—
Uracil	0.88	—	—	—	—	2.39	−2.17
Uridine	—	—	—	—	−2.5	—	—
UMP	—	—	—	—	−2.4	—	—

[a]AMP, adenosine-5′-monophosphate; CMP, cytidine-5′-monophosphate; GMP, guanosine-5′-monophosphate; TMP, thymidine-5′-monophosphate; UMP, uridine-5′-monophosphate, NHE, normal hydrogen electrode; SHE, standard hydrogen electrode; ACN, acetonitrile; DMF, *N,N*-dimethylformamide.

via a thin film of molecularly imprinted polypyrrole. The detection of adenosine was more sensitive at the film-coated electrode compared to the bare GCE.[97] Then, in 2001, Willner *et al.*[98,99] published two articles on the use of very thin (36 ±1 nm) acrylamide-imprinted acrylamidephenylboronic acid copolymer membranes, deposited on ISFETs, to determine nucleotide monophosphate. The results showed that imprinted polymer membranes enhanced both the LOD and sensitivity of the sensing interfaces. To illustrate this result, it was demonstrated that the ISFETs reliably detected nucleotides at very low concentrations in an aqueous buffer solution, with an example LOD equal to 8×10^{-7} M for CMP.[98,99]

In 2006, Kharitonov *et al.*[100] resumed the studies of the Willner group to develop the same FET sensor but dedicated to the determination of nicotinamide adenine dinucleotide (NAD) in a phosphate buffer solution. The reversible interaction of the template with the functional monomer led to the formation of a stable intermolecular complex and to the release of proton near the area of the transistor gate. The signal generated by the proton release was correlated to the concentration of the NAD in the solution.[100] Inspired by the work of Brajter-Toth *et al.*,[97] a Japanese team[101] made an MIP sensitive chemosensor in 2008 by devising an overoxidized imprinted polypyrrole film, selective to ATP. The sensor operated in the triple FIA regime with pulse amperometry, which allowed detection of 5 nM ATP, *i.e.*, a 10^4-fold lower LOD than that found with the bare GCE, without employing any preliminary pre-concentration.[101]

Kugimiya *et al.*[4] selected an ISFET sensor as the type of transducer in cAMP determination where the surface of the transducer was coated with a cAMP-imprinted polymer film used as the recognition element. Application of an artificial polymer receptor as the molecular recognition probe is less expensive than the use of a natural antibody, and less sensitive to chemical and physical stress. Functional monomers, such as 1-allyl-2-thiourea and NIPAM were reliable to distinguish cyclic phosphate of cAMP from similar chemical structure such as adenine, AMP, GMP and cGMP. The linear dynamic concentration range extended from 0.1 to 1.0 mM cAMP in an aqueous buffer solution (pH = 8.0).[4]

The Prasad group[102] evaluated an MIP electrochemical sensor for complex matrices of real samples, such as human plasma. For that, at first in 2009, the team modified an old electrochemical technique engaging a hanging mercury drop electrode.[102] The technique involved modification of the Hg droplet surface with an MIP film prepared by bulk condensation polymerization in DMF. Melamine and chloranil, used as the functional co-monomers to form a non-cross-linked polymer chains, and the 5-FU or uracil template, were mixed with these chains. The goal was to avoid overcrowding and to facilitate unhindered template mass transfer by diffusion across the polymer membrane. The differential pulse cathodic stripping voltammetry runs allowed determination of 5-FU and uracil in human plasma down to 0.34 ng mL^{-1} and 0.26 ng mL^{-1}, respectively.[102] A few years later, in 2012, Prasad *et al.*[103] improved the MIP chemosensor for the determination of 5-FU and uracil devised previously by turning to the synthesis of a hybrid MIP combining a ceramic material and an organic polymer. However, to resolve problems of the use of silica-based MIPs, consisting of polymer shrinkage and cracking, the team used a non-hydrolytic sol–gel (NHSG) formulation without water. The mixture containing 3-(trimethoxysilyl)propyl methacrylate, modified MWCNTs, and the 1,3-diacrylurea template in DMSO was polymerized *via* the activator generated by electron transfer for atom-transfer radical polymerization (AGET-ATRP). MWCNTs were used for their abilities to decrease overpotential, promote electron-transfer kinetics, and minimize electrode fouling. The selectivity to adenine, guanine, thymine, cytosine, barbituric acid, hypoxanthine, caffeine, uric acid, ascorbic acid, dopamine, urea, glucose, and creatine was high for this chemosensor. The compounds sought, namely, 5-FU and uracil, were determined in an aqueous solution of urine or serum. The LOD for uracil and 5-FU in serum was 2.12 ng mL^{-1} and 0.64 ng mL^{-1}, respectively. The sensitivity was lower than the previous one but the measurement was more eco-friendly and the electrode-to-electrode repetability was acceptable.[103]

Before Prasad *et al.*,[102,103] however, the idea of using MWCNTs incorporated in the MIP film for constructing a nucleoside chemosensor was developed by a Chinese team in 2010.[104] These carbon nanostructures were introduced to convert effectively the binding signals of molecular recognition to detectable electrical signals, thus resulting in enhanced electronic transmission and sensitivity. A thin film of a sol–gel silane polymer molecularly imprinted with thymidine was cast on a modified carbon electrode surface by electrochemical deposition. Selectivity of recognition of thymidine and guanosine, cytidine, 2′-deoxyguanosine, adenosine, and 2′-deoxyadenosine by this film was high. The linear dynamic concentration range was 2 to 22 μM for thymidine, in 0.2 M phosphate buffer solution (pH = 7.0). The LOD was 1.6×10^{-9} M. This novel electrochemical sensor showed high sensitivity and rapid response in detection of zidovudine-thymidine tablet samples.[104] Following all these advances, two teams, those of D'Souza and Kutner,[105–109] have worked hard together in the field of MIP chemosensors, dedicated to selective determination of nucleosides, nucleotides, and their analogues. Between 2012 and 2016, five publications in the design, development, characterization, and application of electrochemical sensors in this field marked their collaboration.[105–109]

The first, in 2012,[105] concluded that the LOD of a capacitive impedimetric chemosensor with a film sensitive to ATP was sufficiently low for practical biological applications. For this, the teams employed dedicated functional monomers capable to distinguish ribose, to make complementary Watson–Crick like interactions (functionalized uracil) with the adenosine moiety, and to bind phosphate groups with an amide derivative. These monomers were based on bisbithienyl or thienyl moieties, which provide a very efficient potentiodynamic electropolymerization resulting in the deposition of an MIP-ATP film on an electrode surface. The consequences included a different response of MIP and NIP, and a pronounced selectivity with respect to thymidine triphosphate (TTP), GTP, NAD, ADP, and CTP. Although the LOD of the MIP chemosensor was twofold better (0.2 μM for impedance measurement), it seemed to be more laborious to develop. Capacitive impedimetry (CI) showed a very stable signal compared to those of another transduction used, namely, the piezoelectric microgravimetry (PM) using QCM. Finally, the IF, estimated as 9.4, confirmed a very successful design in the structure of the pre-polymerization complex of ATP and functional monomers, and the complex structure was confirmed by optimization using density functional theory.[105] Based on this first study, the same researchers kept the highly selective recognition mode of the template, involving Watson–Crick interaction. The target molecule was 5-FU, and the bisbithiophene monomer was functionalized with the uracil complementary base, *i.e.*, adenine. Moreover, this pattern allowed for the resulting MIP to become a conducting polymer, electrochemically generated and deposited on the surfaces of transducers. The only problem was that the template could undergo a redox reaction during the potentiodynamic electropolymerization. Then, the MIP would be selective to the electro-oxidation product thus formed, and not to the template initially present in the formulation. Therefore, the electropolymerization was systematically optimized to avoid formation of this undesired product. Working on three different transduction platforms, *i.e.*, differential pulse voltammetry (DPV), CI, and PM, the teams showed that all sensors were selective with respect to the most typical interferences (analogues and metabolites of 5-FU) but only the DPV and CI chemosensors were suitable for determination of 5-FU in human blood plasma (mean concentration of 500 nM), because of their low LOD of 56 nM and 75 nM, respectively. The determinations, therefore, were carried out with CI under flow-injection analysis (FIA) conditions using the 0.1 M KCl–EtOH (1 : 1, *v* : *v*) carrier solution. Physico-chemical parameters, such as the stability constant of the template–(functional monomer) complex formation and the IF were determined as 2.17×10^7 M^{-2} and 3.6, respectively.[106]

The third study involving determination of nucleotides with an MIP chemosensor used C_{60} derivatives as functional monomers.[107] Indeed, the ATP imprinted fullerene polymer showed a high redox conductivity and specific capacity to help detect the ATP analyte with high sensitivity. Moreover, in its derivatized forms, the fullerene played the role as functional monomers with a reductively electroactive behavior. Thus, the template, ATP, an electro-oxidizable molecule, was not impacted when the MIP was prepared by reductive electropolymerization. Accordingly, the MIP was prepared with three different C_{60} functional monomers, one with the uracil moiety to bind selectively to the adenine base of ATP, the second with the carboxyl group to recognize the diol part of the ribose moiety and the third bore the amide group to link the triphosphate chain of the nucleotide. That way, the ATP template was anchored by multi-point interactions, thus increasing its detection selectivity. The MIP-ATP film, containing a Pd(II) acetate dimer and C_{60} as the cross-linking monomers, was deposited on a Pt disk electrode for the

CI measurements. The linear dynamic concentration range for ATP in 0.1 M KF was 0.06 to 1 mM in CI, and the LOD was 31 μM. Moreover, high selectivity to other nucleotides, even to ADP, was demonstrated. The IF and the sorption equilibrium constant were determined as 4 and 6.45×10^2 M^{-1}, respectively.[107] In the next year, the D'Souza and Kutner teams[108] applied their knowledge of electropolymerization aiming at deposition of an MIP thin film on metal electrodes, for determination of another purine nucleoside, 6-thioguanine. The interaction model developed was the same as that in the previous study. It involved Watson–Crick type interactions between an easily polymerizable monomer based on the cytosine linked to (2,2′-bithienyl)-(4-carboxyphenyl) methane ester and the 6-thioguanine template. The cross-linking monomer was 3,3′-bis[2,2′-bis(2,2′-bi-thiophene-5-yl)]thianaphthene because it formed a conductive polymer after electropolymerization. The template-to-(functional monomer) molar ratio was 1 : 2 and the stability constant of the pre-polymerization complex formed was 2.17×10^7 M^{-2}. In CI measurements under FIA conditions, the LOD of 6-thioguanine was 10 μM for 0.5 M KF used as the FIA carrier solution. As an alternative to the capacity measurement, the monitoring of the alternating current of the alternating potential of small amplitude increased the chemosensor selectivity to 2-amino-6-methylmercaptopurine, glucose and 2-aminopurine.[108]

In 2015, Noworyta *et al.*[109] used an extended-gate field-effect transistor (EG-FET) signal-transducing unit, with its gate coated with an MIP film-recognizing unit, to selectively determine inosine.[109] Researchers have employed this technology to decrease the signal instability because of the presence of oxidation products of purine, which are formed on the surface of the electrode with the use of a relatively high potential in classical electrochemical electrode. The MIP was deposited to form a thin film on an Au-coated gate area. For that, the MIP was prepared by potentiodynamic electropolymerization from a solution of two functional monomers, namely, 2-(cytosin-1-yl) ethyl-*p*-bis(2,2-bithien-5-yl)methylbenzolate and 2,2′-bithiophene-5-boronic acid, the inosine template, and the 2,4,5,2′,4′,5′-hexa(thiophen-2-yl)-3,3′-bithiophene cross-linking monomer. Both functional monomers allowed for recognition of inosine *via* the base moiety and by the 2′,3′-diol group of ribose while the bithienyl or bithiophene moieties enabled electropolymerizing and afforded conductivity of the polymer. The linear dynamic concentration range of this chemosensor was 0.5 to 50 μM, and the LOD was 0.62 μM, *i.e.*, higher than that of the PM chemosensor with the same MIP. Moreover, the IF was as high as 29. It was determined as the ratio of sensitivity of the MIP to that of the NIP film-coated EG-FET chemosensor. Selectivity of the proposed MIP film coated EG-FET chemosensor with respect to inosine structural analogues and interferences, including thymine, adenosine, guanosine, and glucose was very high.[109]

Nearly at the same time, in 2014, an Egyptian team[110] was also interested in development of a chemosensor provided with an MIP film for drug determination, a prodrug antiviral guanine analogue.[110] The combination of famciclovir (**30** in Scheme 3.10) MIP with a carbon paste electrode (CPE) allowed

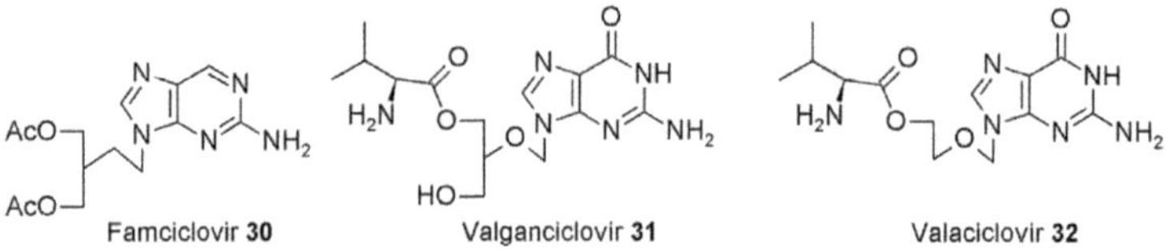

Scheme 3.10 Structural formulas of famciclovir **30**, valganciclovir **31**, and valaciclovir **32**.

for combining high MIP affinity with the sensitivity of electrochemical response originating from electro-oxidation of the 2-aminopurine ring. The MIP was synthetized using a classical solution containing MAA and EGDMA in dichloromethane. This solution was mixed with a graphite powder at 5% in mass. The complex stability constants for the high and low affinity site were ~29×10^4 M^{-1} and ~1.1×10^3 M^{-1}, respectively. The CPE was chosen because of its ease of preparation, wide usable analyte concentration range, and high stability. Moreover, the response was linear over the concentration range of 2.5×10^{-6} to 1×10^{-3} M. Because of the MIP use, the LOQ and LOD was 2.5×10^{-6} and 7.5×10^{-7} M, respectively. The selectivity to lamivudine, ribavirin, caffeine, creatinine, histidine, phenylalanine, tryptophan, and lactose was demonstrated. Finally, the chemosensor was used to quantify famciclovir in dosage formula. The famciclovir recovery from the famvir and propencivir tablets was in the range of 99–104% and 98–102%, respectively.[110] Gholivand *et al.*[5] reported on quantitation of ganciclovir (**4**, in Scheme 3.1) and valganciclovir (**31** in Scheme 3.10) in human serum, by DPV. Here, AuNPs were incorporated in the MIP film to increase the current response *via* facilitating the charge transfer. A sensitive thin MIP film was made by electropolymerizing a mixture of AuNPs, 2,2′-dithiodianiline as the functional monomer, and ganciclovir as the template onto carboxyl-functionalized MWCNTs coupled to the GCE. Thus, the formation of the Au–S bond, and the presence of MWCNTs led to a relatively high conductivity of the system, large specific surface area, and high biocompatibility. Thanks to the MIP, the chemosensor was selective to valganciclovir (**31** in Scheme 3.10) acyclovir (**27** in Scheme 3.8), valaciclovir (**32** in Scheme 3.10), guanine, deoxyguanosine, aniline, and cysteine. The MIP chemosensor allowed for detection of ganciclovir in human serum without any prior separation. Recovery was in the range of 96–103% for this drug in serum, in the concentration range of 0.5 to 1.5 μM. In the phosphate buffer solution (pH = 3.0), the chemosensor presented an LOD equal to 1.5 nM and long-term stability.[5] Another paper of the same group[111] described the same method to prepare an AuNP modified MIP, deposited by electropolymerization onto carboxyl-functionalized MWCNTs, cast on a GCE (GCE/HOOC-MWCNTs/AuNPs-MIP). Selectivity of the resulting MIP chemosensor was investigated with a large number of possible interfering compounds including ganciclovir, acyclovir, valaciclovir, inorganic anions or cations, amino acids, urea, and lactose. These analogues and other possible interferences did not influence the anodic current of oxidation of valganciclovir. For real samples, *e.g.*, for human serum, the recovery of valganciclovir was between 98 and 102% using anodic stripping DPV in the concentration range of 1.5 to 450 nM. The LOD for ganciclovir determination in serum was 1.1 nM.[111]

Before the two papers indicated above appeared, the idea to incorporate AuNPs in the pre-polymerization formulation of an MIP to assay a pyrimidine drug had been developed by the team of Jaffrezic-Renault.[3] Here, a microporous-MOF term was even employed because AuNPs were linked in a 3-D network by *p*-aminothiophenol surrounding the template. These microporous MOFs were tethered to the surface of the gold electrode with the same linker. The sensitive film was prepared by electropolymerization. That was because this electropolymerization allowed controlling the film thickness, thus avoiding the need for rigorous synthesis and film preparation, as opposed to what was required when using the spin-coating or solvent-casting techniques. The MIP film recognized gemcitabine (**2** in Scheme 3.1) *via* hydrogen bonding as well as van der Waals interactions between the amino donor group of *p*-aminothiophenol and the acceptor group of gemcitabine. This MIP chemosensor with embedded AuNPs demonstrated high sensitivity, an enhanced number of accessible binding sites, and fast equilibration with the gemcitabine analyte. The device linearly responded to the gemcitabine concentration over the range of 3.8 fM to 38 nM, with the LOD of 3 fM. The current signal generated by linear sweep voltammetry allowed detection of gemcitabine in spiked calf serum with a recovery between 96 and 102% for concentrations in the range of 10 fM to 1 pM.[3]

Moreover, in 2015, the Prasad group[112] returned to the determination of the now widely used 5-FU and uracil by using an electrochemical sensor with MIP-based nanoarrays.[112] The pre-polymerization mixture, containing the 5-FU template, the *N*-acryloyl-2-mercaptobenzamide functional monomer, and MWCNTs, was spin coated on the tip of the etched nanoporous silver electrode, and then thermal free-radical polymerized. Then, the silver membrane was dissolved in nitric acid, thus leaving the electrode decorated with a nanoarray structure of the MIP. Thanks to advantages of the MIP, such as high mechanical and chemical stability, added to the nanopores of the electrode, and the electroconductivity enhanced with MWCNTs, the LOD of uracil and 5-FU, determined by differential pulse anodic stripping voltammetry in 0.1 M disodium tetraborate buffer solution (pH = 5.6), was 0.50 and 0.33 ng mL^{-1}, respectively. The adsorption coefficient for uracil and 5-FU was 4.5×10^6 and 3.44×10^6 M^{-1}, respectively. Moreover, the IF for uracil and 5-FU was 67 and 77, respectively. Selectivity was examined with respect to adenine, guanine, cytosine, thymine, dopamine, hypoxanthine, barbituric acid, ascorbic acid, caffeine, uric acid, creatine, urea, and glucose. Finally, uracil and 5-FU were determined in complex matrices of blood plasma and pharmaceuticals with assured reliable results, without any matrix effect, cross-reactivity, and false-positive results.[112]

In the beginning of 2016, an innovation was introduced with the use of a natural polymer and a redox dye in the concept of an MIP applied in the sensor electrode.[113] First, a 6-benzylaminopurine template was dissolved in the chitosan solution, a poly-(D-glucosamine) produced by chitin deacetylation. Then, this solution was mixed with a suspension of MWCNTs grafted with SnS_2 NPs. The resulting slurry was dispensed on the GCE surface. High electrocatalytic activity and high conductivity of MWCNT–(SnS_2) dramatically improved the sensitivity. The 6-benzylaminopurine imprinted chitosan provided selective recognition sites of a new chemosensor. Presumably, hydrogen bonding mainly contributed to the interaction of the MIP film with 6-benzylaminopurine because a protic solvent completely removed the template and maintained integrity and stability of the imprinted cavities. For the 0.2 M phosphate buffer solution (pH = 6.5), the LOD was 50 pM. In view of the goal of selective detection of 6-benzylaminopurine in vegetable and fruit samples, including soybean sprout, mung bean sprout, potato, tomato, pear, and apple, the selectivity was investigated with respect to some analogues, including indole-3-acetic acid, 6-benzyloxypurine, 6-benzylguanine, kinetin, *m*-topolin, roscovitine, olomoucine, and bohemine. Finally, in real samples, recovery of the sensor toward 6-benzylaminopurine ranged from 91% to 103%, practically being equal to that of results of LC-MS measurements.[113]

Second, Prasad *et al.*[114] for the first time introduced graphene nanosheets to a pre-polymerization mixture in order to increase the sensitivity of the chemosensor for a nucleoside. That was because the specific surface area of graphene is very large, theoretically 2630 $m^2\ g^{-1}$, and its conductivity is also high, being 550 $S\ cm^{-1}$. Indeed, the sensitivity of an MIP chemosensor is related to the number of active imprinted sites on the sensor surface and, therefore, it is increased by the use of nanomaterials in the film coating the electrode surface. In this study, coupled to the use of electrochemically reduced graphene oxide, neutral red, a phenazine redox dye with π-conjugated structures, served as the functional monomer capable to form a conducting polymer after electropolymerization. The resulting adsorption of 6-thioguanine on the MIP film was strong; the complex stability constant being $176.5 \times 10^6\ M^{-1}$. That was because hydrogen bonding and aromatic π–π interactions were formed because of the presence of the amino group and the aromatic ring, respectively, located on the neutral red. In 0.05 M phosphate buffer solution (pH = 7.5), the MIP film coated pencil graphite electrode, with the differential pulse adsorptive stripping voltammetric technique, allowed determination of the 6-thioguanine analyte in biological and pharmaceutical samples with LOD of 0.07 ng mL^{-1} and 0.04 ng mL^{-1}, for urine and serum, respectively.[114]

3.4.2 Piezoelectric Microgravimetry MIP Chemosensors

As all molecules have mass, mass sensing, in absolute terms, is conceptually more straightforward than chemosensing. However, difficulties arise from a mass measurement variation because this measurement is much less selective than that of a redox potential. Therefore, the importance of using a polymer selective to the target analyte as the recognizing film was addressed in two consecutive papers of 1999 and 2000, by Kanekiyo *et al.*[115,116] In the first one,[115] the gold film electrode of a quartz crystal resonator (Au-QCR) was preliminarily modified with an anionic thiol before deposition of a polyion complex acting as the MIP film. This polyion was imprinted with AMP in a multi-layer adsorption process. The polyion was composed of a mixture of a polycation (poly-*N*,*N*-dimethylpyrrolidine) and a polyanion (boronic acid in a carboxylate polymer), which were hold together through ionic interactions, in the presence of AMP, in 10 mM carbonate buffer solution (pH = 10.2). No cross-linking monomer was added. However, in the second paper,[116] an imprinted hydrogel was anchored to the Au-QCR surface with 2,2′-dithio-bis(acrylamidoethane), and then measurements were performed using 0.1 M carbonate buffer solution (pH = 10.2). In both cases, the hydrogel swelled or shrank, induced by the presence or absence of the nucleotide analyte. In the first study,[115] the LOD for AMP was ~10 μM and selectivity of AMP determination with respect to 2′-deoxyAMP, adenosine, and 2′-deoxyadenosine was demonstrated. In the second work,[116] the LOD was higher, equaling 0.1 mM, but the nucleotides, *viz.*, AMP, ATP, and 2′-deoxyadenosine-5′-monophosphate were discriminated by the hydrogel. Thus, this study has paved the way towards the use of the PM at QCM transduction system with the MIP recognition unit for nucleotide determination. Then, Willner *et al.*[99] coated the Au-QCR with an MIP film by electropolymerization of a mixture of *m*-acrylamidephenylboronic acid, acrylamide, *N*,*N*′-methylenebisacrylamide, $ZnCl_2$ and one nucleotide while developing an ISFET type chemosensor for detection of nucleotides, as described in the previous section. The primary goal of this PM study was to determine the extent of association of the nucleotides with the MIPs. This study demonstrated the interaction of both functional groups of boronic acid and the amide group of functional monomers with different functionalities of the nucleotide. Finally, the response time of QCM was twice as long as than that of an ISFET. Moreover, PM determination of the nucleotides was less sensitive than that with ISFET devices.[99]

Based on these initial results, Yao *et al.*[117] in 2005 devised an Au-QCR chemosensor, where a thin molecularly imprinted sol–gel film selective to cytidine has been deposited by electropolymerization. An amino derivatized silica film was grafted onto Au-NPs, which were then linked to the Au-QCR by a covalent bond with 1,6-hexanedithiol. After electropolymerization, the thickness of the resulting film was 40.7 to 90.1 nm, and resonant frequency linearly decreased with the increase of the cytidine concentration, in 0.1 M phosphate buffer solution (pH = 7.0) over the concentration range of 5.0×10^{-9} to 5.0×10^{-5} M. The selectivity of cytidine determination with respect to guanosine, uridine, thymidine, deoxythymidine, deoxyguanosine, adenosine and AMP was demonstrated. The imprinted film coated Au-QCR was stable over seven measurements for four weeks.[117]

To confirm the previous supposition of the feasibility of application of an MIP film coated Au-QCR for nucleoside determination, the Ersöz team published, in 2008 and 2009, four papers[2,118–120] on determination of 8-OHdG.

In the first paper, Ersöz *et al.*[2] described the use of allyl mercaptan to graft the polymer to the Au electrode *via* the thiol group on the one hand, and to participate in the polymerization by the vinyl group on the other. Moreover, to recognize 8-OHdG selectively, the methacryloylamidohistidine-platinum(II) chelate monomer was selected for ligand exchange because the Pt(II) ion primarily interacted with the guanine base of DNA and the strength of the metal coordination interaction was stronger than that of hydrogen bonding or electrostatic interactions in water. The stability constant of the complex of 8-OHdG with the chelating MIP of the chemosensor was 4.85×10^5 M^{-1}. The LOD was 8.3 nM for 8-OHdG. The linear dynamic concentration range was 11.6 to 750 nM. For 8-OHdG determination in human plasma, the SPE using a cartridge packed with MIP particles was used as the sample pretreatment technique. In effect, the LOD of this new QCM chemosensor was very low, thus allowing for an average 8-OHdG level determination in cancer patient's plasma of 13.8 μM. The LOD of the QCM chemosensor was lower than that of the CE system.[2] In the second work,[118] the team has proposed an imprinting method very close to that described in the first study of 8-OHdG recognition and determination in human plasma.[2] In the second publication,[118] a new metal-chelating monomer, *viz.* methacryloyl aminoantipyrine-Fe(III), was used. The stability constant for the complex of 8-OHdG with the metal-chelating MIP chemosensor was 7.88×10^4 M^{-1}. After MISPE pre-treatment of human plasma, with the same type of MIP, the LOD of this QCM chemosensor was very low equaling 12 nM. The 8-OHdG level in blood serum from a breast cancer patient was 0.34 μM, as determined with this new QCM chemosensor.[118] In the third part of the study on 8-OHdG determination, Ersöz *et al.*[119] combined the effect of the nature of monomers of the two last works to improve the affinity of the QCM chemosensor. The metal-chelating monomers, *viz.*, methacryloyl aminoantipyrine-Fe(III) and methacryloyl histidine-Pt(II), interacted with diols and amino groups, and with N7 and O6 atoms of guanine, respectively. This combination was necessary to reach a high complex stability constant of 1.54×10^5 M^{-1} and sufficiently high selectivity. For 8-OHdG, this selectivity was 32 times higher than that for guanine. That way, MIP SPE of a real sample before guanine determining with the chemosensor was avoided. The LOD was 7.5 nM and the linear dynamic concentration range extended from 10 to 3500 nM, in 0.05 M phosphate buffer solution (pH = 10.0). The 8-OHdG level in plasma from intestinal cancer patients and healthy persons was determined as 0.439 μM and 0.0156 μM, respectively. Finally, because of the use of two different monomers instead of one, cavities better suited for 8-OHdG were generated and, therefore, the selectivity was higher.[119] To complete their extensive work on the MIP-QCM chemosensor for determination of nucleosides, Ersöz *et al.*[120] decided to use a monomer capable of generating interactions of DNA nucleobases with thymine. The synthesized methacryloylamidoadenine functional monomer contained the complementary base of thymine, *i.e.*, adenine. The gold surface of the Au-QCR was grafted with 2-propene-1-thiol, and then an MIP film was deposited on this linker layer. The data were obtained under FIA conditions using a HEPES saline buffer solution (pH = 7.4) as the carrier solution, pumped through the flow cell at 0.1 ml min^{-1}. A linear dynamic concentration range of 0.01 to 0.1 mM and a complex stability constant of 1.0×10^5 M^{-1} for thymine were determined. This MIP chemosensor for thymine was selective with respect to uracil, polythymidylic acid, and polyuridylic acid. The IF for thymine was 3.62. So, Ersöz *et al.*[120] concluded that this chemosensor type was of an increasing interest in detection of DNA of specific sequences, compared to RNA, because the MIP introduced robustness to the recognition part and the device did not require usage of labelled ssDNA or ssRNA.

In 2010, Kutner *et al.*[121] devised a PM chemosensor with an adenine-templated MIP film capable of detecting adenine at an LOD of 5 nM under FIA conditions using an acetonitrile–water (1 : 1, *v* : *v*) mixture as the carrier solution. The adenine-templated MIP film was prepared by electropolymerization of bis(2,2′-bithienyl)-5,5-dimethyl-2-phenyl-[1,3,2]dioxaborinane methane and bis(2,2′-bithienyl)-benzo-[18-crown-6]methane functional monomers, capable of interacting with the nitrogen 3 heteroatom and the protonated amino group of the adenine template, respectively. To avoid adenine electro-oxidation during the electropolymerization step, first a poly(bithiophene) film was deposited on the Pt/quartz electrode. The dynamic linear concentration range of 0.5 to 7 mM for adenine and appreciable selectivity to adenine analogues and possible interferences were determined. The stability constant of the adenine complex with MIP was 1.8×10^5 M^{-1}. Apparently, the use of an MIP film as the recognition unit, integrated with the PM transducer, resulted in a selective and sensitive chemosensor for adenine.[121]

To complete this review of the QCM sensor type, Turner *et al.*[122] returned in 2014 to the basis of the nucleosidic recognition with Watson–Crick pairing of a nucleobase template and a complementary nucleobase functional monomer. Accordingly, they used the 5-(2-carbomethoxyvinyl)-2′-deoxyuridine functional monomer for complementarity with the 2′-deoxyadenosine template. MIP NPs of a mean size of ~20 nm, were prepared by a new synthetic method named the "solid-phase imprinting technique". This method allowed production of MIP NPs of low dispersity and uniform affinity. A layer of molecules of the template and analogues were retained on the Au-QCM resonators by combining the cystamine and glutaraldehyde linkers. Then, MIP NPs were flowed over Au-QCM chips in 0.01 M phosphate buffer solution (pH = 7.2). The more NPs that bound to the Au-QCM, the more the mass was increased and the signal was higher. Concentration of all investigated MIP NPs ranged from 0.125 to 2 μg mL^{-1}. The selectivity with respect to 2′-deoxycytidine and 2′-deoxyguanosine was high. To validate the method, the synthesis of NPs imprinted with analogues of the target analyte was necessary. It was envisioned that this technique could replace biochemical procedures, such as ELISA-like assays, in the near future.[122] On the same principle, the same team, in 2015, fabricated a QCM sensor to recognize and assay a DNA oligomer of a specific sequence. The difference was in the template used. Here, a synthetized oligonucleotide was used as the template, and then grafted to glass beads in order to accomplish a "solid-phase imprinting technique".

Besides, a polymerizable DNA strand, complementary to the template, was used as the functional monomer, thus serving as the recognition element in the MIP NPs composition. With the solid-phase synthesis, the spheroidal oligoMIP NPs were prepared of the size ranging between 5 and 15 nm. The oligonucleotide template was immobilized on the Au-QCM and NPs bearing the template were flowed at different concentrations over the chemosensor. The QCM sensorgrams showed maximum selectivity towards the DNA complementary to the MIP NPs. Thus, these MIP NPs coupled to QCM could potentially be applied for development of *in vivo* therapeutics and diagnostics.[123]

In the field of surface acoustic wave (SAW) devices, Love wave acoustic sensors have received renewed attention for the past two decades. This is because they can operate at high frequencies, which in principle leads to sensitivity in liquid media higher than that of traditional shear-mode oscillating QCM sensors. Therefore, since 2013, French teams[124–126] have been interested in this new acoustic sensor in order to devise a nucleotide selective sensor featuring an AMP MIP film.[124–126] Three publications deal with fabrication of this SAW chemosensor. The first one[124] studied the feasibility of the technique describing spin coating of the pre-polymerization DMSO solution directly on dual-metal layers (Au and Ti) of the AT cut quartz resonator, followed by photopolymerization. Acrylamide, 2-(dimethylamino) ethyl methacrylate, and EGDMA was used as the functional and EGDMA as the cross-linking monomer. The film thicknesses ranged from several hundred nanometres to 2.6 μm and the pore sizes were 1 to 5 μm.[124] The second paper[125] introduced a change in the way of linking the polymer to the support surface. That is, the surface was preliminary activated with the 3-(trimethoxysilyl)propyl methacrylate layer, and then polymer was covalently bound to this layer with the vinyl group present on the linker. The chemosensor was coated with an MIP solution by spin coating, and then polymerization was initiated with UV light. Layer thickness and the pore size range were estimated as 1 μm and 1 to 5 nm, respectively. The LOD was 6 ppm for AMP in 1 mM (acetic acid)-hydroxylamine (pH = 7.0).[125] In the last paper,[126] AMP in an aqueous solution was determined using a microfluidic chip placed above the delay line. The principle was not changed when the target was captured by a sensitive film. In effect, the acoustic wave propagation was disturbed resulting in a decrease of its phase velocity because of the mass addition. Similarly, a frequency shift of the steady-state signal *versus* AMP concentration showed a linear dynamic concentration range of 25 to 600 ppm, and the MIP based sensor demonstrated a real-time frequency shift of 150 Hz for 5 ppm AMP. This new type of gravimetric chemosensor integrated with a microfluidic chip allowed for the detection of nucleotides in real time.[126]

3.4.3 MIP Optical Sensors

Obviously, we cannot close this subchapter without describing optical sensors, as these most often represent the technology currently used in biosensor applications. Among others, SPR spectroscopy is probably one of the best-known label-free optical techniques. The present main shortcoming of this technique is its high cost.

However, this analytical technique assisted by MIP has not been much developed in the field of determination of nucleos(t)ide, because only two publications from the Yoshikawa team[127,128] and one from the Ersöz group[129] reported the use of this technique to this group of compounds. Primarily, SPR is used to measure the strength of interactions between the receptor deposited on the SPR chip and the analyte in solution. However, based on the work of Yoshikawa *et al.*,[127] one can argue that combination of molecular imprinting and SPR sensing is potent for the detection of target nucleos(t)ides in complex matrices. Both papers utilized a 9-methyladenine model compound as the dummy template to prepare two new MIPs capable of recognizing adenosine very selectively. One polymer was aminomethylated polysulfone derivatized with an oligopeptide, capable of recognizing adenosine selectively with an apparent complex stability constant of 1.6×10^4 M^{-1}. The other polymer was polysulfone derivatized with the perillaldehyde moiety. The maximum complex stability constant was 3.31×10^4 M^{-1}. For both devices, linearity was found between the shift in the reflected light angle and the amount of adsorbed adenosine in the investigated concentration range of 0.25 to 2.5 mM. These studies demonstrated that the SPR spectroscopy transduction combined with molecular imprinting recognition contributed to fabrication of chemosensors for nucleos(t)ides.[127,128] The work published by Ersöz *et al.*[129] in 2008 followed the previous study on a QCM chemosensor for 8-OHdG.[2] The authors used the same formulation with two different templates, *viz.*, guanosine and guanine but here the reaction mixture was drop cast on surface of SPR chips. Changes in refractive indices for the guanosine and guanine MIP films coated gold surfaces were recorded when the guanine, guanosine, polyguanylic acid (ssDNA), and dsDNA analytes were adsorbed. Molecular imprinting with guanosine and guanine resulted in the highest affinity for these compounds used as templates and for polyguanylic acid, with a complex stability constant of 3.6×10^4 M^{-1} and 3.7×10^3 M^{-1}, respectively. The signal was proportional to the analyte concentration of 25 to 400 μM. So, combination of this selective recognition element, the MIP, and the sensitive optical sensor, the SPR chip, is a great tool to assay target DNAs *via* hybridization of nucleic acid single strands and mutagenesis.[129]

Fluorescence spectroscopy is one of the most sensitive detection technique used in optical chemosensors. Moreover, it is non-destructive. Therefore, researchers have strived to combine the MIP recognition with fluorescence signal transduction. However, this optical emission sensing is linked to the fluorescence nature of the target analyte, derivatized or not. Unfortunately, intrinsic fluorescence of the nucleos(t)ides is weak, as their quantum yield values show in Table 3.3.

So, in order to use emission spectroscopy to detect nucleobases and their analogues with an optical sensor, a solution is to devise a fluorescent polymer showing a linear signal dependence on concentration of the target analyte in the test solution. This was exactly what Wandelt *et al.*[6]

Table 3.3 Maximum wavelength, λ_{max} and λ_{fluo}, in UV and fluorescence spectroscopy, respectively, as well as absorption coefficient, ε_{max}, and quantum yield, ϕ, at wavelength of emission maximum, for different nucleosides and nucleotides.[a]

Nucleotide or nucleoside	λ_{max}/nm[130]	ε_{max}/mol^{-1} L cm^{-1} [130]	λ_{fluo}/nm[131]	$\phi \times 10^4$ [131]
Adenosine	259.7	14 930	307	0.86
AMP	259.0	15 040		
cAMP	259.0	15 080	306	0.68
Cytidine	270.7	9300	328	0.89
CMP	271.0	8740		
dCMP	271.0	8860	328	1.15
Guanosine	252.7	13 850	334	0.97
GMP	252.0	14 090		
dGMP	252.0	14 230	334	1.09
Uridine	262.0	9820	308	0.40
UMP	262.0	9780		
Thymidine	267.0	9860	330	1.32
TMP	262.0	9490	330	1.54

[a]AMP, adenosine-5′-monophosphate; cAMP, adenosine 3′,5′-cyclic monophosphate; CMP, cytidine-5′-monophosphate; dCMP, deoxycytidine monophosphate; GMP, guanosine-5′-monophosphate; dGMP, deoxyguanosine monophosphate; TMP, thymidine monophosphate; UMP, uridine-5′-monophosphate.

implemented in their first study, in 1998, when developing a biomimetic sensor by combining recognition with a fluorescent MIP with the fluorescent detection. Thus, the MIP served for recognition of cAMP in aqueous solutions. The fluorescent functional monomer used was *trans*-4-[p-(*N*,*N*-dimethylamino)styryl]-*N*-vinylbenzylpyridinium chloride. When it interacted with anionic part of this nucleotide, fluorescence of the polymer was quenched. Another functional monomer, 2-hydroxyethyl methacrylate, and the cross-linking monomer, TRIM, were used in MIP formulation. The stability constant of the MIP complex with cAMP was 3.5×10^5 M^{-1}. The maximal quenching was in the 10 to 100 μM range, and selectivity with respect to cGMP was demonstrated. Therefore, the synthesis of an MIP prepared using a high-quantum-yield monomer in order to fabricate an optical chemosensor of selective emission was demonstrated.[6] In their next study of 2004, Wandelt *et al.*[132] compared the previous results on the cAMP-templated MIP prepared in bulk with the same MIP deposited as a thin film. Here, imprinting in a thin film was performed directly on quartz plates. However, MAA was replaced with 2 hydroxyethyl methacrylate in the formulation. The stability constant of the complex of cAMP with a thin MIP film was 10^6 M^{-1}, as determined from the fluorescence quenching data. The fluorescence quenching for the thin MIP film appeared much more effective than that for the bulk MIP. Selectivity of the thin MIP film to cGMP was not completely satisfactory[132] (Table 3.4).

In the domain of fluorosensors using MIPs to determine nucleos(t)ides, Wandelt *et al.*[132–135] accomplished research between 2004 and 2008 using

Table 3.4 Different nucleobases, nucleosides, nucleotides and oligonucleotides determination using various sensor transduction platforms.[a]

	Electrochemical				Optical		Mass sensitive	
Analyte	FET	Impedimetry	Amperometry	Voltammetry	Fluorescence	SPR	QCM	SAW
Purines & analogues								
2′-deoxyadenosine							122	
6-Benzylaminopurine				113				
6-Mercaptopurine					137			
6-Thioguanine		108		114				
8-Hydroxy-2′-deoxyguanosine							**2**, 118, 119	
Adenine							121	
Adenosine				97	140	128		
AMP	98 and 99	99					99, 115, 116	124–126
ATP		105 and 107	101	96	139		105 and 107	
cAMP	**4**				**6**, 132, 135			
cGMP					133–136			
Famciclovir				110				
Guanine						129		
Guanosine					138	129		
Ganciclovir				**5**				
GMP	98 and 99	99					99	
Inosine	109							
NADH-NAD	100							
Valganciclovir				111				

(continued)

Table 3.4 (*continued*)

Analyte	Electrochemical				Optical		Mass sensitive	
	FET	Impedimetry	Amperometry	Voltammetry	Fluorescence	SPR	QCM	SAW
Pyrimidines & analogues								
5-FU		106		102, 103, 106, 112			106	
Cytidine							117	
CMP	98 and 99	99					99	
Gemcitabine				**3**				
Thymidine		104	104	104				
Thymine							120	
Uracil				102, 103, 112				
UMP	99	99					99	
Nucleic acids								
Poly(dT)							120	
Poly(guanidylic acid)					138	129		
Oligonucleotide							123	

[a]ATP, adenosine-5′-triphosphate; cAMP, adenosine 3′,5′-cyclic monophosphate; cGMP, cyclic guanosine monophosphate; CMP, cytidine-5′-monophosphate; GMP, guanosine-5′-monophosphate; FET, field-effect transistor; 5-FU, 5-fluorouracil; NAD, nicotinamide adenine dinucleotide; NADH, nicotinamide adenine dinucleotide hydride; Poly(dT), polydeoxythymine; SPR, surface plasmon resonance; QCM, quartz crystal microgravimetry; SAW, surface acoustic wave; UMP, uridine-5′-monophosphate.

1,3-diphenyl-6-vinyl-1*H*-pyrazolo[3,4-*b*]quinolone as the fluorescent functional monomer capable to form a complex with cGMP. To summarize the three resulting publications, an optical chemosensor capable of selective cGMP determining in solution was devised by optimizing the formulation and MIP deposition on a quartz plate transducer. For that, first, the plate was coated by spin coating with a thin film of poly(methyl methacrylate) followed by incorporation of the pyrazoloquinoline emitter and the cGMP template. In effect, photoinduced cross-linking in the mixture of polymer chains allowed binding of the fluorescent monomer covalently. The cGMP was determined by analyzing the emission spectra of the MIP chemosensor before and after immersion in the cGMP MeOH solution. The range of fluorescence intensity ratio investigated was 10^{-10} to 10^{-4} M and the complex stability constant for cGMP was 3.7×10^{8} M^{-1}.[133] Then, in 2006, incorporation of the monomer–template complex in PMMA was improved.[134] That is, benzoin ethyl ether, a strong α-cleavage polymerization photoinitiator, was added to the solution for polymerization in order to increase the number of radical centers capable to react with the pre-polymerization mixture. The comparison of methods of deposition of the MIP film on a quartz slide suggested higher homogeneity of the film achieved by the spin-coating technique rather than drop-casting method. Some selectivity to GMP and cAMP was demonstrated but remained moderate.[134]

In the last paper,[135] the team modified the MIP formulation, thus avoiding the use of PMMA. The fluorescent functional monomer, *viz.*, 1,3-diphenyl-6-vinyl-1*H*-pyrazole-[3,4-*b*]-quinoline, was the same as that used in the previous research but it was mixed with 2-hydroxyethyl methacrylate, TRIM, and benzoin ethyl ether in MeOH, with cGMP or cAMP as the template. The solution was drop-cast on quartz slides, and then photopolymerized. The selectivity to cAMP of the cGMP-MIP was 2.38, and it was 0.53 to cGMP of the cAMP-MIP. This latter result indicated that despite the recognition sites being generated by cAMP, cGMP strongly interacted with the 1*H*-pyrazole-[3,4-*b*]-quinoline recognition site in the MIP. Besides, quenching for cGMP-templated MIP film was significant, in contrast to the cAMP-templated MIP film. The quenching effect coming from 10^{-7} M cGMP was not measurable. Moreover, the use of the technique without photocleavable PMMA was successful and the MIP chemosensor based on emission spectroscopy could easily be transposed to analysis of complex samples.[135] Meanwhile, in 2002, Hartell *et al.*[136] also worked on devising an MIP fluorescent chemosensor. In this chemosensor, 2-acrylamidopyridine was used as the fluorescent functional monomer because of its strong fluorescence and the presence of both hydrogen bond donor and acceptor sites capable to form a complex with cGMP with a stability constant of 3×10^{5} M^{-1}. The fluorescence properties of the MIP powder suspension in water were determined using 96-well fluorescence plates. The most important absolute change in fluorescence occurred with the couple of wavelengths of $\lambda_{exc} = 355$ nm and $\lambda_{em} = 460$ nm. The fluorescence quenching linearly correlated with the analyte concentration in the range of 0.1 to 100 μM. Selectivity was determined with respect to GMP and AMP. The polymer formulation and the determination principle could potentially be used for practical fluorescent chemosensors.[136]

Back in 2008, Zhang *et al.*[137] did not use a fluorescent MIP to determine nucleosides but made the target analyte fluorescent once immobilized in an MIP. The pre-polymerization complex solution consisted of the 6-mercaptopurine template as well as the MAA and TRIM functional monomers. An intrinsic emulsion polymerization was chosen by adding to the solution a high volume of MeOH, because of low solubility of the 6-mercaptopurine nucleoside, and by adding poly(vinyl alcohol), PVA. The microtiter plates were coated with MIP microparticles in suspension in PVA, and then 6-mercaptopurine solution was applied. After rinsing, 6-mercaptopurine was oxidized with a solution of H_2O_2 and NaOH. Fluorescence intensities were measured at $\lambda_{exc} = 310$ nm and $\lambda_{em} = 397$ nm. The response was linear in the range of 1.0×10^{-5} to 6.0×10^{-3} g L^{-1} and the LOD was 3.0×10^{-6} g L^{-1}. The (6-mercaptopurine)-templated MIP selectivity to sorbic acid, glucose, urea, uric acid as well as vitamin C and B1 was demonstrated. In human serum samples, for 0.05 to 1 μg mL^{-1} 6-mercaptopurine, recovery was 94–106%. In conclusion, conversion of 6-mercaptopurine to a compound that can be detected with higher sensitivity associated with a selective MIP could be used as an alternative for enzymes and antibodies in sensors.[137] In the same year, Ersöz *et al.*,[138] based on the previously obtained results[2] with their (8-OHdG)-selective MIP, devised a fluorescence chemosensor for guanosine and ssDNA. The metal-chelating functional monomer used was the same as that in the previous study.[2] The novelty involved the use of CdS quantum dots, which incurred native fluorescence to the chemosensor, and a thin MIP film grafted to these dots capable of selective recognition of the target analyte. These dots were nearly spherical with an average diameter of ~45 nm. The increase of the fluorescence intensity was proportional to the guanosine concentration in solution. The recognition sites of the MIP nanoshells on the CdS nanocrystals bound guanosine, guanine, and ssDNA selectively, thus inducing photoluminescence emission from quantum dots. This guanosine imprinted nanoshell chemosensor was more responsive to poly(guanidylic acid) than dsDNA, thus demonstrating the shape-selective cavity presence. So, this type of photoluminescent molecularly imprinted nanosensor could be used for studying DNA defects.[138]

More recently, in 2013, the group of Liao *et al.*[139] used yet another sensor for detecting ATP in urine, *viz.*, a fluorescence type chemosensor. Here, a glass slide was first coated with the AMP templated MIP film, and then an ATP solution was dispensed on top of this MIP. Then, the third phosphate of the adsorbed ATP was coordinated to a uranyl–salophen–fluorescein complex. The latter compound incurred fluorescence to the detection system. Thus, the fluorescence intensity was increased with the increase of the ATP concentration, and hence with the amount of the uranyl–salophen–fluorescein complex immobilized on the sensor. The linear dynamic ATP concentration range was 0.3 to 4.8 μM and the LOD was 0.041 μM. In human urine samples, ATP recovery was 98.5–101.3% in the concentration range of 2 to 4.5 μM.

A new method developed by Liao *et al.* could be applied to the determination of other nucleotides because it was sensitive and the MIP presence brought stability to the sensor at a low cost.[139]

To complete reviewing optical chemosensors responsive in fluorescence, a recent paper developed the field further by incorporating aptamers into the MIP for selective adenosine determination. Indeed, Liu *et al.*[140] used a fragment of biological ligands fluorescently labeled as macromonomers in an adenosine-templated MIP. Two aptamer fragments were labeled on the 5′ end with the acrydite group to copolymerize into the imprinted nanohydrogel. This nanoMIP, with a mean size 170 nm, was prepared by precipitation polymerization in an aqueous solution of acrylamide, *N-iso*propylamide, and *N,N′*-methylene(bis-acrylamide), in the presence of a surfactant. Results for the suspension in an aqueous buffer showed that the relative fluorescence quenching was dependent on the adenosine concentration but not the cytidine concentration. The stability constant of the complex of the aptamer-MIP with adenosine was 6.3×10^4 M^{-1}. Given these first results, the combined use of aptamers and imprinted polymers is, surely, a new way of future devising chemosensors of higher performance.[140]

In conclusion, in view of all these works on the development of new sensors, it is clear that, compared to natural bioreceptors, MIPs offer a number of advantages, which will complement the benefits of sensors. Thus, the development of new sensors provided with the imprinting technology will surely take place in the field of sensors because these sensors are stable after long-term storage, they are re-usable, resistant to microbial damage, cheap, easy to prepare, they show very high affinity and recognition ability for target analytes, and can be used with electrochemical, spectroscopic, and gravimetric transducers, among others.

3.5 Conclusions

Surely, nucleosides and their analogues will have a strong impact, in the future, on human health. Their detection and determination are of great importance for medical diagnostics tools (biomarkers), biological studies, and therapeutic drug monitoring. In view of all the research work on devising and fabricating new MIP chemosensors, it is clear that compared to natural biosensors, the former offer several advantages, which will complement benefits of sensing. For nucleosides and MIPs "the best is still to come!"

List of Abbreviations

3CT	3′-Thiacytidine
8-OHdG	8-Hydroxy-2′-deoxyguanosine
AA	Acrylamide
ABDV	Azobis(4-methoxy-2,4-dimethylvaleronitrile)
ACCN	1,10-Azobis(cyclohexanecarbonitrile)
ADP	Adenosine-5′-diphosphate

ADVN	2,2′-Azobis(2,4-dimethylvaleronitrile)
AGET	Activated generated electron transfer
AIBN	Azobisisobutyronitrile
AMP	Adenosine-5′-monophosphate
APBA	3-Aminophenylboronic acid
APS	Ammonium persulfate
APTEOS	3-Aminopropyltriethoxysilane
ATP	Adenosine-5′-triphosphate
ATRP	Atom transfer radical polymerization
AuNP	Gold nanoparticle
AZT	3′-Azido-3′-deoxythymidine
BAAPy	2,6-Bis(acrylamido)pyridine
cAMP	Adenosine 3′,5′-cyclic monophosphate
CE	Capillary electrophoresis
CEC	Capillary electrochromatography
CE-ECD	Capillary electrophoresis with electrochemical detection
CI	Capacitive impedimetry
CPE	Carbon paste electrode
CTP	Cytidine-5′-triphosphate
dCMP	Cytidine-5′monophosphate
DEGDMA	Diethyleneglycol dimethacrylate
DMF	*N*,*N*-Dimethylformamide
DMSO	Dimethylsulfoxide
DNA	Deoxyribonucleic acid
DPV	Differential pulse voltammetry
dsDNA	double-stranded DNA
DTD-OMI	Dual-template docking oriented molecular imprinting
DVB	Divinylbenzene
EGDMA	Ethyleneglycol dimethacrylate
EG-FET	Extended-gate field-effect transistor
EtOH	Ethanol
FET	Field-effect transistor
FIA	Flow-injection analysis
FT-IR	Fourier-transform infrared (spectroscopy)
5-FU	5-Fluorouracil
GCE	Glassy carbon electrode
GDP	Guanosine-5′-diphosphate
GE	Gel electrophoresis
GMP	Guanosine-5′-monophosphate
GPTES	3-Glycidyloxypropyl triethoxysilane
GTP	Guanosine-5′-triphosphate
HBV	Hepatitis B virus
HEPES	4-(2-Hydroxyethyl)-1-piperazineethanesulfonic acid
HIV	Human immunodeficiency virus
HPLC	High-performance liquid chromatography
HSV	Herpes simplex virus
IF	Imprinting factor

ISFET	Ion-sensitive field-effect transistor
LC-UV	Liquid chromatography coupled ultraviolet detection
LC-MS	Liquid chromatography coupled mass spectrometry
LOD	Limit of detection
LOQ	Limit of quantification
MAA	Methacrylic acid
MAAP	Methacryloyl antipyrine
MAPDIA	5-{[4-(Methacryloyloxy)phenyl]diazenyl}isophthalic acid
MBAA	*N*,*N*′-Methylenebisacrylamide
MeCN	Acetonitrile
MeOH	Methanol
MIP	Molecularly imprinted polymer
MI-MSPD	Molecularly imprinted matrix solid-phase dispersion
MISPE	Molecularly imprinted solid-phase extraction
MNP	Magnetic nanoparticle
MOF	Metal-organic frameworks
MWCNT	Multi-walled carbon nanotube
NHSG	Non-hydrolytic sol–gel
NAD	Nicotinamide adenine dinucleotide
NADH	Nicotinamide adenine dinucleotide hydride
NHE	Normal hydrogen electrode
NIP	Non-imprinted polymer
NIPAM	*N*-Isopropylacrylamide
NMR	Nuclear magnetic resonance
NP	Nanoparticle
ODS	Octadecylsilane
PATP	Pressure assisted thermal processing
PETRA	Pentaerythritol tetra-acrylate
PM	Piezoelectric microgravimetry
PMMA	Poly(methyl methacrylate)
Poly(AN-*co*-MAA)	Poly(acrylonitrile-*co*-methacrylic acid) polymer
Poly(AN-*co*-SMA)	Poly(acrylonitrile-*co*-vinylbenzyl-stearyldimethylamine chloride)
Poly(dT)	Polydeoxythymine
PS	Polystyrene
PTFE	Polytetrafluoroethylene
PVA	Poly(vinyl alcohol)
PVDF	Polyvinylidene-difluoride
QCM	Quartz crystal microbalance
RNA	Ribonucleic acid
ROS	Reactive oxygen species
SAW	Surface acoustic wave
SHE	Standard hydrogen electrode
SMPD	Surface mount power device
SPE	Solid-phase extraction
SPME	Solid-phase microextraction
SPR	Surface plasmon resonance

TEMED	*N*,*N*,*N*′,*N*′-Tetramethylethylenediamine
TEOS	Tetraethoxysilane
TFE	2,2,2-Trifluoroethanol
THF	Tetrahydrofuran
TRIM	Trimethylopropane trimethacrylate
TTP	Thymidine triphosphate
UHPLC-DAD	Ultra HPLC-diode array detector
UTP	Uridine-5′-triphosphate
4-VP	4-Vinylpyridine

References

1. L. A. Agrofoglio and S. R. Challand, *Acyclic, Carbocyclic and L-Nucleosides*, Kluwer Academic Publishers, Dordrecht, 1998.
2. A. Ersöz, S. E. Diltemiz, A. A. Özcan, A. Denizli and R. Say, *Biosens. Bioelectron.*, 2008, **24**, 742.
3. A. Florea, Z. Z. Guo, C. Cristea, F. Bessueille, F. Vocanson, F. Goutaland, S. Dzyadevych, R. Sandulescu and N. Jaffrezic-Renault, *Talanta*, 2015, **138**, 71.
4. A. Kugimiya and K. Kohara, *Mater. Sci. Eng., C*, 2009, **29**, 959.
5. M. B. Gholivand and N. Karimian, *Sens. Actuators, B*, 2015, **215**, 471.
6. P. Turkewitsch, B. Wandelt, G. D. Darling and W. S. Powell, *Anal. Chem.*, 1998, **70**, 2025.
7. W. J. Cheong, S. H. Yang and F. All, *J. Sep. Sci.*, 2013, **36**, 609.
8. L. A. Agrofoglio, I. Gillaizeau and Y. Saito, *Chem. Rev.*, 2003, **103**, 1875.
9. Book: *Modified Nucleosides in Biochemistry, Biotechnology and Medicine*, ed. Piet Herdewijn, Wiley-VCH, 2008.
10. K. J. Shea, D. A. Spivak and B. Sellergren, *J. Am. Chem. Soc.*, 1993, **115**, 3368.
11. D. A. Spivak, M. A. Gilmore and K. J. Shea, *J. Am. Chem. Soc.*, 1997, **119**, 4388.
12. D. A. Spivak and K. J. Kenneth, *Macromolecules*, 1998, **31**, 2160.
13. D. Batra and K. J. Shea, *Mater. Res. Soc. Symp. Proc.*, 2002, **723**, 85.
14. J. Matsui, M. Higashi and T. Takeuchi, *J. Am. Chem. Soc.*, 2000, **122**, 5218.
15. J. Mathew and O. Buchardt, *Bioconjugate Chem.*, 1995, **6**, 524.
16. Y. Kanekiyo, Y. Ono, K. Inoue, M. Sano and S. Shinkai, *J. Chem. Soc., Perkin Trans. 2*, 1999, 557.
17. M. Yoshikawa, M. D. Guiver and G. P. Robertson, *J. Mol. Struct.*, 2005, **739**, 41.
18. K. Sreenivasan, *J. Appl. Polym. Sci.*, 2004, **94**, 2088.
19. K. Sreenivasan, *J. Mater. Sci.*, 2007, **42**, 7575.
20. K. Sreenivasan, *React. Funct. Polym.*, 2007, **67**, 859.
21. E. V. Dmitrienko, R. D. Bulushev, K. Haupt, S. S. Kosolobov, A. V. Latyshev, I. A. Pyshnaya and D. V. Pyshnyi, *J. Mol. Recognit.*, 2013, **26**, 368.
22. I. Chianella, K. Karim, E. V. Piletska, C. Preston and S. A. Piletsky, *Anal. Chim. Acta*, 2009, **559**, 73.

23. L. Longo, G. Vasapollo, A. Scardino, R. A. Picca and C. Malitesta, *J. Porphyrins Phthalocyanines*, 2006, **10**, 1061.
24. L. Longo and G. Vasapollo, *Met.-Based Drugs*, 2008, **5**, 163.
25. L. Longo, S. Scorrano and G. Vasapollo, *J. Polym. Res.*, 2010, **17**, 683.
26. F. Breton, R. Delépée, D. Jégourel, D. Deville-Bonne and L. A. Agrofoglio, *Anal. Chim. Acta*, 2008, **616**, 222.
27. F. Breton, R. Delépée and L. A. Agrofoglio, *J. Sep. Sci.*, 2009, **32**, 3285.
28. A. Lakka and A. Tsakalof, *ChemPlusChem*, 2013, **78**, 808.
29. J. Matsui, J. Nagano, D. Miyoshi, K. Tamaki and N. Sugimoto, *Biosens. Bioelectron.*, 2009, **25**, 563.
30. A. Takata, K. Usui and J. Matsui, *Mol. Imprinting*, 2014, **2**, 93.
31. S. J. Li, S. Pilla and S. Gong, *J. Polym. Sci.*, 2009, **47**, 2352.
32. A. D. Vaughan, S. P. Sizemore and M. E. Byrne, *Polymer*, 2007, **48**, 74.
33. V. D. Salian, A. D. Vaughan and M. E. Byrne, *J. Mol. Recognit.*, 2012, **25**, 361.
34. T. Kusunoki and T. Kobayashi, *J. Appl. Polym. Sci.*, 2010, **117**, 565.
35. K. Wybranska, W. Niemiec, K. Szczubiazka, M. Nowakowska and Y. Morishima, *Chem. Mater.*, 2010, **22**, 5392.
36. A. Plewa, S.-I. Yusa, M. Szuwarzyński, K. Szczubiałka, Y. Morishima and M. Nowakowska, *J. Med. Chem.*, 2012, **55**, 8712.
37. A. Mikulska, M. Inoue, K. Kuroda, A. Iwanowska, S.-I. Yusa, M. Nowakowska and K. Szczubiałka, *Eur. Polym. J.*, 2014, **59**, 230.
38. A. Mujahid, A. Najeeb, A. I. Khan, T. Hussain, M. H. Raza, A. T. Shah, N. Iqbal and M. N. Ahmad, *J. Mater.*, 2015, **2015**, 1.
39. A. J. Hall, P. Manesiotis, J. T. Mossing and B. Sellergren, *Mater. Res. Soc. Symp. Proc.*, 2002, **723**, 11.
40. P. Manesiotis, A. J. Hall and B. Sellergren, *J. Org. Chem.*, 2005, **70**, 2729.
41. Z. Li, J. Ding, M. Day and Y. Tao, *Macromolecules*, 2006, **39**, 2629.
42. T. Muhammad, Z. Nur, E. V. Piletska, O. Yimita and S. A. Piletsky, *Analyst*, 2012, **137**, 2623.
43. J. Abraham and B. Mathew, *Eur. J. Appl. Eng. Sci. Res.*, 2014, **3**, 26.
44. M.-H. Lee, J. L. Thomas, J.-Z. Chen, J.-S. Jan and H.-Y. Lin, *Chem. Commun.*, 2016, **52**, 2137.
45. Y. Liu, L. Wu, X. Zhao and A. Luo, *Int. J. Polym. Anal. Chem.*, 2012, **17**, 38.
46. A. Krstulja, S. Lettieri, A. J. Hall, R. Delépée, P. Favetta and L. A. Agrofoglio, *Anal. Bioanal. Chem.*, 2014, **406**, 6275.
47. D. Jégourel, R. Delépée, F. Breton, A. Rolland, R. Vidal and L. A. Agrofoglio, *Bioorg. Med. Chem.*, 2008, **16**, 8932.
48. S. Scorrano, L. Longo and G. Vasapollo, *Anal. Chim. Acta*, 2010, **659**, 167.
49. A. Krstulja, C. De Schutter, P. Favetta, P. Manesiotis and L. A. Agrofoglio, *J. Chromatogr. A*, 2014, **1365**, 12.
50. S. Vo Duy, I. Lefebvre-Tournier, V. Pichon, F. Hugon-Chapuis, J.-Y. Puy and C. Périgaud, *J. Chromatogr. B*, 2009, **877**, 1101.
51. S. Wu, L. Tan, G. Wang, G. Peng, C. Kang and Y. Tang, *J. Chromatogr. A*, 2013, **1285**, 124.
52. H. Yan, M. Wang, Y. Han, F. Qiao and K. H. Row, *J. Chromatogr. A*, 2014, **1346**, 16.

53. P. Manesiotis, S. Kashani and P. McLoughlin, *Anal. Methods*, 2013, **5**, 3122.
54. P. Mattos dos Santos, A. J. Hall and P. Manesiotis, *J. Chromatogr. B*, 2016, **1021**, 197.
55. M. Shekarchi, M. Pourfarzi, B. Akbari-Adergani, A. Mehramizi, M. Javanbakht and R. Dinarvand, *J. Chromatogr. B*, 2013, **931**, 50.
56. S.-W. Zhang, J. Xing, L.-S. Cai and C.-Y. Wu, *Anal. Bioanal. Chem.*, 2009, **395**, 479.
57. S. Zhang, X. Sun, W. Wang and L. Cai, *J. Sep. Sci.*, 2013, **36**, 752.
58. S. W. Zhang, C. J. Zou, N. Luo, Q. F. Weng, L. S. Cai, C. Y. Wu and J. Xing, *Chin. Chem. Lett.*, 2010, **21**, 85.
59. S. Zhang, X. Song, W. Zhang, N. Luo and L. Cai, *Sci. Total Environ.*, 2013, **450**, 266.
60. Z.-Y. Li, H.-J. Quan, C.-B. Gong, Y.-Z. Yang, Q. Tang, Y.-B. Wei, X.-B. Ma and H.-W. Lam, *Food Chem.*, 2015, **172**, 56.
61. B. B. Prasad, K. Tiwari, M. Singh, P. S. Sharma, A. K. Patel and S. Srivastava, *Biomed. Chromatogr.*, 2009, **23**, 499.
62. Z. Terzopoulou, M. Papageorgiou, G. Z. Kyzas, D. N. Bikiaris and D. A. Lambropoulou, *Anal. Chim. Acta*, 2016, **913**, 63.
63. Y. Han, H. Yan, X. Cheng, G. Yang and B. Li, *Anal. Methods*, 2013, **5**, 3285.
64. M. C. Cela-Pérez, L. Barbosa-Pereira, X. Vecino, M. Pérez-Ameneiro, A. L. Latorre, J. M. López-Vilariño, M. V. González Rodríguez, A. B. Moldes and J. M. Cruz, *Talanta*, 2015, **135**, 58.
65. Y. Chen, X. Li, D. Yin, D. Li, Z. Bie and Z. Liu, *Chem. Commun.*, 2015, **51**, 10929.
66. C. J. Allender, K. R. Brain, C. Ballatore, D. Cahard, A. Siddiqui and C. McGuigan, *Anal. Chim. Acta*, 2001, **435**, 107.
67. S. Aoki, A. Jikiba, K. Takeda and E. Kimura, *J. Phys. Org. Chem.*, 2004, **17**, 489.
68. J. Haginaka, H. Tabo and H. Matsunaga, *Anal. Chim. Acta*, 2012, **748**, 1.
69. Y.-C. Huang, C.-C. Lin and C.-Y. Liu, *Electrophoresis*, 2004, **25**, 554.
70. M. Ogiso, N. Minoura, T. Shinbo and T. Shimizu, *Biomaterials*, 2006, **27**, 4177.
71. M. Ogiso, N. Minoura, T. Shinbo and T. Shimizu, *Biosens. Bioelectron.*, 2007, **22**, 1974.
72. S. A. Piletsky, I. Ya Dubey, D. M. Fedoryak and V. P. Kukhar, *Biopolym. Cell*, 1990, **6**, 55.
73. S. A. Piletsky, Y. P. Parhometz, N. V. Lavryk, T. L. Panasyuk and A. V. El'skaya, *Sens. Actuators, B*, 1994, **18–19**, 629.
74. J. Mathew-Krotz and K. J. Shea, *J. Am. Chem. Soc.*, 1996, **118**, 8154.
75. V. Kochkodan, N. Hilal, P. J. Windsor and E. Lester, *Chem. Eng. Technol.*, 2003, **26**, 463.
76. N. Hilal and V. Kochkodan, *J. Membr. Sci.*, 2003, **213**, 97.
77. N. Hilal, V. Kochkodan, G. Busca, O. Kochkodan and B. P. Atkin, *Sep. Purif. Technol.*, 2003, **31**, 281.

78. M. Yoshikawa, J.-I. Izumi, M. D. Guiver and G. P. Robertson, *Macromol. Mater. Eng.*, 2001, **286**, 52.
79. M. Yoshikawa, K. Kawamura, A. Ejima, T. Aoki, S. Sakurai, K. Hayashi and K. Watanabe, *Macromol. Biosci.*, 2006, **6**, 210.
80. H. Y. Wang, S. L. Xia, H. Sun, Y. K. Liu, S. K. Cao and T. Kobayashi, *J. Chromatogr. B*, 2004, **804**, 127.
81. S. Xia, H. Sun, H. Wang, K. Takeda and T. Kobayashi, *AZojomo*, 2007, 3.
82. Q. Zhang, T. Kusunoki, Q. Xu, H. Wang and T. Kobayashi, *Anal. Bioanal. Chem.*, 2007, **388**, 665.
83. T. Kobayashi, S. S. Leong and Q. Zhang, *J. Appl. Polym. Sci.*, 2008, **108**, 757.
84. X.-J. Qu, C.-B. Chen, J. Zhou and C.-H. Wu, *Chin. J. Chem.*, 2007, **25**, 213.
85. X. J. Qu, Q. X. Meng, S. Y. Ai, J. Zhou and L. S. Zhu, *J. Anal. Chem.*, 2008, **63**, 999.
86. S. Scorrano, L. Mergola, M. P. Di Bello, M. R. Lazzoi, G. Vasapollo and R. Del Sole, *Int. J. Mol. Sci.*, 2015, **16**, 13746.
87. I. Tabushi, K. Kurihara, K. Naka, K. Yamamura and H. Hatakeyama, *Tetrahedron Lett.*, 1987, **28**, 4299.
88. L. Andersson, C. F. Mandenius and K. Mosbach, *Tetrahedron Lett.*, 1988, **29**, 5437.
89. E. Hedborg, F. Winquist, I. Lundstrom, L. I. Andersson and K. Mosbach, *Sens. Actuators, A*, 1993, **37–38**, 196.
90. S. A. Piletsky, Y. P. Parhometz, N. V. Lavryk, T. L. Panasyuk and A. V. El'skaya, *Sens. Actuators, B*, 1994, **18–19**, 629.
91. S. V. Jovanovic and M. G. Simic, *J. Phys. Chem.*, 1986, **90**, 974.
92. K. Lewis, K. Copeland and G. Hill, *Int. J. Quantum Chem.*, 2014, **114**, 1678.
93. A. M. Oliveira-Brett, J. A. P. Piedade, L. A. Silva and V. C. Diculescu, *Anal. Biochem.*, 2004, **332**, 321.
94. C. A. Schroeder, E. Pluhařová, R. Seidel, W. P. Schroeder, M. Faubel, P. Slavíček, B. Winter, P. Jungwirth and S. E. Bradforth, *J. Am. Chem. Soc.*, 2015, **137**, 201.
95. C. A. M. Seidel, A. Schulz and M. H. M. Sauer, *J. Phys. Chem.*, 1996, **100**, 5541.
96. A. Boyle, E. M. Genies and M. Lapkowsk, *Synth. Met.*, 1989, **28**, C769.
97. L. D. Spurlock, A. Jaramillo, A. Praserthdam, J. Lewis and A. Brajter-Toth, *Anal. Chim. Acta*, 1996, **336**, 3746.
98. M. Zayats, M. Lahav, A. B. Kharitonov and I. Willner, *Tetrahedron*, 2001, **22**, 815.
99. N. Sallacan, M. Zayats, T. Bourenko, A. B. Kharitonov and I. Willner, *Anal. Chem.*, 2002, **74**, 702.
100. O. M. Petrukhin, S. P. Pogorelova, A. B. Kharitonov and E. V. Shipulo, *J. Anal. Chem.*, 2007, **62**, 894.
101. S. Takeda, H. Yagi, S. Mizuguchi, H. Funahashi, H. Shiigi and T. Nagaoka, *J. Flow Injection Anal.*, 2008, **25**, 77.

102. B. B. Prasad, S. Srivastava, K. Tiwari and P. S. Sharma, *Sens. Mater.*, 2009, **21**, 291.
103. B. B. Prasad, D. Kumar, R. Madhuri and M. P. Tiwari, *Electrochim. Acta*, 2012, **71**, 106.
104. Z.-H. Zhang, Y.-F. Hua, H.-B. Zhang, L.-J. Luo and S.-Z. Yao, *J. Electroanal. Chem.*, 2010, **644**, 7.
105. T.-P. Huynh, A. Pietrzyk-Le, K. C. C. Bikram, K. R. Noworyta, J. W. Sobczak, P. S. Sharma, F. D'Souza and W. Kutner, *Biosens. Bioelectron.*, 2013, **41**, 634.
106. T.-P. Huynh, P. Pieta, F. D'Souza and W. Kutner, *Anal. Chem.*, 2013, **85**, 8304.
107. P. S. Sharma, M. Dabrowski, K. Noworyta, T.-P. Huynh, B. K. C. Chandra, J. W. Sobczak, P. Pieta, F. D'Souza and W. Kutner, *Anal. Chim. Acta*, 2014, **844**, 61.
108. T.-P. Huynh, A. Wojnarowicz, M. Sosnowska, S. Srebnik, T. Benincori, F. Sannicolò, F. D'Souza and W. Kutner, *Biosens. Bioelectron.*, 2015, **70**, 153.
109. Z. Iskierko, M. Sosnowska, P. S. Sharma, T. Benincori, F. D'Souza, I. Kaminska, K. Fronc and K. Noworyta, *Biosens. Bioelectron.*, 2015, **74**, 526.
110. N. A. ElGohary, A. Madbouly, R. M. El Nashar and B. Mizaikoff, *Biosens. Bioelectron.*, 2015, **65**, 108.
111. M. B. Gholivand and M. Torkashvand, *Mater. Sci. Eng., C*, 2016, **59**, 594.
112. B. B. Prasad and A. Kumar, *J. Mater. Chem.*, 2015, **3**, 5864.
113. T. Gan, Z. Lu, Y. Sun, Z. Shi, J. Sun and A. Zhao, *J. Appl. Electrochem.*, 2016, **46**, 389.
114. B. B. Prasad, R. Singh and A. Kumar, *Carbon*, 2016, **102**, 86.
115. Y. Kanekiyo, K. Inoue, Y. Ono, M. Sano, S. Shinkai and D. N. Reinhoudt, *J. Chem. Soc., Perkin Trans. 2*, 1999, 2719.
116. Y. Kanekiyo, M. Sano, R. Iguchi and S. Shinkai, *J. Polym. Sci.*, 2000, **38**, 1302.
117. Z. Zhang, L. Nie and S. Yao, *Talanta*, 2006, **69**, 435.
118. A. Ersöz, S. E. Diltemiz, A. A. Özcan, A. Denizli and R. Saya, *Sens. Actuators, B*, 2009, **137**, 7.
119. R. Say, A. Gültekin, A. A. Özcan, A. Denizli and A. Ersöz, *Anal. Chim. Acta*, 2009, **640**, 82.
120. S. E. Diltemiz, D. Hür, A. Ersöz, A. Denizli and R. Say, *Biosens. Bioelectron.*, 2009, **25**, 599.
121. A. Pietrzyk, S. Suriyanarayanan, W. Kutner, R. Chitta, M. E. Zandler and F. D'Souza, *Biosens. Bioelectron.*, 2010, **25**, 2522.
122. A. Poma, H. Brahmbhatt, J. K. Watts and N. W. Turner, *Macromolecules*, 2014, **47**, 6322.
123. H. Brahmbhatt, A. Poma, H. M. Pendergraff, J. K. Watts and N. W. Turner, *Biomater. Sci.*, 2016, **4**, 281.
124. N. Lebal, H. Hallil, C. Dejous, B. Plano, A. Krstulja, R. Delepée, D. Rebière and L. A. Agrofoglio, *IEEE-NEMS*, 2013, 550.

125. N. Lebal, H. Hallil, C. Dejous, B. Plano, A. Krstulja, R. Delepée, L. Agrofoglio and D. Rebière, *Micro Nano Lett.*, 2013, **8**, 563.
126. N. Lebal, H. Hallil, V. Raimbault, J.-L. Lachaud, A. Krstulja, R. Delépée, L. A. Agrofoglio, C. Dejous and D. Rebière, *Proceeding of 29th SBMicro, September 2014*, Brazil, 2014.
127. K. Taniwaki, A. Hyakutake, T. Aoki, M. Yoshikawa, M. D. Guiver and G. P. Robertson, *Anal. Chim. Acta*, 2003, **489**, 191.
128. M. Yoshikawa, M. D. Guiver and G. P. Robertson, *J. Appl. Polym. Sci.*, 2008, **110**, 2826.
129. S. E. Diltemiza, A. Denizli, A. Ersöz and R. Say, *Sens. Actuators, B*, 2008, **133**, 484.
130. M. J. Cavaluzzi and P. N. Borer, *Nucleic Acids Res.*, 2004, **32**, e13.
131. D. Onidas, D. Markovitsi, S. Marguet, A. Sharonov and T. Gustavsson, *J. Phys. Chem. B*, 2002, **106**, 11367.
132. B. Wandelt, A. Mielniczak and P. Cywinski, *Biosens. Bioelectron.*, 2004, **20**, 1031.
133. P. Cywinski, B. Wandelt and A. Danel, *Adsorpt. Sci. Technol.*, 2004, **22**, 719.
134. P. Cywinski, M. Sadowska, A. Danel, W. J. Buma, A. M. Brouwer and B. Wandelt, *J. Appl. Polym. Sci.*, 2007, **105**, 229.
135. M. Sadowska and B. Wandelt, *Mol. Cryst. Liq. Cryst.*, 2008, **486**, 203.
136. N. T. Kim Thanh, D. L. Rathbone, D. C. Billington and N. A. Hartell, *Anal. Lett.*, 2002, **35**, 2499.
137. L. Wang and Z. Zhang, *Talanta*, 2008, **76**, 768.
138. S. E. Diltemiz, R. Say, S. Büyüktiryaki, D. Hüra, A. Denizli and A. Ersöz, *Talanta*, 2008, **75**, 890.
139. M. Yang, L. Liao, G. Zhang, X. Xiao, Y. Lin and C. Nie, *Anal. Bioanal. Chem.*, 2013, **405**, 7545.
140. Z. Zhang and J. Liu, *ACS Appl. Mater. Interfaces*, 2016, **8**, 6371.

CHAPTER 4

Application of Nanomaterials to Molecularly Imprinted Polymers

ALESSANDRA MARIA BOSSI*, LUCIA CENCI AND RICCARDO TOGNATO

Department of Biotechnology, University of Verona, Strada Le Grazie 15, 371234 Verona, Italy
*E-mail: alessandramaria.bossi@univr.it

4.1 Introduction

The molecular imprinting of polymers (MIP) refers to the technique that obtains biomimetic polymeric materials by exploiting a template assisted synthesis.[1,2] The preparation of the MIP materials involves the polymerization of monomer molecules in the presence of a template molecule that is afterwards extracted, leaving complementary cavities in the material formed. Thus, the MIPs possess an affinity for the template molecule and have the ability to rebind it.

Remarkable affinity and selectivity have been reported for MIPs, making them attractive as recognition elements, in substitution of the natural ones, in the areas of sensing and separating.[3] Moreover, the possibility to produce MIPs in various formats, such as bulk materials, micro- and nano-particles and -surfaces, render them suitable to be integrated in a variety of applications, from analytical methods to drug delivery.

Polymer Chemistry Series No. 28
Molecularly Imprinted Polymers for Analytical Chemistry Applications
Edited by Wlodzimierz Kutner and Piyush Sindhu Sharma

Published by the Royal Society of Chemistry, www.rsc.org

In addition to their recognition properties, imparted by the imprinting technique, the MIPs are characterized by the inherent physical properties of their constituent material. The majority of MIPs are made of poly-acrylates and -methacrylates. Alternatively, MIPs based on silica and on cellulose have been reported, as indicated by browsing a website dedicated to the collection of the MIP literature.[4] The advantages gained by using each of these constituent materials can be summarized in the protocols for their preparation and manipulation, which are well established. A significant variety of functional monomers and cross-linking monomers are available on the market allowing targeting of templates bearing different chemical groups. The resulting polymers offer mechanical stability and the possibility to be prepared in different forms. However, these MIP materials present also some inherent limitations, *e.g.*, they do not exhibit optical properties and are characterized by electrical insulating properties. Both aspects could limit the use of the resulting MIPs.

To overcome some of these limitations, semiconducting MIP materials have been described based either on semiconducting[5] or later on versatile monomers.[6] As for the introduction of optical properties into MIPs, the co-polymerization of fluorescent monomers was proposed.[7] These preparation strategies widened considerably the potential of the MIPs. More recently, with the research area on nanomaterials reaching a significant level of maturity, the potential of most of the nanomaterials to be coupled to MIPs was revealed. In fact, nanomaterials manifest extremely attractive and advantageous physico-chemical properties, such as high surface-to-area ratio, high homogeneity, and particular physical properties, such as the optical bandgap, outstanding conductivity, magnetization, thermal resistance, which can be exploited in variety of applications, ranging from micro-electronics to nano-medicine.

The integration of nanomaterials and MIPs, with particular attention paid to magnetic and conducting nanomaterials, represents a new frontier of the MIP-materials research, especially in the fields of analytical measurements, sensing, and diagnostics. The main achievements and strategies relating to the MIP/nanomaterial integration are here discussed and speculations are attempted over their roles and implications on the future of MIP technology.

4.2 Introduction to Magnetic Nanoparticles

Magnetic nanoparticles (MNPs) are materials whose behavior is susceptible to the presence of external magnetic field gradients. Among MNPs, iron oxide nanoparticles are characterized by diameters of *ca.* 1–300 nanometres and are found typically in the form of magnetite (Fe_3O_4) or in the oxidized form called maghemite (γ-Fe_2O_3). In terms of atomic structure, iron oxides are characterized by an inverse spinel structure, in which the oxygen forms a cubic crystal system where the tetrahedral sites are occupied by Fe^{3+} and the octahedral sites are occupied by both Fe^{3+} and Fe^{2+}. The unpaired electrons in the 3d shell of the iron atoms give rise to strong magnetic moments, thus magnetite can behave as a ferromagnetic, antiferromagnetic or ferrimagnetic material.[8] Most interestingly, when considering magnetite in a nano-format, *i.e.* of the size < 20 nm, superparamagnetism is observed (Figure 4.1). As outlined in Figure 4.1a, in the case of superparamagnetism

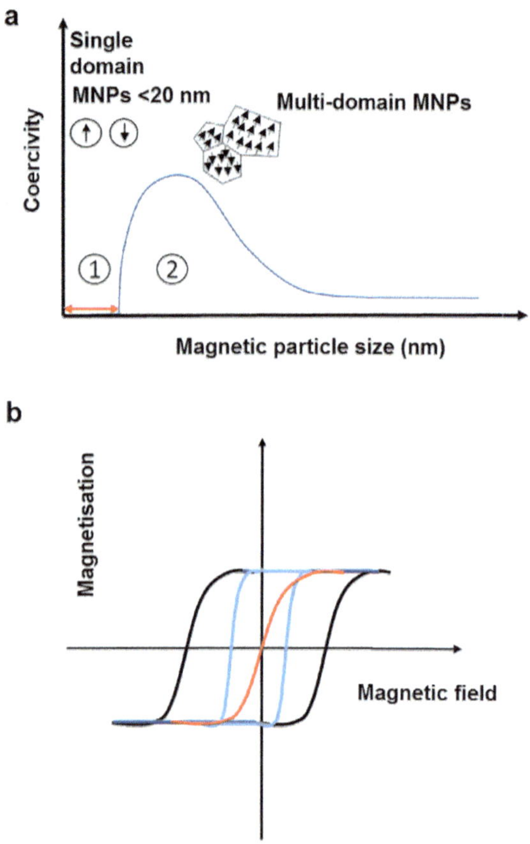

Figure 4.1 The magnetization of iron oxide particles is dependent on their sizes. Superparamagnetic MNPs of a size <20 nm comprise rotating crystals that align with direction of an external magnetic field applied. Crystal reorientation provides high magnetic susceptibility and magnetization saturation. (a) Coercivity (H_c), *i.e.* the ability of a material to withstand an external magnetic field without becoming demagnetized, as a function of the particle size: (1) single-domain MNPs of the size of <20 nm; (2) multi-domain MNPs. (b) Magnetization curve: the magnetization (M) is plotted as a function of the applied magnetic field (H). In red, the superparamagnetic MNPs exhibit nearly zero value for coercivity and nearly zero magnetic remanence (*i.e.* the magnetization remaining after an external magnetic field is removed), while for nanoparticles of larger sizes (black – "hard" ferromagnetic, blue – "soft" ferromagnetic) hysteresis loops are observed, as the alignment of the magnetic moments for the multi-domain material is less homogeneous and faster than that for MNPs of the size < 20 nm.

each nanoparticle has a large magnetic moment and behaves as an enormous single paramagnetic atom, yielding to a fast response and a coherent movement of all the spins, when the magnetic field is applied (Figure 4.1b, magnetization curve).

The superparamagnetic properties are at the basis of the applications of MNPs in nano-medicine and nano-biotechnology. The use of superparamagnetic MNPs in medicine and biotechnology has been steadily increasing over the years: MNPs offer a fast response to the applied magnetic fields, advantages in the separation process, the possibility to magnetize directly the crude biological samples in which the MNPs are suspended, good dispersibility and fast binding kinetics.[9] Established areas of application are targeted drug delivery, magnetic resonance imaging (MRI) contrast enhancement, proteomics, purification of proteins and cells.[10–12]

4.3 Methods for the Preparation of Molecularly Imprinted MNPs

The combination of magnetic properties to molecular imprinting yields nanomaterials capable of both magnetic-field-induced separation and targeted recognition properties. Such a combination is foreseen to provide, in principle, a powerful analytical tool having the characteristics of ease of operation, flexibility, and selectivity. Indeed, as a pioneering example to prove the concept, magnetic iron oxide superparamagnetic composite MIP beads, prepared using suspension polymerization in perfluorocarbon, were proposed for the first time. The resulting material, although having an average diameter of 13 μm, was responsive to magnetization.[13]

In general, early works proceeded by carrying out the polymerization of the MIPs within a suspension of MNPs, which led to MIPs-MNPs often composed of aggregates of MNPs embedded in the MIP matrix. In effect, their size grew relatively large (with diameters > 500 nm), and this growth severely undermined their targeting selectivity in biomedical applications, their magnetic properties and their penetration ability into tissues. As an advance in the MNPs-MIP preparation techniques, individual ultra-small MNPs coated with thin MIP layers were conceived, with the advantages of (i) control of the NPs sizes, (ii) higher surface-to-volume ratios, and (iii) higher binding capacity and faster binding kinetics.[14]

In general, all the synthetic strategies proceeded at the modification of the MNPs through the addition of a coating layer to the magnetic core, such as a silica layer,[15] so to provide chemical functionalities as anchorage points suitable for further covalent chemistry couplings. The subsequent step takes advantage of the introduced chemical functionalities and proceeds through three main routes, as summarized in Scheme 4.1: (i) the immobilization of the template,[16,17] (ii) the introduction of polymerizable double bonds onto the MNP,[18] (iii) the employment of reversible addition–fragmentation chain transfer initiators for a controlled "grafting-from" MIP synthesis.[19,20] Alternatively, the use of aryl diazonium salt chemistry, to grow very thin MIP layers,

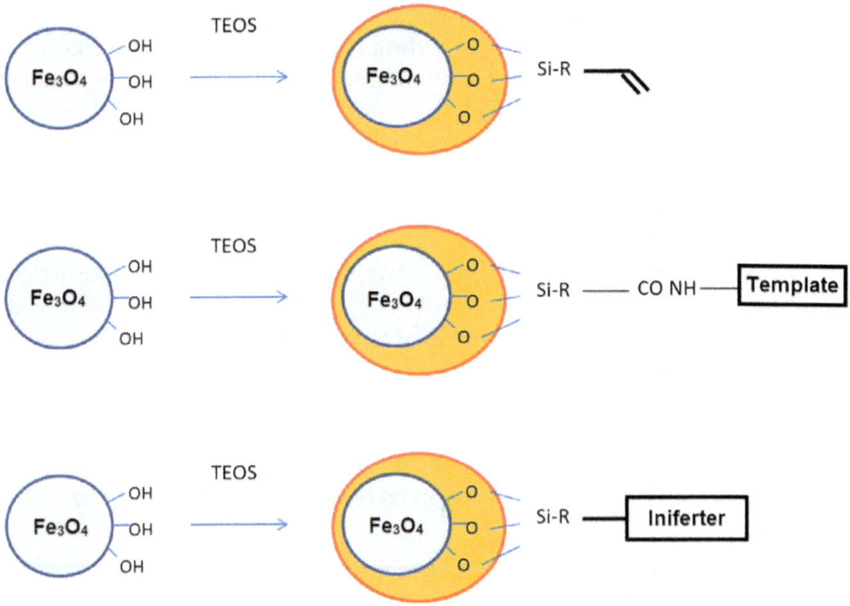

Scheme 4.1 Schematic representation of the strategies for the preparation of MNP-MIPs. MNP cores are often derivatized with silane derivatives, such as tetraethyl orthosilicate (TEOS), and further modified so to covalently attach either the template, or the iniferter, or a vinyl funcitonality.

was successfully reported[21] and more recently click chemistry protocols were exploited onto MNPs coated with silica shells, which were modified with terminal azide groups, for obtaining highly controlled and few nanometre thick MIP layers.[22]

4.4 Applications of MNPs-MIPs

In MNP-MIPs, the magnetic separation properties are conjugated to the ability of the MIPs to target selectively a defined analyte. In this combination, the magnetic separation allows a straightforward enrichment, while the MIP confers attuned selection. Thus, MIP-MNPs offer the possibility to be separated by an external magnetic field with relative rapidity, easy protocols, cost effectiveness and high efficiency.

Elective areas of application of such combined technology are the determinations of analytes within highly complex sample matrices. In fact, the quali/quantification of a target analyte in these samples becomes feasible only upon the introduction of a separation step that frees the analyte from the many interfering components of the matrix and that can be achieved with either on-line or off-line strategies. Clinical, environmental, food and feed analytical areas, despite great differences in the composing matrixes, all share the problem of high complexity of the starting sample and all have been considered as niches of interest for the MNP-MIPs.

4.4.1 Clinically Directed MNP-MIPs

In clinical research, MPN-MIP targets mostly selected proteins present in the biological fluids. Yet, nowadays to understand a biological process, such as the onset and the progression of a pathology, at the molecular level, so as to gain information on the regulation and cooperation of multiple protein subunits,[23] the mere approach of descriptive proteomics, *i.e.* the list of expressed proteins in a cell or tissue or organ in a defined time and condition, has been demonstrated to be not exhaustive. Thus such information shall be accompanied by the analysis of protein complexes and protein interaction networks both in time and in space.[23] For these reasons, it becomes evident that additional technologies are needed to bridge the gap between 'interactomics' and classical proteomics, and structural biology, so as to gain detailed information about dynamics and functions of biomacromolecules. These envisaged technologies should be easily integrated to state-of-art proteomic platforms, such as mass spectrometry.[24] Experimental-wise insights into functional proteomic start by gathering target proteins, single or classes, from the biological fluids. For this reason, the selective enrichment of defined analytes is a goal that is suitable for a series of MNP-MIPs works, which are dedicated to accomplishing selection towards a defined protein or towards selected protein classes or protein interaction-complexes. From the early research of MNP-MIP, Chau and colleagues proposed MNP-MIPs of ~500 nm in size for the recovery of albumin.[16] Hemoglobin was also targeted by core–shell MNP-MIPs, either based on aminophenylboronate moyeties,[15,25] or as an alternative, hemoglobins of different sources were distinguished by MNP-MIP based on polydopamine.[26] More recently, copper-mediated MNP-MIP polymers for hemoglobin were prepared by surface imprinting. In this synthetic process, copper ions were used to chelate carboxyl groups of Fe_3O_4@COOH and histidine moieties of bovine hemoglobin. The resultant imprinted MNP-MIPs were ~50 nm in diameter, thus the MNP-MIPs displayed high mono-dispersion, high crystallinity, the desired superparamagnetic susceptibility, together with rapid separation, fast adsorption kinetics, high binding capacity as well as favorable reproducibility and reusability. Bovine hemoglobin was successfully recovered from a standard protein mixture and from real bovine blood samples.[27]

Lysozyme was also a highly studied protein target. Core-shell MNP-MIP composites targeted it with a reported imprinting factor, IF > 8.[28] Lysozyme was also an object of interest in a paper[29] where MNP-MIPs were obtained through the immobilization of the protein-template onto the core–shell MNPs and the MIP layer was prepared by self-polymerization of dopamine (DA). The MNP-MIP-DA material demonstrated the ability to selectively enrich lysozyme from diluted egg white, by a simple incubation followed by elution of the recovered protein on the target plate of a MALDI-TOF MS and followed by the mass spectrum analysis, showing a workflow and a combination of extraction and analytical methodology of high interest for clinical purposes and for targeted proteome analyses.[29]

As the necessity of clinical measurements and proteomics is not restricted to a single target analyte, instead it often invokes multiple targets, an attempt to recover multiple analytes was proposed. For that, MNP-MIPs were synthesized from poly(ethylene-*co*-ethylene alcohol) *via* phase inversion, placing both target molecules and hydrophobic magnetic nanoparticles in a polymer solution. The resulting MNP-MIPs were down in size to 56 nm, with the advantage of a more effective response to the magnetic field with respect to greater dimensions. Moreover, they demonstrated the ability to treat urine samples and to recover the following panel of analytes: lysozyme, albumin, urea and creatinine.[30] Moreover, multiple targets were the object of interest of the work on dual protein removal based on a surface imprinting technique onto MNPs and on a two-step immobilized template method: the magnetic separation allowed simultaneous selective extraction of bovine serum albumin and bovine hemoglobin.[31]

Glycoproteins, glyco-modifications and glycome represent a topic of increasing importance when evaluating phato-physiology and when assessing clinical status.[32] Glycoproteins were therefore targeted with MNPs coated with gold (Au) nanoparticles derivatized with aminothiophenol and co-polymerized with aminophenylboronic acid so to produce a glyco-recognitive imprinted shell. The resulting multifunctional MIP nanofibers, with micrometre length and sections of about 160 nm, showed high loading capacity, excellent selectivity and fast binding kinetics towards the target glycol-proteins, plus, the semi-conductive properties of the nanocomposite, allowed for the electrochemical detection of the bound protein in a linear concentration range of 0.01 to 0.30 mg mL^{-1}.[33]

Among the MNP-MIPs used for protein recognition and recovery, an original perspective on the potential uses of the composite materials is proposed by the work of Liu and colleagues.[34] In fact, they fabricated magnetic antibody-like nanoparticles *via* a one-step surface-initiated *in situ* molecular imprinting over silica coated magnetite core–shell nanocomposites. As a result, the MNP-MIPs had an overall size of ~80 nm in diameter showing excellent aqueous dispersion stability together with the predetermined selectivity to the target protein and high biocompatibility. Rapid, efficient, selective and optically tractable *in situ* sequestration of target proteins was demonstrated directly within living cells. This work, by now, represents a new concept as it is the first example of fully artificially engineered multifunctional MNP-MIPs for intracellular protein-sequestration within the cell.

Finally, in the field of protein imprinting, the additional novelty is represented by a greener approach for the preparation of MNP-MIPs: the derivatization of MNPs with polysaccharide-imprinted polymers prepared by a layer-by-layer assembly, and then surface imprinting was proposed.[35] Results were a core–shell MNP-MIP able of the recognition of ovalbumin. Tested types of natural polysaccharides were sodium alginate and chitosan. The MNP-MIP exhibited a magnetic susceptibility (45.30 emu g^{-1}) convenient for efficient separation, thanks to their thin imprinted layer of just ~8 nm, and offered high water compatibility, given the selected hydrophilic monomers.[35]

Beside the issues posed by proteomics, the determination of hormones, drugs and any signaling molecule of clinical relevance were considered as matter of interest for the development of clinically directed MNP-MIP quali-/quantitative measurement strategies.

The spectrofluorimetric properties were exploited in the case of surface-initiated reversible addition–fragmentation chain transfer polymerization of ofloxacin (OFX) as the template for the preparation of a nanomaterial suitable for the separation of the class of fluoroquinolones (FQs). The resulting MNP-MIP exhibited high selectivity toward the family of FQs (ofloxacin, pefloxacin, enrofloxacin, norfloxacin, gatifloxacin) in a competitive binding assay. The successful direct enrichment of the five FQs from human urine was demonstrated on spiked human urine samples. In the proposed method, the recoveries ranged from 83.1% to 103.1% and the relative standard deviation (RSD) was 0.8 to 8.2%.[36]

Moreover, the analytical measurements benefitted greatly from MNP-MIPs, especially when these latter were used as solid phases for extraction, as a new class of sorbent medium,[37] and were therefore coupled to (liquid chromatography)-(mass spectrometry) analysis, such as in the case of cocaine metabolites in urine samples.[38] As a further example of (MNP-MIP)-based solid-phase extraction (SPE), the power of the procedure was evident in the analysis of the antioxidant metabolites of the plant Polygonum cuspidatum directly in rat plasma in combination with LC/Q-TOF MS.[39]

In another attempt, an unprecedented combination of an MIP nanomaterial was proposed: magnetic graphene (MG) was chosen as the supporting substrate, exhibiting both conducting and magnetic properties, and served to absorb the targeted imprinted molecule, *i.e.* bovine hemoglobin. Acrylamide-*co*-bisacrylamide was *in situ* polymerized to the MG-MNP. The so-prepared MG-MIP showed a remarkable maximum adsorption capability that was 186.73 mg g^{-1} and also an adequate imprinting factor of ~2.[40]

Instead, when MNPs were used as an integral part of chemosensors, electrochemical transduction was mostly exploited. A highly sensitive and selective chemosensor based on (magnetic field)-induced self-assembly of MNPs-polyaniline-MIPs, directed at the clinical determination of creatinine in human plasma and in urine samples, was proposed with a limit of detection, LOD = 0.35 nmol L^{-1}, average recovery of 90.8 to 104.9% and RSD below 2.7%.[41] Photoelectrochemistry was used for the fabrication of a bilirubin chemosensor where MNPs-hydroxyapatite-MIP nanoparticles were attached to the surface of the magnetic glassy carbon electrode (MGCE). The developed chemosensor was highly sensitive to bilirubin in solutions with a linear concentration range of 0.1 to 17 μM and LOD of 0.007 μM. Moreover, the chemosensor exhibited outstanding selectivity while used in coexisting systems containing various interferences with high concentration.[42]

Interleukin-8 (IL-8), given its correlation with cancer, was the target analyte of an electrochemical sensor based on MNPs coated with graphene oxide (GO) co-MIPs.[43] The IL-8 was surface imprinted. The Fe_3O_4@GO@MIP was self-assembled onto a gold film electrode under magnetic field. Remarkable achievements were reported as the sensor exhibited superb resistance to non-specific binding and a very low LOD of 0.04 pM for IL-8. This successful detection of IL-8 in saliva samples led to the foreseeing of applications in cancer marker detection.[43]

The lamotrigine anti-convulsal drug, used nowadays to treat epilepsy and other neurological disorders, was imprinted to form MNP-MIP lamotrigine nanoparticles, which were then the basis for an electrochemical sensor that allowed the trace determination of lamotrigine in biological samples.[44]

A bi-metallic sensor was prepared for the detection of two major components of vitamin B6, *i.e.* pyridoxine (Py) and pyridoxal-5′-phosphate (PLP). In this case, acrylic acid modified zero valent iron nanoparticles were combined with copper nanoparticles. This combination resulted in bimetallic Fe/Cu magnetic nanoparticles, modified with vinyl groups, synthesized on the surface of vinyl silane pencil graphite electrodes, with the LOD for the target analytes of 0.040 $\mu g\ L^{-1}$ and 0.043 $\mu g\ L^{-1}$ for Py and PLP, respectively.[45]

In the case of optical sensing, the combined separation and concentration of phenylalanine in urine, given the role of such analyte as precursor of catecholamines, was proposed. Poly(ethylene-*co*-vinyl alcohol)s-MNPs were imprinted with phenylalanine and the composite nanomaterial was used to separate and sense the target phenylalanine in urine by fluorescence spectroscopy and Raman scattering microscopy.[46]

In conclusion, some attempts, quite recent, are investigating the potential of MNPs as a magnetic delivery nanomaterial for a triggered cancer therapy: MNPs-MIP showed the possibility of active control over the drug release by using an alternative magnetic field. *In vitro* and *in vivo* release of doxorubicin was investigated, demonstrating a massive doxorubicin release under the alternative magnetic field without a temperature increase in the medium. These results open new routes for targeted therapies.[47] Quercetin was also imprinted onto Fe_3O_4@SiO_2 nanoparticles using vinylimidazolium as the monomer and Tragacanth gum as the cross-linking matrix, demonstrating MNP-MIP nanogels that possessed biocompatibility, thanks to the choice of the cross-linking monomer, and showed the ability to release drug by diffusion.[48] Moreover, aspirin,[49] meloxicam[50] and mitomycin[51] were imprinted onto composite MNP-MIP for controlled release.

4.4.2 Water Contaminant Recovery and Analysis by MNP-MIPs

A completely different scenario that harnesses the benefits of the MNPs-MIPs is the monitoring of water pollutants.[52] Here, the sample matrix is essentially an aqueous solution with debris and large contaminants; in such a context the magnetic separation allows for a straightforward enrichment, as the analytes of interest are present at high dilutions. In this respect, many papers appeared for the determination of endocrine disruptive chemicals[53] and some dedicated efforts are in particular towards estradiol.[54–56] The interest was also in the sorption and removal of dye molecules, such as azo-dyes[57] and Sudan dye.[58] Chlorophenols also represented a class of pollutants suitable target for MNP-MIPs. A sensitive and selective determination of 4-chlorophenol (4-CP), 2,4-dichlorophenol (2,4-DCP) and 2,4,6-trichlorophenol (2,4,6-TCP) in an environmental water sample was proposed by molecularly imprinted magnetic nanoparticles. The MNP-MIP extraction was coupled to capillary electrophoresis (CE). The results, under optimal concentration conditions, showed remarkable signal enhancement for the CP analytes (*i.e.* from 781 to 8916 fold increase). The average recovery of the analytes in spiked water samples was 99.33 to 104.5% with RSD below 6.8%. The LODs were from 10^{-3} to 10^{-5} μg mL^{-1}.[59] Alternatively, the issue of measuring CPs was also tackled by fabrication of imprinted silica nanomaterials co-loaded with Fe_3O_4 nanoparticles and ZnS : Mn^{2+} quantum dots (QDs). The combination allowed separation of pentachlorophenol with an external magnetic field and directly monitoring the uptake by the changes of fluorescence intensity of ZnS : Mn^{2+} QDs.[60]

4.4.3 Food and Feed-related MNP-MIPs

Recently, the issue of food and feed[61] quality is becoming central for health policy in many countries, because the wellbeing and the health of the individual strongly correlates and have been demonstrated to depend on the diet.[62] Yet, the latest research seems to suggest significant dietary impact of the microbiome.[63–65] In this scenario, MPN-MIPs could be invaluable technical tools to enrich the target analytes prior to their analytical assessment. In fact, the extraction of estradiol in feed[66] and in milk[67] has been reported. Estrogens were extracted and measured from milk powders.[68] Enrofloxacin imprinted organic-inorganic hybrid mesoporous sorbent from nanomagnetic polyhedral oligomeric silsesquioxanes allowed the selective extraction of fluoroquinolones in milk samples.[69]

Protocatechuic acid (PCA) was assessed, being the metabolite of antioxidant polyphenols, and its potential role in cancer cells was indicated. Magnetic hollow porous MIPs with high binding capacity, fast mass transfer, and easy magnetic separation were fabricated. Hollow porous MIPs were firstly synthesized using 4-vinylpyridine and glycidilmethacrylate (GMA) as the co-monomer, on a sacrificial support. After that, the epoxide ring of GMA was opened for chemisorbing Fe_3O_4 nanoparticles. The results indicated that magnetic hollow porous MIPs exhibited an outstanding large specific surface area (548 m^2 g^{-1}) with hollow porous structure and magnetic sensitivity. The recognition demonstrated high selectivity towards PCA with respect to analogues. The results of the real sample analysis confirmed the superiority of the proposed magnetic HPMIPs for selective and efficient enrichment of trace PCA from complex matrices.[70]

The extraction of bisphenol in food by MNP-MIPs was also proposed[71] and of chologenic acid from fruit juices.[72] Moreover, the determination of acrylamide in potato chips was a matter of research,[73] as well as macrolidic antibiotics[74] and sulphonamides in chicken[75] and in poultry feeds,[76] resveratrol in wine,[77] and dicophol in tea.[78] Ochratoxin A was a target for quality control[79] and vanillin too,[80] along with ion determination regarded copper ions[81] and traces of mercury(II) ions in fish samples.[82]

The integration of personal portable, real time monitoring meters, such as the glucose meter, together with new additional functionalities aimed at controlling the quality of nutrition and of nutrients was proposed, by developing an antibody-free sandwich assay, (MNP-MIP)-based, directed at the determination of CP in food.[83]

4.5 Carbon Nanotubes

A relatively novel kind of nanomaterial with inherent interesting characteristics of promising potentials is the so-called carbon nanotube (CNT). CNTs are building blocks suitable for many nanotechnological applications,[84] and nowadays MIP-related technologies foresee future development perspectives using them.

In terms of their physical aspect, CNTs are cylindrical, fiber-like nanostructures, formed from the roll of a graphene planar atomic sheet into a continuous tubular structure (Scheme 4.2). Yet in the process, a single-walled carbon nanotube (SWCNT) can be formed, which are characteristic single-tube entities or, alternatively, multi-walled carbon nanotubes (MWCNTs) can be formed. These latter consist of multiple concentric nanotubes nested one within the other. They exhibit telescoping properties so

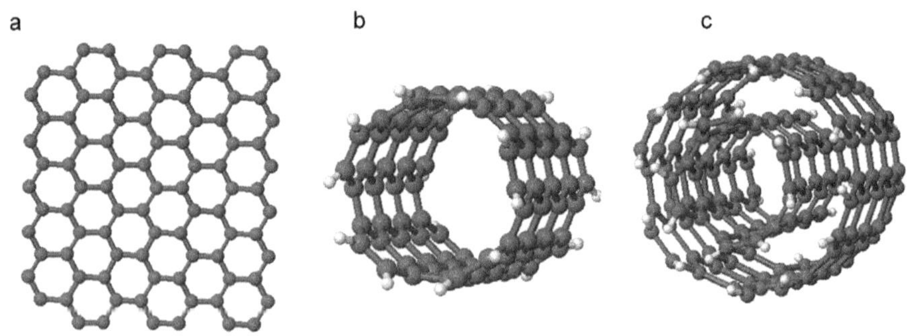

Scheme 4.2 (a) Structure of planar graphene, a two-dimensional (2-D) network with a hexagonal lattice, where carbon atoms are in sp^2 hybridization, except for a small number of sp^3 hybridized atoms at edges or defects. (b) When a graphene layer is induced to roll into a tube, it forms a continuous single-walled carbon nanotube (SWCNT). (c) Multiple SWCNTs, packed in a tight concentric structure, give rise to multi-walled carbon nanotubes (MWCNT).

that the inner nanotube may slide, almost without friction, within its outer nanotube shell, thus creating an atomically perfect linear or rotational system.[84] In terms of dimensions, SWCNTs have diameters in the range of 0.8 to 5 nm and a variable length from a few nanometres to millimetres. MWCNTs have outer diameters in the range of 3 to 300 nm and lengths from a few nanometres to millimetres, most commonly ~200 μm. The core of CNT is hollow. Fully dispersed SWCNTs have all their atoms exposed at the surface and their calculated specific surface area is quite remarkably expanded, being ~1600 $m^2 g^{-1}$.

In terms of physical properties, the highly delocalized π-electrons in the structure of a CNT induce the formation of strong interactions among atoms present in the same layer, whereas the interactions between adjacent layers are much weaker. This difference in interactions leads to drastically different properties in and out of the graphene plane, and to highly anisotropic properties. The result is that the electrical, thermal, mechanical, and electrochemical properties of CNTs vary by orders of magnitude between in-plane and cross-plane measurements.[85] Yet, as the properties of carbon materials correlate to the dimension and organization of the anisotropic graphitic domains in their internal structures, it is of critical importance to organize CNTs into proper configurations and architectures so to achieve the desired high performance.

Given π-orbital electrons delocalized across their hexagonal atomic arrays and the geometrical arrangement of the hexagonal arrays, CNTs exhibit high electron mobility and hence conductivity, and behave as metallic or semiconducting materials. The arc-produced MWCNTs sustain high current density (>100 A cm^{-2}), a property that is of particular interest in electrochemical sensing.[86]

CNTs present also some minuses, such as their marked nonpolar character and a high length-to-diameter ratio, both leading to CNTs insolubility in water and to their spontaneous aggregation, with the consequent decrease of the favorable surface-to-area ratio, although dimethyl formamide, dimethyl acetamide and dimethyl pyrolidone are reported as sufficient dispersants for CNTs.[87]

The sidewall of SWCNTs and MWCNTs is hydrophobic and chemically inert so the chemical functionalization of CNTs can be based on weak bonds, such as π–π stacking of conjugate molecules or encapsulation in surfactants.[88] Alternatively, when CNTs are strongly oxidized, especially at the defects on the walls and edges of the sidewall, hydroxyl or carboxyl groups can be grafted onto the pipe walls and to the pipe orifices so to later perform covalent functionalization. A comprehensive review on covalent functionalization of CNTs was published earlier.[89]

In terms of applications, CNTs are exploited for innovative conduction metrologies, represent the basis for micro-fabricated platforms for sensing and can be used as scanning thermal probes.[90] Owing to their high (surface area)-to-volume ratio and excellent physisorption performance, CNTs find niches of application also in SPE and for the absorption of chemicals.[91] CNTs have been successfully used as "molecular wires" to realize direct electron transfer between redox enzymes and electrode surfaces, which can fabricate new kinds of mediator-free biosensors.[91–94]

4.6 Nanotubes Coupled to MIPs: Derivatization Strategies

Given the conducting performances of CNT nanomaterials, their inclusion into MIPs to obtain CNT-MIP composites was a matter of interest. In fact, it opened up perspectives of increased sensing signals. The early attempts account for the deposition of an MIP film on preformed areas of conducting carbon nanomaterials, the filling of CNTs with MIPs and the addition of graphene flakes to the MIPs during the polymerization.[95] However, uninterrupted MIP areas confer insulating properties and the lack of a direct path for electron conduction between the MIP and the CNT. To overcome such problems, the surface of CNTs was modified both covalently and non-covalently (Scheme 4.3).

The hydrophobic character of the MIP was used to immobilize aromatic organic units such as vinyl anthracene non-covalently, by interactions based on π–π stacking.[88] The derivatization of the surface of CNTs with polar groups offered the possibility of covalent modifications: iniferter[96] and vinyl groups where anchored prior to the MIP synthesis (Scheme 4.3).[97] The best results were achieved when "grafting from" procedures were employed, as this allowed a controlled polymerization to take place and avoided the formation of bulk polymer in the solution. Both atom transfer radical polymerization (ATRP) and reversible addition–fragmentation chain transfer (RAFT) were successfully used. Vinyl silane was also used for the covalent modification of CNTs, by the condensation of the –OH groups at the defect sites with silanes (Scheme 4.3). Further acrylic acid-based monomers were used to grow the MIP film.[97]

4.7 Nanotubes Coupled to MIPs: Applications

The performance of CNT-MIPs has been put to the test in a variety of applications and several reviews are devoted to the subject.[98–101] In particular, the high loading capacity of the CNT-MIP composite has been exploited for SPE of typical pollutants of the environmental, biological, food and pharmaceutical samples prior to chromatographic analyses.[102] The CNT-MIP is the basis for columns[103] or steel frits,[104] which are mainly combined off-line to the separation method, whereas in some cases on-line combination was proposed.[102]

The excellent conductive properties of the CNT-MIPs are at the basis of their uses in chemical- and biochemical sensors, especially for niches of applications characterized by the necessity to quali/quantify an analyte

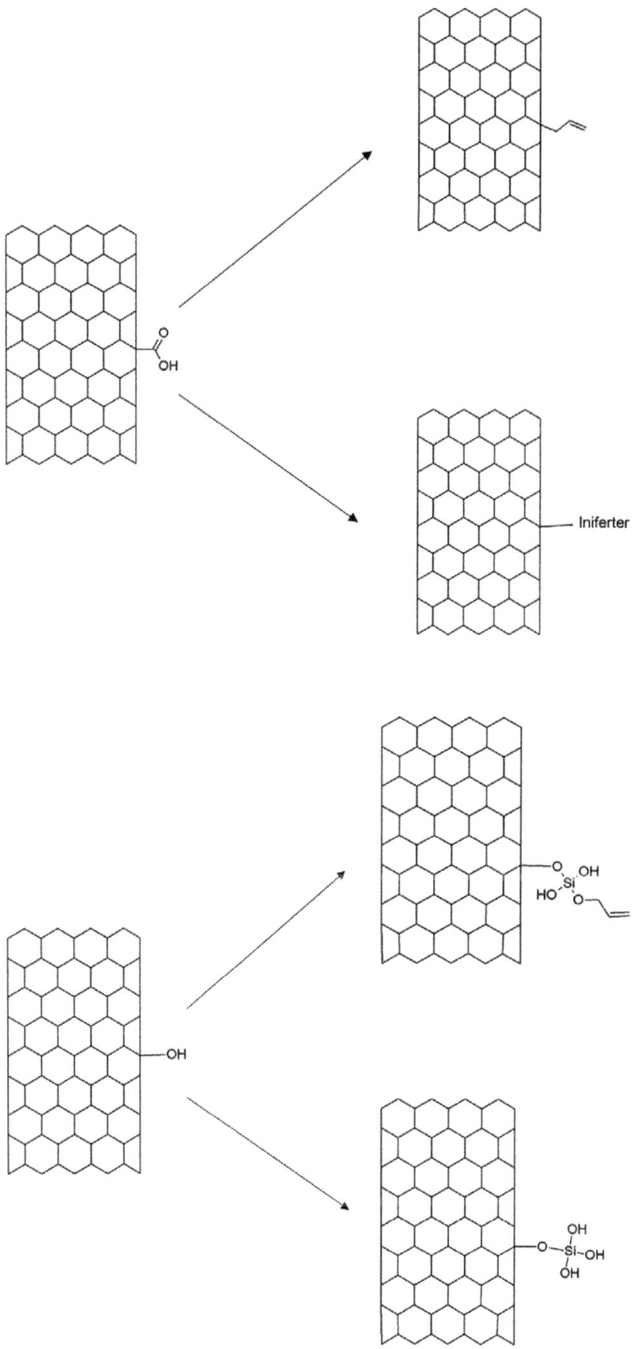

Scheme 4.3 Representation of strategies for derivatization of CNTs in view of the preparation of composite CNT-MIPs: a general procedure starts from an oxidizing treatment, which leads to the formation of carboxyl and hydroxyl groups on the CNT walls, especially at sites characterized by defects of junctions. Carboxylic acid sites on the CNT are derivatized mainly with polymerizable double bonds or iniferters, in a "grafting from" strategy aimed at maximizing the control over the composite CNT-MIP, especially in the case of polymethacrylate MIPs. Alternatively, the hydroxyl groups are modified with silane derivatives so to prcparc silica bascd MIPs.

Table 4.1 Performance of CNT-MIP nanomaterials in analytical measurements.[a]

Analyte	Linear range/M	LOD	CNTs-MIPs types	Ref.
Guanine-rich DNA	5.0×10^{-8}–3.0×10^{-5}	7.52 nM	MWCNTs-MIPs	105
1,4-Dihydroxy anthraquinone	1×10^{-8}–1×10^{-4}	4.15 nM	MWCNTs-MIPs	106
γ-Hexachloro-cyclohexane	1×10^{-9}–1×10^{-3}	0.1 nM	MWCNTs-MIPs	107
Dioctylphthalate	n.d.	2.3 ng L^{-1}	MWCNTs-MIPs	108
Ganciclovir	0.05×10^{-6}–500×10^{-6}	1.5 nM	MIPs-Au-MWCNTs	109
Metronidazole	1.71×10^{-4}–2.05 10^{-1} mg L^{-1}	49.2 μM	MWCNTs-MIPs	110
Triamterene	80.0×10^{-9}–265×10^{-6}	3.35 nM	MWCNTs-conductive MIPs	111
Dinitrotoluene	2.2×10^{-9}–1.0×10^{-6}	1 nM	MWCNTs-MIPs	112
Amoxicillin	1.0×10^{-9}–1.0×10^{-6}	0.89 nM	MWCNTs-MIPs-Pt-Pd dendrite	113
Dopamine	n.d.	1 nM	MWCNTs-MIPs	114
Hydrochloro-thiazide	n.d.	0.1 nM	MWCNTs-MIPs	114
Tartrazine	0.03×10^{-3}–5.0×10^{-3}	8 nM	MIP-MWCNTs-IL@PtNPs	115

[a]LOD - limit of detection, n.d. - not determined.

at trace or ultratrace levels, for hazardous substances, or for monitoring analytes in complex matrices. The list reported in Table 4.1 compares LODs and the linear dynamic concentration ranges for analytes measured recently by CNT-MIP approaches: the remarkable LODs are confirmed in many cases, the variety of the analytes recognized confirms once more that the MIP technology offers a general approach for the target selection, thus overall the results described fully support the strength of these nanocomposites.

4.8 Conclusions

The conjugation of MIPs to nanomaterials yields composites of superior qualities which could overcome the main actual limits inherent to the MIPs and, in particular, the insulating properties of the constituent polymeric material. In fact, the nanocomposites retain the affinity and selectivity typically achieved by the MIP technology, including seemingly to antibodies and biological receptors, while adding several benefits including high mechanical strength, increased surface areas, fast equilibria, chemical stability, conductivity and a magnetic response of the produced nanocomposite.

In particular, (magnetic nanoparticles)-MIP composites exhibit superior recovery properties, suitable for greatly easing the extraction processes and the recovery of selective removal of analytes from complex samples, while (carbon nanotube)-MIPs greatly enhance the conductivity to the benefit of sensing measurements. The potentials of the MIP nanocomposites are quite well expressed by the actual state of art of the technology. Yet there are still challenges for the future of MIP nanocomposites: further advancements are expected when solutions to improve our mastering of the MIP recognition ability and of the selection of properties are found. Moreover, the mastering of the manipulation of nanocomposites, such as tools and protocols for ordered single- and multi-layer deposition, as well as for the growth of two- and three-dimensional nanocomposite architectures, will be achieved and transferred into technological processes suitable for mass fabrication. These are the forthcoming challenges in the MIP research and technology.

List of Abbreviations

ATRP	Atom transfer radical polymerization
CE	Capillary electrophoresis
CNT	Carbon nanotube
CNT-MIP	Carbon nanotube molecular imprinted polymer
CP	Chrolophenol
4-CP	4-Chlorophenol
DA	Dopamine
2,4-DCP	2,4-Dichlorophenol
FQ	Fluoroquinolone
GMA	Glycidilmethacrylate
GO	Graphene oxide
LC/Q-TOF MS	Liquid chromatography quadrupole time of flight mass spectrometry
LOD	Limit of detection
MALDI-TOF MS	Matrix assisted laser desorption ionization time of flight mass spectrometry
MWCNT	Multi-walled carbon nanotube
MGGE	Magnetic glassy carbon electrode
MG	Magnetic graphene
MG-MNP	Magnetic graphene magnetic nanoparticles
MIP	Molecular imprinted polymer
MNPs-MIP	Magnetic nanoparticles molecular imprinted polymer
MNP	Magnetic nanoparticle
MRI	Magnetic resonance imaging
OFX	Ofloxacin
PCA	Protocatechuic acid
QD	Quantum dot
RAFT	Reversible addition–fragmentation chain transfer

RSD	Relative standard deviation
SPE	Solid phase extraction
SWCNT	Single-walled carbon nanotube
2,4,6-TCP	2,4,6-Trichlorophenol
TEOS	Tetraethyl orthosilicate

References

1. G. Wulff and A. Sarhan, *Angew. Chem., Int. Ed.*, 1972, **11**, 341.
2. R. Arshady and K. Mosbach, *Macromol. Chem. Phys.*, 1981, **182**, 687.
3. M. J. Whitcombe, N. Kirsch and I. A. Nicholls, *J. Mol. Recognit.*, 2014, **27**, 297.
4. www.mipdatabase.com.
5. P. S. Sharma, M. Dabrowski, F. D'Souza and W. Kutner, *TrAC, Trends Anal. Chem.*, 2013, **51**, 146.
6. D. Lakshmi, A. M. Bossi, M. J. Whitcombe, I. Chianella, S. A. Fowler, S. Subrahmanyam, E. V. Piletska and S. A. Piletsky, *Anal. Chem.*, 2009, **81**, 3576.
7. W. Wan, S. Wagner and K. Rurack, *Anal. Bioanal. Chem.*, 2016, **408**, 1753.
8. A. H. Lu, E. L. Salabas and F. Schüth, *Angew. Chem., Int. Ed.*, 2007, **46**, 1222.
9. K. S. Lee and I. S. Lee, *Chem. Commun.*, 2008, **6**, 709.
10. A. K. Gupta and M. Gupta, *Biomaterials*, 2005, **26**, 3995.
11. S. Bucak, D. A. Jones, P. E. Laibinis and T. A. Hatton, *Biotechnol. Prog.*, 2003, **19**, 477.
12. I. Brigger, C. Dubernet and P. Couvreur, *Adv. Drug Delivery Rev.*, 2002, **54**, 631.
13. R. J. Ansell and K. Mosbach, *Analyst*, 1998, **123**, 1611.
14. N. Griffete, H. Li, A. Lamouri, C. Redeuilh, K. Chen, C. Z. Dong, S. Nowak, S. Ammar and C. Mangeney, *J. Mater. Chem.*, 2012, **22**, 1807.
15. N. Li, X. He, L. Chen and Y. Zhang, *Chem.–Asian J.*, 2009, **4**, 286.
16. J. T. Chau, G. C. Hong, H. K. Kwee and W. T. Yen, *Anal. Chem.*, 2008, **80**, 683.
17. R. Gao, X. Kong, X. Wang, X. He, L. Chen and Y. Zhang, *J. Mater. Chem.*, 2011, **21**, 17863.
18. T. Jing, H. Du, Q. Dai, H. Xia, J. Niu, Q. Hao, S. Mei and Y. Zhou, *Biosens. Bioelectron.*, 2010, **26**, 301.
19. Y. Li, C. Dong, J. Chu, J. Qi and X. Li, *Nanoscale*, 2011, **3**, 280.
20. Y. Li, X. Li, J. Chu, C. Dong, J. Qi and Y. Yuan, *Environ. Pollut.*, 2010, **158**, 2317.
21. Y. J. Dai, J. Pan, L. Xu, X. Li, Z. Zhou, R. Zhang and Y. Yan, *J. Hazard. Mater.*, 2012, **205–206**, 179.
22. C. Xu, X. Shen and L. Ye, *J. Mater. Chem.*, 2012, **22**, 7427.
23. B. Alberts, *Cell*, 1998, **92**, 291.
24. A. J. Heck, *Nat. Methods*, 2008, **5**, 927.
25. L. Li, X. He, L. Chen and Y. Zhang, *Sci. China, Ser. B: Chem.*, 2009, **52**, 1402.

26. R. Ouyang, J. Lei and H. Ju, *Nanotechnology*, 2010, **21**, 185502.
27. R. Gao, X. Cui, Y. Hao, G. He, M. Zhang and Y. Tang, *Talanta*, 2016, **150**, 46.
28. S. Yang, X. Zhang, W. Zhao, L. Sun and A. Luo, *J. Mater. Sci.*, 2016, **51**, 937.
29. W. Wan, Q. Han, X. Zhang, Y. Xie, J. Sun and M. Ding, *Chem. Commun.*, 2015, **51**, 3541.
30. M. H. Lee, J. L. Thomas, M. H. Ho, C. Yuan and H. Y. Lin, *ACS Appl. Mater. Interfaces*, 2010, **2**, 1729.
31. R. Gao, S. Zhao, Y. Hao, L. Zhang, X. Cui, D. Liu, M. Zhang and Y. Tang, *J. Sep. Sci.*, 2015, **38**, 3914.
32. H. Vlassara, R. Bucala and L. Striker, *Lab. Invest.*, 1994, **70**, 138.
33. Y. Li, M. Hong, Miaomiao, Q. Bin, Z. Lin, Z. Cai and G. Chen, *J. Mater. Chem. B*, 2013, **1**, 1044.
34. Y. Liu, S. Fang, J. Zhai and M. Zhao, *Nanoscale*, 2015, **7**, 7162.
35. R. Gao, S. Zhao, Y. Hao, L. Zhang, X. Cui, D. Liu and Y. Tang, *RSC Adv.*, 2015, **5**, 88436.
36. Y. He, Y. Huang, Y. Jin, X. Liu, G. Liu and R. Zhao, *ACS Appl. Mater. Interfaces*, 2014, **6**, 9634.
37. J. Płotka-Wasylka, N. Szczepańska, M. de la Guardia and J. Namieśnik, *TrAC, Trends Anal. Chem.*, 2016, **77**, 23.
38. J. Sánchez-González, M. Jesús Tabernero, A. María Bermejo, P. Bermejo-Barrera and A. Moreda-Piñeiro, *Talanta*, 2016, **147**, 641.
39. Z. Yang, Q. Cai, N. Chen, X. Zhou and J. Hong, *RSC Adv.*, 2016, **6**, 12193.
40. J. Guo, Y. Wang, Y. Liu, C. Zhang and Y. Zhou, *Talanta*, 2015, **144**, 411.
41. T. Wen, W. Zhu, C. Xue, J. Wu, Q. Han, X. Wang, X. Zhou and H. Jiang, *Biosens. Bioelectron.*, 2014, **56**, 180.
42. Z. Yang, X. Shang, C. Zhang and J. Zhu, *Sens. Actuators, B*, 2014, **201**, 167.
43. P. Tang, H. Zhang, J. Huo and X. Lin, *Anal. Methods*, 2015, **7**, 7784.
44. M. K. Bojdi, H. Behbahani, G. Hesam and M. H. Mashhadizadeh, *RSC Adv.*, 2016, **6**, 32374.
45. S. Patra, E. Roy, R. Das, P. Karfa, S. Kumar, R. Madhuri and P. K. Sharma, *Biosens. Bioelectron.*, 2015, **73**, 234.
46. C. Y. Hsu, M. H. Lee, J. L. Thomas, C. P. Shih, T. L. Hung, T. J. Whang and H. Y. Lin, *Nanotechnology*, 2015, **26**, 305502.
47. N. Griffete, J. Fresnais, A. Espinosa, C. Wilhelm, A. Bée and C. Ménager, *Nanoscale*, 2015, 7, 18891.
48. K. Hemmati, A. Masoumi and M. Ghaemy, *Carbohydr. Polym.*, 2016, **136**, 630.
49. X. Kan, Z. Geng, Y. Zhao, Z. Wang and J. J. Zhu, *Nanotechnology*, 2009, **20**, 165601.
50. S. Azodi-Deilami, M. Abdouss and D. Kordestani, *Appl. Biochem. Biotechnol.*, 2014, **172**, 3271.
51. D. Türkmen, N. Bereli, M. E. Çorman, H. Shaikh, S. Akgöl and A. Denizli, *Artif. Cells, Nanomed., Biotechnol.*, 2014, **42**, 316.
52. Prachi, P. Gautam, D. Madathil and A. N. Brijesh Nair, *Int. J. ChemTech Res.*, 2013, **5**, 2303.

53. Y. Li, X. Li, J. Chu, C. Dong, J. Qi and Y. Yuan, *Environ. Pollut.*, 2010, **158**, 2317.
54. X. Xia, E. P. C. Lai and B. Ormeci, *Polym. Eng. Sci.*, 2012, **52**, 1775.
55. R. Gao, Y. Hao, L. Zhang, X. Cui, D. Liu and Y. Tang, *Anal. Methods*, 2014, **6**, 9791.
56. Y. Hao, R. Gao, L. Shi, D. Liu, Y. Tang and Z. Guo, *J. Chromatogr. A*, 2015, **1396**, 7.
57. C. Y. Chen, J. C. Chang and A. H. Chen, *J. Hazard. Mater.*, 2011, **185**, 430.
58. X. Xie, L. Chen, X. Pan and S. Wang, *J. Chromatogr. A*, 2015, **1405**, 39.
59. S. Qi, H. Zhang, Q. Zhu, H. Chen, Y. Dong, L. Zhou, C. Ren and X. Chen, *Anal. Methods*, 2014, **6**, 1219.
60. M. Yang, A. Han, J. Duan, Z. Li, Y. Lai and J. Zhan, *J. Hazard. Mater.*, 2012, **237–238**, 63.
61. L. Pinotti, A. Krogdahl, I. Givens, C. Knight, A. Baldi, V. Baeten, L. V. Raamsdonk, S. Woodgate, D. P. Marin and J. Luten, *Biotechnol., Agron., Soc. Environ.*, 2014, **18**, 471.
62. G. Fattore and C. Agostoni, *Crit. Rev. Food Sci. Nutr.*, 2016, **56**, 1960.
63. S. Tamburini, N. Shen, H. C. Wu and J. C. Clemente, *Nat. Med.*, 2016, **22**, 713.
64. P. E. Wischmeyer, D. McDonald and R. Knight, *Curr. Opin. Crit. Care*, 2016, **22**, 347.
65. J. A. Wargo, S. M. Reddy, A. Reuben and P. Sharma, *Curr. Opin. Immunol.*, 2016, **41**, 23.
66. S. Wang, Y. Li, M. Ding, X. Wu, J. Xu, R. Wang, T. Wen, W. Huang, P. Zhou, K. Ma, X. Zhou and S. Du, *J. Chromatogr. B: Anal. Technol. Biomed. Life Sci.*, 2011, **879**, 2595.
67. R. Gao, X. Cui, Y. Hao, L. Zhang, D. Liu and Y. Tang, *Food Chem.*, 2016, **194**, 1040.
68. H. Lan, N. Gan, D. Pan, F. Hu, T. Li, N. Long and L. Qiao, *J. Chromatogr. A*, 2014, **1331**, 10.
69. H. B. He, C. Dong, B. Li, J. P. Dong, T. Y. Bo, T. L. Wang, Q. W. Yu and Y. Q. Feng, *J. Chromatogr. A*, 2014, **1361**, 23.
70. H. Li, X. Hu, Y. Zhang, S. Shi, X. Jiang and X. Chen, *J. Chromatogr. A*, 2015, **1404**, 21.
71. Z. Xu, L. Ding, Y. Long, L. Xu, L. Wang and C. Xu, *Anal. Methods*, 2011, **3**, 1737.
72. Y. Hao, R. Gao, D. Liu, G. He, Y. Tang and Z. Guo, *Food Chem.*, 2016, **200**, 215.
73. M. Arabi, A. Ostovan, M. Ghaedi and M. K. Purkait, *Talanta*, 2016, **154**, 526.
74. Y. Zhou, T. Zhou, H. Jin, T. Jing, B. Song, Y. Zhou, S. Mei and Y. I. Lee, *Talanta*, 2015, **137**, 1.
75. M. Karimi, F. Aboufazeli, H. R. L. Z. Zhad, O. Sadeghi and E. Najafi, *Food Anal. Methods*, 2014, **7**, 73.
76. X. Kong, R. Gao, X. He, L. Chen and Y. Zhang, *J. Chromatogr. A*, 2012, **1245**, 8.

77. F. F. Chen, X. Y. Xie and Y. P. Shi, *J. Chromatogr. A*, 2013, **1300**, 112.
78. H. Yan, X. Cheng and N. Sun, *J. Agric. Food Chem.*, 2013, **61**, 2896.
79. E. Turan and F. Şahin, *Sens. Actuators, B*, 2016, **227**, 668.
80. F. Ning, H. Peng, L. Dong, Z. Zhang, J. Li, L. Chen and H. Xiong, *J. Agric. Food Chem.*, 2014, **62**, 11138.
81. M. Kaykhaii, M. Khajeh and S. H. Hashemi, *J. Anal. Chem.*, 2015, **70**, 1325.
82. E. Najafi, F. Aboufazeli, H. R. L. Z. Zhad, O. Sadeghi and V. Amani, *Food Chem.*, 2013, **141**, 4040.
83. S. Chen, N. Gan, H. Zhang, F. Hu, T. Li, H. Cui, Y. Cao and Q. Jiang, *Anal. Bioanal. Chem.*, 2015, **407**, 2499.
84. J. Li and G. P. Pandey, *Annu. Rev. Phys. Chem.*, 2015, **66**, 331.
85. H. Dai, D. Xiao, H. He, H. Li, D. Yuan and C. Zhang, *Microchim. Acta*, 2015, **182**, 893.
86. C. A. Santini, P. M. Vereecken, A. Volodin, G. Groeseneken, S. De Gendt and C. V. Haesendonck, *Nanotechnology*, 2011, **22**, 39520.
87. M. Okamoto, T. Fujigaya and N. Nakashima, *Adv. Funct. Mater.*, 2008, **18**, 1776.
88. X. L. Zhang, Y. Zhang, X. F. Yin, B. B. Du, C. Zheng and H. H. Yang, *Talanta*, 2013, **105**, 403.
89. K. Balasubramanian and M. Burghard, *Small*, 2005, **1**, 180.
90. S. Zhu and G. Xu, *Nanoscale*, 2010, **2**, 2538.
91. J. X. Wang, M. X. Li, Z. J. Shi, N. Q. Li and Z. N. Gu, *Anal. Chem.*, 2002, **74**, 1993.
92. J. Wang, *Electroanalysis*, 2005, **17**, 7.
93. J. J. Gooding, R. Wibowo, J. Q. Liu, W. Yang, D. Losic, S. Orbon, F. J. Mearns, J. G. Shapter and D. B. Hibbert, *J. Am. Chem. Soc.*, 2003, **125**, 9006.
94. F. Patolsky, Y. Weizmann and I. Willner, *Angew. Chem., Int. Ed.*, 2004, **43**, 2113.
95. J. C. C. Yu, A. Hirdina, C. Mancini and E. P. C. Lai, *J. Nanosci. Nanotechnol.*, 2007, **7**, 3095.
96. H.-Y. Lee and B. S. Kim, *Biosens. Bioelectron.*, 2009, **25**, 587.
97. X. Kan, Y. Zhao, Z. Geng, Z. Wang and J.-J. Zhu, *J. Phys. Chem. C*, 2008, **112**, 4849.
98. C. Baggiani, C. Giovannoli and L. Anfossi, *Toxins*, 2015, **7**, 4083.
99. J. G. Yu, X. H. Zhao, H. Yang, X. H. Chen, Q. Yang, L. Y. Yu, J. H. Jiang and X. Q. Chen, *Sci. Total Environ.*, 2014, **482–483**, 241.
100. A. L. Capriotti, C. Cavaliere, A. Laganà, S. Piovesana and R. Samperi, *TrAC, Trends Anal. Chem.*, 2013, **43**, 53.
101. E. A. Souza Silva, S. Risticevic and J. Pawliszyn, *TrAC, Trends Anal. Chem.*, 2013, **43**, 24.
102. Y. Wen, L. Chen, J. Li, D. Liu and L. Chen, *TrAC, Trends Anal. Chem.*, 2014, **59**, 26.
103. Z. Zhang, H. Zhang, Y. Hu and S. Yao, *Anal. Chim. Acta*, 2010, **661**, 173.
104. J. C. C. Yu and E. P. C. Lai, *React. Funct. Polym.*, 2006, **66**, 702.

105. M. You, S. Yang, F. Jiao, L. Z. Yang, F. Zhang and P. G. He, *Electrochim. Acta*, 2016, **199**, 133.
106. A. Nezhadali, S. Senobari and M. Mojarrab, *Talanta*, 2016, **146**, 525.
107. T. S. Anirudhan and S. Alexander, *Biosens. Bioelectron.*, 2015, **64**, 586.
108. J. Du, R. Gao and H. Mu, *Food Anal. Methods*, 2016, **9**, 2026.
109. M. B. Gholivand and N. Karimian, *Sens. Actuators, B*, 2015, **215**, 471.
110. Y. Liu, J. Liu, H. Tang, B. Xu, F. Yu and Y. Li, *Sens. Actuators, B*, 2015, **206**, 647.
111. A. Nezhadali and M. Mojarrab, *J. Electroanal. Chem.*, 2015, **744**, 85.
112. D. Nie, Z. Han, Y. Yu and G. Shi, *Sens. Actuators, B*, 2015, **224**, 584.
113. G. Yang and F. Zhao, *Electrochim. Acta*, 2015, **174**, 33.
114. T. S. Anirudhan, S. Alexander and A. Lilly, *Polymer*, 2014, **55**, 482.
115. L. Zhao, B. Zeng and F. Zhao, *Electrochim. Acta*, 2014, **146**, 611.

CHAPTER 5

Molecularly Imprinted Polymer-based Materials for Quantifying Pharmaceuticals

D. MACIEJEWSKA*, M. SOBIECH AND P. LULIŃSKI

Medical University of Warsaw, Department of Organic Chemistry, Faculty of Pharmacy, Banacha 1, 02-097 Warsaw, Poland
*E-mail: dmaciejewska@wum.edu.pl

5.1 Introduction to Molecularly Imprinted Polymers as Materials for Separation

Separation of substances from the samples is a very important step in analytical procedures, especially in environmental, pharmaceutical and medical determinations. Solid-phase extraction (SPE), solid-phase microextraction, matrix solid-phase dispersion, stir-bar sorptive extraction and different kinds of chromatography are the most frequently used techniques for isolation and pre-concentration of analytes from the complex matrices.[1–3] The commercial sorbents used as the stationary phases are not-selective which leads to co-separation of the matrix components and to an insufficient isolation of the target molecule. One of the ways to overcome this problem is the preparation of new materials which can mimic natural receptors or enzymes (in which selectivity is very high). Molecularly imprinted polymers (MIPs), which are highly cross-linked synthetic polymers characterized by high recognition properties towards predetermined molecules, can be

Polymer Chemistry Series No. 28
Molecularly Imprinted Polymers for Analytical Chemistry Applications
Edited by Wlodzimierz Kutner and Piyush Sindhu Sharma

Published by the Royal Society of Chemistry, www.rsc.org

the answer. The technique of molecular imprinting (formation of imprints of molecule in polymeric matrix) has received attention for a long time. It is difficult to say when the beginning was. The studies of silica gel adsorption properties can be regarded as the beginning.[4,5] The adsorbents revealing selective affinities for predetermined substances were formed by their synthesis in the presence of the suitable compound. Dickey[6] formed silica gels in the presence of methyl orange and its derivatives, and obtained sorbents which preferably adsorbed the dye used in the preparation of the adsorbent. The separation of compounds having even small structural differences was possible. The scientists compared the functions of the imprinted polymers to enzymes and receptors, and analyzed the problems with the introduction of functional groups into the polymer in a definite arrangement. Wulff and co-workers[7–9] had an idea to prepare polymers with appropriate functional groups by using compounds which were polymerizable vinyl derivatives of the suitable template molecule. Their first template was D-glyceric acid to which 4-vinylanilin was bound as an amide, and 4-vinylphenylboronic acid as a diboronate. The functional groups, which covalently bind the template, should undergo an easily reversible interaction. After template removal, the functional groups should remain exactly in the predefined steric arrangement, and the polymer should preferentially recognize and interact with the template. The described approach was named reversible covalent imprinting. Another approach to the preparation of MIPs was named non-covalent. It relies on prearrangement of molecules of functional monomers around the template molecule by electrostatic, hydrophobic, π–π, or hydrogen bonding interactions.[10–12] During the polymerization, the steric prearrangement of molecules is preserved, and the cross-linked polymer matrix is created. After extracting the template, a three-dimensional cavity is left in the polymer. The cavity is complementary in its shape and in molecular electrostatic potential on its surface to the template molecule. The non-covalent approach is simple from the synthetic point of view, and the kinetics of the non-covalent binding during adsorption favorably compares with the reversible binding necessary in the covalent approach. A mixed approach, so called semi-covalent, is also very useful when the covalent binding is used for polymer preparation and non-covalent interactions for analyte adsorption.[13] In Scheme 5.1 the imprinting idea is illustrated. An interesting strategy was proposed to prepare the synthetic host molecules containing one binding site each. It involved monomolecular imprinting inside dendrimers with the porphyrin template using covalent attachment of dendrons to the template, and then removal of the template by hydrolysis.[14] Promptly, the imprinting concept was patented together with the polymerization and the imprinting methods.[11] The recognition process was scrutinized and, especially, separation of enantiomers was examined.[12,15]

The crucial moment in the preparation of sorbents is quantitative removal of the template from polymer matrices. The template, which is left in the

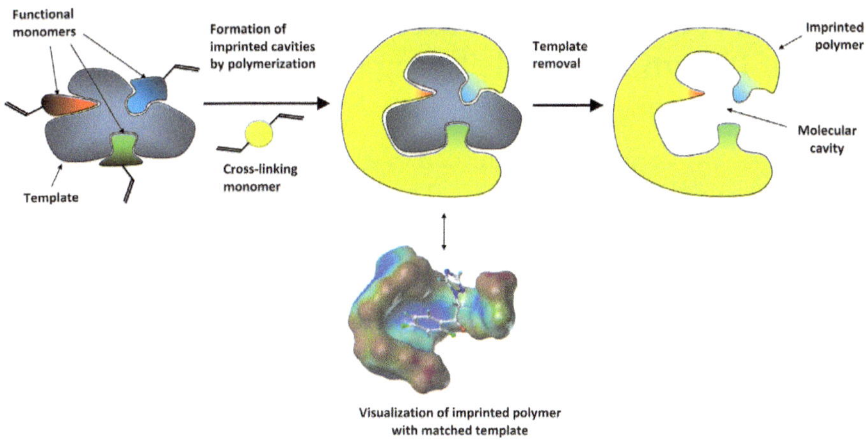

Scheme 5.1 Illustration of the imprinting/adsorption processes.

polymer, can be removed during sample analysis (so-called leakage or bleeding) resulting in false findings (especially if the analyte concentration is low). The problem can be solved by using a structural analogue of the template, *i.e.*, a dummy template, for polymerization.

MIPs can be applied in different formats. After bulk polymerization and laborious preparation, irregular shape particles are packed into cartridges and used successfully in off-line SPE. Spherical particles obtained during precipitation, suspension, and multi-step swelling polymerizations can be used efficiently in both off-line and on-line SPE. To obtain MIP in a capillary format, MIP particles can be packed between two frits in a capillary column.[16] MIPs can be synthesized by *in situ* polymerization inside appropriate supports to obtain surface-coated capillaries, pipette tips or monolithic columns.[17,18] High-performance liquid chromatography (HPLC) is one of the cost-effective methods of separation of compound mixtures at short time. A selective separation approach to the HPLC technique concentrates on molecularly imprinted sorbents used for monolithic or conventional particle-packed columns. Prepared *in situ* monoliths reveal integrated structures and high porosity, which enables rapid mass transport. The main disadvantages of monolith application as the stationary phases include peak broadening and tailoring, and therefore, SPE is proposed as their mode of application.[19]

Although MIPs can be applied in different areas, such as separation techniques, chemical sensors, catalysis, drug delivery systems, or cell culture, the number of commercially available MIPs is smaller than one could expect taking into account the number of publications on imprinted materials or development of the imprinting technique. Therefore, in Sections 5.2 and 5.3, below, we present examples of the determination of pharmaceuticals using patented commercial imprinted sorbents.

5.2 Validated Analytical Methods for Separation of Pharmaceuticals Using Commercial MIP Sorbents

Commercial MIPs are mainly dedicated to separation because they are used in analytical methods involving SPE. They allow for an increase of the target analyte concentration and, hence, the method sensitivity. Moreover, by using imprinted materials one can avoid the derivatization step and that way shorten sample preparation time. There are several companies manufacturing commercially available MIPs for SPE, *e.g.*, MIP Technologies, which now is a part of Biotage, Affinisep and Chrysalis Scientific Technologies, Inc. Commercial MIP SPE cartridges are dedicated to different kinds of analytes including pharmaceuticals (*e.g.*, fluoroquinolones, chloramphenicol, nonsteroidal anti-inflammatory drugs (NSAIDs), β-agonists, β-blockers, estrogens, amphetamines, tetracyclines, catecholamines, antidepressants, aminoglycosides, nitroimidazoles, and riboflavin), herbicides (triazines, picolinic acid derivatives, glyphosate), pesticides (pyrethroids, polycyclic aromatic hydrocarbons), tobacco-specific nitrosamines, mycotoxins, bisphenols, phenolic compounds. Moreover, suppliers of the MIP sorbents offer procedures for isolation of analytes. However, the majority of researchers propose new optimized and validated methods for the analytes to be determined.

Quantification of pharmaceuticals is very important because of the impact of drugs on the health quality of the human population. There are several examples of determination of pharmaceuticals in different matrices including biological samples (*e.g.*, urine, plasma, and blood), food samples (*e.g.*, milk, honey, eggs, meat, and baby food) or environmental samples (*e.g.*, natural waters and sewage).

Antimicrobial agents are used for treatment of different human infections and in veterinary practice. The first example includes quinolones. These are chemotherapeutics for animals. Quinolones cumulate in foodstuff, which can result in adverse reactions in humans, *e.g.*, allergy or antibiotic resistance. The quinolone level in food of animal origin is monitored and the European Union has set maximum residues limits (MRLs) of quinolones. Pretreatment of food samples is necessary before quantification of quinolones. In this step, SPE is most often used and commercial SupelMIP® SPE-Fluoroquinolones cartridges are employed by different researcher groups. Lombardo-Agüí and co-workers[20] used the extraction cartridges (SupelMIP® SPE-Fluoroquinolones cartridge) for SPE of four different fluoroquinolones (ciprofloxacin, enrofloxacin, danofloxacin, sarafloxacin) from bovine raw milk and pig kidney samples. The authors optimized the molecularly imprinted solid-phase extraction (MISPE) procedure by simplifying commercially proposed protocol. Finally, they used methanol (1 mL), ultra-pure water (2 mL) and extraction buffer (0.5 mL) for the conditioning/equilibrating step. After loading appropriately prepared samples (1 mL), the cartridges were washed with ultra-pure water (3 mL), and then acetonitrile (1 mL), and subsequently the analytes were eluted using the methanol–water (50/50, *v*/*v*) mixture with 3% ammonium hydroxide addition (3 mL). The extract was evaporated under nitrogen at 35 °C, then the residue was dissolved in acetonitrile supporting electrolyte solution (25/75, *v*/*v*, 400 μL), and afterward filtered and analyzed by capillary electrophoresis coupled with laser induced fluorescence. That way, very clean extracts were obtained. Limits of quantification (LOQs) were much lower than MRLs (*e.g.*, LOQ and MRL was 1.78 mg kg^{-1} and 100 mg kg^{-1}, respectively, for ciprofloxacin in a milk sample). The recovery (85.2–98.6%) and precision were satisfactory. The procedure results showed high sensitivity of the MIP to fluoroquinolone residues in food samples. Kaur and co-workers[21] used the MISPE procedure also with SupelMIP® SPE-Fluoroquinolones cartridges for milk samples treatment before micelle-enhanced and terbium-sensitized spectrofluorimetric determination of danofloxacin. The protocol of the MISPE procedure recommended by the manufacturer was evaluated, and then modified. For that, authors used methanol (1 mL) and ultra-pure water (2 mL) for conditioning. After sample loading (2 mL), the cartridge was washed, consecutively, with ultra-pure water (3 mL), acetonitrile (1 mL), 15% acetonitrile in ultra-pure water (1 mL), 0.5% acetic acid in acetonitrile (1 mL) and 0.1% ammonia in ultra-pure water (1 mL). For the elution step, the mixed solvent solution of methanol and acetic acid (50/50, *v*/*v*, 1 mL) was applied after testing different solutions. The authors analyzed interferences of foreign substances. The interferences of metal ions, other pharmaceuticals or organic substances were not found, but impact of other coexisting fluoroquinolones was significant. Enrofloxacin, enoxacin, ofloxacin, and norfloxacin were studied for their effect on danofloxacin quantification. It appeared that only enoxacin and norfloxacin with a molar ratio to danofloxacin equal or exceeding 1 : 1 could interfere with the determination of the target analyte. Recovery of the MISPE procedure was high (96–105%). The proposed method was simple, rapid, and cost effective, as compared with that involving HPLC. It can be used for determination of danofloxacin in pharmaceutical and environmental samples.

Rodriguez and co-workers[22] successfully used a modified commercial MISPE procedure with SupelMIP® SPE-Fluoroquinolones cartridges before determination of seven fluoroquinolones in baby foods by liquid chromatography (LC). The eluent obtained after applying the MISPE method[22] was cleaned additionally using Oasis MAX cartridges before final determination of three fluoroquinolones in animal meat products by (isotope dilution)–(liquid chromatography)–(tandem mass spectrometry)(LC-MS/MS).[23]

Lombardo-Agüí and co-workers[24] compared different sample treatments for determination of quinolones in milk by capillary liquid chromatography with laser-induced fluorescence detection. They tested two procedures, namely, Quick, Easy, Cheap, Effective, Rugged and Safe (QuEChERS) and MISPE for extraction and cleanup of a whole cow milk sample from seven different quinolones (ciprofloxacin, danofloxacin, oxolinic acid, flumequine, difloxacin, enrofloxacin and sarafloxacin). For executing the MISPE procedure, the same SupelMIP® SPE-Fluoroquinolones cartridges, as mentioned above, were used and the procedure was similar to that described above.[20] Despite that authors tested different washing steps, they encountered difficulties with high losses of oxolinic acid and flumequine as well as with removing interferences that produced signals overlapping with those of difloxacin and oxolinic acid. Although cleaner extracts were obtained with the MISPE procedure, the authors could not quantify two analytes and had problems with low recovery of two quinolones, whereas with the QuEChERS procedure these difficulties were absent. The MISPE procedure appeared effective only for four quinolones whereas the QuEChERS procedure was faster and allowed for determination of all tested quinolones.

Blasco and Picó[25] used the MISPE procedure with SupelMIP® SPE-Fluoroquinolones cartridges and column–(liquid chromatography)–(electrospray ionization)–(tandem mass spectrometry) for trace analysis of eleven quinolones in eggs. According to earlier publications,[20,22] they modified commercial procedures to optimize conditions for MISPE. Moreover, the authors used the next extraction procedures including SPE with the Oasis HLB column and solvent extraction for comparison of all techniques. The MISPE recovery was 10% higher than those reached by SPE and 20% higher than those obtained by solvent extraction. Sensitivity of the MISPE procedure was higher than those of SPE and solvent extraction (5 to 30 times). Moreover, accuracy and precision of the former procedure was higher. Additionally, no matrix effect was observed with MISPE whereas SPE and solvent extraction showed this effect. Besides, in view of a quite high cost of the commercial MIP in comparison with the cost of Oasis HLB and the cost of solvent extraction, the MISPE procedure could be successfully used for determination of quinolones in eggs.

Furthermore, the SupelMIP® SPE-Fluoroquinolones columns were used as sorbents for construction of an in-line SPE analyte concentrator in capillary electrophoresis coupled with mass spectrometry for the determination of eight veterinary quinolones in bovine milk.[26] The effect of different factors, such as sample pH, injection pressure, injection time, volume and composition of elution medium, on the analyte concentrator performance was studied. After method optimization, repeatability of individual peak areas and signal reproducibility were analysed using three different analyte concentrators. Repeatability and reproducibility studies revealed that their relative standard deviation (RSD) values ranged between 3.4–9.8% and between 5.1–10.4%, respectively. These results showed satisfactory precision of the method, even though the analyte concentrator had to be renovated. Limits of detections (LODs) of the procedures examined were below MRLs (for example 1.1 *vs.* 100 μg kg^{-1} for ciprofloxacin in a milk sample) and lower or at the same level, as compared with other analytical techniques connected with different off-line extraction methods. Recovery exceeding 70% and high precision made this method potentially applicable for determination of fluoroquinolones in different samples.

An interesting example shows that commercial SupelMIP® SPE-Fluoroquinolones could be utilized to fabricate a magnetic composite as the sorbent for extraction of fluoroquinolones from milk samples.[27] A simple method was proposed for synthesis of a magnetic MIP. That is, magnetic nanoparticles and commercial MIP, with the mass ratio of 1 : 1, in methanol were co-mixed and vortexed spontaneously forming aggregates of magnetic composites. The extraction procedure involved the following steps: washing magnetic MIPs with 2 mL of ultra-pure water, then vortexing, followed by discarding the supernatant with an external magnet, subsequent loading of 2 mL of the sample solution, and then vortexing, separation of fluoroquinolones adsorbed on magnetic MIPs with an external magnet, and then washing consecutively with water, acetonitrile, 15% acetonitrile in water, elution of fluoroquinolones from magnetic MIP with methanol, which was 7% in ammonia (*v*/*v*), separation of the eluent from magnetic MIPs with a magnet, evaporation to dryness, dissolution of the residue in a mobile phase, and finally the HPLC analysis. The MISPE procedure of determination of quinolones was optimized by comparing results for different milk samples. That is, optimization was performed through the investigation of several parameters, such as sample pH, the amount of sorbent, extraction time as well as desorption solution composition and time to achieve the highest extraction efficiency. LODs reached in this method were in the range of 1.8–3.2 ng g^{-1} and were comparable to those of other methods for the determination of fluoroquinolone residues, however the analysis time was shorter. Moreover, satisfactory recovery (94.0–124.4%) and precision of the described method made them a good reason for further exploration of co-mixing materials.

The second example of an antimicrobial drug, used in veterinary medicine, revealing a toxicological effect in humans is chloramphenicol. Its use is banned in food of animal origin production in many countries. The MRL for chloramphenicol is set at 0.3 ng mL^{-1}. There are different commercial MIP sorbents dedicated to adsorb chloramphenicol selectively including SupelMIP SPE-Chloramphenicol, MIP4SPE dedicated to chloramphenicol, MIP™ Chloramphenicol. Moreover, different separation and determination methods have been studied for quantification of chloramphenicol in food or body fluids.

Gaugain and co-workers[28] used SupelMIP® SPE-Chloramphenicol as the sorbent for extraction and cleanup of human urine samples before LC-MS/MS determination of chloramphenicol. The MISPE procedure was, as follows. Methanol (1 mL) and water (1 mL) were used during conditioning step, and then urine (1 mL) was loaded on the cartridge. Next, the cartridge was washed, consecutively, with water (2 × 1 mL), a solution of acetonitrile in acetic acid 0.5% (1 mL), 1% ammonia solution (2 × 1 mL), dried, washed with 2% acetic acid in dichloromethane (2 × 1 mL), dried and eluted with methanol (2 × 1 mL). No matrix effect was observed and recovery was high (93.3–104.6%). Moreover, this method was used for stability studies of chloramphenicol glucuronide in urine.

Rejtharová and Rejthar[29] used the same SupelMIP® SPE-Chloramphenicol as the sorbent for cleanup samples of pig and cow urine, feed water, as well as milk and honey, before chloramphenicol determination and the results obtained compared with those for routinely used C18 HPLC columns. The MISPE procedure was quite similar, as described earlier,[28] with differences in volume of conditioning solvents (5 mL of methanol and 5 mL of deionised water), the washing solvent (dichloromethane) and the eluting solvent (a mixture of dichloromethane, acetic acid and methanol, 89/1/10, *v*/*v*/*v*). The SPE procedure involving the C18 column was different than those devised with the use of the MIP. The results obtained from both procedures were quite similar, but the extracts obtained after MISPE were sufficiently clean for gas chromatography analyses (low matrix effect). The differences arose when contaminated urine samples were examined. The range of concentrations, which could be detected with MISPE varied from 0.1 to 50.0 ng mL^{-1}. The procedure was simple, low-time-consuming, and robust, thus offering the possibility to use it for routine screening and confirmatory analyses.

Mohamed and co-workers[30] provided another example where different sample pretreatments were compared. In this work, MISPE using a commercial MIP4SPE adsorbent dedicated for chloramphenicol, or classical SPE with Oasis HLB columns (followed by liquid–liquid extraction, LLE), were employed before LC-EI-MS/MS of milk samples containing chloramphenicol. The MISPE procedure was quite similar to that described above,[28] however, with some modifications. After the first water washing, washing solutions subsequently applied included a mixture of water and acetonitrile (95/5, *v*/*v*) with addition of 5% acetic acid (1 mL), water (2 mL), a water/acetonitrile (80/20, *v*/*v*) mixture which was 1% in a 25% ammonia solution (1 mL), dichloromethane (3 mL), and the elution solution was a water/methanol mixture (10/89, *v*/*v*) with 1% acetic acid (2 mL). Higher recovery of chloramphenicol was obtained after MISPE, and data quality were higher (in terms of sample preparation, extractability of target analyte, and selectivity) than after using a classical procedure. Moreover, it enabled detection of the analyte at a level ten time less than MRL. Similar comparison studies with quite similar conclusions were accomplished for the MISPE and LLE methods during chloramphenicol determination in honey, urine, milk, and plasma samples using high-performance liquid chromatography–tandem mass spectrometry (HPLC-MS/MS).[31]

An interesting study was performed by Li and co-workers:[32] the matrix effect was evaluated for quantitative determination of chloramphenicol residues in the milk powder using isotope dilution mass spectrometry. The authors tested four methods of sample treatment, namely, LLE, HLB-SPE, MCX-SPE, and MISPE (MIP™ Chloramphenicol). Interestingly, the MISPE extracts caused a larger impact on different MS instruments coupled with different ion sources.

Non-steroidal anti-inflammatory drugs are another pharmaceuticals group of medicines widely used in medical treatment of humans and animals. In view of their high consumption, their ubiquity in the environment needs to be monitored. There are commercial MIP sorbents dedicated to these substances, *e.g.*, SupelMIP® SPE-NSAIDs and Affinilute MIP-NSAIDs.

Phenylbutazone and flunixin meglumine were determined in equine plasma using electrochemical method coupled with MIPs.[33] Determination of level of NSAIDs in racehorses is important because of anti-doping control. SupelMIP® SPE-NSAIDs was used as the sorbent during execution of the MISPE procedure. This procedure was performed according to the commercial data sheet with modification. That is, acetonitrile (1 mL), methanol (1 mL), and a buffer solution (1 mL) were applied during conditioning step, then after the loading step (2 mL of equine plasma appropriately prepared), the cartridge was washed with water (1 mL), then consecutively dried, washed with the acetonitrile–water (40/60, *v*/*v*) solution (1 mL) dried, and twice eluted with the solution of acetic acid (1%) in an acetone–methanol (20/80, *v*/*v*) mixture. High recovery (95.2–98.7%) was obtained. So, the MISPE procedure could successfully be used for pretreatment of biological samples.

The same MISPE procedure was applied for evaluating the kinetics of elimination of phenylbutazone from plasma after its intravenous and oral administration in healthy horses.[34] The MISPE procedure was optimized in detail for the SupelMIP® SPE-NSAIDs sorbent during determination of acidic pharmaceuticals (clofibric acid, naproxen, diclofenac, and ibuprofen) in wastewater.[35] Different solutions for each step, different pH of the sample, and different breakthrough volume of MISPE were tested. The final procedure was very similar to that described above[33,34] with a difference in the washing step where additionally, a non-polar solvent, *viz.*, a mixture of toluene and heptane (40/60, *v*/*v*, 1 mL) was used. The breakthrough volume was 25 mL. The MISPE provided clean extracts, without any matrix effect, and it was employed before the LC-MS/MS analysis. As a result, both inter- and intra-day precision (below 10%) and recovery (84–100%) were high with low LOQs (below 8.0 ng L^{-1}).

Another example of MISPE application for NSAID analyses of environmental wastewater samples was performed by Gilart and co-workers.[36] They used the Affinilute MIP NSAIDs SPE sorbent. The procedure was the same as that described earlier[33,34] with higher volume of solutions (5 mL) because of the higher amount of sorbent used. Recovery for six acidic pharmaceuticals in effluent and influent wastewater was between 62 and 103%. A matrix-matched calibration curve was required to balance the suppression/enhancement effect. After the MISPE procedure was completed, tandem mass spectrometry showed a low LOD level (0.05–0.10 μg L^{-1}). The same authors compared the MISPE method for pretreatment of samples during NSAIDs analysis in wastewater (using the same sorbent as that described above) with SPE procedures employing three different commercially available sorbents; namely, Oasis HLB (hydrophilic-lipophilic balance), Oasis MAX (strong anion exchanger) and Oasis WAX (weak anion exchanger).[37] The MISPE procedures were optimized with respect to sample pH, washing and elution solvents as well as the maximum loading volume. After extraction, liquid chromatography coupled with ultraviolet (LC-UV) detection and LC-MS/MS analyses were developed for effluent wastewater. Only the extracts obtained using MIP were highly cleaned and did not show any matrix effect. The Affinilute MIP-NSAIDs SPE sorbent was chosen for the analysis of real samples, although recovery was slightly lower than that for other sorbents (63–85%). LODs were low for both these analytical techniques *i.e.*, they were 0.15–1.0 μg L^{-1} and 0.5–2.0 ng L^{-1} for LC-UV and for LC-MS/MS, respectively. Moreover, repeatability and reproducibility in consecutive days were high.

The next work was performed by Brogat and co-workers,[38] in which commercial MIP (SupelMIP® SPE-NSAIDs) and other sorbents (10 commercially available cartridges) were compared in the SPE/UV analysis of micropollutants in pure and in river water. The same SPE procedure was employed for all tested sorbents and the SupelMIP® SPE-NSAIDs appeared insufficiently selective under extraction conditions used because it adsorbed not only the NSAIDs drugs but also other substances, such as caffeine, atrazine, 1,7α-ethinylestradiol.

Another group of pharmaceuticals, which needs to be monitored in food of animal origin, includes β-agonists. The reason for that is their common use for growth promotion in cattle, which is banned in the European Union. The commercial MIPs were used for their determination. Blomgren and co-workers[39] applied the MIP4SPE column dedicated for clenbuterol for clenbuterol extraction from calf urine before HPLC-UV determination. After optimization, the MISPE procedure was as follows. The columns were conditioned consecutively with methanol (1 mL), water (1 mL), and the solution of 25 mM ammonium acetate pH = 6.7 (1 mL). Next, a 5 mL sample of pretreated urine was loaded. After that, the columns were washed with water (1 mL), the acetonitrile-(acetic acid) (98/2, *v*/*v*) solution (1 mL), 0.5 M ammonium acetate buffer, pH = 5.0 (1 mL), and 70% acetonitrile in water (1 mL). Elution was performed using the methanol–trifluoroacetic acid (99/1, *v*/*v*) solution (2 × 1 mL). Although selectivity of the tested sorbent was higher, recovery after reusing of columns was low. This difficulty was eliminated by using MIP cartridges only once. Very clean extracts obtained after MISPE made possible clenbuterol quantification down to 0.5 ng mL^{-1} by HPLC-UV. The MIP4SPE β-agonist columns were used for extraction of twelve different β-agonists from bovine muscles before LC-MS quantification.[40] The conditioning step of the MISPE procedure was the same as that described above. The washing step consisted of acetonitrile (1 mL), 0.5% acetic acid (1 mL), 50 mM ammonium acetate (1 mL) and an acetonitrile–water (60/40, *v*/*v*) mixture. For elution two portions (2 × 5 mL) of a methanol–(concentrated acetic acid) (90 : 10, *v*/*v*) were employed. Results showed that only eight pharmaceuticals (cimaterol, ractopamine, clenproperol, clenbuterol, brombuterol, mabuterol, mapenterol, and isoxsuprine) could be quantified by the tested method. It was impossible to measure the level of salbutamol and terbutaline, and for cimbuterol and zilpaterol quantitative determination was only possible. Van Hoof and co-workers[41] compared the MIP4SPE β-agonist columns with the non-imprinted commercial clean screen Dau (CSD) columns used for extraction and cleanup of urine sample before multi-residue LC-MS/MS analysis of β-agonists. Although the same sorbent was used, the washing step in the

MISPE procedure was not the same. The authors applied solutions of water (1 mL), 1% acetic acid in acetonitrile (1 mL), 50 mM ammonium acetate buffer (1 mL), 60% acetonitrile in water (1 mL). The results showed that cleanup using MIP is more selective than using CSD columns. The high selectivity of MIP4SPE β-agonists columns enabled determining concentration of salbutamol and terbutaline.

β-Blockers are another group of pharmaceuticals, which are one of the most frequently prescribed all over the world (they are used in hypertension treatment, suppression of cardiac arrhythmia, or heart failure), and they need to be controlled during anti-doping tests. The following commercial MIPs are available: SupelMIP® SPE-β-Blocker or MIP4SPE™ β-Blockers. Morante-Zarcero and Sierra[42,43] used SupelMIP® SPE-β-Blocker cartridges for treatment of river water samples before HPLC separation of β-blockers in comparative study using chiral HPLC stationary phases. The MISPE procedure was as follows. Methanol (1 mL) and water (1 mL) were used during a conditioning step, and then a 100 mL sample of river water was loaded onto the column. After that, cartridges were consecutively treated, as follows. They were consecutively washed with water (2 × 1 mL), dried, washed with acetonitrile (1 mL), dried, washed with dichloromethane (1 mL), dried and eluted with methanol which was 10% in acetic acid (2 × 1 mL). The authors optimized the flow rate and the breakthrough volume. Acceptable extraction recovery (89–113%) and precision (RSD less than 10%) were obtained. LODs (0.4–22.0 μg L^{-1}) made this method suitable for controlling of these pharmaceuticals in natural waters.

Comparative studies of MISPE using MIP4SPE™ β-Blockers and SPE using Oasis HLB cartridges were performed during development of an analytical procedure of determining eight β-blockers in waste waters.[44] The MISPE procedure was the same as that described above[42] with the only difference being the volume of the sample (25 mL). Although recovery reached for both sorbents was quite similar ranging from 40–110%, the MIP provided lower LODs (0.2–6.5 ng L^{-1}). Moreover, competitive selectivity tests and matrix effect studies revealed high selectivity and a lower matrix effect for MIP4SPE™ β-Blockers. Liquid chromatography–quadrupole-linear ion trap mass spectrometry was applied after MISPE for determination of the tested analytes.

In another work, Boonjob and co-workers[45] compared different sorbents for SPE of β-blockers. The authors tested seven different commercial sorbents including Oasis HLB, Bond Elut Plexa, Oasis MAX, Plexa PAX, Oasis MCX, Plexa PCX, and SupelMIP® SPE-β-Blocker. The MISPE procedure used was the same as that described earlier[42,44] with small differences in loading volume of the urine sample (5 mL), the volume of dichloromethane during washing step (0.5 mL) and the volume of solution used for elution (3 × 0.5 mL). Recovery for the SupelMIP® SPE-β-Blocker sorbent was ~95% for the standard solution and was higher than those for other sorbents. The MISPE strategy was confirmed to be the best and, therefore, chosen for analysis of real samples. The authors validated the analytical method. For determination of β-blockers in urine samples using HPLC-UV, recovery ranged from 94% to 106%, LOD ranged from 0.7 to 2.0 ng mL^{-1}, whereas the maximum concentrations of β-blockers in human urine during doping control could be on the level of 500 ng mL^{-1}.

Other pharmaceuticals banned by the European Union, but used as animal growth promotion factors, are steroid hormones. The problem with them is connected to their ability to interact with the endocrine system of humans and animals. Even small amounts present in the environment may cause problems with reproduction in humans and animals. Therefore, their concentration in water and environment samples needs to be monitored. For that, the following two commercial MIP sorbents are available, among others: Affinimip® SPE Estrogens and Affinimip® SPE Zearalenone. González-Sálamo and co-workers[46] compared effectiveness of both these commercial MIP sorbents in SPE of natural, synthetic, and mycoestrogens from environmental water samples before the analysis of liquid chromatography combined with mass spectrometry. Different MISPE procedures were applied. In both cases, the conditioning step was the same using acetonitrile (3 mL) and water (3 mL). After sample loading, water (3 mL) and the solution of acetonitrile–water (40/60, *v*/*v*, 3 mL) were used in the washing step for Affinimip® SPE Estrogens or a solution of water–acetonitrile–(acetic acid) (58/40/2, *v*/*v*/*v*) was employed in the washing step for Affinimip® SPE Zearalenone. The elution step was performed with methanol (5 mL) and methanol–(acetic acid) (98/2, *v*/*v*, 5 mL) for the Affinimip® SPE Estrogens and the Affinimip® SPE Zearalenone, respectively. Both sorbents could extract the majority of estrogenic compounds, however, with some exceptions. Affinimip® SPE Estrogens could not extract mycoestrogens, and Affinimip® SPE Zearalenone could not extract two estrogens from waste water samples. The recovery obtained for both sorbents exceeded 58% with RSD below 20% and LODs ranged between 0.01 and 0.44 μg L^{-1}. The cross-reactivity effect was lower for Affinimip® SPE Estrogens.

Lucci and co-workers[47] compared the MISPE procedure using Affinimip Estrogens SPE with the C18-SPE method in determination of seven natural and synthetic estrogens in aqueous samples before the LC-MS analysis. The MISPE protocol used was different than that described above.[46] The conditioning step was performed with acetonitrile (5 mL) and water (5 mL). The washing step employed a solution of water and acetonitrile (80/20, *v*/*v*, 4 mL), water (2 mL), acetonitrile (2 mL), and an acetonitrile–methanol solution (95/5, *v*/*v*, 2 mL), and the elution step was accomplished with methanol (3 × 1 mL). The recovery for the MIP (82–106%) was higher than that for C18. Moreover, low LODs (4.5–9.8 ng L^{-1}) and high repeatability were obtained for MIP.

Matějíček and co-workers[48] compared performance of a mixture of Affinimip® SPE Estrogens, Affinimip® SPE Bisphenol A, and Affinimip® SPE Phenolics sorbents with that of the Oasis WAX cartridge in multicomponent determination of endocrine-disrupting compounds in water and sediments. The authors concluded that SPE on the Oasis WAX sorbent is not suitable for quantitative analysis of target analytes without elimination of ion suppression which was possible when MIPs were used. Moreover, the influence of the sample volume was smaller in the case of MIPs. The same researchers successfully applied Affinimip® SPE Estrogens in the on-line MISPE coupled to LC-MS/MS for the determination of hormones in water and sediments.[49]

Another group of drugs which has to be controlled in environmental and biological samples are amphetamine derivatives. The reason for that is quite obvious because amphetamine-like compounds are widely abused and, very often, illicit drugs with stimulant and hallucinogenic properties. Kumazawa and co-workers[50] used SupelMIP™ SPE-Amphetamine cartridges in the SPE procedure of determination of seven different amphetamines in human whole blood before gas chromatography–mass spectrometry analysis. After optimization of the MISPE procedure, the conditions were, as follows. The cartridges were conditioned with methanol (1 mL), then equilibrated with 10 mmol L^{-1} ammonium acetate, pH = 8.0 (1 mL), washed with water (2 × 1 mL), an acetonitrile–water (60/40, *v*/*v*, 1 mL) solution, an acetic acid–acetonitrile (1/100, *v*/*v*, 1 mL) solution, eluted with a formic acid–methanol (1/100, *v*/*v*, 2 mL) solution. Recovery was high (89.1–100%) and the LOQ of 0.25–1.0 ng $(200\ \mu L)^{-1}$ was low. Therefore, this method could be a useful tool for determination of amphetamines in clinical and toxicological studies.

Wongniramaikul and co-workers[51] compared the MISPE procedure using the SupelMIP™ SPE-Amphetamine sorbent with the conventional SPE method using Super-select HLB cartridges applied for extraction of amphetamine and methamphetamine from river water. The target analytes were more retained on MIPs leading to a cleaner chromatogram. Moreover, MIPs revealed a higher breakthrough volume and a lower LOD. González-Mariño and co-workers[52] performed another comparative analysis of three different sorbents, namely, SupelMIP™ SPE-Amphetamine, Oasis HLB, and Oasis MCX. They tested amphetamine drug extraction from waste water for LC-MS/MS determination. Molecularly imprinted sorbent showed higher selectivity, lower matrix effect, lower LODs, as well as higher accuracy and precision than classical sorbents tested.

More examples of the use of commercially available MIPs dedicated to different pharmaceuticals involve quantification of aminoglycosides,[53] antidepressants,[54] riboflavin[55] and biogenic amines.[56,57] SupelMIP® SPE-Aminoglycosides cartridges were used for extraction and cleanup of honey samples before capillary electrophoresis tandem mass spectrometry determination of nine aminoglycosides.[53] SupelMIP® SPE-Antidepressant cartriges were applied for extraction of seven antidepressant drugs from environmental waters, and subsequent cleanup followed by ultra high-performance liquid chromatography coupled to triple quadrupole mass spectrometry.[54] The SupelMIP® SPE-Riboflavin sorbent was employed for determination of riboflavin in foodstuffs using the hyphenated MISPE-bead injection-lab-on-wave procedure. This step was performed for making the MISPE to serve as the front end to LC.[55]

Affinimip® SPE Catecholamines & Metanephines cartridges were applied for extraction of dopamine, 3-methoxytyramine, and serotonin from urine before capillary electrophoresis determination.[56] Luliński and co-workers[57] used Affinimip® SPE Catecholamines as a comparative sorbent for MIPs obtained by the authors for dopamine quantification in human urine. This study revealed low dopamine recovery (24.5%) for the commercial sorbent.

In the presented context, commercial molecularly imprinted sorbents can be used for rapid isolation and determination of pharmaceuticals important for human health protection and therapy. However, the SPE steps should be carefully optimized towards the target analytes and the matrices.

5.3 Inventions and Patents Concerning Molecularly Imprinted Sorbents

In this section, recent inventions and patents will be presented in relation to the pharmaceutical analysis and quantification of drugs using MIP technology. These inventions and patents have been dedicated to separation and determination of different classes of pharmaceuticals including antibiotics, sulfonamides, antihistamine drugs, alkaloids, tricyclic antidepressants, anti-inflammatory drugs, expectorants, antiretroviral medications, β_2-agonists, local anesthetics, and antimalarial drugs. Selected inventions disclosed in the last decade are presented below.

A vast majority of patents were devoted to the preparation of MIPs for separation or determination of antibiotics. The increase in prescription of antibiotics as well as the common use of antibiotics in animal husbandry has caused a significant presence of these drugs in human environment. In order to protect human health, restricted MRLs of drugs in foodstuff or environmental samples have been implemented in many countries across the world. Therefore, several inventions devoted to determination of antibiotics with the use of MIPs are described below.

Lei and co-inventors[58] patented a method for preparation and application of uniform MIP particles dedicated to the separation of chloramphenicol. In the synthetic process, the molecule of chloramphenicol served as the template and different reagents were claimed as functional monomers (4-vinylpiridyne or 1-allylpiperazine), cross linking monomers (ethylene glycol dimethacrylate-EGDMA- or divinylbenzene) and porogens (chloroform or ethyl acetate). Different dispersing agents (Tween 20, Tween 60, Tween 80, or polyvinyl alcohol) were applied in a continuous aqueous phase. The ratio of reagents was optimized. The invention disclosed the use of a microfluidic device made of glass or quartz coated by a layer with multiple holes that served as microchannels to facilitate the formation of emulsion. The use of this device ensured the formation of uniform MIP particles. The pre-polymerization solution, under inert gas atmosphere, was transferred with the predefined flow rate from the syringe pump to the microchannels, and then to the continuous phase to form the emulsion in the subsequent polymerization step.

In another invention, Guo and co-inventors[59] disclosed the method for preparation of a selective electrode dedicated to the determination of tetracycline. The electrode comprised a quartz tube of 1–3 mm in diameter and 20–50 mm in length containing molecularly imprinted monolith. The pre-polymerization solution contained tetracycline as the template, methacrylic acid as the functional monomer, EGDMA as the cross-linking monomer and a mixed solvent solution of methanol, toluene, and dodecanol as the porogens. The homogeneous degassed pre-polymerization solution was placed in the quartz tube before polymerization under UV light initiation. After the polymerization was finished, the glass tube with monolith inside was placed in the Soxhlet apparatus for 36 h to remove tetracycline in continuous extraction with methanol. Moreover, the invention disclosed a procedure of preparation of a homogeneous electroactive mixture attached to the lower end of the electrode rod. A platinum wire or Ag/AgCl electrode served as the internal reference electrodes.

He and co-inventors[60] patented a detecting device of a molecularly imprinted membrane and the method for preparation and detection of tetracycline. The detecting device was described as a reactor with the working and reference electrodes. The working electrode had a molecularly imprinted membrane for detection of tetracycline. Fabrication of the tetracycline molecularly imprinted membrane comprised the formation of the pre-polymerization solution of tetracycline (the template), styrene (the functional monomer), EGDMA (the cross-linking monomer), and methanol (the porogen), followed by the photopolymerization after ultrasound dissolving of all components.

Furthermore, Zhao and co-inventors[61] claimed the fabrication of an instrument for detection of tetracycline based on the imprinting technology. Bao and co-inventors[62] disclosed the preparation of gatifloxacin MIP sorbent. The invention was related to an organic polymer, which provided high selectivity and adsorption of gatifloxacin. The invention involved thermal precipitation polymerization of gatifloxacin (the template), methacrylic acid (the functional monomer), EGDMA (the cross-linking monomer) and chloroform (the porogen) to generate porous spherical particles of the specific surface area exceeding 90 $m^2 g^{-1}$. The maximum binding capacity was established for different polymer examples presented in this invention. For instance, particles prepared at the stoichiometry of the template to the functional monomer to the cross-linking monomer of 1 : 4 : 20 and 1 : 8 : 20 revealed maximum binding capacities of 246 and 292 $\mu mol\ g^{-1}$, respectively. This invention provided a simple method for preparation of low-cost uniform MIP microspheres with a large pore volume, high adsorption selectivity, and high absorption rate. The convenient and simple procedure patented did not require any tedious post-polymerization treatment. Thus, the invented precipitation polymerization of molecular imprinting microspheres could have broad application prospects.

An interesting invention was presented by Liu and co-inventors.[63] In their patent, a method of fabrication of a levofloxacin imprinted polymer was disclosed. The polymer was prepared by polymerization of methacrylic acid, divinylbenzene, and acrylamide in the presence of the levofloxacin template as well as surfactants and foaming agents. The limited scale-up capability was the main disadvantage of the polymerization commonly used in fabrication of imprinted polymers. Thus, the frontal polymerization propagation through the reaction vessel was disclosed. To initiate the front, an external electric iron was used. The polymerization was easily controlled and revealed a predictable polymer structure, low cost, and environmental safety.

Fang and co-inventors[64] patented a method for preparing an MIP for enrofloxacin. This invention described a surface molecular imprinting, which provided an inorganic material that was fabricated on activated silica beads. The reagents included also enrofloxacin (the template), *N*,*N*-dimethylformamide (the porogen) and tetraethyl orthosilicate (the cross-linking monomer). The polymerization was carried out using an aqueous ammonia solution at 46 °C for 14 h before post-polymerization treatment which involved rinsing with methanol, then with hydrochloric acid, and then with methylene chloride. Final template removal was carried out in a mixture of methanol and glacial acetic acid in the Soxhlet apparatus for 10–12 h. Other patent examples include inventions related to determination of valnemulin.[65] The invention involves methods of preparation of an MIP. The valnemulin antibiotic molecule has an octagonal cyclic carbon system and one unsaturated bond in its structure. To prepare an MIP for valnemulin determination, a valnemulin precursor was used as the template. The next patent described a method for determining traces of terramycin, in which structure of the polycarbocyclic system is present.[66] The invention comprises the use of oxytetracycline molecularly imprinted immunosensor.

Sulfonamides are antibacterial or bacteriostatic drugs. Application of sulfonamides in humans is nowadays declining. However, there are still many veterinary sulfonamides that are commonly used and very often misused in farming. Here, one example of an invention by Zhang and co-inventors[67] disclosed the preparation of a core–shell fluorescent imprinted material and application of this material for sulfanilamide determination. Preparation of this material involved synthesis of a fluorescent functional monomer, *viz.* lysine methyl ester. The first step involved protecting susceptible groups, and then reacting them with methacryloyl chloride, followed by purification to obtain the required monomer as a derivative of *N*-methacryloyl-lysine methyl ester. Next, the core–shell particles were prepared in the presence of ethanol, water, ammonia as well as tetraethyl orthosilicate and 3-(trimethoxysilyl) propyl methacrylate and the previously obtained monomer. The molecule of sulfamethyldiazine acted as the template as well as the structural analogue of targeted compound, sulfanilamide. Finally, the template was removed by

Soxhlet extraction with the methanol–(acetic acid) solution. Another invention claimed the method for purification of sulfanilamide drugs with an MIP prepared using a mixture of templates.[68]

An interesting invention by Liu and co-inventors[69] was dedicated to the preparation of an MIP monolithic column for separation of an antihistamine drug, cetirizine. The invention comprised a method for the fabrication of an MIP monolith. In this method, an ionic liquid, 1-butyl-3-methylimidazolium tetrafluoroborate (the porogen), cetirizine (the template), 4-vinylpyridine (the functional monomer) and EGDMA (the cross-linking monomer) were mixed with choline chloride to obtain a eutectic solvent. The polymerization was carried out in the presence of *N*,*N*-dimethylformamide inside a stainless-steel HPLC column. After the polymerization and subsequent template removal, the molecularly imprinted stationary phase served as the separation medium. Moreover, a reference non-imprinted monolith was prepared. By comparing retention factors for cetirizine standards on both imprinted and non-imprinted stationary phases, the imprinting factor was calculated. It was equal to 32, thus indicating very high selectivity and high ability of a new MIP material for cetirizine separation.

Other two inventions disclosed methods for preparation of MIPs dedicated to the determination of alkaloids, strychnine,[70] and reserpine.[71] Polymerization methods claimed were devoted to the fabrication of imprinted stationary phases for liquid chromatography separations. The inventions disclosed the use of allyl-β-cyclodextrin as the functional monomer and strychnine or reserpine as the templates. EGDMA was selected as the cross-linking monomer and dimethyl sulfoxide acted as the porogen. The bulk polymerization technique was employed. After post-polymerization grinding and sieving, the MIP particles were wet packed into the chromatographic column. Moreover, the control non-imprinted polymers were prepared. The methods allowed using dual recognition abilities of both β-cyclodextrin and the MIP.

Zhang and co-inventors[72] patented a method for selective separation of an antiepileptic carbamazepine drug and preparation of a magnetic MIP used for that purpose. Carbamazepine is widely used in treatment of neurological disorders. It is often detected in analysis of municipal sewage. The presence of carbamazepine in the environment can bring some potential risk for human health. Therefore, new advanced materials for separation of carbamazepine are devised to enable determining this compound in complex environment samples. The synthesis protocol involved the application of chitosan (sufficiently soluble in acetic acid solution) and magnetic nanoparticles to form magnetic-shell polysaccharide nanoparticles. Then, the carbamiazepine template was imprinted by precipitation polymerization using acrylic acid or methacrylic acid or methyl methacrylate or 2- and 4-vinylpyridine as functional monomers as well as EGDMA or divinylbenzene or pentaerythritol acrylate as cross-linking monomers. The disclosed polymerization was carried out using acetonitrile or methanol or toluene or methylene chloride or chloroform or carbon tetrachloride. Polyvinylpyrrolidone was added as the dispersing agent. The invented method could be used for fast, easy and selective separation of carbamazepine on the magnetic MIP nanoparticles from complex environmental samples by simple application of the external magnet.

The next interesting example of an invention of Wei and co-inventors[73] disclosed a method for preparation and fluorescent determination of aspirin by using a quantum-dot imprinted chemosensor. This method involved the following steps. (i) Preparation of a solution of the NaHTe precursor from sodium borohydride, tellurium powder and water under sonication. (ii) Injection of the NaHTe precursor solution into a cadmium chloride solution under inert gas atmosphere in the presence of thioglycollic acid. (iii) Reacting the mixture prepared at 100–110 °C to obtain quantum dots of different sizes governed by different backflow time intervals. (iv) A film was deposited on the surface of quantum dots to act as an element recognizing aspirin. This MIP was synthesized by a sol–gel method from (3-aminopropyl) triethoxysilane as the functional monomer and tetraethyl orthosilicate as the cross-linking agent in the presence of aspirin as the template. In summary, importantly, the MIP-QD system revealed high optical and pH stability, and an aspirin selective-identification function.

Different other examples of inventions can be found. Claims involve both preparation and applications of MIPs. Amongst them, there are method for preparation of MIPs dedicated to separation of the fudosteine expectorant drug,[74] the lamivudine antiviral agent,[75] the salbutamol β_2-adrenergic receptor agonist,[76] the lidocaine local anesthetic,[77] and the artemisin antimalarial drug.[78] To enhance the imprinting effect in their invention, Cao and co-inventors disclosed[78] a method for preparation of silica gel particles modified by calix[4]arene to produce a surface with artemisim imprinted polymer.

Summing up, the recently disclosed inventions presented in this section revealed a vast range of applications of MIPs as advanced separators or selective chemosensors for quantification of pharmaceuticals. The methods developed for preparation of imprinted materials show their unique utility and versatility as advanced determination tools in modern analytical chemistry.

5.4 Conclusions

The commercial molecularly imprinted sorbents are very useful for effective isolation of pharmaceuticals from the complex matrices before their determinations by various detectors. However, MISPE procedures should be carefully optimized towards the target analytes in appropriate matrices. There are still a lot of challenges which can be taken up by researchers working on MIPs including selectivity of sorbents, simplicity of procedures, binding capacity of polymers, controlled release of analyte from polymer matrices, and new formats of polymers.

List of Abbreviations

EGDMA	Ethylene glycol dimethacrylate
HPLC	High-performance liquid chromatography
HPLC-UV	High-performance liquid chromatography coupled with ultraviolet spectroscopy detection
LC-MS/MS	Liquid chromatography-tandem mass spectrometry
LC-UV	Liquid chromatography coupled with ultraviolet spectroscopy
LLE	Liquid–liquid extraction
LOD	Limit of detection
LOQ	Limit of quantification
MIP	Molecularly imprinted polymer
MISPE	Molecularly imprinted solid-phase extraction
MRL	Maximum residue limit
NSAID	Nonsteroidal anti-inflammatory drug
QuEChERS	Quick, easy cheap, effective, rugged and safe
RSD	Relative standard deviation
SPE	Solid-phase extraction

References

1. D. E. Raynie, *Anal. Chem.*, 2010, **82**, 4910.
2. S. A. Barker, *J. Biochem. Biophys. Methods*, 2007, **70**, 151.
3. F. David and P. Sandra, *J. Chromatogr. A*, 2007, **1152**, 54.
4. M. V. Polyakov, P. M. Stadnik, M. V. Patrizkii, I. M. Malkin and F. S. Dukhina, *Zh. Fiz. Khim.*, 1933, **4**, 454.
5. M. V. Polyakov, L. Kuleshina and I. Neimark, *Zh. Fiz. Khim. Akad. SSSR*, 1937, **10**, 100.
6. F. H. Dickey, *Proc. Natl. Acad. Sci. U. S. A.*, 1949, **35**, 227.
7. G. Wulff, A. Sarhan and K. Zabrocki, *Tetrahedron Lett.*, 1973, **44**, 4329.
8. G. Wulff, *ACS Symp. Ser.*, 1986, **308**, 186.
9. G. Wulff, *Microchim. Acta*, 2013, **180**, 1359.
10. L. Andersson, B. Ekberg and K. Mosbach, *Tetrahedron Lett.*, 1985, **26**, 3623.
11. B. Sellergren, B. Ekberg and K. Mosbach, *J. Chromatogr.*, 1985, **347**, 1.
12. B. Sellergren, M. Lepisto and K. Mosbach, *J. Am. Chem. Soc.*, 1988, **110**, 5853.
13. M. J. Whitcombe, M. E. Rodriguez, P. Villar and E. N. Vulfson, *J. Am. Chem. Soc.*, 1995, **117**, 7105.
14. G. Wullf and A. Sarhan, US Pat. 4127730 A, 1978.
15. D. S. Spivak, *Adv. Drug Delivery Rev.*, 2005, **57**, 1779.
16. V. Pichon and F. Chapui-Hugon, *Anal. Chim. Acta*, 2008, **622**, 48.
17. J. Tan, Z.-T. Jiang, R. Li and X.-P. Yan, *Trends Anal. Chem.*, 2012, **39**, 207.
18. J. Matsui, T. Kato, T. Takeuchi, M. Suzuki, K. Yokoyama, E. Tamiya and I. Karube, *Anal. Chim. Acta*, 1993, **65**, 2223.

19. M. Diaz-Bao, R. Barreiro, J. M. Miranda, A. Cepeda and P. Regal, *Chromathography*, 2015, **2**, 79.
20. M. Lombardo-Agüí, A. M. García-Campaña, L. Gámiz-Gracia and C. Cruces Blanco, *J. Chromatogr. A*, 2010, **1217**, 2237.
21. K. Kaur, S. Singh Saini, A. Kumar Malik and B. Singh, *Spectrochim. Acta, Part A*, 2012, **96**, 790.
22. E. Rodriguez, M. C. Moreno-Bondi and M. D. Marazuela, *Food Chem.*, 2011, **127**, 1354.
23. S. Lee, B. Kim and J. Kim, *J. Chromatogr. A*, 2013, **1277**, 35.
24. M. Lombardo-Agüí, L. Gámiz-Gracia, C. Cruces Blanco and A. M. García-Campaña, *J. Chromatogr. A*, 2011, **1218**, 4966.
25. C. Blasco and Y. Picó, *J. Agr. Food Chem.*, 2012, **60**, 11005.
26. D. Moreno-González, F. J. Lara, L. Gámiz-Gracia and A. M. García-Campaña, *J. Chromatogr. A*, 2014, **1360**, 1.
27. H.-B. Zheng, J.-Z. Mo, Y. Zhang, Q. Gao, J. Ding, Q.-W. Yu and Y.-Q. Feng, *J. Chromatogr. A*, 2014, **1329**, 17.
28. M. Gaugain, M.-P. Chotard, D. Hurtaud-Pessel and E. Verdon, *J. Chromatogr. B*, 2016, **1011**, 145.
29. M. Rejtharová and L. Rejthar, *J. Chromatogr. A*, 2009, **1216**, 8246.
30. R. Mohamed, J. Richoz-Payot, E. Gremaud, P. Mottier, E. Yilmaz, J.-C. Tabet and P. A. Guy, *Anal. Chem.*, 2007, **79**, 9557.
31. B. Boyd, H. Björk, J. Billing, O. Shimelis, S. Axelsson, M. Leonora and E. Yilmaz, *J. Chromatogr. A*, 2007, **1174**, 63.
32. X. Q. Li, Z. Yang, Q. H. Zhang and H. M. Li, *Anal. Chim. Acta*, 2014, **807**, 75.
33. V. Meucci, M. Vanni, M. Sgorbini, R. Odore, M. Minunni and L. Intorre, *Sens. Actuators, B*, 2013, **179**, 226.
34. V. Meucci, G. Luci, M. Vanni, M. Sgorbini and L. Intorre, *J. Equine Vet. Sci.*, 2015, **35**, 906.
35. S. Zorita, B. Boyd, S. Jönsson, E. Yilmaz, C. Svensson, L. Mathiasson and S. Bergström, *Anal. Chim. Acta*, 2008, **626**, 147.
36. N. Gilart, R. M. Marcé, N. Fontanals and F. Borrull, *Talanta*, 2013, **110**, 196.
37. N. Gilart, R. M. Marcé, F. Borrull and N. Fontanals, *J. Sep. Sci.*, 2012, **35**, 875.
38. M. Brogat, A. Cadiere, A. Sellier, O. Thomas, E. Baures and B. Roig, *Microchem. J.*, 2013, **108**, 215.
39. A. Blomgren, C. Berggren, A. Holmberg, F. Larsson, B. Sellergren and K. Ensing, *J. Chromatogr. A*, 2002, **975**, 157.
40. P. R. Kootstra, C. J. P. F. Kuijpers, K. L. Wubs, D. van Doorn, S. S. Sterk, L. A. van Ginkel and R. W. Stephany, *Anal. Chim. Acta*, 2005, **529**, 75.
41. N. van Hoof, D. Courtheyn, J.-P. Antignac, M. Van de Wiele, S. Poelmans, H. Noppe and H. De Brabander, *Rapid Commun. Mass Spectrom.*, 2005, **19**, 2801.
42. S. Morante-Zarcero and I. Sierra, *J. Pharm. Biomed. Anal.*, 2012, **62**, 33.
43. S. Morante-Zarcero and I. Sierra, *Chirality*, 2012, **24**, 860.

44. M. Gros, T.-M. Pizzolato, M. Petrović, M. J. López de Alda and D. Barceló, *J. Chromatogr. A*, 2008, **1189**, 374.
45. W. Boonjob, H. Sklenářová, F. J. Lara, A. M. García-Campaña and P. Solich, *Anal. Bioanal. Chem.*, 2014, **406**, 4207.
46. J. González-Sálamo, B. Socas-Rodríguez, J. Hernández-Borges, M. del Mar Afonso and M. A. Rodríguez-Delgado, *J. Sep. Sci.*, 2015, **38**, 2692.
47. P. Lucci, O. Núñez and M. T. Galceran, *J. Chromatogr. A*, 2011, **1218**, 4828.
48. D. Matějíček, A. Grycová and J. Vlček, *J. Sep. Sci.*, 2013, **36**, 1097.
49. D. Matějíček, J. Vlček, A. Burešová and P. Pelcová, *J. Sep. Sci.*, 2013, **36**, 1509.
50. T. Kumazawa, C. Hasegawa, K. Hara, S. Uchigasaki, X.-P. Lee, H. Seno, O. Suzuki and K. Sato, *J. Sep. Sci.*, 2012, **35**, 726.
51. W. Wongniramaikul, A. Choodum, L. Dennany and N. N. Daeid, *J. Sep. Sci.*, 2012, **35**, 3332.
52. I. González-Mariño, J. B. Quintana, I. Rodríguez, R. Rodil, J. González-Peñas and R. Cela, *J. Chromatogr. A*, 2009, **1216**, 8435.
53. D. Moreno-González, F. J. Lara, N. Jurgovská, L. Gámiz-Gracia and A. M. García-Campaña, *Anal. Chim. Acta*, 2015, **891**, 321.
54. K. Demeestere, M. Petrović, M. Gros, J. Dewulf, H. van Langenhove and D. Barceló, *Anal. Bioanal. Chem.*, 2010, **396**, 825.
55. H. M. Oliveira, M. A. Segundo, J. L. F. C. Lima, M. Miró and V. Cerdà, *Anal. Bioanal. Chem.*, 2010, **397**, 77.
56. B. Claude, R. Nehmé and P. Morin, *Anal. Chim. Acta*, 2011, **699**, 242.
57. P. Luliński, M. Bamburowicz-Klimkowska, M. Dana, M. Szutowski and D. Maciejewska, *J. Sep. Sci.*, 2016, **39**, 895.
58. J. Lei, G. Ma, Z. Su, X. Kou and Y. Zhai, Cn. Pat. 102336871 A, 2013.
59. Z. Guo, P. Gai, H. Zhang, J. Duan, S. Wang and D. Wei, Cn. Pat. 102116758 A, 2011.
60. F. Wang, S. He, X. Wang and X. Liao, Cn. Pat. 101650335 A, 2010.
61. H. Zhao, H. Wang, X. Quan and Y. Xie, Cn. Pat. 201965114 A, 2011.
62. L. Bao, J. Fan, Y. Gao, C. Jin, X. Su, H. Wang, J. Wang, Z. Xie and G. Zhu, Cn. Pat. 101829550 A, 2010.
63. X. Liu, X. Li, Y. Huang and Z. Liu, Cn. Pat. 102212160 A, 2011.
64. G. Fang, M. Pan, J. Wang and S. Wang, Cn. Pat. 101747506 A, 2010.
65. L. He, H. Guo, B. Fang, Y. Liu, J. Zhang, R. Liu and L. Chen, Cn. Pat. 101921370 A, 2010.
66. F. Jiang, J. Li and X. Wei, Cn. Pat. 101710116 A, 2012.
67. H. Zhang, B. Yu, J. Zhang, Y. Yang and K. Zhang, Cn. Pat. 104277190 A, 2015.
68. S. Song, Z. Zhao and C. Yao, Cn. Pat. 102344527 A, 2012.
69. Z. Liu, X. Sun, X. Li and Y. Huang, Cn. Pat. 104209104 A, 2014.
70. D. Chen, Cn. Pat. 102443119 A, 2012.
71. D. Chen, Cn. Pat.102504142 A, 2012.
72. Y. Zhang, C. Dai, J. Zhang, Z. Gao, X. Zhou and S. Liu, Cn. Pat. 102827321 A, 2013.

73. X. Wei, J. Dai, H. Li, L. Gao, J. Wang, Z. Yu, Z. Zhou and Y. Yan, Cn. Pat. 104165874 A, 2014.
74. D. Liu, F. Yan, J. Dong, L. Dai and A. Li, Cn. Pat. 102060958 A, 2013.
75. M. Tuerhong, H. Zeng and K. Nuermaimaiti, Cn. Pat. 102008946 A, 2011.
76. M. Yang, R. Shen, X. Chu and D. Wang, Cn. Pat. 101962420 A, 2014.
77. M. Tuerhong and C. Liu, Cn. Pat. 101905150 A, 2011.
78. X. Cao and B. Yang, Cn. Pat. 101831076 A, 2012.

CHAPTER 6

Micro and Nanofabrication of Molecularly Imprinted Polymers

FRANK BOKELOH[a,b], CÉDRIC AYELA*[b] AND KARSTEN HAUPT*[a]

[a]Sorbonne Universités, Université de Technologie de Compiègne, CNRS Laboratory for Enzyme and Cell Engineering, Rue Roger Couttolenc, CS 60319, 60203 Compiègne, France; [b]Laboratoire de L'Intégration du Matériau au Système, Université de Bordeaux, 351 Cours de la Libération, 33405 Talence cedex, France
*E-mail: karsten.haupt@utc.fr, cedric.ayela@ims-bordeaux.fr

6.1 Introduction

Affinity technology, *i.e.*, the use of molecular recognition for the selective and sensitive detection of target molecules for different applications, is a field of great interest in the scientific community. Antibodies and enzymes are therefore intensively studied objects for selective target recognition with high affinity. An alternative solution is offered by molecularly imprinted polymers (MIPs).[1,2] These synthetic receptors, often referred to as artificial antibodies,[3] are synthesized by copolymerizing functional and cross-linking monomers around a template molecule. After template extraction, the polymer matrix comprises imprinted cavities bearing recognizing sites which are complementary in size and shape to the template molecule (Scheme 6.1).[4] These

Polymer Chemistry Series No. 28
Molecularly Imprinted Polymers for Analytical Chemistry Applications
Edited by Wlodzimierz Kutner and Piyush Sindhu Sharma

Published by the Royal Society of Chemistry, www.rsc.org

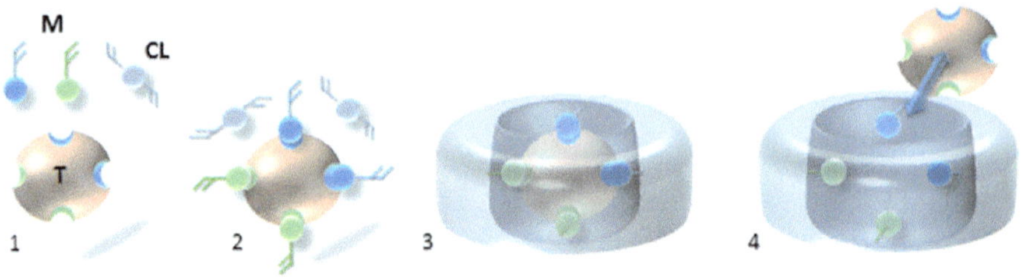

Scheme 6.1 General principle of molecular imprinting. A template molecule (T), functional monomers (M) and cross-linking monomers (CL) initially form a pre-polymerization complex by self-assembly (1,2), which is then polymerized (3), resulting in a polymer with molecularly imprinted cavities, complementary to the template in terms of size, shape, and position of recognizing sites, *e.g.*, functional groups. Target molecules can selectively bind to these sites (4).

artificial biomimetic receptors show, unlike their natural counterparts, a superior chemical and physical stability, which facilitates their storage and handling, but also their engineering and integration into standard processes. A large variety of different targets can be addressed by MIPs and they can be regenerated for potential reuse. These advantages make MIPs an interesting tool for many applications, such as solid-phase extraction, antibody mimics in immunoassays,[5,6] controlled drug release,[7–9] protection of molecules from further modifications[10] and recognition elements in biosensors.[11–14]

Especially for the latter application, MIPs offer an interesting alternative, because these polymers are compatible with a wide range of microfabrication methods and, therefore, can overcome limitations associated with less stable biological recognition elements. During the past decades, several techniques of micro-structuring and fabrication of MIPs have been developed, which are reviewed in the following section.

6.2 Methods of MIP Microfabrication

The development of modern micro- and nanotechnology occurred in surprisingly short periods of time. In 1975 Gordon E. Moore described for computer technology a doubling of density of transistors in an integrated circuit every two years.[15] This trend refers today as "Moore's law" and became possible by improving existing and developing new microfabrication techniques. The trend of miniaturisation was not only developed for computer technology but in many other fields as well, such as chemical sensors and biochips, where microfabrication techniques of polymers play a key role. The most popular fabrication scheme is probably photolithography, which uses light to cure a defined pattern of polymeric photoresist.[16,17] But also other technologies have been developed for the defined and localized deposition of polymers and can be classified into electrical methods, optical methods including photolithography or laser-based structuring of polymers, and mechanical patterning techniques, such as soft-lithography and printing techniques.

In this context MIPs offer a great potential as sensitive layer for biochips or sensing elements since their polymer matrix is accessible to these modern microfabrication techniques. During the past decades, sensing schemes based on MIPs with defined structures at the micro- and nanoscale have been developed thanks to the use of modern microfabrication tools or new approaches.

6.2.1 Electropolymerization

Electropolymerization of MIPs was introduced in the late 90s,[18,19] as an alternative approach to the traditionally employed free radical polymerization. This method gave, for the first time, high control over film thickness and morphology, which is governed by the amount of transferred charges and composition of the electrolyte/solvent system in which the polymerization is carried out. For the development of MIPs used as the recognition element in sensing applications, this method offers an interesting alternative to the direct deposition and patterning of thin films on the transducer surface with high mass transfer and accessibility.[20]

The development of electrochemical synthesis of MIPs can be seen as a parallel development in the field of MIPs, since traditionally used vinyl and acrylic monomers are not electropolymerizable and hence have to be replaced by suitable electroactive monomers. The resulting polymer films can be classified in conducting and non-conducting films. Among the conducting polymers polypyrrole,[18,21–23] polyaniline,[24,25] polythiophene,[26,27] polycarbazole,[20] and their derivatives are well studied. For non-conducting films *o*-phenylenediamine, phenol[28,29] and over-oxidized pyrrole[18] are well established monomers for electrosynthesized MIPs. In addition, the design of novel electroactive monomers with functional groups for selective target recognition has made considerable progress.

The method is based on the generation of free radical species by oxidizing monomers on an electrode due to an applied potential. These free radical species adsorb strongly on the electrode surface and subsequently form a polymer network.[30] By adjusting the current or potential, thin polymer films of controlled thickness and density can easily be deposited,[31] which has special interest in the field of MIPs and potential sensing application.[18,19,28] Detailed review articles about electropolymerization of MIPs have been published by the groups of Kutner and Piletsky.[32,33]

6.2.2 Optical Methods

Photoinduced polymerization of MIPs is probably the most versatile and popular method to pattern defined micro- and nanostructures. It opens the field to many modern fabrication methods starting from standard microfabrication techniques, such as photolithography,[34,35] to more advanced technologies, such as two-photon-stereolithography.[36]

One simple approach for generating thin films of controlled thickness and morphology can be achieved by spin-coating a precursor mixture on a substrate, followed by polymerization. This fabrication method was not only applied directly to sensing schemes[37] but is also the initial fabrication step for further structuring by UV-photolithography, a technology which enables, due to parallel processing, large-scale fabrication of micro devices. For standard UV-photolithography, a photoresist is spin coated on a substrate, baked and placed under a mask aligner to transfer a geometric pattern from a photomask to the photoresist. In 2002, Byrne *et al.*[38] adapted this microfabrication technique for the photo-patterning of a polymer gel containing poly(ethylene glycol) dimethacrylate and acrylamide imprinted with D-glucose. Different geometrical configurations at the micro-scale with a thickness of ~13 μm were achieved in this study. The first MIP based chemosensor fabricated by photolithography was introduced by Huang *et al.*[39] In this study, a sensitive MIP layer (directed against albuterol) with a feature size of 50 μm was patterned using photolithography on a gold electrode. This approach of combining MIPs with photolithography was achieved by using a chemically modified photoresist instead of commonly used vinyl and acrylic functional monomers. In another approach combining photolithography and MIPs, the same group patterned acrylic monomers on a three-electrode cell.[40] For this study, an MIP pattern made of copolymerized benzyl methacrylate methacrylic acid and 2-hydroxyl ethyl methacrylate with a feature size of 20 μm could be fabricated. In 2009, Guillon and co-workers[41] proposed the use of photomasks for the direct large scale fabrication of MIP micro-biochips. Different shapes, such as lines, spirals, circle matrices, squares, or hexagons, were patterned on a standard 4-inch silicon wafer with resolution reaching down to 1.5 μm. Fluorescent model analytes were used in this study to characterize the binding properties of polymerized micro structures. UV photolithography was used as well to fabricate a sandwich type MIP structure.[42] In an initial step, a thin MIP film was fabricated on a substrate. On top of this first layer, another MIP film with a periodic grid pattern was fabricated by UV-photolithography. Different MIPs were directly compared by this fabrication method.

As an alternative to photolithography, projection lithography is, as the name suggests, projecting an image from a photomask to the photoresist assisted by an optical system. Appreciable resolutions can be achieved while a direct contact between mask and resist is avoided. The resolution of this setup is limited by diffraction of the light at the edges of the photomask. Linares *et al.*[43] have developed a custom made projection lithography setup for MIP patterning. Concretely, the field diaphragm of a fluorescence microscope was replaced by a mask with 0.5 mm squared features (pitch 2 mm). The system used the excitation light source of the microscope and reduced the pattern size by the magnification factor of the objective. MIP patterns were fabricated using a 10× objective resulting in a feature size of 70 μm. Furthermore, the system was used to polymerize an MIP precursor on a nanoporous aluminium oxide membrane, resulting in a microarray of dots consisting of nanofilaments 150 nm in diameter and 4 μm long. Thus, the surface-to-volume aspect ratio of these structures was drastically increased, improving the accessibility to the molecular cavities bearing recognizing sites for the imprinted analyte (myoglobin). Mask-less lithography methods including electron beam lithography, microstereolithography and two-photon-stereolithography have in general a lower throughput than that of the described UV-photolithography systems where a photomask is used, enabling parallel processing of multiple devices. However, mask-less optical lithography methods are excellent tools for rapid prototyping at the micro and nano scales, since the patterns are directly written with a focused beam and no external mask is needed. Electron beam lithography is probably the most popular maskless method, which can generate arbitrary patterns with a nanometre resolution. This extraordinary resolution can be achieved by scanning a focused electron beam over the surface of an electron-sensitive photoresist. Carrasco *et al.*[44] have used electron beam lithography for synthesizing an MIP, with a sub-micrometre resolution (Figure 6.1), targeting the rhodamine model template.

In this study, the MIP consisted of a linear copolymer, which contained recognition groups for complexing the target molecule and triggerable cross-linkable groups. The material showed, depending on the electron dose, dual-tone behaviour. That is, negative-tone behaviour was reported for low doses while for high doses the copolymer showed positive-tone characteristics (*i.e.* irradiated regions are removed after development).

Another example of sequential writing of MIPs is micro-stereolithography that uses a sharply focused laser beam to induce local photopolymerization and enables the fabrication of 3-D micro-structures. For the synthesis of MIPs by this method, Conrad *et al.*[45] achieved 3-D microstructures by polymerizing layer-by-layer an MIP precursor mixture with 9-dansyl adenine used as the template molecule. Design and architecture of the pattern remain limited because the focal point of a laser cannot be focused enough to fabricate real 3-D structures with under cuts, thus the structures are referred to as 2.5-D architectures. This limitation was recently overcome by Gomez *et al.*,[36] using a two-photon stereolithography set-up comprising a microscope and a femtosecond laser to synthesize real 3-D micro- and nanoarchitectures based on MIPs. The polymerization in this work was initiated by a two-photon absorption, which is a highly non-linear process and further amplifies the nonlinear chemical response of the photoresist during polymerization. These two effects together induce photopolymerization on a very confined focal point (typically less than 1 μm^3) in a 3-D volume (voxel) (Figure 6.3). An interesting aspect in this work was that polymerization was performed in the absence of additional solvents, which are used for standard bulk polymerization as a porogen to guarantee access to the imprinted cavities bearing recognizing sites but can interfere with both template and functional monomers. Thus, polymerization of small voxels seemed to generate a sufficient porosity that enhanced the binding properties of the MIP. Using this technology, MIP structures with

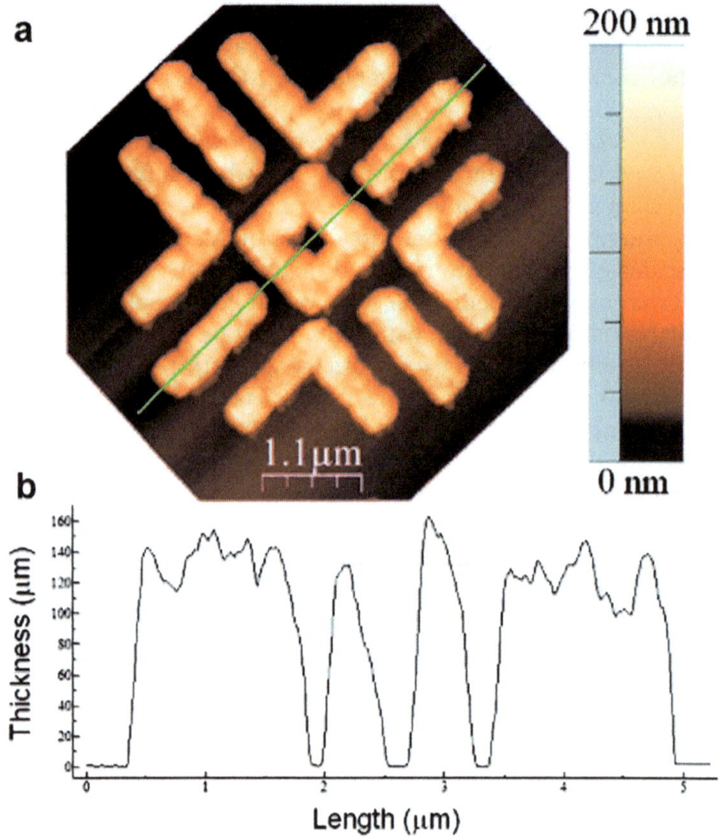

Figure 6.1 (a) AFM image of an MIP nanostructure fabricated by electron-beam lithography (EBL). Electron radiation dose: 750 mC cm^{-2} (positive tone resist). (b) Cross-sectional profile along the green line shown in the top image. The RMS value equals 30 nm. Reproduced from ref. 44 with permission from The Royal Society of Chemistry.

a line width of 250 nm could be achieved. Furthermore, this technology paves the way for the direct writing of new sensing devices, since shapes are not limited anymore to 2- or 2.5-D structures but allows for a single-step fabrication of real 3-D architectures. One example was given by fabricating free-standing cantilever microelectromechanical system (MEMS) sensors (Figure 6.2). These devices measured 20 × 60 μm^2 and were written in a single step on a glass substrate, including their clamped base. Detection of the target analyte by these gravimetric sensors was than evaluated by analysing the resonance frequency, which shifts due to mass variations (analyte binding or extraction) of the system.

Other mask-less methods to photopattern MIPs were reported by Fuchs *et al.*[46,47] using holographic techniques. These allow fabrication of diffractive optical structures, which can be integrated in label-free sensor elements.

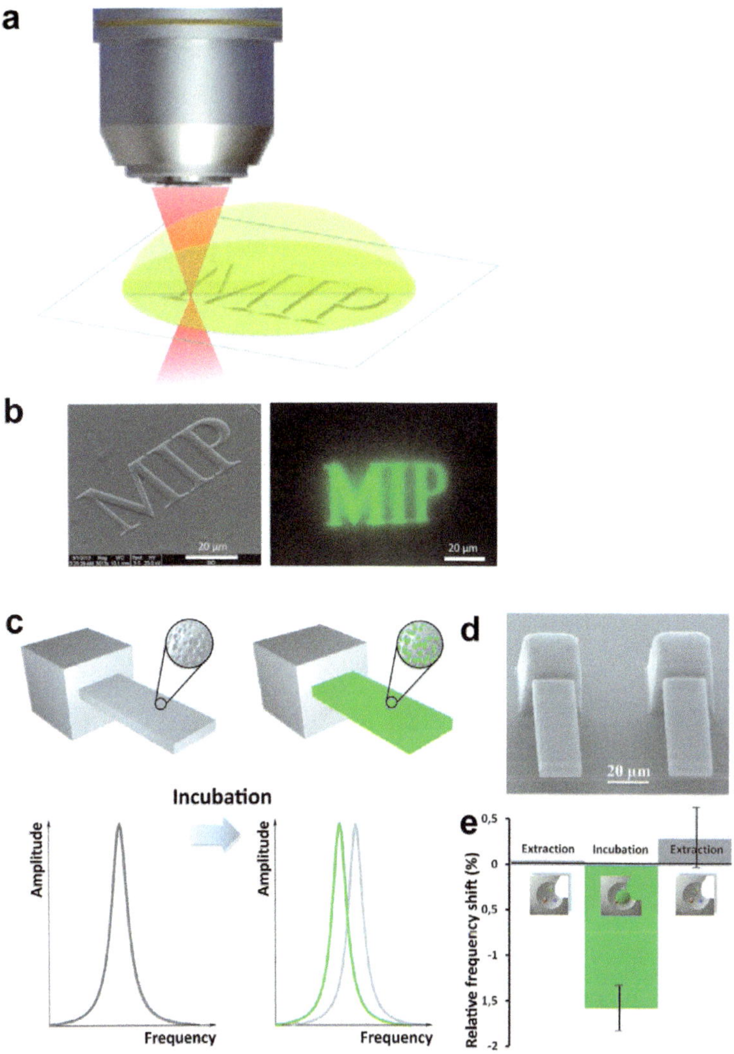

Figure 6.2 (a) Schematic representation of the direct laser writing of MIPs, by displacing an NIR femtosecond laser focused beam inside the MIP precursor solution. (b) Examples of MIP microstructures imaged by SEM and fluorescence microscopy after binding a fluorescent derivative of the template (dansyl-L-Phe). (c) Schematic representation of the MIP cantilever fabricated by two-photon stereolithography and of the resonance frequency variation before and after contacting the target Z-L-Phe solution. (d) SEM images of two MIP cantilevers. (e) A relative resonance frequency shift of the cantilevers after extraction, contacting Z-L-Phe and second extraction. Error bars correspond to the average of four measurements performed on four different microcantilevers. Adapted from L. P. C. Gomez, A. Spangenberg, X.-A. Ton, Y. Fuchs, F. Bokeloh, J.-P. Malval, B. Tse Sum Bui, D. Thuau, C. Ayela, K. Haupt and O. Soppera, Rapid Prototyping of Chemical Microsensors Based on Molecularly Imprinted Polymers Synthesized by Two-Photon Stereolithography, *Adv. Mater*,[36] John Wiley and Sons, © 2016 WILEY-VCH Verlag GmbH & Co. KGaA, Weinheim.

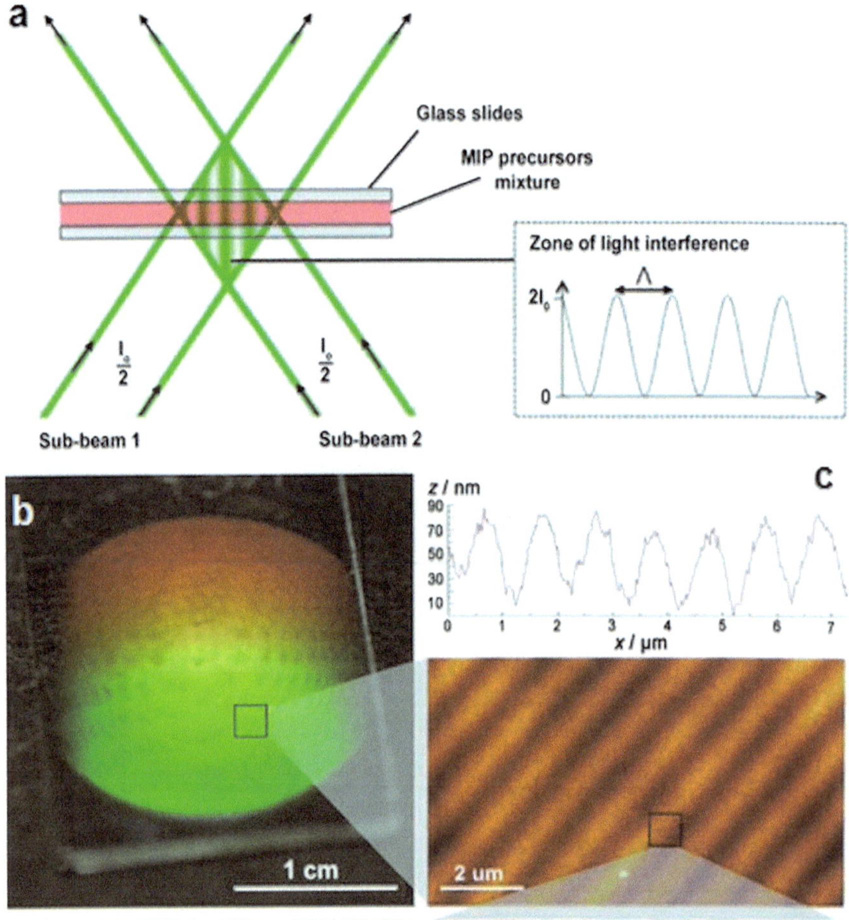

Figure 6.3 (a) Scheme showing the *in situ* MIP microstructuring by photopolymerization with two interfering laser beams at 532 nm. (b) Holographic MIP film supported on a glass slide. (c) the AFM image of the surface of a holographic MIP film (10 µm). Reproduced from Y. Fuchs, O. Soppera, A. G. Mayes and K. Haupt, Holographic Molecularly Imprinted Polymers for Label-Free Chemical Sensing, Adv. Mater,[46] John Wiley and Sons, © 2016 WILEY-VCH Verlag GmbH & Co. KGaA, Weinheim.

In a first approach,[46] a metal-less hologram was generated directly during the synthesis of the MIP by interference lithography (Figure 6.3). This technique combines two coherent laser beams in order to create a stable pattern of light through interference. As a consequence, a sinusoidal light intensity distribution facilitates an inhomogeneous photopolymerization of the MIP precursor solution, thus generating a periodic polymer pattern, which was used as a diffraction sensing element. Moreover, the same authors fabricated silver-halide holographic structures to generate reflection holograms based on MIPs.[47] Here, a homogeneous film of an MIP for testosterone was first deposited on a flat substrate by photopolymerization. Next, silver bromide was infused into the MIP film with a photosensitizer, and reflection holograms were recorded by passing a collimated 532 nm laser beam through the film backed by a mirror. Testosterone binding was then followed *via* wavelength changes of the holographic reflection peak as a function of testosterone concentration.

Evanescent fields are non-propagating electromagnetic waves, characterized by an exponential decay.[48] For many applications, *e.g.* surface plasmon resonance (SPR), near-field scanning optical microscopy (NSOM/SNOM), attenuated total reflection spectroscopy (ATR), the interaction of this field with matter is of particular interest because optical interactions are on sub-wavelength scales and entail physical properties different from propagating electromagnetic waves.[49] Evanescent fields appeared suitable for polymerization as well.[50,51] An evanescent field is generated on a glass surface by total reflection of an incident laser beam, which polymerises a photosensitive monomer mixture. The thickness of the resulting polymer film was determined by the penetration depth of the evanescent field. Fuchs *et al.*[52] adapted this method and photopatterned a MIP targeting *Z*-L-phenylalanine on a glass slab. The optimized precursor solution was photopolymerized locally by total reflection of a laser beam (405 nm), resulting in sub 100 nm thick polymer spots. More recently, a fibber-based approach was reported,[53] using an evanescent field for polymerizing a homogeneous MIP core-shell around the polymer fibber. The evanescent field was generated by coupling the fibber to a 410 nm light and used not only for fabrication but also for the readout of the generated MIP chemosensor, too. This readout was achieved by incorporating a fluorescent functional monomer whose fluorescence was triggered by protonation. In this context, the evanescent field was also used for the excitation of this fluorophore.

For the efficient transport of light optical elements, such as waveguides and optical fibbers, these are standard tools in the field of telecommunication. These elements guide an incident light beam through a core with a refractive index higher than that of the surrounding media. Recently, this approach has been used to fabricate defined polymer tips in the micrometre range[54,55] and was adapted by Ton *et al.*[56] for the development of a fluorescent based MIP sensing system. An MIP tip was fabricated at the end of an 8 μm core telecommunication optical fibre by placing a precursor droplet at the cleaved end of the fibber and guiding a 375 nm laser through the core. The shape was controlled by a self-guiding effect of the growing polymer tip (Figure 6.4). For some applications, photoinduced polymerization of MIPs might be not suitable. In this case, the use of a focused infrared laser to induce locally thermal polymerization is an interesting concept. The group of Piletsky used a standard MIP recipe targeting dansyl-L-phenylalanine (dansyl-L-phe) to generate MIP microdots in a microfluidic device by thermal polymerization with a CO_2 laser.[57]

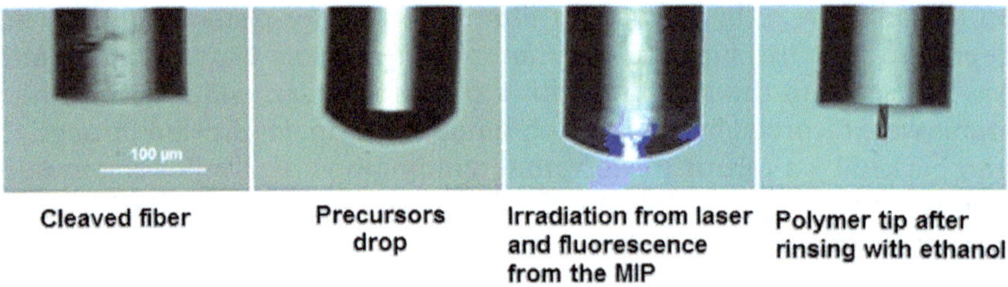

Figure 6.4 Microtip fabrication on an optical fiber (optical microscope images). Reproduced from X. A. Ton, B. Tse Sum Bui, M. Resmini, P. Bonomi, I. Dika, O. Soppera and K. Haupt, A Versatile Fiber-Optic Fluorescence Sensor Based on Molecularly Imprinted Microstructures Polymerized *in situ*, *Angew. Chem. Int. Ed.*,[56] John Wiley and Sons, © 2012 WILEY-VCH Verlag GmbH & Co. KGaA, Weinheim.

6.2.3 Mechanical Patterning

Mechanical approaches allow precise depositing and patterning of small liquid volumes on a substrate. The most popular mechanical spotting device is probably inkjet printing, which reproduces digital images by precise, contactless deposition of small ink droplets on a substrate (*i.e.* paper). Today, inkjet printers are widely spread and used not only by industry but also by private customers. Another example of commercially available mechanical spotting systems is nano-spotters, which are intensively used in the field of biotechnology to generate miniaturised test assays, for example enzyme immunoassays. These test assays consist of micro-droplets with volumes down to the picolitre range, of which each droplet contains one complete assay, providing a cost efficient and rapid screening method for expensive biological samples.

In this context, the development of microsystems for simple and reliable sample deposition was continuously investigated or improved over the last decades. One of these developed techniques is based on microcantilevers. Although these free standing structures were initially used as transducer elements for sensing applications, cantilevers can also be used for the transfer of small liquid volumes by using a direct-contact deposition technique.[58] Therefore an ink, loaded with a functional material, is transferred directly to the surface of a substrate through an etched channel on the cantilever. The system consists of a chip with 12 cantilevers, 10 for deposition and 2 for positioning, each cantilever is ~1500 µm long, ~120 µm wide and ~5 µm thick, with a fluidic channel incorporated in the cantilever tip. This array of cantilevers was used for the deposition of a pattern of MIP precursors on a glass slide, followed by *in situ* polymerization under UV light. Binding of the target molecule to the MIP dots was analysed by fluorescence microscopy.

For features with a sub-micrometre resolution, deposition techniques based on atomic force microscopy (AFM) were developed.[59,60] Nanofoutain pen writing is one of these methods. The AFM cantilever is replaced in this application by capillary nanopipettes of sub-micrometre size. Liquids can be transferred through this pipette onto a substrate by capillary forces, which can be used to deposit proteins, enzymes and pre-polymer solutions polymerized *in situ* after deposition.[61] Belmont *et al.*[62] were able to apply this technology to generate MIP patterns with a feature size in the low micrometre range. A nanofountain pen with an aperture of 100–600 nm was filled with an MIP precursor solution and mounted on an AFM system, thus enabling the precise deposition of the loaded sample. For demonstrating the feasibility of this deposition technique, a pattern of MIP and NIP microdots directed against fluorescein was synthesized and binding was analyzed by fluorescence microscopy. However, writing dots and lines with this technique is not trivial because several parameters, such as substrate wettability, viscosity of the MIP precursor and writing speed, had to be optimised. The method based on sequential writing is slow and, therefore, more suitable for prototype development.

The above-mentioned inkjet printers operate, in contrary to the above given examples of cantilever and nanofountain pen based system, without contacting the substrate. Spray coating through a shadow mask is a similar non-contact deposition technique that can be used for the patterning of microstructured polymer features. Recently, Ayela *et al.*[63] have published a study, which picked up this idea of non-contact deposition technique. Microcantilever sensors based on MIPs were fabricated by propelling an MIP precursor solution with the help of a spray coater on a substrate. A precise shape and dimension of a patterned MIP were triggered by micro-stencils that defined the area on which the sprayed aerosol should remain. One clear advantage of this approach is the simultaneous deposition and patterning of the MIP layer. Cantilever-shaped structures were fabricated with this technique and attached to a support of epoxy resist (SU-8) to generate a free-standing microgravimetric chemosensor (Figure 6.5). Typical dimensions of the cantilever beam were 500 µm in length, 200 µm in width and 12 µm in thickness, as determined by the stencil dimensions. The analyte propranolol was recognized selectively, as analysed with the resonance frequency change of fabricated MIP based cantilevers.

6.2.4 Soft Lithography

Soft lithography is a fabrication method which combines the precision of photolithography with the cost efficiency and versatility of mechanical deposition methods. The technology, initially introduced in the late 1990s for micro transfer moulding,[64] is now a well-established micro- and nanofabrication method, which comprises the two approaches of micro-moulding and micro-contact printing (µCP). Both techniques rely on the similar principle, in which a stamp, generally made of polydimethylsiloxane (PDMS), is used to transfer micro-patterns onto a substrate. The PDMS stamp is generally fabricated from a master mould (which in turn is fabricated in an initial step

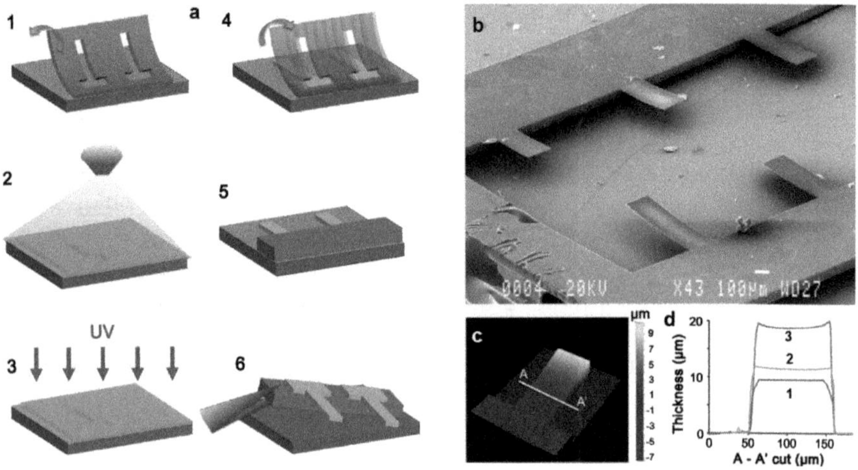

Figure 6.5 (a) Fabrication process of MIP cantilevers (1) the SU-8 stencil containing apertures of the desired shape of the cantilevers is deposited on a silicon wafer coated with a sacrificial Omnicoat layer, followed by (2) spray-coating the MIP precursors, (3) photopolymerization, (4) lift-off of the stencil, and then (5) a thick SU-8 support is overlaid by photolithography, finally (6) lifting-off of the cantilevers after dissolution of the sacrificial layer. (b) A matrix of MIP cantilevers of different lengths (from 300 to 500 μm) after their release from the silicon support. (c) Initial bending of one MIP cantilever of 500 μm in length, 200 μm in width and 12 μm in thickness imaged by optical interferometry. (d) Thickness profile of three cantilevers fabricated with: (*1*) 8, (*2*) 12, and (*3*) 20 μm thick stencils. Adapted from C. Ayela, G. Dubourg, C. Pellet and K. Haupt, All-Organic Microelectromechanical Systems Integrating Specific Molecular Recognition – A New Generation of Chemical Sensors, *Adv. Mater*,[63] John Wiley and Sons, © 2013 WILEY-VCH Verlag GmbH & Co. KGaA, Weinheim.

by photolithography or related techniques), and carries the negative image of the master mould. Soft lithography methods allow depositing of numerous materials including self-assembled monolayers of alkanethiolates,[65] polymers[66,67] and biomolecules.[68]

Micromoulding in capillaries (MIMIC) was the first soft lithography method that was successfully applied to MIPs by Yan and Kapua.[69] MIP microstructures, templated by 2,4-dichlorophenoxyacetic acid (2,4-D), were fabricated by placing a PDMS stamp on a silicon wafer and filling the empty cavities between the wafer and the stamp with an MIP precursor. A negative image of the micro-structured PDMS stamp made of MIP was obtained after polymerization, with a feature size of 20×20 μm^2. After isolating the microstructures from the substrate, binding of the MIP microstructures was analysed using radio-labelled 2,4-D. The method aimed at microfabrication of micro monoliths, which can be directly characterized, instead of mechanically ground and sieved bulk polymers. In contrast, micro-contact printing focuses on the direct patterning of two-dimensional features. It was applied to imprint a surface-immobilized template.[70] The technique of immobilizing a template on a surface was initially introduced to increase the accessibility of potential binding sites for the analyte.[71] In this approach, theophylline was covalently attached to the surface of the PDMS stamp and subsequently brought into contact with recognition monomers and anchoring monomers. The PDMS stamp was then pressed gently against a substrate before polymerization. After removing the stamp, a surface imprinted polymer array with a line width of 25 μm was obtained on the substrate surface. Binding assays with fluorescent analytes revealed selectivity for theophylline. In both examples, the feature size exceeded 20 μm, which is quite large. How MIP pattern can be generated with a submicron resolution using soft lithography was demonstrated by Lalo *et al.*[72] They created arrays of nanopatterns by micro-contact printing. In order to achieve this high resolution, the surface of the PDMS stamp was rendered hydrophilic by oxygen plasma treatment to increase its wettability. The stamp was then inked with an MIP precursor solution templating *N*-Boc-L-phenylalanine, dried with nitrogen and brought in contact with a silanized glass substrate. Photopolymerization was induced using a 365 nm light source resulting in a periodic pattern of lines (400×400 μm^2) with a line width of 660 nm, a height of 140 nm and a pitch of 1 μm. Recognition of MIP cavities was analysed using the fluorescently labelled phenylalanine derivative dansyl-L-phe.

A similar approach was published shortly afterwards, generating a periodic pattern by nanoimprint lithography.[73] This lithography relies on a similar principle as micro-contact printing but provides a resolution down to a few tens of nanometre (Figure 6.6). It was used to generate a pattern of periodic lines with a lateral resolution of 100 nm. The interest of this small feature size is, in addition to improved mass transport of analytes to potential molecular cavities bearing recognizing sites, the investigation of solvent effects. Traditionally, MIPs are synthesized in the presence of a porogen (solvent) in order to generate access to the molecular cavities. However, solvents that are used to generate these pores can interfere with the template or the functional monomer and thus with the recognizing performance of the synthesized MIP. With polymer features in the nanometre range, pores do not have to be introduced anymore to the system because enough imprinted cavities should be available on the developed polymer surface.

The above-mentioned soft lithography methods used a stamp generated from a master mould to print active features on a substrate. This printing can be a critical step because the stamp has to be brought into contact with the substrate with a certain pressure. This pressure, if not adjusted carefully, might lead to distorted features. As an alternative approach, the master mould can be used to directly fabricate the features instead of the use of a PDMS stamp. Using this approach, a diffractive grating chemosensor based on MIPs was described more recently by Moreno-Bondi's

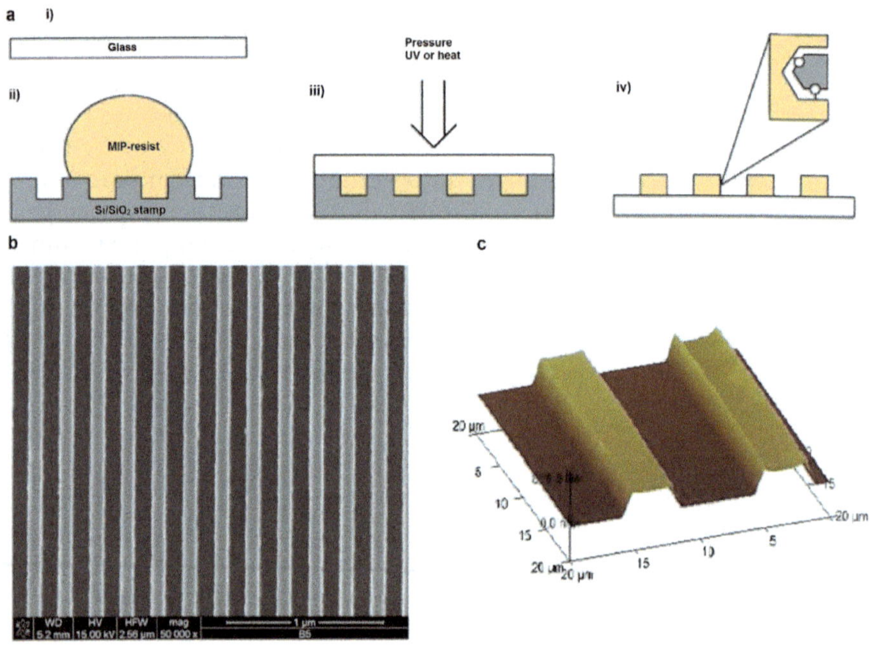

Figure 6.6 (a) Patterning MIP nanostructures with reactive nanoimprint lithography (NIL). A monomer mixture containing the molecular template is deposited on a patterned stamp. A substrate is pressed against the stamp causing the reaction mixture to fill the empty space defined by the stamp pattern. The monomers are polymerized by UV or thermal excitation. Releasing the stamp (unmolding) reveals an MIP pattern containing predefined sites. (b) SEM images of a patterned MIP fabricated by UV-NIL. (c) The AFM image of the MIP pattern. Reproduced from ref. 73 with permission from the Royal Society of Chemistry.

group.[74] An MIP precursor solution was cast on a master mould with a periodic pattern of squares, coated with a Mylar® film and polymerized under argon atmosphere. The resulting MIP film exhibited a diffractive behaviour and could be used for the sensitive and selective recognition of enrofloxacin (Figure 6.7).

6.3 MIP Nanomaterials and Their Fabrication

Nanomaterials and their unique physical and chemical properties have opened a broad range of possible applications not only in academia but also in commercially available mass products. Therefore, it is not surprising that the synthesis of nanostructured MIPs and nanocomposites gained increasing interest among the research community. Synthesizing MIPs at the nanoscale comprises several advantages over the classical macroporous bulk monoliths.[75] The surface-to-volume ratio is much higher for the former nanomaterials and because accessible imprinted cavities are mainly present

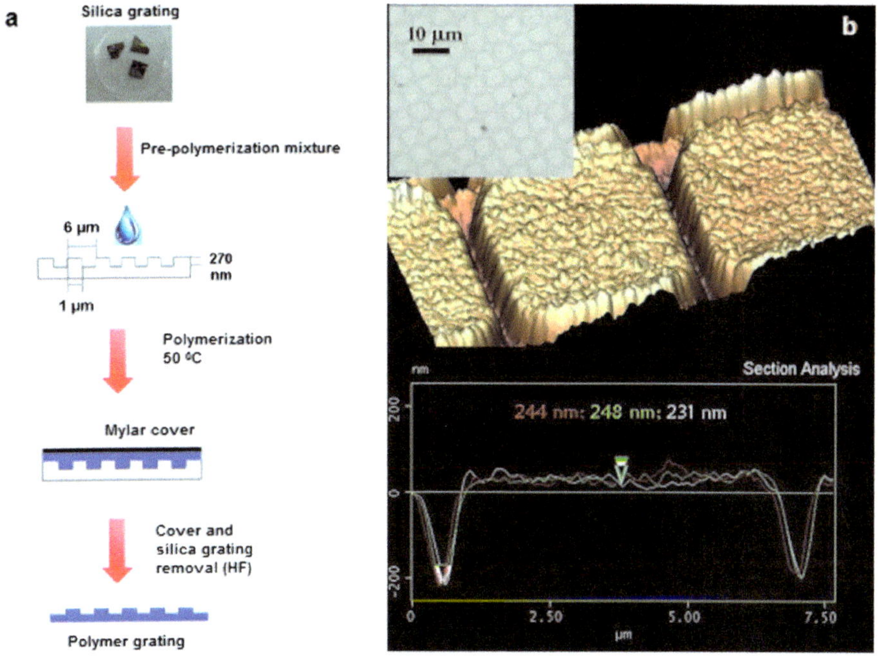

Figure 6.7 (a) Micropatterning of MIP films using SiO_2/Si grating molds. (b) AFM image and cross-section scan of a representative portion of an MIP 2-D grating. Inset: optical microscope photograph of the micropatterned MIP film. Adapted from *Biosensors and Bioelectronics*, **26**(5), C. A. Barrios, C. Zhenhe, F. Navarro-Villoslada, D. López-Romero and M. C. Moreno-Bondi, Molecularly imprinted polymer diffraction grating as label-free optical bio(mimetic)sensor, 2801–2804,[74] Copyright (2011) with permission from Elsevier.

on the surface of the MIP, template removal can be more efficient, thus leading to more cavities available for target analytes. Next to the imprint efficiency, binding kinetics of MIPs at the nanoscale is improved significantly, due to lower mass-transfer resistance.[76–78] Furthermore, nano-MIPs open the door towards applications in the field of biomedicine, diagnostics and theranostics, and they have already been applied for *in vitro* studies, such as cell imaging[79] and *in vivo* applications.[80,81] In this context, the design of nanostructured MIPs is of great interest and continuously improved, adapted and developed.

6.3.1 MIP Nanoparticles

6.3.1.1 Precipitation Polymerization

Precipitation polymerization is probably the most straightforward method to synthesize MIP nanoparticles with homogenous size distribution. In this method, monomers are dissolved in an excess of suitable solvent (near-θ solvents), from which seed particles precipitate and grow by subsequently addition of monomers and oligomers from the solution. The resulting monodispersed spherical particles can be tuned in size and porosity by changing the polymerization conditions.[82–84] Pioneering work of Ye *et al.*[75,85] introduced the concept of precipitation polymerization to the field of MIPs. Theophylline, and 17β-estradiol were successfully imprinted in the first published study,[85] resulting in nanoparticles with a size of 200 nm and 300 nm, respectively.

In several interesting publications, the group of Zhang[86] focused on the synthesis of spherical particles by precipitation polymerization using controlled radical polymerization, such as atom transfer radical polymerization (ATRP), reversible addition-fragment transfer (RAFT) polymerization[87–89] and iniferter-based polymerization.[90] The main advantage of using controlled polymerization conditions is the possibility of further functionalization of these particles due to the "living" fragments on the polymer surface.

Precipitation polymerization can be seen, due to the simplicity of this one-pot, one-step synthesis method, as a standard fabrication scheme of nanoparticles. Therefore, it found its way into many applications, such as commercially available solid-phase extraction kits,[91,92] bead-based sensors,[53] cell-imaging,[79] and cosmetic applications.[10]

6.3.1.2 *Suspension and Emulsion Polymerizations*

Suspension and emulsion polymerizations are other fabrication schemes to obtain nano-scaled polymers by polymerizing the components of the desired polymer in individual 'nanoreactors'. Mayes and Mosbach[93] have introduced a suspension polymerization to MIPs in 1996. In this work perfluorocarbons, which are immiscible with most common liquid organic compounds, were used as the continuous phase. Microdroplets, containing the MIP precursor, were stabilised by a polymeric surfactant containing a perfluorinated part, and then homogeneously suspended in the inert phase. The size of the resulting polymer beads ranged between 5 and 50 μm. The polymer particles obtained by this fabrication scheme are not of the nanoscale size, but they paved the way for introducing emulsion polymerization to the field of MIPs. Emulsion polymerization is based on the same principle as that of suspension polymerization but suspended droplets are scaled down to the size of several hundred nanometres. Vaihinger *et al.*[94] used a miniemulsion polymerization to synthesize MIPs with a particle size of ~200 nm. Water with sodium dodecyl sulphate as the surfactant was used in this study as an inert phase. Nanodroplets consisting of the MIP precursor were stabilised by hydrophobic hexadecane to prevent Ostwald ripening during miniemulsion polymerization. Other authors[95–97] used emulsion polymerization to polymerize a core made of methyl methacrylate or styrene and prepared an MIP-shell around. The group of Haupt[97] used a two-step approach (Scheme 6.2) starting with the emulsion polymerization of a polystyrene core and subsequent

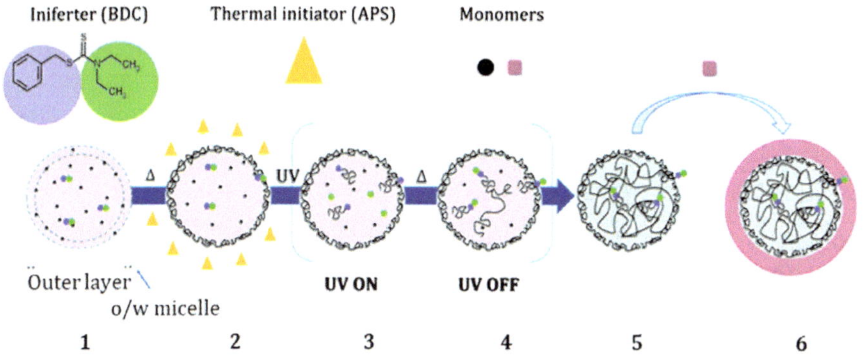

Scheme 6.2 Formation of stable oil in water (o/w) micelles, (2) thermal polymerisation in the outer layer of the micelles with the initiator ammonium persulfate (APS), (3) iniferter *N*,*N*-diethyldithiocarbamate (BDC) activation, (4 and 5) total formation of latex particles, and (6) re-initiation from the iniferter seed under UV irradiation in the presence of desired monomers. Reproduced from ref. 97 with permission from the Royal Society of Chemistry.

polymerization of a shell by re-initiation from an iniferter, which was already present in the mixture of precursors of the seed particles, and subsequently activated.

6.3.1.3 *Solid-phase Synthesis of MIPs*

Solid-phase synthesis of MIPs is based on the immobilization of the template molecule on a solid surface before imprinting. This immobilization leads to the major advantage that the imprinted cavities are always accessible because the MIP is synthesized directly at the interface with the template support, and all cavities are located on the surface of the polymer material. Furthermore, this immobilization leads to a controlled and uniform orientation of the template, which in turn should lead to a more homogeneous cavity population. In a first approach[71] of oriented, immobilised templates, the 8-carboxypropyl derivate of theophylline was linked *via* an amide bond to a silica gel. In a second step, the immobilized template was imprinted with trifluoromethacrylic acid and the divinylbenzene cross-linking monomer. The support was dissolved after polymerization leaving MIP particles behind with imprinted cavities at, or near to the polymer surface. Similar approaches have been published for the solid-phase synthesis of MIPs for other small molecules[98] and macromolecules, such as proteins.[99] More recently, Poma *et al.*[100] reported a fully automated synthesis of MIP nanoparticles based on a solid-phase method. The resulting nanoparticles were collected by increasing the temperature to 60 °C, thus allowing reusing of the immobilised template over 30 times. In the same year, the group of Haupt[101] developed a method to design soluble MIP nanoparticles for protein recognition by solid-phase synthesis. Oriented immobilization of the target protein was achieved by attaching an affinity ligand of the protein to the solid support. Thermoresponsive MIP nanoparticles were synthesized on this support and released just by a temperature change.[102]

6.3.2 Nanocomposite MIPs

Core-shell MIPs at the nanoscale present interesting additional features, because the cores of this material allow the inclusion or association of a second material, which can provide additional useful features, such as magnetic susceptibility, fluorescence or plasmonic enhancement. For the fabrication of these nanocomposites, different strategies have been developed to graft MIP shells around a functional core.

6.3.2.1 *Free-radical Polymerization*

For designing MIP nanocomposites, different strategies have been developed including involvement of a magnetic core. One of the first magnetic MIP core–shell structure was reported in the late 90s.[95] The author of this work used a core-shell emulsion polymerization to form an imprinted shell around a ferrofluidic seed. Resulting core-shell particles were 50 and 100 nm in size and imprinted with a sacrificial spacer, cholesteryl (4-vinyl)phenyl carbonate, attached to target cholesterol. The magnetic properties of the particles have been used for sedimentation of the MIP nanoparticles after extraction steps or binding assays. Other authors used grafting techniques to functionalize magnetic Fe_3O_4 nanoparticles.[103–105] In a first report, magnetic nanoparticles were functionalised with vinyl groups, which were subsequently grafted by a copolymerized methacrylic acid and ethylene glycol dimethacrylate (EGDMA) polymer imprinted with 4-nitrophenol.[103] In a more recent work,[104] the same authors reported on an approach in which the same template (4-nitrophenol) was immobilised on silica coated Fe_3O_4 nanoparticles functionalized with 3-aminopropyltriethoxysilane (APTES). In a second step, EGDMA was linked to the APTES and polymerized, resulting in an imprinted 25 nm thick film. Presented studies use free-radical polymerization for the fabrication of the functional MIP shell, a technique which is difficult to control and may lead to an inhomogeneous distribution of molecular cavities with recognizing sites.

6.3.2.2 *Controlled Radical Polymerization*

The introduction of controlled radical polymerization (CRP) techniques was an important step for the development of nanocomposite MIPs. CRP initiators can be easily grafted onto substrates and give control over polymer thickness directly through monomer conversion. Multifunctional nanomaterials and nanocomposites can be prepared, which otherwise were difficult or impossible to obtain. For the controlled design of MIP nanoparticles with magnetic features, CRP techniques, ATRP and RAFT polymerization have been employed by several groups.[106–110] A detailed review of controlled polymerization was recently published by Beyazit *et al.*[111] For RAFT polymerization, a transfer agent competes with the propagation of generated macro radicals and, therefore, decreases the probability of termination. For polymerization, an azoinitiator is used while the reversible transfer agent reacts with generated radicals due to their higher reactivity with respect to ordinary vinyl monomers.[111] Gonzato *et al.*[107] used this polymerization technique for the synthesis of magnetic core-shell MIPs. The RAFT agent was grafted to the surface of amino-functionalized Fe_3O_4 nanoparticles, which enabled preparation by polymerization of thin MIP shells of controlled size. In addition, the living character could be used for the polymerization of a further outer layer, thus enabling the fine tuning of surface properties of composite MIPs. Similar approaches of immobilised RAFT initiators on solid cores have been reported,[112,113] resulting in reproducible and homogeneous core-shell structures with improved properties compared with azo- or iniferter-based initiation.[113]

Another approach to controlling polymerization conditions can be achieved by ATRP. Free-radical species and oxidised metals are generated by a redox reaction involving a transition metal complex and an organic halide. The generated free radical propagates only until a halide back transfer from the oxidised metal complex stops the chain growth. This reaction scheme can be described as a metal-catalyzed radical polymerization. It was used for the fabrication of core-shell nanocomposites as well as for immobilization of the initiator on a solid surface.[106,109,114] One of the drawbacks is that a high amount of transition-metal catalyst is needed for ARTP and has to be removed subsequently. Matyjaszewski and co-workers[115] developed, therefore, an "activator regenerated by electron transfer radical polymerization" (ARGET ATRP). This technique constantly regenerates the Cu(I) ATRP activator by reducing inactive Cu(II) species to active Cu(I) species in the presence of a reducing agent. Thus, the concentration of the transition-metal catalyst could be decreased to only a few ppm. This technique was applied to graft an MIP shell on the SiO_2 functionalized, magnetic ZnO nanorods using ascorbic acid as the reducing agent.[116] Another drawback of ATRP for MIP synthesis is that acidic monomers, such as widely used methacrylic acid (MAA), are not compatible with the standard ATRP protocols. However, very recently this limitation was overcome by introducing a more robust catalyst to the ATRP process.[117] Adali-Kaya *et al.*[118] later applied this approach to MIPs demonstrating that the synthesis of MAA containing MIPs by ATRP was possible. The living character of the ATRP process could be proven by post-modification of the MIP nanoparticles. Polymer brushes were grafted to the surface and allowed the modification of surface charges.

Stable free-radical polymerization (SFRP) allows precise control of reaction kinetics and can be applied by initiating polymerization with an iniferter or by nitroxide-mediated radical polymerization. The control mechanism of both methods depends on a dynamic equilibrium between the propagating species and alkoxyamine species. In the field of MIPs, many applications can be found in which organic sulphides are used as photoiniferters (*e.g.* dithiocarbamate) to control polymerization[38,113,119–122] and for applications in which a re-initiation is desired.[97] By combining emulsion polymerization and iniferter initiated polymerization, magnetic[95] and polymer-polymer[97] core-shell particles were prepared.

6.3.2.3 *Core Induced Polymerization*

The cores of presented structures (*e.g.* magnetic core-shell MIPs) present additional features for sample treatment (extraction of particles with an external magnet, for instance). Recent research in the area of multifunctional, core-shell MIPs focused on fabrication strategies by using intrinsic properties of the core material. Panagiotopoulou *et al.*[123] recently used individual internal light sources of green- and red-emitting InP/ZnS quantum dots (QDs) to render the core hydrophilic before preparing an MIP shell by polymerization in a second step (Figure 6.8). Polymerization was initiated with suitable initiators sensitive to the emitted light of the quantum dots (550 and 660 nm) but not to the excitation light of 365 nm. Hydrophilic core-shell particles were obtained with a 14 nm poly-HEMA first shell stabilizing the QDs, and a 30 nm thick outer MIP layer. These particles have been applied for cell imaging. The energy of the external light source (365 nm) was in this case converted down to the visible spectra (550 nm and 660 nm respectively). In contrast, upconverting nanoparticles (UCPs) are excited in the near-infrared region and emit light in the visible or UV region. The same method as that for QDs has been employed to render the core of UCPs hydrophilic or hydrophobic by photoinduced polymerization of suitable monomers. An MIP shell was then polymerized by re-initiation using, again, the emitted light of the UCPs.[124]

6.3.3 Nanotube and Nanowire-based MIPs

6.3.3.1 *Carbon Nanotubes*

MIPs were not only grafted onto spherically shaped cores, such as magnetic nanoparticles, quantum dots, or UCPs, but also on surfaces,[107] rod shaped particles,[116] cubic nanoparticles[125] and carbon nanotubes CNTs.[126–129] In this case, the surface of the substrate is modified (*e.g.* by silanization) to link a CRP initiator to the surface and to start deposition of a functional polymer film from there. By grafting MIPs onto CNTs, imprinted nanofilaments can be obtained with a high surface-to-volume ratio and interesting features resulting from the CNT substrate, such as fluorescence and conducting properties. Lee *et al.*[127] designed a field-effect transistor by grafting an MIP targeting theophylline and deposited a 10 nm thick MIP film on the surface of the iniferter modified CNTs. In a similar way, an ATRP initiator was grafted on the surface of multiwalled CNTs,[126] through the acylation of carboxyl groups,

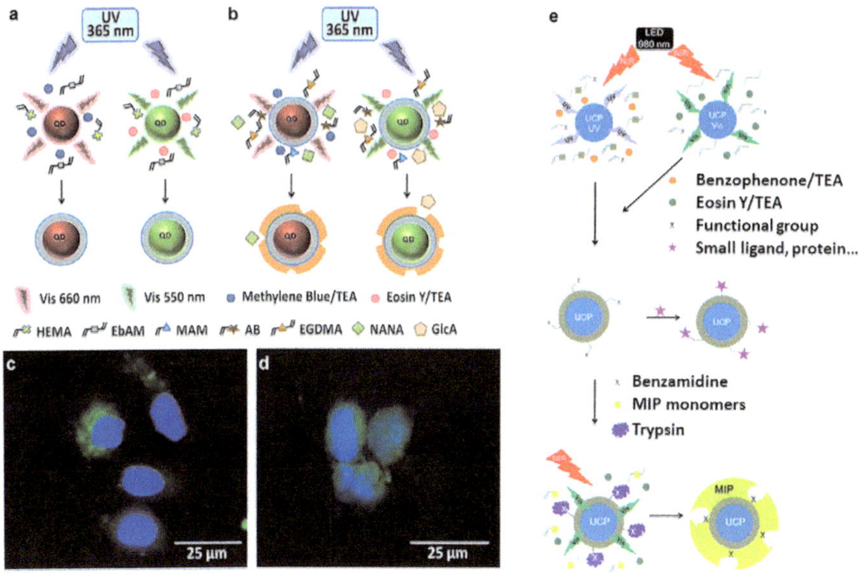

Figure 6.8 (a) Red or green light emitted from InP/ZnS quantum dots (QDs), excited by UV irradiation, is used to synthesize *in situ* a polymeric shell around the particles by photopolymerization. Methylene blue/triethylamine (TEA) is used as the initiator system for red QDs and eosin Y/TEA for green-QDs. (b) A second shell of the molecularly imprinted polymer (MIP) is synthesized by re-initiation in the presence of functional and cross-linking monomers. Adapted from M. Panagiotopoulou, T. Salinas, S. Beyazit, S. Kunath, L. Duma, E. Prost, A. G. Mayes, M. Resmini, B. Tse Sum Bui and K. Haupt, Molecularly Imprinted Polymer Coated Quantum Dots for Multiplexed Cell Targeting and Imaging, *Angew. Chemie Int. Ed.*,[123] John Wiley and Sons, © 2016 WILEY-VCH Verlag GmbH & Co. KGaA, Weinheim. (d) UV or visible light, emitted from upconverting nanoparticles excited in the near-infrared region with a light-emitting diode (LED) (980 nm), is used to create *in situ* a polymeric shell around the particles by photopolymerization. Benzophenone and eosin Y are used as UV and visible initiators, respectively. The incorporation of functional monomers enables subsequent attachment of small ligands, proteins, or other moieties. A second shell, *e.g.*, an MIP, can be synthesized by re-initiation in the presence of different monomers and a molecular template (trypsin). Adapted from S. Beyazit, S. Ambrosini, N. Marchyk, E. Palo, V. Kale, T. Soukka, B. Tse Sum Bui and K. Haupt, Versatile Synthetic Strategy for Coating Upconverting Nanoparticles with Polymer Shells through Localized Photopolymerization by Using the Particles as Internal Light Sources, Angew. Chem. Int. Ed.,[124] John Wiley and Sons, © 2014 WILEY-VCH Verlag CmbH & Co. KGaA, Weinheim.

previously introduced *via* oxidation using a HNO_3/H_2SO_4 mixture. The subsequent reaction with 2-hydroxylethyl-2′-bromo-isobutyrate covalently bound the ATRP initiator to the CNTs. Theophylline was then imprinted using HEMA and EGDMA. The resulting CNT nanocomposites had a 5 nm thick MIP shell with binding kinetics improved compared to that of an analogous bulk MIP.

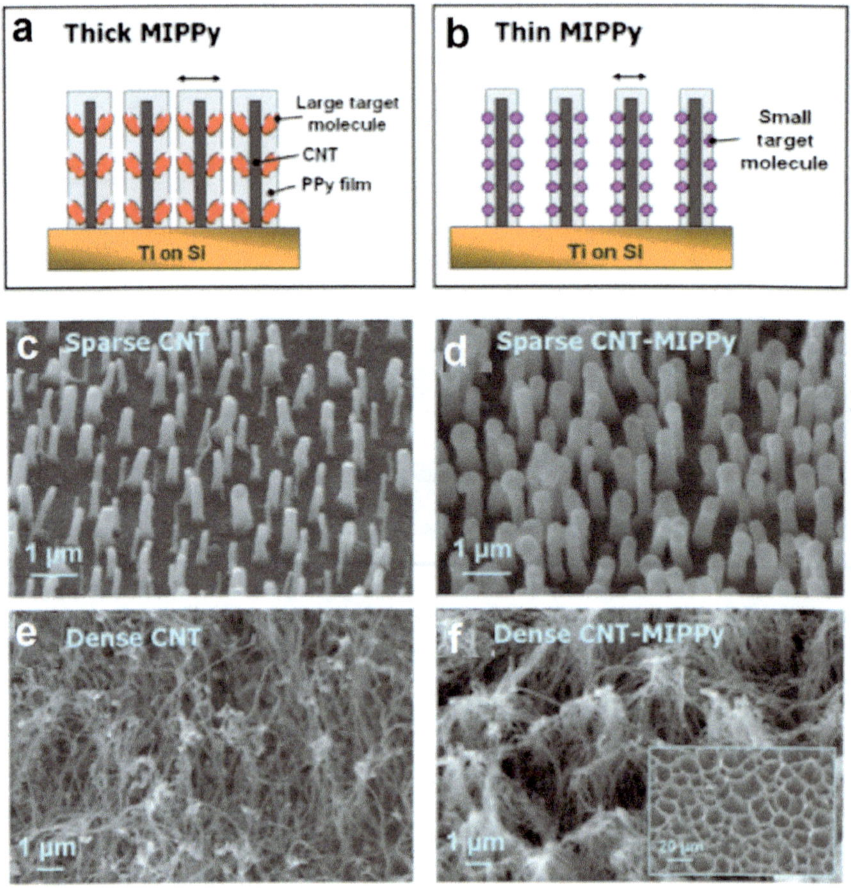

Figure 6.9 (a) Schematic illustration of the concept of high density, surface molecular imprinting technique using a CNT array. Thick molecularly imprinted polypyrrole (MIPPy) film grafted around each CNT to accommodate large target molecules whereas (b) thin MIPPy film was deposited for small molecules. SEM images of a sparse CNT array (c) before and (d) after caffeine imprinting electropolymerization. SEM images of a dense CNT array (e) before and (f) after caffeine imprinting electropolymerization. The inset of (F) shows the macroscopic view of the densely packed CNTs after collapsing into honeycomb structure when the CNT samples were withdrawn from the HNO_3 solution during pre-treatment. All the SEM images were taken at 30° tilted view. Reprinted from *Biosensors and Bioelectronics*, 25(3), C. Choong, J. S. Bendall and W. I. Milne, Carbon nanotube array: A new MIP platform, 652–656,[128] Copyright (2009) with permission from Elsevier.

A different approach was suggested by the group of Milne.[128] It used the conductive properties of CNTs to coat an MIP shell by electropolymerization on the surface of an array of free-standing, vertically aligned CNTs (Figure 6.9). Therefore a CNT array was fabricated on a silicon substrate by plasma-enhanced chemical vapour deposition (PECVD), and subsequently grafted with a polypyrrole thin film imprinted with the target molecule. The thickness of the MIP shell could be adjusted from 10 nm for small imprinted targets to 100 nm for larger molecules.

6.3.3.2 Nanofilaments Obtained Through Porous Alumina Membranes

The presented nanofabrication techniques can be described so far as bottom-up approaches in order to obtain an MIP material at a nanoscale. An interesting alternative is the design of MIPs by nanomoulding, which can be seen as a top-down approach because the procedure can be compared with the soft lithography technique described in Section 6.2.4. Nanoporous aluminium oxide (alumina) substrates can be obtained by anodic oxidation in an acid electrolyte.[130,131] This interesting substrate was used in several reports as sacrificial mould or template for the synthesis of MIPs at the nanoscale.[43,132–137]

Li *et al.*[137] have synthesized magnetic MIP nanowires in alumina membranes. The 8-carboxypropyltheophylline template was immobilized on the membrane surface in the alumina nanopores and imprinted by a standard free radical polymerization technique using a thermal initiator. In a final step, the alumina support was removed leaving long imprinted filaments behind. The inclusion of superparamagnetic $MnFe_2O_4$ nanocrystallites could facilitate the separation from a suspension. The group of Haupt[132,138] has used alumina membranes to synthesize layers of surface bound nanofilaments based on MIPs. These MIP filaments were fabricated on a silanized glass slide by pressing an alumina membrane on an MIP precursor droplet. After dissolving the alumina mould, surface bound MIP nanofilaments were obtained. Moreover, the technique of nanomoulding with alumina as a sacrificial mould was combined with projection lithography[43] to control the shape at the microscale of the nanofilament array. This example shows the flexibility of MIPs and, therefore, feasibility for micro- and nanofabrication techniques.

Other authors reported the use of alumina to produce very thin MIP sol–gel nanotubes.[136] In this case the aluminium oxide membrane was not dissolved but used as a scaffold for a silica-monomer imprinted molecule complex. The covalently imprinted estrone formed a thermally cleavable urethane bond with the cross-linkable silica monomer, thus resulting in a very thin imprinted sol–gel layer in the nanopores of the alumina.

6.4 Conclusions

The great potential of molecularly imprinted polymers resides in their stability and their accessibility to micro- and nanofabrication techniques. These features make these artificial receptors excellent candidates not only for their integration in chemosensors and biochips but also for many other applications. Thus, it is not surprising that research in this area is still expanding.

Several standard microfabrication schemes are suitable to process MIPs, allowing precise patterning in 2-D and 3-D with sub-micrometre resolution. The possibility of photopolymerizing MIPs might be one of the key features of this development. In parallel, the development of nanomaterials based on MIPs has made major improvements and allow these days the precise design of nano-tools based on MIPs. Thereby, the adaptation of controlled polymerization processes, such as iniferter polymerization, RAFT polymerization and ATRP can be seen as the main approaches that pushed the precision and versatility of nanofabricated MIPs towards new standards. The developments of micro- and nanofabrication presented in this chapter could pave the way for commercial applications in the field of chemical and biochemical detection, monitoring, and screening, in cosmetics, and even in medical treatment.

List of Abbreviations

μCP	Micro-contact printing
2,4-D	2,4-Dichlorophenoxyacetic acid
AFM	Atomic force microscopy
APS	Ammonium persulfate
APTES	3-Aminopropyltriethoxysilane
ARGET ATRP	Activator regenerated by electron transfer radical polymerization
ATR	Attenuated total reflection
ATRP	Atom transfer radical polymerization
BDC	*N,N*-diethyldithiocarbamate
CNT	Carbon nanotube
CRP	Controlled radical polymerization
dansyl-L-phe	Dansyl-L-phenylalanine
EGDMA	Ethylene glycol dimethacrylate
HEMA	2-Hydroxyethyl methacrylate
InP	Indium phosphide
MAA	Methacrylic acid
MEMS	Microelectromechanical system
MIMIC	Micromoulding in capillary
MIP	Molecularly imprinted polymer
MIPPy	Molecularly imprinted polypyrrole
NIL	Nanoimprint lithography
NIP	Non-imprinted polymer
NSOM/SNOM	Near-field scanning optical microscopy
PDMS	Polydimethylsiloxane
PECVD	Plasma-enhanced chemical vapour deposition
QDs	Quantum dots
RAFT	Reversible addition-fragment transfer
SFRP	Stable free-radical polymerization
SPR	Surface plasmon resonance

TEA	Triethylamine
UCP	Upconverting nanoparticle
Z-L-phe	*Z*-L-phenylalanine
UV	Ultraviolet

References

1. G. Wulff and A. Sarhan, *Angew. Chem., Int. Ed.*, 1972, **46**, 334.
2. R. Arshady and K. Mosbach, *Makromol. Chem.*, 1981, **182**, 687.
3. G. Wulff, *Angew. Chem., Int. Ed.*, 1995, **34**, 1812.
4. K. Haupt, *Anal. Chem.*, 2003, **75**, 376A.
5. G. Vlatakis, L. I. Andersson, R. Muller and K. Mosbach, *Nature*, 1993, **361**, 645.
6. L. I. Andersson, *Anal. Chem.*, 1996, **68**, 111.
7. C. Alvarez-Lorenzo and A. Concheiro, *J. Chromatogr. B: Anal. Technol. Biomed. Life Sci.*, 2004, **804**, 231.
8. B. Sellergren and C. J. Allender, *Adv. Drug Delivery Rev.*, 2005, **57**, 1733.
9. M. C. Norell, H. S. Andersson and I. A. Nicholls, *J. Mol. Recognit.*, 1998, **11**, 98.
10. S. Nestora, F. Merlier, S. Beyazit, E. Prost, L. Duma, B. Baril, A. Greaves, K. Haupt and B. Tse Sum Bui, *Angew. Chem., Int. Ed.*, 2016, 6360.
11. L. Ye and K. Haupt, *Anal. Bioanal. Chem.*, 2004, **378**, 1887.
12. K. Dario, O. Ramström, A. Svensson and K. Mosbach, *Anal. Chem.*, 1995, **67**, 2142.
13. T. A. Sergeyeva, S. A. Piletsky, A. A. Brovko, E. A. Slinchenko, L. M. Sergeeva and A. V. El'skaya, *Anal. Chim. Acta*, 1999, **392**, 105.
14. S. A. Piletsky, Y. P. Parhometz, N. V. Lavryk, T. L. Panasyuk and A. V. El'skaya, *Sens. Actuators, B*, 1994, **19**, 629.
15. G. E. Moore, *Proc. IEEE*, 1998, **86**, 82.
16. A. del Campo and C. Greiner, *J. Micromech. Microeng.*, 2007, **17**, R81.
17. M. Madou, *Fundamentals of Microfabrication the Science of Miniaturisation*, Taylor & Francis Ltd, 2nd edn, 2002.
18. L. D. Spurlock, A. Jaramillo, A. Praserthdam, J. Lewis and A. Brajter-Toth, *Anal. Chim. Acta*, 1996, **336**, 37.
19. T. Panasyuk, V. Campo Dall'Orto, G. Marrazza, A. El'skaya, S. Piletsky, I. Rezzano and M. Mascini, *Anal. Lett.*, 1998, **31**, 1809.
20. D. C. Apodaca, R. B. Pernites, R. Ponnapati, F. R. Del Mundo and R. C. Advincula, *Macromolecules*, 2011, **44**, 6669.
21. A. Ramanaviciene and A. Ramanavicius, *Biosens. Bioelectron.*, 2004, **20**, 1076.
22. A. Dupont-Filliard, M. Billon, T. Livache and S. Guillerez, *Anal. Chim. Acta*, 2004, **515**, 271.
23. T. Darmanin, H. Bellanger, F. Guittard, P. Lisboa, M. Zurn, P. Colpo, D. Gilliland and F. Rossi, *RSC Adv.*, 2012, **2**, 1033.
24. K. Sreenivasan, *J. Mater. Sci.*, 2007, **42**, 7575.
25. B. Deore and M. S. Freund, *Analyst*, 2003, **128**, 803.

26. R. B. Pernites, R. R. Ponnapati and R. C. Advincula, *Macromolecules*, 2010, **43**, 9724.
27. A. Pietrzyk, S. Suriyanarayanan, W. Kutner, R. Chitta and F. D'Souza, *Anal. Chem.*, 2009, **81**, 2633.
28. T. L. Panasyuk, V. M. Mirsky, S. A. Piletsky and O. S. Wolfbeis, *Anal. Chem.*, 1999, **71**, 4609.
29. E. Granot, R. Tel-Vered, O. Lioubashevski and I. Willner, *Adv. Funct. Mater.*, 2008, **18**, 478.
30. A. S. N. Murthy, *Electrochim. Acta*, 1982, **28**, 473.
31. *Electropolymerization*, ed. E. Schab-Balcerzak, InTech, 2011, ISBN: 978-953-307-693-5.
32. C. Malitesta, E. Mazzotta, R. A. Picca, A. Poma, I. Chianella and S. A. Piletsky, *Anal. Bioanal. Chem.*, 2012, **402**, 1827.
33. P. S. Sharma, A. Pietrzyk-Le, F. D'Souza and W. Kutner, *Anal. Bioanal. Chem.*, 2012, **402**, 3177.
34. R. H. Schmidt and K. Haupt, *Chem. Mater.*, 2005, **17**, 1007.
35. R. H. Schmidt, A. S. Belmont and K. Haupt, *Anal. Chim. Acta*, 2005, **542**, 118.
36. L. P. C. Gomez, A. Spangenberg, X.-A. Ton, Y. Fuchs, F. Bokeloh, J.-P. Malval, B. Tse Sum Bui, D. Thuau, C. Ayela, K. Haupt and O. Soppera, *Adv. Mater.*, 2016, **28**, 5931.
37. A. S. Belmont, S. Jaeger, D. Knopp, R. Niessner, G. Gauglitz and K. Haupt, *Biosens. Bioelectron.*, 2007, **22**, 3267.
38. M. E. Byrne, E. Oral, J. Zachary Hilt and N. A. Peppas, *Polym. Adv. Technol.*, 2002, **13**, 798.
39. H. C. Huang, C. I. Lin, A. K. Joseph and Y. Der Lee, *J. Chromatogr., A*, 2004, **1027**, 263.
40. H. C. Huang, S. Y. Huang, C. I. Lin and Y. Der Lee, *Anal. Chim. Acta*, 2007, **582**, 137.
41. S. Guillon, R. Lemaire, A. V. Linares, K. Haupt and C. Ayela, *Lab Chip*, 2009, **9**, 2987.
42. R. I. Boysen, L. J. Schwarz, S. Li, J. Chowdhury and M. T. W. Hearn, *Microsyst. Technol.*, 2014, **20**, 2037.
43. A. V. Linares, A. Falcimaigne-Cordin, L. A. Gheber and K. Haupt, *Small*, 2011, **7**, 2318.
44. S. Carrasco, V. Canalejas-Tejero, F. Navarro-Villoslada, C. A. Barrios and M. C. Moreno-Bondi, *J. Mater. Chem.*, 2014, 1400.
45. P. G. Conrad, P. T. Nishimura, D. Aherne, B. J. Schwartz, D. Wu, N. Fang, X. Zhang, M. J. Roberts and K. J. Shea, *Adv. Mater.*, 2003, **15**, 1541.
46. Y. Fuchs, O. Soppera, A. G. Mayes and K. Haupt, *Adv. Mater.*, 2013, **25**, 566.
47. Y. Fuchs, S. Kunath, O. Soppera, K. Haupt and A. G. Mayes, *Adv. Funct. Mater.*, 2014, **24**, 688.
48. C. Girard and A. Dereux, *Rev. Prog. Phys.*, 1996, **59**, 657.
49. Y. Fuchs, O. Soppera and K. Haupt, *Anal. Chim. Acta*, 2012, **717**, 7.

50. C. Ecoffet, A. Espanet and D. J. Lougnot, *Adv. Mater.*, 1998, **10**, 411.
51. A. Espanet, C. Ecoffet and D. J. Lougnot, *J. Polym. Sci., Part A: Polym. Chem.*, 1999, **37**, 2075.
52. Y. Fuchs, A. V. Linares, A. G. Mayes, K. Haupt and O. Soppera, *Chem. Mater.*, 2011, **23**, 3645.
53. X. A. Ton, V. Acha, P. Bonomi, B. Tse Sum Bui and K. Haupt, *Biosens. Bioelectron.*, 2015, **64**, 359.
54. C. Ecoffet, R. Bachelot, D. Deloeil, P. Royer and D. J. Lougnot, *Synth. Met.*, 2001, **124**, 29.
55. S. Jradi, O. Soppera, D. J. Lougnot, R. Bachelot and P. Royer, *Opt. Mater. (Amsterdam, Neth.)*, 2009, **31**, 640.
56. X. A. Ton, B. Tse Sum Bui, M. Resmini, P. Bonomi, I. Dika, O. Soppera and K. Haupt, *Angew. Chem., Int. Ed.*, 2013, **52**, 8317.
57. O. Y. F. Henry, S. A. Piletsky and D. C. Cullen, *Biosens. Bioelectron.*, 2008, **23**, 1769.
58. P. Belaubre, M. Guirardel, G. Garcia, J. B. Pourciel, V. Leberre, A. Dagkes-samanskaia, E. Trévisiol, J. M. François and C. Bergaud, *Appl. Phys. Lett.*, 2003, **82**, 3122.
59. K. Kantarovich, I. Tsarfati, L. A. Gheber, K. Haupt and I. Bar, *Anal. Chem.*, 2009, **81**, 5686.
60. R. D. Piner, J. Zhu, F. Xu, S. Hong and C. A. Mirkin, *Science*, 1999, **283**, 661.
61. A. Lewis, Y. Kheifetz, E. Shambrodt, A. Radko, E. Khatchatryan and C. Sukenik, *Appl. Phys. Lett.*, 1999, **75**, 2689.
62. A.-S. Belmont, M. Sokuler, K. Haupt and L. A. Gheber, *Appl. Phys. Lett.*, 2007, **90**, 193101.
63. C. Ayela, G. Dubourg, C. Pellet and K. Haupt, *Adv. Mater.*, 2014, **26**, 5876.
64. X. Zhao and Y. Xia, *Adv. Mater.*, 1995, **8**, 837.
65. J. C. Love, D. B. Wolfe, M. L. Chabinyc, K. E. Paul and G. M. Whitesides, *J. Am. Chem. Soc.*, 2002, **124**, 1576.
66. T. W. Odom, J. C. Love, D. B. Wolfe, K. E. Paul and G. M. Whitesides, *Langmuir*, 2002, **18**, 5314.
67. D. Qin, Y. Xia and G. M. Whitesides, *Nat. Protoc.*, 2010, **5**, 491.
68. H. Lalo, J. C. Cau, C. Thibault, N. Marsaud, C. Severac and C. Vieu, *Microelectron. Eng.*, 2009, **86**, 1428.
69. M. Yan and A. Kapua, *Anal. Chim. Acta*, 2001, **435**, 163.
70. R. Voicu, K. Faid, A. A. Farah, F. Bensebaa, R. Barjovanu, C. Py and Y. Tao, *Langmuir*, 2007, **23**, 5452.
71. E. Yilmaz, K. Haupt and K. Mosbach, *Angew. Chem., Int. Ed.*, 2000, **39**, 2115.
72. H. Lalo, C. Ayela, E. Dague, C. Vieu and K. Haupt, *Lab Chip*, 2010, **10**, 1316.
73. D. Forchheimer, G. Luo, L. Montelius and L. Ye, *Analyst*, 2010, **135**, 1219.
74. C. A. Barrios, C. Zhenhe, F. Navarro-Villoslada, D. López-Romero and M. C. Moreno-Bondi, *Biosens. Bioelectron.*, 2011, **26**, 2801.

75. L. Ye, R. Weiss and K. Mosbach, *Macromolecules*, 2000, **33**, 8239.
76. J. Wackerlig and R. Schirhagl, *Anal. Chem.*, 2016, **88**, 250.
77. F. Canfarotta, M. J. Whitcombe and S. A. Piletsky, *Biotechnol. Adv.*, 2013, **31**, 1585.
78. E. Moczko, A. Poma, A. Guerreiro, I. Perez de Vargas Sansalvador, S. Caygill, F. Canfarotta, M. J. Whitcombe and S. Piletsky, *Nanoscale*, 2013, **5**, 3733.
79. S. Kunath, M. Panagiotopoulou, J. Maximilien, N. Marchyk, J. Sänger and K. Haupt, *Adv. Healthcare Mater.*, 2015, **4**, 1322.
80. Y. Hoshino, H. Koide, T. Urakami, H. Kanazawa, T. Kodama, N. Oku and K. J. Shea, *J. Am. Chem. Soc.*, 2010, **132**, 6644.
81. E. Asadi, M. Abdouss, R. M. Leblanc, N. Ezzati, J. N. Wilson and D. Kordestani, *Polymer*, 2016, **97**, 226.
82. K. Li and H. D. H. Stover, *J. Polym. Sci., Part A: Polym. Chem.*, 1993, **31**, 3257.
83. J. Wang, P. A. G. Cormack, D. C. Sherrington and E. Khoshdel, *Angew. Chem., Int. Ed.*, 2003, **42**, 5336.
84. K. Yoshimatsu, K. Reimhult, A. Krozer, K. Mosbach, K. Sode and L. Ye, *Anal. Chim. Acta*, 2007, **584**, 112.
85. L. Ye, P. A. G. Cormack and K. Mosbach, *Anal. Commun.*, 1999, **36**, 35.
86. B. Zu, G. Pan, X. Guo, Y. Zhang and H. Zhang, *J. Polym. Sci., Part A: Polym. Chem.*, 2009, **47**, 3257.
87. G. Pan, B. Zu, X. Guo, Y. Zhang, C. Li and H. Zhang, *Polymer*, 2009, **50**, 2819.
88. G. Pan, Y. Zhang, Y. Ma, C. Li and H. Zhang, *Angew. Chem., Int. Ed.*, 2011, **50**, 11731.
89. Y. Ma, Y. Zhang, M. Zhao, X. Guo and H. Zhang, *Chem. Commun.*, 2012, **48**, 6217.
90. J. Li, B. Zu, Y. Zhang, X. Guo and H. Zhang, *J. Polym. Sci., Part A: Polym. Chem.*, 2010, **48**, 3217.
91. L. Chimuka, M. van Pinxteren, J. Billing, E. Yilmaz and J. Å. Jönsson, *J. Chromatogr., A*, 2011, **1218**, 647.
92. S. Zorita, B. Boyd, S. Jönsson, E. Yilmaz, C. Svensson, L. Mathiasson and S. Bergström, *Anal. Chim. Acta*, 2008, **626**, 147.
93. A. G. Mayes and K. Mosbach, *Anal. Chem.*, 1996, **68**, 3769.
94. D. Vaihinger, K. Landfester, I. Krauter, H. Brunner and G. E. M. Tovar, *Macromol. Chem. Phys.*, 2002, 1965.
95. N. Pérez, M. J. Whitcombe and E. N. Vulfson, *J. Appl. Polym. Sci.*, 2000, **77**, 1851.
96. N. Pérez, M. J. Whitcombe and E. N. Vulfson, *Macromolecules*, 2001, **34**, 830.
97. N. Marchyk, J. Maximilien, S. Beyazit, K. Haupt and B. Tse Sum Bui, *Nanoscale*, 2014, **6**, 2872.
98. M. M. Titirici, A. J. Hall and B. Sellergren, *Chem. Mater.*, 2002, **14**, 21.
99. A. Nematollahzadeh, W. Sun, C. S. A. Aureliano, D. Lütkemeyer, J. Stute, M. J. Abdekhodaie, A. Shojaei and B. Sellergren, *Angew. Chem., Int. Ed.*, 2011, **50**, 495.

100. A. Poma, A. Guerreiro, M. J. Whitcombe, E. V. Piletska, A. P. F. Turner and S. A. Piletsky, *Adv. Funct. Mater.*, 2013, **23**, 2821.
101. S. Ambrosini, S. Beyazit, K. Haupt and B. Tse Sum Bui, *Chem. Commun.*, 2013, **49**, 6746.
102. J. Xu, S. Ambrosini, E. Tamahkar, C. Rossi, K. Haupt and B. Tse Sum Bui, *Biomacromolecules*, 2016, **17**, 345.
103. A. Mehdinia, T. Baradaran Kayyal, A. Jabbari, M. O. Aziz-Zanjani and E. Ziaei, *J. Chromatogr., A*, 2013, **1283**, 82.
104. A. Mehdinia, S. Dadkhah, T. Baradaran Kayyal and A. Jabbari, *J. Chromatogr., A*, 2014, **1364**, 12.
105. S. Wang, Y. Li, M. Ding, X. Wu, J. Xu, R. Wang, T. Wen, W. Huang, P. Zhou, K. Ma, X. Zhou and S. Du, *J. Chromatogr., B: Anal. Technol. Biomed. Life Sci.*, 2011, **879**, 2595.
106. Q. Q. Gai, F. Qu, Z. J. Liu, R. J. Dai and Y. K. Zhang, *J. Chromatogr., A*, 2010, **1217**, 5035.
107. C. Gonzato, M. Courty, P. Pasetto and K. Haupt, *Adv. Funct. Mater.*, 2011, **21**, 3947.
108. Y. Li, X. Li, J. Chu, C. Dong, J. Qi and Y. Yuan, *Environ. Pollut.*, 2010, **158**, 2317.
109. Z. Xu, L. Ding, Y. Long, L. Xu, L. Wang and C. Xu, *Anal. Methods*, 2011, **3**, 1737.
110. J. Dai, J. Pan, L. Xu, X. Li, Z. Zhou, R. Zhang and Y. Yan, *J. Hazard. Mater.*, 2012, **205–206**, 179.
111. S. Beyazit, B. Tse Sum Bui, K. Haupt and C. Gonzato, *Prog. Polym. Sci.*, 2016, **62**, 1.
112. A. Zengin, E. Yildirim, U. Tamer and T. Caykara, *Analyst*, 2013, **138**, 7238.
113. M. R. Halhalli, C. S. A. Aureliano, E. Schillinger, C. Sulitzky, M. M. Titirici and B. Sellergren, *Polym. Chem.*, 2012, **3**, 1033.
114. Y. Liu, Y. Huang, J. Liu, W. Wang, G. Liu and R. Zhao, *J. Chromatogr., A*, 2012, **1246**, 15.
115. W. Jakubowski, K. Min and K. Matyjaszewski, *Macromolecules*, 2006, **39**, 39.
116. L. Xu, J. Pan, J. Dai, Z. Cao, H. Hang, X. Li and Y. Yan, *RSC Adv.*, 2012, **2**, 5571.
117. B. P. Fors and C. J. Hawker, *Angew. Chem., Int. Ed.*, 2012, **51**, 8850.
118. Z. Adali-Kaya, B. Tse Sum Bui, A. Falcimaigne-Cordin and K. Haupt, *Angew. Chem., Int. Ed.*, 2015, **54**, 5192.
119. F. Barahona, E. Turiel, P. A. G. Cormack and A. Martín-Esteban, *J. Polym. Sci., Part A: Polym. Chem.*, 2010, **48**, 1058.
120. A. D. Vaughan, S. P. Sizemore and M. E. Byrne, *Polymer*, 2007, **48**, 74.
121. S. Subrahmanyam, A. Guerreiro, A. Poma, E. Moczko, E. Piletska and S. Piletsky, *Eur. Polym. J.*, 2013, **49**, 100.
122. C. Baggiani, P. Baravalle, G. Giraudi and C. Tozzi, *J. Chromatogr., A*, 2007, **1141**, 158.
123. M. Panagiotopoulou, Y. Salinas, S. Beyazit, S. Kunath, L. Duma, E. Prost, A. G. Mayes, M. Resmini, B. Tse Sum Bui and K. Haupt, *Angew. Chem., Int. Ed.*, 2016, **55**, 8244.

124. S. Beyazit, S. Ambrosini, N. Marchyk, E. Palo, V. Kale, T. Soukka, B. Tse Sum Bui and K. Haupt, *Angew. Chem., Int. Ed.*, 2014, **53**, 8919.
125. P. K. Ivanova-Mitseva, A. Guerreiro, E. V. Piletska, M. J. Whitcombe, Z. Zhou, P. A. Mitsev, F. Davis and S. A. Piletsky, *Angew. Chem., Int. Ed.*, 2012, **51**, 5196.
126. X. Kan, Y. Zhao, Z. Geng, Z. Wang and J.-J. Zhu, *J. Nanosci. Nanotechnol.*, 2011, **11**, 1217.
127. H. Y. Lee and B. S. Kim, *Biosens. Bioelectron.*, 2009, **25**, 587.
128. C. L. Choong, J. S. Bendall and W. I. Milne, *Biosens. Bioelectron.*, 2009, **25**, 652.
129. X. Kan, Y. Zhao, Z. Geng, Z. Wang and J. J. Zhu, *J. Phys. Chem. C*, 2008, **112**, 4849.
130. G. Thompson, R. C. Furneaux, G. C. Wood, J. A. Richardson and J. S. Goode, *Nature*, 1978, **272**, 433.
131. F. Li, L. Zhang and R. M. Metzger, *Chem. Mater.*, 1998, **10**, 2470.
132. A. V. Linares, F. Vandevelde, J. Pantigny, A. Falcimaigne-Cordin and K. Haupt, *Adv. Funct. Mater.*, 2009, **19**, 1299.
133. F. Vandevelde, A. S. Belmont, J. Pantigny and K. Haupt, *Adv. Mater.*, 2007, **19**, 3717.
134. J. Zdunek, E. Benito-Peña, A. Linares, A. Falcimaigne-Cordin, G. Orellana, K. Haupt and M. C. Moreno-Bondi, *Chem.–Eur. J.*, 2013, **19**, 10209.
135. H. H. Yang, S. Q. Zhang, F. Tan, Z. X. Zhuang and X. R. Wang, *J. Am. Chem. Soc.*, 2005, **127**, 1378.
136. H. H. Yang, S. Q. Zhang, W. Yang, X. L. Chen, Z. X. Zhuang, J. G. Xu and X. R. Wang, *J. Am. Chem. Soc.*, 2004, **126**, 4054.
137. Y. Li, X. Yin, F. Chen, H. Yang, Z. Zhuang and X. Wang, *Macromolecules*, 2006, **39**, 4497.
138. F. Vandevelde, T. Leïchlé, C. Ayela, C. Bergaud, L. Nicu and K. Haupt, *Langmuir*, 2007, **23**, 6490.

CHAPTER 7

Theoretical and Computational Strategies in Molecularly Imprinted Polymer Development

IAN A. NICHOLLS*[a], GUSTAF D. OLSSON[a], BJÖRN C. G. KARLSSON[b], SUBRAMANIAN SURIYANARAYANAN[a] AND JESPER G. WIKLANDER[a]

[a]Bioorganic & Biophysical Chemistry Laboratory, Linnæus University Centre for Biomaterials Chemistry, Department of Chemistry & Biomedical Sciences, Linnæus University, SE-391 82 Kalmar, Sweden; [b]Physical Pharmacy Laboratory, Linnæus University Centre for Biomaterials Chemistry, Department of Chemistry & Biomedical Sciences, Linnæus University, SE-391 82 Kalmar, Sweden

*E-mail: ian.nicholls@lnu.se

7.1 Introduction

The molecular imprinting concept[1–5] is founded on the role that template–monomer interactions present in the pre-polymerization mixture have on the structure and recognition properties of the resulting molecularly imprinted polymer (MIP). As the characteristics of the recognition sites is a direct function of the interactions present in the pre-polymerization mixture, *e.g.*

Polymer Chemistry Series No. 28
Molecularly Imprinted Polymers for Analytical Chemistry Applications
Edited by Wlodzimierz Kutner and Piyush Sindhu Sharma

Published by the Royal Society of Chemistry, www.rsc.org

template–monomer, monomer–monomer, solvent–template/monomer, *etc.*, an appreciation of the physical rules governing the formation of monomer–template complexes is essential for understanding the imprinting process. This insight into molecular recognition processes is necessary if we are to achieve true rational design of polymerization systems for producing materials with predetermined recognition properties.

The molecular-level events underlying the synthesis of MIPs and the polymer–ligand recognition properties of an imprinted material can in principle be described in thermodynamic terms.[6] Moreover, with the aid of computational techniques, models can be developed to describe the molecular-level events present in the pre-polymerization mixture and even to characterize polymer–ligand interactions.[7–9] In this chapter, we first provide a background to the thermodynamic factors that influence the molecular imprinting process before moving to the use of computational methods for studying molecular imprinting systems.

7.2 Molecular Imprinting from a Thermodynamic Perspective

Understanding the physical factors governing whether or not two molecules interact has attracted the interest of researchers for decades. Of note are Jenck's paradigms that were employed by a number of groups in attempts to achieve this goal.[10,11] The independently developed semi-empirical approaches of Andrews[12] and Williams[13–15] each attempted to define the physical terms that regulate binding (recognition) events, however it is perhaps Williams' more detailed treatment,[13,16] eqn (7.1), that is best used to describe aspects of molecular imprinting systems as reflected in a series of studies;[6,17,18]

$$\Delta G_{\mathrm{bind}} = \Delta G_{\mathrm{t+r}} + \Delta G_{\mathrm{r}} + \Delta G_{\mathrm{h}} + \Delta G_{\mathrm{vib}} + \Sigma\Delta G_{\mathrm{p}} + \Delta G_{\mathrm{conf}} + \Delta G_{\mathrm{vdW}} \tag{7.1}$$

where the Gibbs free energy changes are: ΔG_{bind}, complex formation; $\Delta G_{\mathrm{t+r}}$, translational and rotational; ΔG_{r}, restriction of rotors upon complexation; ΔG_{h}, hydrophobic interactions; ΔG_{vib}, residual soft vibrational modes; $\Sigma\Delta G_{\mathrm{p}}$, the sum of interacting polar group contributions; ΔG_{conf}, adverse conformational changes; and ΔG_{vdW}, unfavourable van der Waals interactions.

The simplicity of the molecular imprinting concept, as reflected in the highly schematic representations of the imprinting process depicted in many papers, belies the actual complexity of these polymerization systems.

As the molecular memory of MIPs arises from the formation of template–functional monomer solution adducts in the pre-polymerization reaction mixture, a system under thermodynamic control, the position of the equilibrium for formation of self-assembled solution adducts between templates and attendant monomers is reflected in the free energy of binding, ΔG_{bind}. The position of the equilibrium also reflects the number of resultant sites, and the degree of receptor site heterogeneity. Accordingly, the more stable and regular the ensemble of template–functional monomer complexes is, the greater the number, and fidelity, of resultant MIP receptors. The extent of "optimal" template coordination by a functional monomer is, in turn, dependent on the nature of the chemical components in the polymerization mixture, in addition to the physical environment (temperature and pressure). Notably, for examples employing reversible covalent functional monomer–template interactions, although the complex stability is guaranteed with respect to the template and functional monomer, a certain degree of van der Waals derived shape complementarity and even electrostatic interactions should be expected from interactions with cross-linking monomers.

Direct evidence for the formation of non-covalent monomer–template interactions, the extent of which is reflected in the total binding term, ΔG_{bind}, was first presented in an elegant NMR study by Sellergren *et al.*[19] In this seminal paper, a combination of line-broadening and chemical shift arguments was used by the authors to confirm the formation of functional monomer (methacrylic acid, MAA)–template (phenylalanine anilide) complexes. Their results also suggested the minor presence of template self-association and higher order complexes. More recent molecular dynamics (MD)-based computational studies[20] have helped confirm, in part, this hypothesis (see below).

Subsequently, numerous studies deploying spectroscopic methods have been used to calculate the dissociation constants for the solution adducts and the relative concentration of fully complexed template in polymerization mixtures.[21–27] Importantly, spectroscopic methods have also been used to investigate other aspects of the events present in pre-polymerization mixtures, such as cross-linker interactions and self-association.[27,28]

In spectroscopic titration studies, such as those cited above, it is observed that moving to higher (functional monomer)-template ratios, in order to push the equilibrium towards complex formation, leads to increased numbers of randomly oriented functional groups, which in turn leads to increased levels of non-specific binding in resultant MIPs. Much higher ratios, a non-imprinted polymer being the extreme case, lead to lower site population densities in the polymer. The need for further studies of such fundamental aspects of molecular imprinting is reflected in the results from an early, detailed study by Mayes and Lowe[29] where they demonstrated that the number and quality of high affinity sites selective for morphine was similar to that in polymers prepared using just 2% of the amount of template used in a standard protocol.[30]

Turning to the other terms in eqn (7.1), the ΔG_{t+r} term is the change in translational and rotational Gibbs free energy associated with combining two or more free entities in a complex, a process that is entropically unfavourable. This term carries implications for the order (number of components) of complexes that may be formed. It can be stated that functional monomer systems capable of multiple simultaneous interactions of correct geometry, relative to multiple single point interacting monomers, should yield higher concentrations of complexed template, due to a reduction in the adverse loss of translational and rotational free energy, an entropically unfavourable process. From a semantic perspective, the use of multi-dentate monomers in molecular imprinting may be seen as lying between the concepts of traditional non-covalent molecular imprinting (mono-dentate monomers) and supramolecular chemistry. Over recent years, some very elegant examples have appeared in the literature, in particular from the groups of Takeuchi[31] and Sellergren and Hall.[32,33] The power of such monomers is reflected in their use in the selective recognition of, *e.g.*, phosphorylated peptides in biological fluids.[34]

The selectivity demonstrated for MIPs prepared with rigid templates are superior to those of less rigid structures, which is a direct consequence of the ΔG_r term. The origin of this effect lies in the fact that the more rigid the structure, the lower the number of possible solution conformations and thus solution complexes. This, in turn, renders a narrower site distribution in the resultant polymer. That the highest MIP–ligand affinities thus far observed have been for rigid templates, such as the alkaloids morphine[30] and yohimbine,[35] provides support for this statement. This is further underscored by the affinity of the morphine MIP for its template being superior to that of the opioid receptor binding neuropeptide leucine enkephalin for an enkephalin MIP, even though the peptide has a greater number of potential electrostatic interaction points.

The number and relative strengths of different template–functional monomer interactions, as given by the $\Sigma\Delta G_p$ term, dictates the degree of selectivity of the resultant polymer.[36] In terms of electrostatic interactions, several reports have been made of enhanced selectivity through the use of more strongly interacting functional monomers that favour template–monomer complex stability. The Takeuchi group has presented a series of studies using trifluoromethylacrylic acid as the functional monomer[37–40] and demonstrated that it can be used to augment template recognition. The lower pK_a of this monomer enhances ion pairing and ion–dipole interactions.

Metal ions have been successfully used to coordinate templates in a range of studies. The possibility of multiple interactions with the template, in conjunction with the greater relative strengths of template-metal ion interactions, lends to their use in highly polar solvents, such as methanol, as exemplified by an early work from the Arnold group's use of Cu(II) coordination of imidazole-containing templates.[41] It is of interest that metal ions have been successfully deployed in catalytic MIPs, were the metal ion's capacity to engage in multiple simultaneous interactions is used to coordinate the reacting moieties, *e.g.* Co(II),[42,43] Ti(IV),[44] and Zn(II).[45] The use of crown ethers by Andersson and Ramström provided an interesting approach for solubilizing highly polar, cationic, *e.g.* Zwitter ionic, species at low polarities.[46] The value of this approach has been seen through the use of polymerizable crown ether derivatives.[47]

An examination of the literature shows that the vast majority of reported MIP systems are prepared and evaluated using organic (non-polar) media. The ligand selectivity that is often demonstrated reflects the strengths of electrostatic interactions in non-polar media. However, for applications in environmental engineering or biomedical analysis, where the targets of interest are often water soluble and incompatible with many conventional polymerization systems *e.g.* peptides, proteins, oligonucleotides, and sugars, other approaches are necessary. The use of water as the solvent of polymerization opens up hydrophobic moiety selective functional monomers, which can contribute to solution adduct formation through the hydrophobic effect, the contribution of which is reflected in the ΔG_h term of eqn (7.1). Alternatively, or as a complement, as mentioned above, functional monomers capable of sustaining electrostatic interactions at such high polarities, *e.g.* metal ion chelation, can be employed. With this in mind, a number of groups have developed water-compatible polymerization systems based upon hydrophobic moiety selective functional monomers, in particular based upon cyclodextrin-based monomers, where the hydrophobic inside of the cyclodextrin toroidal structure has been used to efficiently recognize the hydrophobic domains of, most predominantly, steroids[48,49] and amino acids.[50,51]

To summarize, each of the thermodynamic terms contributing to the stability and uniformity of the solution adducts formed during the molecular imprinting pre-arrangement play a role in determining ΔG_{bind}. Conclusively, the exclusive use of non-covalent template–functional monomer interactions is inherently limited by the stability of the template–(functional monomer) complexes. The subsequent heterogeneity of the template-selective imprinted cavities bearing recognition sites, their polyclonal nature, is a direct consequence of the extent of template complexation. A small number of studies have attempted to combine the versatility of non-covalent interaction based molecular imprinting with the benefits afforded by the constraint of functional monomer and template through reversible covalent bonds,[52–55] which in a number of cases has been described as "stoichiometric imprinting". This approach, although requiring the synthesis of template derivatives, offers much scope for enhancing the fidelity of template selective imprinted cavities. How these terms influence the non-covalent MIP-ligand recognition shall be discussed in the following Section 7.3.

7.3 Computational Strategies for Studying and Developing Molecular Imprinting Systems

7.3.1 Introduction

The development of molecular imprinting, and not least the increasing number of applications of molecularly imprinted materials has driven the establishment of a variety of *in silico* tools for the study of various aspects of MIP systems and for MIP design. These tools can provide molecular-level insights concerning many aspects of MIP systems ranging from events in the pre-polymerization stage, through polymer morphology, to factors regulating polymer–ligand interactions.

Historically, results from empirical studies of ligand-polymer recognition characteristics have been used to deduce the nature of molecular-level events underlying MIP behaviour. Later, thermodynamic models of the molecular imprinting process[6,17,18,56,57] were used to gain insights into aspects of both events in the pre-polymerization stage and of polymer-ligand recognition, as discussed earlier in this chapter. A more recent contribution to this area was the use of stochastic simulations of interactions of pre-polymerization mixture components,[58] where monomer–template binding affinities were used in a stochastic algorithm[59] to position monomer–template units in a lattice matrix. A cross-linking monomer was subsequently added, and template models then removed to leave imprinted cavity models that were assessed with respect to their heterogeneity. The MIPs studied using the simulated methods demonstrated performance comparable to the corresponding synthesized polymeric materials.

The primary factor underlying the last decade's rapid growth in the use of computational tools in molecular imprinting science and technology is the improvement in computing power and general availability of suitable software.[60] Together, these developments have made possible the use of multivariate analyses, quantum mechanical calculations and large-scale MD studies. Here, we review the current status of computational strategies in molecular imprinting and we focus on methods used in the study of electronic structure, for MD and other, *e.g.* statistics-based approaches, where each section includes a brief general presentation of the underlying techniques.

7.3.2 Electronic Structure Methods

Computational electronic structure-based methods, *e.g. ab initio*, semi-empirical, and density functional strategies, are becoming common in conjunction with both designing and evaluating MIPs. Methods and basis sets are generally developed to provide an *acceptably correct* description of a system. While high degrees of accuracy are desirable, this goal is always in balance with the cost of computation. These methods describe the electronic structure of a molecular species and can therefore provide the basis for accurate and highly detailed descriptions of potential non-covalent interactions between the studied species. In the majority of reports using electronic structure-based methods, the primary purpose has been to describe putative interactions present in the MIP pre-polymerization stage.

On account of the limits imposed by the demands on both hardware and time to perform such calculations, the most common use of electronic structure determining methods is for solvation and characterization of putative complexes of a template with monomers. Of the electronic structure determining methods, the semi-empirical strategies are less demanding on computational resources. For example, studies of a complex of (*S*)-nilvadipine with 4-vinylpyridine (4VP)[61] have been performed using the AM1 semi-empirical method, as well as studies involving the optimization of a 2,4,6-trichlorophenol–4VP complex in a 1:4 stoichiometry.[62] The approach of studying relative binding energies obtained from calculations on complexes between the template and one or a set of different functional monomers and/or different stoichiometry ratios has become a more frequently used strategy for choosing functional monomers. The AM1 method was used in a study by Holdsworth *et al.*,[63] where calculated binding energies of template-monomer complexes of cocaine with 1–14 monomers of either MAA or 4VP were used to optimize polymerization mixture composition *i.e.*, the template–(functional monomer) ratio. The results from template–monomer stoichiometry optimization studies were validated by both NMR and analyte binding studies of subsequently synthesized MIPs, confirming the viability of these approaches. Ogawa *et al.*[64] included AM1 based calculations of template–monomer complexes, where structural parameters obtained from performed calculations were correlated to experimental results from chromatographic evaluation, in the production of quinine selective MIPs. In a slightly different approach, Lai and Feng[65] used molecular structures of buffer acids and bases derived from AM1 based calculations to show a correlation between MIP binding of the template metformin in different media based on an assumed mechanism for competitive binding. In another approach, Meng *et al.*[66] used AM1 based calculations to derive interaction energies between functional monomers and putative reaction intermediates in a screening approach to select monomers for the design of a MIP-based transesterification catalyst. Further examples of complex studies in MIP production, characterization or optimization include the use of AM1 based calculations in the design of a MIP selective for *N*,*O*-dibenzylcarbamate,[67] as well as studies targeting theophylline[68] in combination with density functional theory (DFT) calculations at the B3LYP/6-31+G**//B3LYP/3-21G level of theory. Voshell and Gagné[69] employed a combination of semi empirical AM1 and *ab initio* HF/6-31G* calculations to determine the conformational rigidity of a dendritic system in efforts to produce BINOL MIPs. An optimal rigid dendrimeric structure was identified that enhanced polymer enantioselectivity and decreased heterogeneity of imprinted cavities bearing recognizing sites.

The PM3 method, another semi-empirical method, was used in the description of two complexes formed between (*S*)-naproxen and one or two molecules of acrylamide (AAM).[70] Similar template–monomer interaction energy screening approaches were adopted by Luliński *et al.*,[71] in a study aimed at designing a dopamine MIP using PM3, by Boysen *et al.*[72] in the pre-production optimization of composition and stoichiometry of template and functional monomers as well as in a slightly modified approach by Wu and Li[73] where PM3 calculations were used to produce an equilibrated pre-polymerization complexes of the MAA functional monomer with nicotinamide and *iso*-nicotinamide templates. By restraining the positions of monomers and removing the template, a binding site model was created that was probed with ligand structures to determine interaction energies, producing results with strong correlation to experimentally determined retention factors. Wu *et al.*[74] used a combination of PM3 and MP2/6-31G//HF/6-31G calculations to demonstrate a relationship between capacity factors determined experimentally and calculated binding energies derived from computational studies of various (functional monomer)-template complexes. In a novel study, Rathbone *et al.*[75] used PM3 based calculations in the design of a cytochrome CYP2D6 mimic where templates to be imprinted were chosen based on superposition studies of candidate molecule geometries on known CYP2D6 substrates. PM3 calculations were also implemented in the design of an ester hydrolysis-catalysing polymer, adding support for the theory that the template used in MIP synthesis is an analogue of the transition-state structure in the reaction to be catalysed.[76] Wu and Li[77] managed to produce an explanation for the failed attempt to produce picolinamide MIPs based on results from PM3 calculations, and subsequently developed a Cu(II) complex-based system displaying selectivity towards the target, a strategy that was later extended towards recognition of small organic acids.[78]

The approach of template–monomer complex studies has been utilized in several investigations using different methods, basis sets and levels of theories. He *et al.*[79] screened a functional monomer candidate through evaluation of modelled template-(functional monomer) complexes using HF/6-31G(d, p) calculations where functional monomers were one of AAM, *p*-vinylbenzoic acid or 4VP in production of baicalein selective MIPs. A similar approach was employed utilizing density functional methods by Pietrzyk *et al.*[80] in the study of a model complex between melamine and three functional monomer molecules using DFT at the B3LYP/3-21G(d) level of theory.

DFT calculations using the B3LYP functional are, now, one of the more commonly adopted methods in electronic studies for MIP evaluation and design, due to their high accuracy at a still relatively low computational cost. There are a growing number of examples where the B3LYP functional has been implemented in modelling of template–monomer complexes. These examples include a study by Demircelik *et al.*[81] where the 6-31G(d,p) basis set was implemented in the preparation of polymeric beads with selectivity towards Al^{3+} ions for plasma extraction/removal, by Azimi *et al.*,[82] where MIPs targeting hydroxyzine were prepared after evaluation of (cross-linking monomer)/(functional monomer)–template complex stoichiometry using a combination of PM3, HF/6-31G and B3LYP/6-31G based calculations and by Riahi *et al.*,[83] using a 6-311+G(d,p) basis set in the production of polymeric materials selective for chlorphenamine. This latter study also accounted for solvent effects through inclusion of a polarizable continuum model (PCM).

A clear limitation associated with electronic structure methods is the difficulty associated with handling the large number of atoms necessary to provide a holistic model of the pre-polymerization or polymer system. This difficulty often leads to the exclusion of components deemed "non-interacting". One component omitted from many calculations is solvent, though this has been shown to be problematic.[84,85] Treating solvent explicitly is generally not achievable in electronic structure calculations due to the limitations of computational time and resources. To address this problem, methods such as the PCM were developed providing the possibility of solvent representation without inclusion of explicit solvent molecules,[86] where the effect of the solvent is approximated by surrounding the model(s) with a surface that is polarizable according to the electric permittivity of the solvent, like a polarizable cavity. The accuracy of this approach has also been questioned[84] since not all solvents can be modelled adequately using a PCM model. This constitutes a major drawback of the method, stemming from the inability of a PCM to produce representations of hydrogen bonds between solvent molecules and structures that would be competing with hydrogen bond interactions between the template and functional monomers. This was exemplified in a study by Liu *et al.*[87] where the solvation energies obtained from PCM calculations were compared to calculated template–solvent interaction energies, using a B3LYP/6-31+G(d,p) basis set and level of theory. Dong *et al.*[88] presented an alternative way of using the energies of solvation of a template and functional monomers in different solvents as a measure of potential competition for interaction with the solvent using B3LYP/6-31+G(d,p) calculations including a PCM model. In an interesting alternative approach, Pardeshi *et al.*[89] employed DFT based B3LYP calculations using a 6-31G(d,p) basis set and a self-consistent (reaction field)-PCM in studies of solvent effects on the stability of template–monomer complexes in optimization of MIP compositions in the design of gallic acid MIPs. Wu *et al.*[90] used a similar approach in screening the stability of equilibrated monomer–template complexes in different solvents including PCM models representative of different solvents using MP2/6-311+G*//B3LYP/6-311G* calculations. It was shown that, as long as the solvent is aprotic, this approach could be used to predict solvent induced effects on template retention and selectivity characteristics of MIPs. In extensions of this approach, Diñeiro *et al.*[91,92] performed B3LYP/6-311+G**//B3LYP/6-31G* calculations including solvent implicitly through PCMs in an effort to screen and select both the functional monomer and porogen for pre-production optimization in the synthesis of homovanillic acid MIPs. Moreover, de Barros *et al.*[93] reported on the use of DFT B3LYP/6-31G(d) calculations to screen both potential functional monomers as well as porogenic solvents through inclusion of PCMs in an elaborate study aiming to produce fenitrothion selective MIPs. Dong *et al.*[94] developed another approach for monomer and porogen selection in which B3LYP/6-31G(d) calculations of candidate systems were performed after first being evaluated with MD simulations (discussed below). Similarly, Luliński *et al.*[95] used a combination of MD and DFT B2LYP/6-311+G(d,p) calculations and a PCM, pre-equilibrating template-(functional monomer)–(cross-linking monomer) complexes using MD and evaluating association energies using DFT, in their work on tyramine-selective MIPs.

Further examples of studies that have made use of a PCM to reproduce solvent effects have been performed by Ahmadi *et al.*[96,97] using DFT B3LYP/6-31G+(d,p) calculations and a PCM in the production of enantioselective (*S*)-warfarin targeting MIPs and in the design of MIPs for selective extraction of metaproterenol from human blood plasma for implementation in drug-screening using a combination of PM3 and DFT based B3LYP/6-31G(d,p) calculations. Huynh *et al.*[98] employed B3LYP/3-21Gn calculations of a template cavity model in evaluations of novel functional monomers for production of nicotine MIPs and Liu *et al.*[99] used B3LYP based calculations at a 6-31G(d,p) level of theory in the stoichiometry optimization of template–(functional monomer) composition in the production of enrofloxacin selective MIPs. Qi *et al.*[100] used electronic structure methods in the design of MIPs targeting a group of six carbamate pesticides, applying DFT B3LYP/6-31G(d) calculations in a screening protocol of different monomers against a template structure in different solvents through inclusion of PCMs. Qin *et al.*[101] employed DFT B3LYP/6-311++G(d,p) calculations in selection of an optimal functional monomer for production of imprinted dibenzothiophene polymers on the surface of carbon microspheres. Tadi *et al.*[102] used B3LYP/6-31+G(d,p)//B3LYP/6-31G(d) DFT calculations in the selection of a functional monomer through monomer–template complex screening in the production of MIPs targeting sulphanilamide. Yang *et al.*[103] implemented DFT B3LYP/6-311G calculations screening both monomers and stoichiometry in the pre-production optimization of MIPs for selective removal of quinolone from octane. Alizadeh[104] included B3LYP/6-31G calculations in the development of a pyridoxine MIP. Li *et al.*[105] based the selection of a functional monomer for the production of chlorophenol MIPs on results from B3LYP/6-31G(d,p) calculations. Moreover, harmane selective MIPs were fabricated based on functional monomer screening using B3LYP/6-31G(d,p) calculations by Kowalska *et al.*[106] A recognition mechanism was suggested by Wang *et al.*[107] based on results from B3LYP/6-311G** calculations, explaining the selectivity of an *N*-(4-isopropylphenyl)-*N*′-butyleneurea MIP. The mechanism of selective MIP recognition of the theophylline template was proposed by Che *et al.*[108] based on B3LYP/6-31G(d,p) calculations in combination with 2-D IR-spectroscopy. The MIP-ligand interaction mode was studied and described by Christoforidis *et al.*,[109] for semiquinone MIPs using a combination of electron spin echo envelope modulation (ESEEM) spectroscopy and a combination of B3LYP and MP2 calculations.

A number of studies using approaches other than purely semi-empirical or DFT based calculations have been reported. These include, as seen above, combinations of both *ab initio*, semi-empirical and DFT based calculations. Examples of approaches based on *ab initio* calculations include a study by Gholivand *et al.*[110] aiming to produce furosemide selective MIPs using HF/6-31G(d) calculations with a PCM of a 1:3 template–monomer stoichiometry of different monomers, by El Gohary *et al.*[111] engaged in the fabrication of MIPs targeting famiciclovir to be included in a biosensor application using HF/6-31G(d) calculations with a PCM, and by Yao *et al.*[112] using MP2/6-31++G(d) calculations to examine mechanisms in an aniline templated MIP system. Moreover, Appell *et al.*[113] included template–monomer complex optimization and binding energy studies using *ab initio* based MP2/6-31+G* calculations in efforts to develop and optimize fusaric acid selective imprinted materials. Efforts at preparing MIPs targeting nicotinamide were undertaken by Del Sole *et al.*[114] using a range of calculations including different density functional methods and *ab initio* MP2 with several basis sets. Mukawa *et al.*[115] explained the relative selectivity of an allyl phenyl disulfide containing MIP towards phenol and thiophenol by comparing differences in strengths of the hydrogen bonds formed between the MIP and ligands, calculated at the HF/6-31G* level of theory.

Most polymer systems studied to date have been organic polymers, though with some interesting alternatives. A seminal case was the work by Tada *et al.*[116] with a silica-based system where Rh–amine complex imprinted materials displayed shape-selectively in the catalysis of hydrogenation reactions. In this study, the Perdew–Wang correlation (PWC) functional was used in combination with double-zeta polarized numerical basis set (DNP) based calculations to characterize the complexation of the metal centre, with the catalytically active species being studied in subsequent experiments. In another study, Azenha *et al.*[117] worked with β-damascenone selective silica-based materials, where they performed *ab initio* HF and DFT B3LYP calculations using different basis sets.

The frequency of reports of quantum chemical calculations in MIP design and characterization is increasing rapidly. To date, most approaches have primarily focused on sets of interactions in the pre-polymerization mixtures or isolated non-solvated molecular complexes *in vacuo*. However, the constantly increasing availability of computational power in conjunction with novel mixed approaches where electronic structure and MD simulations are combined should lead to an increased use of electronic structure determining methods in the study and design of molecular imprinting systems.

7.3.3 Molecular Dynamics

In comparison to the high-level electronic structure methods described above, MD simulations place a lower demand on computational resources, making them more accessible to researchers. MD simulations are force field-based techniques[118] with the foundation for the current methodologies cemented in the pioneering work of Alder and Wainwright in 1957[119] in their study of a gaseous argon model system. MD simulations typically do not allow direct study of processes involving bond formation and breaking because electrons are not represented explicitly, though as a trade-off, they do allow for studies of much larger molecular systems compared to *ab initio* and semi-empirical methods, thus enabling their use in studies of the ensembles of molecular complexes that are formed between components in the pre-polymerization mixtures. The increasing number of application areas for MD studies has contributed to a rapid development of numerous software packages and force fields aimed at studying different types of systems, such as proteins and protein folding,[120,121] DNA conformal changes,[122,123] phospholipid bilayer systems,[124,125] and membrane transport of pharmaceutically active compounds.[126] Moreover, MD simulations have successfully been used for physical characterization of solvents,[127] surfaces[128,129] and biomolecular interactions.[130–132] Examples of frequently used force fields include AMBER,[133] GAFF,[134] CHARMM,[135] OPLS[136] and GROMOS.[137,138]

The use of MD simulations in the study of molecularly imprinted systems has primarily focused on events in the pre-polymerization mixture, and on the assumption that the recognition properties of MIPs are generated during the pre-polymerization stage.[4,57,139–141] These include simulation of the nature, frequency and strength of monomer-template interactions correlated to the quality of imprints and final material performance. Studies of the molecular interplay present in the pre-polymerization stage, a liquid system, can provide insights that can be extrapolated and correlated to final MIP properties. Based on the comparably small computational demands required by (force field)-based calculations, as compared to electronic structure determination methods, MD simulations have an advantage due to the fact that they can allow studying of several thousands of molecules, including explicit solvent models.

The first examples were presented by the Piletsky group,[142] who in 2001 described a composition optimization approach where a virtual library of functional monomers was first screened against an ephedrine enantiomer in order to identify the most stable complexes. Next, a simulated annealing protocol was developed and used in investigations of the effect of solvent–monomer composition on template pre-polymerization complexation. Low-energy ensembles of solvated multimolecular complexes of ephedrine and functional- and cross-linking monomers were generated, including explicit solvent molecules. This strategy was later included in the successful preparation of MIPs targeting simazine,[140] cocaine, methadone and morphine,[143,144] creatinine,[145] and biotin.[146] Furthermore, it was used in the production of microcystin-LR MIPs[147,148] where the produced materials demonstrated recognition characteristics close to those of antibodies. Similar library screening approaches has been adopted by other researchers, including Wei *et al.*[149] In their study, 17-β-estradiol MIPs were fabricated after computational evaluation of solvated (explicit solvent models of acetone or chloroform) 1 : 1 and 8 : 1 monomer-template complexes to include monomer dimerization effects. A library of nine different functional monomers was screened in this manner, identifying MAA, methacrylamide and 2-(diethylamino)ethyl methacrylate as promising candidate monomers based on strong hydrogen bond interactions with the template. The results were validated through MIP-analyte batch binding experiments. Further, Bakas *et al.*[150,151] implemented a form of the functional monomer library screening approach based on the leapfrog algorithm to identify promising functional monomers for the production of MIP based chemosensors for organophosphate insecticides. Similarly, Toro *et al.*[152] used functional monomer library screening simulations to identify promising functional monomers in the imprinting of the hexanizone herbicide for production of a biomimetic chemosensor and Abdin *et al.*[153] in the selection of monomers for the *in silico* design of a nano MIP chemosensor targeting lipopolysaccharide endotoxins.

Although the polymerization itself cannot be described using MD simulations, studies involving polymeric chains have been performed. Pavel and Lagowski[154,155] presented results from a study where energy differences were calculated for clusters of functional monomers in the absence and presence of a template in a search for potent functional monomer candidates in efforts to produce MIPs selective for theophylline and analogues. What made this study especially interesting was the inclusion of a MIP model in the evaluation, where pre-constructed polymer chains were simulated in a similar manner to evaluate effects of polymer growth on template complex stability while screening for an optimal functional monomer. Later, the approach was successfully implemented in the virtual screening of a number of warfare agents.[156] These studies aimed at characterization of the non-covalent (functional monomer)–template complex interactions, thus highlighting the importance of electrostatic interactions in driving the monomer-template complexation. Lv *et al.*[157] studied homo-polymer and template interactions in an effort to characterize the adsorption properties of dimethoate MIPs, identifying an optimal functional monomer through screening of different candidates. The results in this study were supported through chromatographic evaluations.

There are other examples of studies where attempts to simulate aspects of polymerized material have been attempted, *e.g.*, by including polymer-matrix models. Srebnik *et al.*[158–160] investigated analyte binding mechanisms and the importance of pores in the MIP matrix for template recognition and binding through a series of MD studies. Yungerman and Srebnik[161] investigated factors contributing to heterogeneity of imprinted cavities of MIPs using coarse-grained MD simulations. Results from simulations with polymer matrix models showed that a commonly observed low yield of quality imprints was a result of the pores and that, considering both size and shape of the template, the best performing MIPs should have a high degree of cross-linking. Moreover, low quality imprints could even originate from template aggregation during the pre-polymerization stage. Zhao *et al.*[162] included a cubic, pre-constructed MIP matrix model representative of a hydrogel network in the simulations performed. They investigated the dynamics of the hydrogel model, in the presence and absence of template, and its interaction with solvent, revealing a highly ordered water structure solvating the hydrogel. Apparently, modification of the carboxyl group content in the MIP matrix could be used to control the water structure and diffusion of water throughout the MIP network with a strong correlation between experimental template diffusion data and computational modelling results. Following up on this publication, the effect of the functional monomer charge was evaluated through modifications to the charge of a carboxyl group in the pre-constructed cubic MIP matrix model,[163] suggesting that analyte binding and recognition solely depended upon the mesh size of the network and that modifications to functional group charges would rather affect the polymer structure and solvation properties.

In efforts to elucidate the mechanisms underlying MIP selectivity and analyte binding properties in aqueous solutions, Molinelli *et al.*[164] studied the stability of (2,4-dichlorophenoxyacetic acid)-4VP template–(functional monomer) complexes in different explicitly represented solvents (water or chloroform). Based on the results of the characterization of hydrogen bond interactions in chloroform, combined with the observed π–π stacking interactions in water, the authors were able to produce a model describing the pre- and post-polymerization interaction pattern responsible for the formation of selective sites and subsequent selective analyte binding from aqueous solution.

The combination of different computational approaches has shown potential in minimizing the demand on resources, as well as producing descriptive results. Monti *et al.*[165] used a combination of MD, docking and site mapping procedures to identify a potent functional monomer through an elaborate screening protocol in the production of theophylline MIPs with results correlating to experimental data. Although the computational demand posted by studies on multimolecular representations of pre-polymerization mixtures including multiple copies of monomers, template and explicit solvent often limits the use of electronic structure methods directly, there are several examples where quantum mechanics (QM) calculations have been integrated in combination with MD methods in MIP related studies. Dong *et al.*[94] combined MD and QM based methods in an elegant screening study for monomer selection in the production of acetochlor MIPs. Initially, MD simulations of a functional monomer and template were performed using an explicit solvent to estimate interaction energies of molecular complexes. These estimations were followed by evaluation of complex energies using DFT B3LYP/6-31G(d) calculations, identifying three potential functional monomers. MIPs were then synthesized and evaluated, confirming the validity of the mixed MD/QM method. This procedure has since been implemented in the imprinting of both rhodamine B[166] and sulfadimidine.[167] Liu *et al.*[168] has recently combined MD-based investigations of (functional monomer)–template stoichiometry ratios with interaction energies obtained using semi-empirical AM1-PM3 calculations in the production of estrone and meolcarb MIPs.

Different polymerization formats and factors known to influence polymerization reactions, *e.g.* temperature and pressure,[169–171] can be incorporated into MD simulations. For instance, Yoshida *et al.*[172] investigated template–monomer complex stability by examining complex dynamics *in vacuo* and at a toluene–water interface in an emulsion based MIP production protocol. This examination was performed in an attempt to characterize (functional monomer)–template interactions in the production of surface imprinted polymers targeting tryptophan methyl ester using phenyl phosphonic acid monododecyl ester (*n*-DDP) to form the recognition site in a water-in-oil emulsion. The results indicated that the observed imprinting effect was based on the stability of *n*-DDP–tryptophan methyl ester complexes formed in the pre-polymerization emulsion. Toorisaka *et al.*[173] reported on the stability of a template–monomer complex at a water–toluene interface in the preparation of a catalytically active MIP through studies of the interaction of cobalt ion and one equivalent of alkyl imidazole and a substrate analogue, *N*α-*t*-Boc-L-histidine. More recently, Olsson *et al.*[174] presented the first all component MD simulation of a mini-emulsion pre-polymerization mixture used to produce bisphenol A (BPA) MIP nanoparticles. The simulation results were used to explain emulsion and particle properties, as well as the analyte binding behaviour of materials produced.

The past decade has led to a growing awareness of the importance of considering the entire ensemble of interactions present in a pre-polymerization mixture if accurate predictions are required. In efforts to explain observed unexpected behaviour during analyte binding experiments of nicotine MIPs,[36] the influence of template and monomer dimerization was investigated using spectroscopic methods.[23] Katz and Davis[175] also proposed template dimerization during the pre-polymerization stage as a reason for the formation of analyte selective sites and for the subsequent binding of the phenylalanine anilide analyte, proposing the imprinting of multimolecular template–template complexes based on experimental evaluation, a theory that was later further investigated using MD by Olsson *et al.*[20] Ansell *et al.*[24–26] investigated the impact of functional monomer dimerization in ephedrine recognition of MIPs, demonstrating experimental support for the importance of computationally approaching the pre-polymerization mixtures as all-component systems. This approach was further supported by Zhang *et al.*[176] who investigated monomer–monomer interactions and pointed to the importance of monomer dimerization at the pre- and post-polymerization stage during imprinting. In this study, the dimerization was evaluated based on statistical simulations, an area that will be addressed below.

To address the need for holistic treatments of pre-polymerization mixtures, a series of all-component and all-atom studies were presented, firstly by O'Mahony *et al.*[84] who used MD simulations to evaluate template–monomer interactions and the effect of template dimerization on the performance of naproxen MIPs, though excluding the initiator from simulated systems. The results from these studies showed a high degree of (functional monomer)–template interaction and a significant involvement of the cross-linking monomer in template complexation. O'Mahony *et al.*[177] also investigated the impact of template dimerization on MIP performance for quercetin MIPs, identifying sheet-like structures of template–monomer complexes that displayed a low dependence on the cross-linking monomer content. More recently, and building upon the observed template–template interactions, O'Mahony *et al.*[178] included template dimerization as an aspect of imprint formation in the development of a MIP for extracting BPA from milk.

Karlsson *et al.*[85] later published studies where all MIP components were represented in the performed simulations in stoichiometric ratios directly correlated to experimental mixtures. Thorough evaluations of performed simulations resulted in a comprehensive description of all non-covalent interactions in the pre-polymerization mixtures of bupivacaine MIPs, describing a correlation between computational and experimental results as well as an explanation for the recognizing site heterogeneity in the produced MIPs. This approach has more recently been used to investigate mechanisms underlying recognition in MIP catalysts for a Diels–Alder reaction,[179] MIPs selective for 17-β-estradiol[180,181] and oseltamivir.[182] The general outline of this approach is presented in Figure 7.1.

Further investigations into the origin of the often observed recognizing site heterogeneity and possible correlation between pre-polymerization mixture events and final MIP morphology, as well as analyte binding properties of corresponding MIPs were undertaken by Golker *et al.*[183,184] Here, MD simulation data were correlated to the physical characterization of polymer

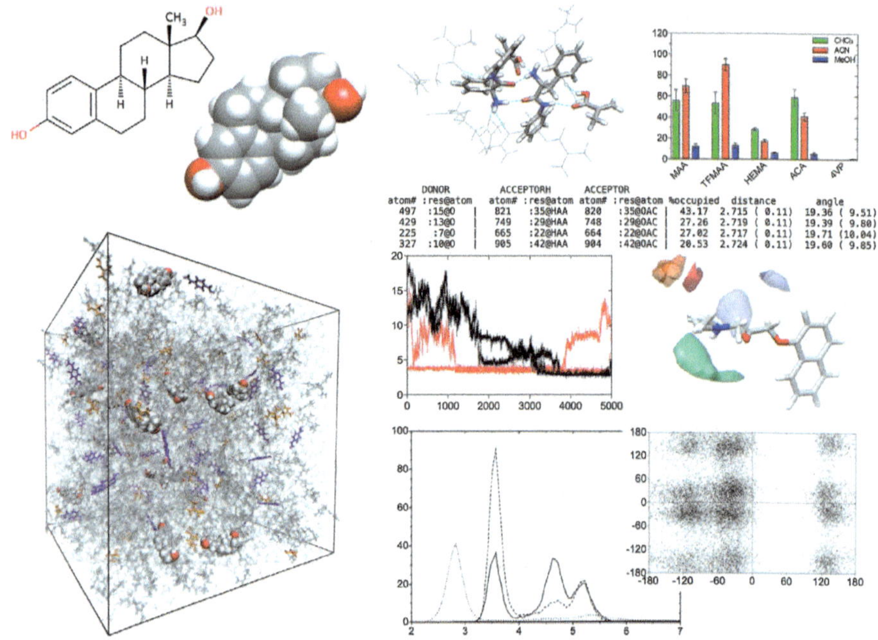

Figure 7.1 A schematic overview of the general strategy used for all-component MD simulation of pre-polymerization mixtures; illustrated with examples from studies of imprinting systems targeting 17-beta-estradiol,[180,181] phenylalanine anilide[20] and (*S*)-propranolol.[28] (Top left) *In silico* molecular models are prepared and parameterized for the various species present in the pre-polymerization mixture. (Bottom left) Systems are constructed that include copies of all MIP pre-polymerization components (molecular models) in stoichiometry ratios corresponding to compositions used for synthesis. Systems are generated using randomized starting geometries. Initially, energy minimization steps are performed followed by equilibration under conditions of NVT and then NPT. Subsequently, collection of data is undertaken during so called production-phase runs under conditions of NVT or NPT. Generally, replicate system simulations are run in parallel as a control to ensure statistical reliability. (Right) Statistical analysis is used to identify molecular interactions and motions using analytical tools, such as radial distribution function and hydrogen bond analysis routines, distance plots, dihedral angle analyses, grid density analyses and more, producing a molecular level view of interactions and motions as well as statistical averages.

morphology, as well as evaluation of analyte binding properties of the corresponding MIPs. There was a correlation between the nature and extent of non-covalent interactions in the pre-polymerization mixtures and fabricated MIPs. The influence of the cross-linking monomer in these studies motivated further investigations into the influence of cross-linking monomer selection on template interactions prior to polymerization. This was subsequently evaluated in a study of propranolol MIP pre-polymerization mixtures using MD.[28]

Another interesting study evaluated the influence of π–π stacking interactions in the MIP pre-polymerization mixture on the imprinting and analyte binding properties in MIPs imprinted with polychlorinated benzenes.[185]

Xerogels are another polymer format used in molecular imprinting. They have also been subjected to MD studies. Azenha *et al.*[186] examined the role of polyethylene glycol (PEG) additions to modify the porosity of the MIP matrix, as well as the sol–gel phase separation of components in xerogel pre-gelification mixtures in the production of damascenone imprinted materials. The inclusion of PEG had an adverse effect on template–(functional monomer) complex formation and would not improve the imprinting effect, in accordance with experimental findings. It was suggested, since the PEG content seemingly had no effect on network structuring, that PEG was excluded from the other sol–gel components into the aqueous-methanolic sol phase when added to the pre-gelification mixture.

The dramatic increase in reports on the use of MD simulations in MIP studies can be attributed to the often strong correlation of these studies to experimental results, especially in all-atom, all-component studies. Nonetheless, the new and promising reactive force fields that are emerging, and that potentially allow for the simulation of bond formation and breaking in force field based simulations, should allow researchers to bridge the gap between the pre-polymerization mixture and MIP matrix, thus providing invaluable insights regarding polymer properties.

7.3.4 Multivariate Analysis and Other Computational Strategies

Chemometrics, the application of mathematical and statistical methods to chemical data,[187–189] is another form of modelling employed within the molecular imprinting field. It has been used as a tool for the selection and optimization of experimental parameters. In order to use chemometric strategies, an experiment needs to be designed and defined and the factors affecting the outcome or success should be identified. Relevant factors should be systematically and simultaneously varied, thus changing simultaneously over a series of experiments and allowing for interpretation of the outcome using mathematical models. In this manner, a minimal number of experiments needed to extract the maximal amount of information can be performed.

Examples of different experimental designs that have been applied in analysis of MIP systems include fractional factorial,[190–195] full factorial,[196–200] central composite,[194,197,198,200,201] Box-Behnken[202] and Doehlert[192,193,195,199,203] methods. In essence, the number of experiments required, and the manner in which factors are varied, is what differs between the design methods (illustrated in Figure 7.2). When selecting the number of factors to vary in an evaluation, there is no limit to how many can be included, though seeing how the computational approaches are intended to help in designing and minimizing trial and error it is neither practical nor desirable to include more than is needed. Just as a PCM can be included to relieve the computational

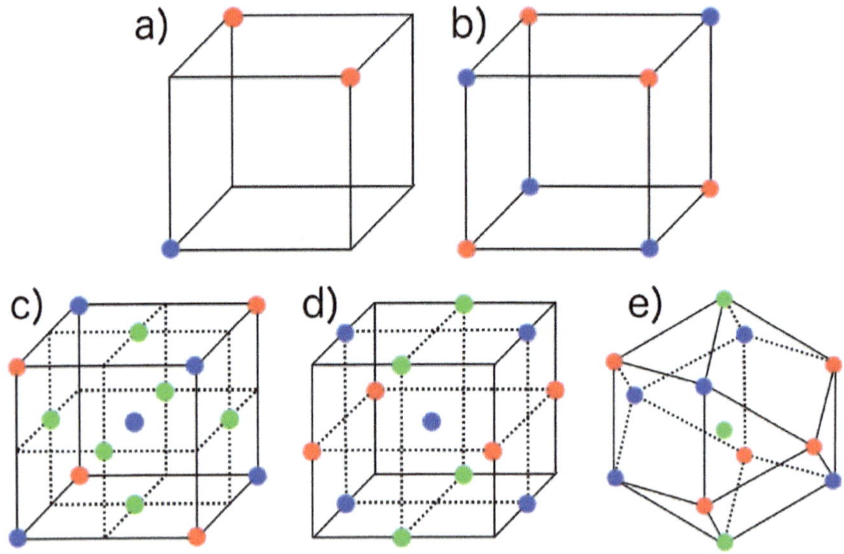

Figure 7.2 Illustration of schematic experimental designs that have been used in statistical evaluation of MIPs. The variables that can be controlled (*e.g.* amount of functional monomer, cross-linking monomer and porogen) in the analysis are called variables. Points represent experimental runs of a three-factor (a) fractional factorial; (b) full factorial; (c) central composite; (d) Box-Behnken and (e) Doehlert design.

demand of including solvent, or the force field description of molecular and atomic motions and interactions in MD allows study of larger systems than QM based methods, the key in chemometrics is to find the appropriate number of factors and experimental design methods to extract accurate and relevant information. When performing screening studies, the full and the fractional designs are primarily used, while Behnken and Doehlert methods are implemented for surface modelling. If the intention is to perform both optimization and surface modelling, the central composite design method, being an extension of fractional designs, is the method of choice.

During recent years, statistical modelling and chemometric methods have become increasingly popular, as reflected in several literature reports. The intention here is to provide a brief description of the use and potential of these methods. Traditional MIP analysis is generally univariate in nature, where one parameter is evaluated and optimized and then used to optimize the next parameter and so forth. The problem with using univariate analysis is that the optimum identified might be false. However, MIP composition (both choice of components and stoichiometry), production parameters (*e.g.* pressure, temperature, and polymerization methods), analyte binding properties and chromatographic performance as well as physical characteristics of materials produced (porosity and pore volume, accessible surface area and so forth) can be evaluated and optimized using chemometric approaches.

Chemometric strategies can provide a powerful complement to traditional MIP analysis and effort has been made to include chemometric studies in the optimization of polymer composition[190,191,196–198,203–205] and binding parameters[192–195,199–202,206–209] albeit the frequency of which multivariate statistical analyses are included is still relatively limited, as compared to MD-based reports.

There are several statistical methods available for analysing experimental data, where the analysis of variance (ANOVA) is a commonly used method for analysing observations dependent on the variation of one or more factors. The ANOVA cannot account for covariance of different variables because it is a univariate method. Another method that can be used in data classification is the principal component analysis (PCA). It is used to identify so-called principal components (PCs), which can then be represented as vectors. An approximation of the maximum variance direction in the data is obtained from the first PC while the second PC is chosen orthogonal to the first, thus approximating the second maximum variance directions, and so forth. In this manner, the PCs describing the largest variation can be identified, revealing patterns in the data and identifying the factors that influence the variance to the greatest extent. There are several examples of methods that can be used in multivariate calibration of data, where partial least squares regression (PLSR) and principal component regression (PCR) are two examples that handles covariance rather well, while a method called multiple linear regression (MLR) is the preferred regression method if the variables included are independent.[187] The artificial neural network (ANN) construction functions as an associative memory by programming itself using experimental data.[210] The ANN is comprised of different layers where the input layer receives input, a hidden layer performs processing and transformation of the input, and the output processes the final results. Weights and numerical values describing the relative strength of the input are assigned to connections between nodes of different layers. These layers are adjusted with respect to the error calculated from the difference between the actual and predicted values. This calculation is commonly performed using the so-called back-propagation of error (BPE) algorithm, which is a supervised learning method used in ANN. This method requires both input and output to be known in advance.

There are reports where optimization of the MIP composition has been performed through chemometric approaches. Davies *et al.*[196] highlighted difficulties upon introduction of new templates when using common synthetic protocols, and developed a chemometric approach for designing sulfamethazine MIPs to circumvent this issue. The study was based on data from a HPLC multi-analyte competition rebinding assay. For a given amount of functional and cross-linking monomer in the MIPs, a three-level full fractional design was selected for experimental setup combined with a quadratic regression model containing squared terms for data fitting, thus validating the proposed model through ANOVA. In a second case, Nezhadali and Shadmehri[211] implemented an ANN approach in the production optimization of pantoprazole MIPs with regard to polymer composition, stoichiometry, preparation method and production parameters. Functional monomer candidates were initially selected through computational screening using RHF/6-31G* calculations of interaction energies of template monomer complexes including a PCM. A Plackett–Burman fractional design method was then used to produce data for the identification and optimization of significant factors during the electropolymerization. A Pareto plot was evaluated to identify key factors further investigated using central composite design and ANN modelling combined with generic algorithm optimization. Navari-Villoslada *et al.*[190] used PLSR analyses in evaluation of a first order model in composition optimization of a BPA MIP. Both composition and stoichiometry of the MIP pre-polymerization mixtures varied in the production of a series of mini-MIPs that were evaluated through analyte binding experiments. The model was used to predict the polymer composition, stoichiometry and preparation method to identify the variable leading to maximal selective binding; this MIP was prepared on a larger scale to validate the predictive power of the model. In this study, only the optimum was validated, where the sparse validation of the model presents a weakness of the approach. Navarro-Villoslada, this time in collaboration with Takeuchi,[191] also performed a screening and evaluation of a small library of small batch piroxicam MIPs using a fractional factorial design approach for polymer compositions by, again, varying stoichiometric and molecular composition of the MIP pre-polymerization mixtures and polymerization method, where a first order calibration curve was used for experimental data fitting and cross-validation. Similarly, Salimraftar *et al.*[212] implemented a three-level full-factorial design to investigate effects of MIP composition with regard to functional and cross-linking monomer on analyte binding capacity and MIP selectivity in an effort to investigate the optimal material composition for maximal selectivity and capacity in the production of diclofenac sodium MIPs. ANOVA was subsequently performed to estimate the significance and to validate the models. In an early study, Kempe and Kempe[197] utilized a central composite design method in efforts to optimize the composition of propranolol imprinted MIP beads using a regression MLR method to analyse variable stoichiometric compositions. Propranolol adsorption was subsequently evaluated through a binding assay and the resulting quadratic model validated. Kempe *et al.*[213] later used force field based calculations in combination with semi-empirical AM1 computations to derive molecular descriptors that were used in multivariate data analysis, correlating derived molecular descriptors and MIP nanocarrier ligand uptake through PLSR in the production of MIP based erythromycin nanocarriers. The predicted sum of squares (PRESS) for each model dimension and residual sum squares (RSS) of previous dimension were calculated and compared for cross validation of the model. In a similar study, Baggiani *et al.*[206] derived molecular descriptors from AM1 optimized structures, and then used them as variables in chemometric studies of the correlation between HPLC column selectivity and the calculated molecular descriptors using PCA and PCR. The cross-selectivity of the pentachlorophenol-MIP towards several related phenols was explained through steric and electronic molecular descriptors.

The effect of binding media on template uptake of bupivacaine MIPs at various temperatures was investigated by Rosengren *et al.*[207] through equilibrium binding studies. Binding could be described in terms of temperature and electric permittivity of the solvent mixture through PLSR analysis, following a third degree equation with cross-term calculations. Independent binding data from a separate polymer batch was used for model validation, thus highlighting the necessity of a robust validation. This validation was based on the observed complexity of the described dependences, when constructing chemometric prediction models. Tarley *et al.*[192] used a flow pre-concentration system coupled to amperometric detection for studying MIP rebinding of chloroguaiacol. This study included a fractional design to establish the influence of the mobile phase physical properties (pH, flow rate, KCl concentration, elution flow rate, and eluent volume) on MIP uptake. The factors with the highest influence on analyte binding were identified and a quadratic model with cross-terms derived from a Doehlert design used to optimize this binding by accounting for the most important factors (pH and KCl concentration). In a similar work, Nantasenamat *et al.*[208,214] used literature data as a basis for chemometric studies using ANN to correlate the reported selectivity of MIPs from diverse HPLC studies with molecular descriptors and mobile phase compositions. Mobile phase descriptors were obtained from the literature or measured and the BPE algorithm was used to calculate the model. The results, interestingly, indicated that an observed difference in data obtained depends on the regularity of particle size of MIPs.

In another interesting study, stochastic simulations were used in examination of interactions between pre-polymerization mixture components.[58] That is, monomer–template binding affinity was used in a stochastic algorithm[59] to construct a lattice matrix with monomer and template units selectively positioned. A cross-linking monomer was added, the template removed and the binding site models analysed for heterogeneity where results correlated with evaluation of experimentally produced MIPs.

In conclusion, while multivariate statistical analyses offer a description of covariance and a higher probability of finding an optimized composition and set of production variables compared to univariate analytical methods, they do not guarantee the identification of an optimally experimental MIP design. This is illustrated by cases where the optimum is found in a corner of the designed experimental region, where extension of the region is needed until a local optimum emerges.[190,191,196,203–205] On the other hand, one of the great strengths of the chemometric methods lies in their potential to deal with a large number of variables, so we argue that chemometric approaches still offer a great, though as yet unrealized, potential for accelerating the development of MIP systems.

7.4 Conclusions

During the past decade, the number of molecular imprinting reports appearing in the literature that involve some form of computational studies or modelling have grown most dramatically. This trend does not seem to be abating and with ever more powerful and affordable computer software and hardware we should see a further increase in the impact of these methods on the development of molecular imprinting.

List of Abbreviations

4VP	4-Vinylpyridin
AAM	Acrylamide
ANN	Artificial neural network
ANOVA	Analysis of variance
BPA	Bisphenol A
BPE	Back-propagation of error
DFT	Density functional theory
MAA	Methacrylic acid
MD	Molecular dynamics
MIP	Molecularly imprinted polymer
MLR	Multiple linear regression
n-DDP	Phenyl phosphonic acid monododecyl ester
PC	Principal component
PCA	Principal component analysis
PCR	Principal component regression
PEG	Polyethylene glycol
PLSR	Partial least squares regression
PCM	Polarizable continuum model
QM	Quantum mechanics

References

1. G. Wulff and A. Sarhan, *Angew. Chem.*, 1972, **84**, 364.
2. R. Arshady and K. Mosbach, *Makromol. Chem.*, 1981, **182**, 687.
3. *Molecularly Imprinted Polymers: Man-made Mimics of Antibodies and Their Application in Analytical Chemistry*, ed. B. Sellergren, Elsevier, Amsterdam, 2001.
4. C. Alexander, H. S. Andersson, L. I. Andersson, R. J. Ansell, N. Kirsch, I. A. Nicholls, J. O'Mahony and M. J. Whitcombe, *J. Mol. Recognit.*, 2006, **19**, 106.
5. M. J. Whitcombe, N. Kirsch and I. A. Nicholls, *J. Mol. Recognit.*, 2014, **27**, 297.
6. I. A. Nicholls, *J. Mol. Recognit.*, 1998, **11**, 79.
7. I. A. Nicholls, H. S. Andersson, C. Charlton, H. Henschel, B. C. G. Karlsson, J. G. Karlsson, J. O'Mahony, A. M. Rosengren, K. J. Rosengren and S. Wikman, *Biosens. Bioelectron.*, 2009, **25**, 543.
8. I. A. Nicholls, H. S. Andersson, K. Golker, H. Henschel, B. C. G. Karlsson, G. D. Olsson, A. M. Rosengren, S. Shoravi, S. Suriyanarayanan, J. G. Wiklander and S. Wikman, *Anal. Bioanal. Chem.*, 2011, **400**, 1771.

9. I. A. Nicholls, B. C. G. Karlsson, G. D. Olsson and A. M. Rosengren, *Ind. Eng. Chem. Res.*, 2013, **52**, 13900.
10. M. I. Page and W. P. Jencks, *Proc. Natl. Acad. Sci. U. S. A.*, 1971, **68**, 1678.
11. W. P. Jencks, *Proc. Natl. Acad. Sci. U. S. A.*, 1981, **78**, 4046.
12. P. R. Andrews, D. J. Craik and J. L. Martin, *J. Med. Chem.*, 1984, **27**, 1648.
13. D. H. Williams, J. P. L. Cox, A. J. Doig, M. Gardner, U. Gerhard, P. T. Kaye, A. R. Lal, I. A. Nicholls, C. J. Salter and R. C. Mitchell, *J. Am. Chem. Soc.*, 1991, **113**, 7020.
14. M. S. Searle, D. H. Williams and U. Gerhard, *J. Am. Chem. Soc.*, 1992, **114**, 10697.
15. S. E. Holroyd, P. Groves, M. S. Searle, U. Gerhard and D. H. Williams, *Tetrahedron*, 1993, **49**, 9171.
16. D. H. Williams, E. Stephens, D. P. O'Brien and M. Zhou, *Angew. Chem., Int. Ed.*, 2004, **43**, 6596.
17. I. A. Nicholls, *Chem. Lett.*, 1995, **24**, 1035.
18. I. A. Nicholls, K. Adbo, H. S. Andersson, P. O. Andersson, J. Ankarloo, J. Hedin-Dahlström, P. Jokela, J. G. Karlsson, L. Olofsson, J. Rosengren, S. Shoravi, J. Svenson and S. Wikman, *Anal. Chim. Acta*, 2001, **435**, 9.
19. B. Sellergren, M. Lepistö and K. Mosbach, *J. Am. Chem. Soc.*, 1988, **110**, 5853.
20. G. D. Olsson, B. C. G. Karlsson, S. Shoravi, J. G. Wiklander and I. A. Nicholls, *J. Mol. Recognit.*, 2012, **25**, 69.
21. M. J. Whitcombe, L. Martin and E. N. Vulfson, *Chromatographia*, 1998, **47**, 457.
22. J. Svenson, H. S. Andersson, S. A. Piletsky and I. A. Nicholls, *J. Mol. Recognit.*, 1998, **11**, 83.
23. J. Svenson, J. G. Karlsson and I. A. Nicholls, *J. Chromatogr. A*, 2004, **1024**, 39.
24. R. J. Ansell and K. L. Kuah, *Analyst*, 2005, **130**, 179.
25. R. J. Ansell, D. Wang and J. K. L. Kuah, *Analyst*, 2008, **133**, 1673.
26. R. J. Ansell and D. Wang, *Analyst*, 2009, **134**, 564.
27. H. S. Andersson and I. A. Nicholls, *Bioorg. Chem.*, 1997, **25**, 203.
28. S. Shoravi, G. D. Olsson, B. C. G. Karlsson and I. A. Nicholls, *Int. J. Mol. Sci.*, 2014, **15**, 10622.
29. A. G. Mayes and C. R. Lowe, in *Methodological Surveys in Bioanalysis of Drugs*, ed. E. Reid, H. M. Hill and I. D. Wilson, Royal Society of Chemistry, Cambridge, 1998, pp. 28–36.
30. G. Vlatakis, L. I. Andersson, R. Müller and K. Mosbach, *Nature*, 1993, **361**, 645.
31. K. Tanabe, T. Takeuchi, J. Matsui, K. Ikebukuro, K. Yano and I. Karube, *J. Chem. Soc., Chem. Commun.*, 1995, 2303.
32. A. J. Hall, P. Manesiotis, M. Emgenbroich, M. Quaglia, E. De Lorenzi and B. Sellergren, *J. Org. Chem.*, 2005, **70**, 1732.
33. M. Emgenbroich, C. Borrelli, S. Shinde, I. Lazraq, F. Vilela, A. J. Hall, J. Oxelbark, E. De Lorenzi, J. Courtois, A. Simanova, J. Verhage, K. Irgum, K. Karim and B. Sellergren, *Chem.–Eur. J.*, 2008, **14**, 9516.

34. J. Chen, S. Shinde, M.-H. Koch, M. Eisenacher, S. Galozzi, T. Lerari, K. Barkovits, P. Subedi, R. Krüger, K. Kuhlmann, B. Sellergren, S. Helling and K. Marcus, *Sci. Rep.*, 2015, **5**, 11438.
35. J. Berglund, I. A. Nicholls, C. Lindbladh and K. Mosbach, *Bioorg. Med. Chem. Lett.*, 1996, **6**, 2237.
36. H. S. Andersson, J. G. Karlsson, S. A. Piletsky, A.-C. Koch-Schmidt, K. Mosbach and I. A. Nicholls, *J. Chromatogr. A*, 1999, **848**, 39.
37. J. Matsui, Y. Miyoshi and T. Takeuchi, *Chem. Lett.*, 1995, **24**, 1007.
38. J. Matsui and T. Takeuchi, *Anal. Commun.*, 1997, **34**, 199.
39. J. Matsui, I. A. Nicholls and T. Takeuchi, *Tetrahedron: Asymmetry*, 1996, **7**, 1357.
40. J. Matsui, O. Doblhoff-Dier and T. Takeuchi, *Anal. Chim. Acta*, 1997, **343**, 1.
41. S. Vidyasankar, P. K. Dhal, S. D. Plunkett and F. H. Arnold, *Biotechnol. Bioeng.*, 1995, **48**, 431.
42. A. Leonhardt and K. Mosbach, *React. Polym., Ion Exch., Sorbents*, 1987, **6**, 285.
43. J. Matsui, I. A. Nicholls, I. Karube and K. Mosbach, *J. Org. Chem.*, 1996, **61**, 5414.
44. B. P. Santora, A. O. Larsen and M. R. Gagné, *Organometallics*, 1998, **17**, 3138.
45. J. Liu and G. Wulff, *J. Am. Chem. Soc.*, 2008, **130**, 8044.
46. H. S. Andersson and O. Ramström, *J. Mol. Recognit.*, 1998, **11**, 103.
47. S. Lay, X. Ni, H. Yu and S. Shen, *J. Sep. Sci.*, 2016, **39**, 2321.
48. H. Asanuma, M. Kakazu, M. Shibata and T. Hishiya, *Chem. Commun.*, 1997, 1971.
49. K. Sreenivasan, *J. Appl. Polym. Sci.*, 1998, **70**, 15.
50. S. A. Piletsky, H. S. Andersson and I. A. Nicholls, *J. Mol. Recognit.*, 1998, **11**, 94.
51. S. A. Piletsky, H. S. Andersson and I. A. Nicholls, *Macromolecules*, 1999, **32**, 633.
52. B. Sellergren and L. Andersson, *J. Org. Chem.*, 1990, **55**, 3381.
53. S. E. Byström, A. Börje and B. Åkermark, *J. Am. Chem. Soc.*, 1993, **115**, 2081.
54. M. J. Whitcombe, M. E. Rodriguez, P. Villar and E. N. Vulfson, *J. Am. Chem. Soc.*, 1995, **117**, 7105.
55. M. Lübke, M. J. Whitcombe and E. N. Vulfson, *J. Am. Chem. Soc.*, 1998, **120**, 13342.
56. V. S. Pande, A. Y. Grosberg and T. Tanaka, *Phys. D*, 1997, **107**, 316.
57. S. A. Piletsky, T. L. Panasyuk, E. V. Piletskaya, I. A. Nicholls and M. Ulbricht, *J. Membr. Sci.*, 1999, **157**, 263.
58. X. Wu, W. R. Carroll and K. D. Shimizu, *Chem. Mater.*, 2008, **20**, 4335.
59. M. Veitl, U. Schweiger and M. L. Berger, *Comput. Biomed. Res.*, 1997, **30**, 427.
60. I. A. Nicholls, H. S. Andersson, K. Golker, H. Henschel, B. C. G. Karlsson, G. D. Olsson, A. M. Rosengren, S. Shoravi, J. G. Wiklander and

S. Wikman, in *Molecular Imprinting - Principles and Applications of Micro- and Nanostructured Polymers*, ed. Y. Lei, Pan Stanford Publishing, Singapore, 2013, pp. 71–104.
61. Q. Fu, H. Sanbe, C. Kagawa, K.-K. Kunimoto and J. Haginaka, *Anal. Chem.*, 2003, **75**, 191.
62. L. Schwarz, C. I. Holdsworth, A. McCluskey and M. C. Bowyer, *Aust. J. Chem.*, 2004, **57**, 759.
63. C. I. Holdsworth, M. C. Bowyer, C. Lennard and A. McCluskey, *Aust. J. Chem.*, 2005, **58**, 315.
64. T. Ogawa, K. Hoshina, J. Haginaka, C. Honda, T. Tanimoto and T. Uchida, *J. Pharm. Sci.*, 2005, **94**, 353.
65. E. P. C. Lai and S. Y. Feng, *Microchem. J.*, 2003, **75**, 159.
66. Z. Meng, T. Yamazaki and K. Sode, *Biosens. Bioelectron.*, 2004, **20**, 1068.
67. C. Baggiani, L. Anfossi, P. Baravalle, C. Giovannoli and C. Tozzi, *Anal. Chim. Acta*, 2005, **531**, 199.
68. W. Dong, M. Yan, M. Zhang, Z. Liu and Y. Li, *Anal. Chim. Acta*, 2005, **542**, 186.
69. S. M. Voshell and M. R. Gagné, *Organometallics*, 2005, **24**, 6338.
70. P. Li, F. Rong, Y. Xie, V. Hu and C. Yuan, *J. Anal. Chem.*, 2004, **59**, 939.
71. P. Luliński, D. Maciejewska, M. Bamburowicz-Klimkowska and M. Szutowski, *Molecules*, 2007, **12**, 2434.
72. R. I. Boysen, L. J. Schwarz, S. Li, J. Chowdhury and M. T. W. Hearn, *Microsyst. Technol.*, 2014, **20**, 2037.
73. L. Wu and Y. Li, *J. Mol. Recognit.*, 2004, **17**, 567.
74. L. Wu, B. Sun, Y. Li and W. Chang, *Analyst*, 2003, **128**, 944.
75. D. L. Rathbone, A. Ali, P. Antonaki and S. Cheek, *Biosens. Bioelectron.*, 2005, **20**, 2353.
76. T. Sagawa, K. Togo, C. Miyahara, H. Ihara and K. Ohkubo, *Anal. Chim. Acta*, 2004, **504**, 37.
77. L. Wu and Y. Li, *Anal. Chim. Acta*, 2003, **482**, 175.
78. L. Wu and Y. Li, *Anal. Chim. Acta*, 2004, **517**, 145.
79. H. He, X. Gu, L. Shi, J. Hong, H. Zhang, Y. Gao, S. Du and L. Chen, *Anal. Bioanal. Chem.*, 2014, **407**, 509.
80. A. Pietrzyk, W. Kutner, R. Chitta, M. E. Zandler, F. D'Souza, F. Sannicolò and P. R. Mussini, *Anal. Chem.*, 2009, **81**, 10061.
81. A. H. Demircelik, M. Andac, C. A. Andac, R. Say and A. Denizli, *J. Biomater. Sci., Polym. Ed.*, 2009, **20**, 1235.
82. A. Azimi and M. Javanbakht, *Anal. Chim. Acta*, 2014, **812**, 184.
83. S. Riahi, F. Edris-Tabrizi, M. Javanbakht, M. R. Ganjali and P. Norouzi, *J. Mol. Model.*, 2009, **15**, 829.
84. J. O'Mahony, B. C. G. Karlsson, B. Mizaikoff and I. A. Nicholls, *Analyst*, 2007, **132**, 1161.
85. B. C. G. Karlsson, J. O'Mahony, J. G. Karlsson, H. Bengtsson, L. A. Eriksson and I. A. Nicholls, *J. Am. Chem. Soc.*, 2009, **131**, 13297.
86. S. Miertuš, E. Scrocco and J. Tomasi, *Chem. Phys.*, 1981, **55**, 117.

87. Y. Liu, F. Wang, T. Tan and M. Lei, *Appl. Biochem. Biotechnol.*, 2008, **160**, 328.
88. W. Dong, M. Yan, Z. Liu, G. Wu and Y. Li, *Sep. Purif. Technol.*, 2007, **53**, 183.
89. S. Pardeshi, R. Dhodapkar and A. Kumar, *Compos. Interfaces*, 2013, **21**, 13.
90. L. Wu, K. Zhu, M. Zhao and Y. Li, *Anal. Chim. Acta*, 2005, **549**, 39.
91. Y. Diñeiro, M. I. Menéndez, M. C. Blanco-López, M. J. Lobo-Castañón, A. J. Miranda-Ordieres and P. Tuñón-Blanco, *Anal. Chem.*, 2005, **77**, 6741.
92. Y. Diñeiro, M. I. Menéndez, M. C. Blanco-López, M. J. Lobo-Castañón, A. J. Miranda-Ordieres and P. Tuñón-Blanco, *Biosens. Bioelectron.*, 2006, **22**, 364.
93. L. A. de Barros, L. A. Pereira, R. Custódio and S. Rath, *J. Braz. Chem. Soc.*, 2014, **25**, 619.
94. C. Dong, X. Li, Z. Guo and J. Qi, *Anal. Chim. Acta*, 2009, **647**, 117.
95. P. Luliński, M. Sobiech, T. Żołek and D. Maciejewska, *Talanta*, 2014, **129**, 155.
96. F. Ahmadi, E. Yawari and M. Nikbakht, *J. Chromatogr. A*, 2014, **1338**, 9.
97. F. Ahmadi and E. Karamian, *Iran. J. Pharm. Res.*, 2014, **13**, 1.
98. T.-P. Huynh, C. B. KC, M. Sosnowska, J. W. Sobczak, V. N. Nesterov, F. D'Souza and W. Kutner, *Biosens. Bioelectron.*, 2015, **64**, 657.
99. J. Liu, Z. Dai, B. Li, S. Tang and R. Jin, *J. Mol. Model.*, 2014, **20**, 2456.
100. P. Qi, X. Wang, X. Wang, H. Zhang, H. Xu, K. Jiang and Q. Wang, *J. Sep. Sci.*, 2014, **37**, 2955.
101. L. Qin, W. Liu, Y. Yang and X. Liu, *Monatsh. Chem.*, 2014, **146**, 449.
102. K. K. Tadi, R. V. Motghare and V. Ganesh, *Electroanalysis*, 2014, **26**, 2328.
103. W. Yang, P. Ma, T. Fan, Z. Zhou, H. Liu and W. Xu, *J. Appl. Polym. Sci.*, 2014, **132**, 41730.
104. T. Alizadeh, *Anal. Chim. Acta*, 2008, **623**, 101.
105. Y. Li, X. Li, C. Dong, Y. Li, P. Jin and J. Qi, *Biosens. Bioelectron.*, 2009, **25**, 306.
106. A. Kowalska, A. Stobiecka and S. Wysocki, *J. Mol. Struct.: THEOCHEM*, 2009, **901**, 88.
107. J. Wang, R. Guo, J. Chen, Q. Zhang and X. Liang, *Anal. Chim. Acta*, 2005, **540**, 307.
108. A.-F. Che, L.-S. Wan, J. Ling, Z.-M. Liu and Z.-K. Xu, *J. Phys. Chem. B*, 2009, **113**, 7053.
109. K. C. Christoforidis, M. Louloudi, A. W. Rutherford and Y. Deligiannakis, *J. Phys. Chem. C*, 2008, **112**, 12841.
110. M. B. Gholivand, M. Khodadadian and F. Ahmadi, *Anal. Chim. Acta*, 2010, **658**, 225.
111. N. A. El Gohary, A. Madbouly, R. M. El Nashar and B. Mizaikoff, *Biosens. Bioelectron.*, 2015, **65**, 108.
112. J. Yao, X. Li and W. Qin, *Anal. Chim. Acta*, 2008, **610**, 282.
113. M. Appell, M. A. Jackson, L. C. Wang, C.-H. Ho and A. Mueller, *J. Sep. Sci.*, 2013, **37**, 281.

114. R. Del Sole, M. R. Lazzoi, M. Arnone, F. D. Sala, D. Cannoletta and G. Vasapollo, *Molecules*, 2009, **14**, 2632.
115. T. Mukawa, T. Goto, H. Nariai, Y. Aoki, A. Imamura and T. Takeuchi, *J. Pharm. Biomed. Anal.*, 2003, **30**, 1943.
116. M. Tada, T. Sasaki and Y. Iwasawa, *J. Phys. Chem. B*, 2004, **108**, 2918.
117. M. Azenha, P. Kathirvel, P. Nogueira and A. Fernando-Silva, *Biosens. Bioelectron.*, 2008, **23**, 1843.
118. A. R. Leach, *Molecular Modelling: Principles and Applications*, Prentice Hall, Harlow, 2001, pp. 353–409.
119. B. J. Alder and T. E. Wainwright, *J. Chem. Phys.*, 1957, **27**, 1208.
120. Y. Duan, *Science*, 1998, **282**, 740.
121. K. Pervushin, R. Riek, G. Wider and K. Wuthrich, *Proc. Natl. Acad. Sci. U. S. A.*, 1997, **94**, 12366.
122. T. E. Cheatham III and P. A. Kollma, *J. Mol. Biol.*, 1996, **259**, 434.
123. T. E. Cheatham III, J. L. Miller, T. Fox, T. A. Darden and P. A. Kollman, *J. Am. Chem. Soc.*, 1995, **117**, 4193.
124. O. Berger, O. Edholm and F. Jähnig, *Biophys. J.*, 1997, **72**, 2002.
125. P. van der Ploeg, *J. Chem. Phys.*, 1982, **76**, 3271.
126. B. L. de Groot, *Science*, 2001, **294**, 2353.
127. A. R. van Buuren, S. J. Marrink and H. J. C. Berendsen, *J. Phys. Chem.*, 1993, **97**, 9206.
128. Q. C. Hsu, C. D. Wu and T. H. Fang, *Comput. Mater. Sci.*, 2005, **34**, 314.
129. B. J. Garrison, A. Delcorte and K. D. Krantzman, *Acc. Chem. Res.*, 2000, **33**, 69.
130. K. M. Masukawa, P. A. Kollman and I. D. Kuntz, *J. Med. Chem.*, 2003, **46**, 5628.
131. H. Yang and A. H. Elcock, *J. Am. Chem. Soc.*, 2003, **125**, 13968.
132. X. Y. Li and L. A. Eriksson, *Holzforschung*, 2005, **59**, 253.
133. J. Wang, P. Cieplak and P. A. Kollman, *J. Comput. Chem.*, 2000, **21**, 1049.
134. J. Wang, R. M. Wolf, J. W. Caldwell, P. A. Kollman and D. A. Case, *J. Comput. Chem.*, 2004, **25**, 1157.
135. B. R. Brooks, R. E. Bruccoleri, B. D. Olafson, D. J. States, S. Swaminathan and M. Karplus, *J. Comput. Chem.*, 1983, **4**, 187.
136. W. L. Jorgensen, D. S. Maxwell and J. Tirado-Rives, *J. Am. Chem. Soc.*, 1996, **118**, 11225.
137. H. J. C. Berendsen, D. van der Spoel and R. van Drunen, *Comput. Phys. Commun.*, 1995, **91**, 43.
138. W. R. P. Scott, P. H. Hünenberger, I. G. Tironi, A. E. Mark, S. R. Billeter, J. Fennen, A. E. Torda, T. Huber, P. Krüger and W. F. van Gunsteren, *J. Phys. Chem. A*, 1999, **103**, 3596.
139. A. Mayes and M. Whitcombe, *Adv. Drug Delivery Rev.*, 2005, **57**, 1742.
140. E. Piletska, N. Turner, A. Turner and S. A. Piletsky, *J. Controlled Release*, 2005, **108**, 132.
141. S. Wei, M. Jakusch and B. Mizaikoff, *Anal. Chim. Acta*, 2006, **578**, 50.
142. S. A. Piletsky, K. Karim, E. V. Piletska, A. P. F. Turner, C. J. Day, K. W. Freebairn and C. Legge, *Analyst*, 2001, **126**, 1826.

143. E. V. Piletska, M. Romero-Guerra, A. R. Guerreiro, K. Karim, A. P. F. Turner and S. A. Piletsky, *Anal. Chim. Acta*, 2005, **542**, 47.
144. E. V. Piletska, M. Romero-Guerra, I. Chianella, K. Karim, A. P. F. Turner and S. A. Piletsky, *Anal. Chim. Acta*, 2005, **542**, 111.
145. S. Subrahmanyam, S. A. Piletsky, E. V. Piletska, B. Chen, K. Karim and A. P. F. Turner, *Biosens. Bioelectron.*, 2001, **16**, 631.
146. E. Piletska, S. Piletsky, K. Karim, E. Terpetschnig and A. Turner, *Anal. Chim. Acta*, 2004, **504**, 179.
147. I. Chianella, M. Lotierzo, S. A. Piletsky, I. E. Tothill, B. Chen, K. Karim and A. P. F. Turner, *Anal. Chem.*, 2002, **74**, 1288.
148. I. Chianella, S. A. Piletsky, I. E. Tothill, B. Chen and A. P. F. Turner, *Biosens. Bioelectron.*, 2003, **18**, 119.
149. S. Wei, M. Jakusch and B. Mizaikoff, *Anal. Bioanal. Chem.*, 2007, **389**, 423.
150. I. Bakas, N. Ben Oujji, G. Istamboulié, S. Piletsky, E. Piletska, E. Ait-Addi, I. Ait-Ichou, T. Noguer and R. Rouillon, *Talanta*, 2014, **125**, 313.
151. I. Bakas, A. Hayat, S. Piletsky, E. Piletska, M. M. Chehimi, T. Noguer and R. Rouillon, *Talanta*, 2014, **130**, 294.
152. M. J. U. Toro, L. D. Marestoni and M. D. P. T. Sotomayor, *Sens. Actuators, B*, 2015, **208**, 299.
153. M. J. Abdin, Z. Altintas and I. E. Tothill, *Biosens. Bioelectron.*, 2015, **67**, 177.
154. D. Pavel and J. Lagowski, *Polymer*, 2005, **46**, 7528.
155. D. Pavel and J. Lagowski, *Polymer*, 2005, **46**, 7543.
156. D. Pavel, J. Lagowski and C. J. Lepage, *Polymer*, 2006, **47**, 8389.
157. Y. Lv, Z. Lin, T. Tan, W. Feng, P. Qin and C. Li, *Sens. Actuators, B*, 2008, **133**, 15.
158. S. Srebnik, O. Lev and D. Avnir, *Chem. Mater.*, 2001, **13**, 811.
159. S. Srebnik and O. Lev, *J. Chem. Phys.*, 2002, **116**, 10967.
160. S. Srebnik, *Chem. Mater.*, 2004, **16**, 883.
161. I. Yungerman and S. Srebnik, *Chem. Mater.*, 2006, **18**, 657.
162. Z.-J. Zhao, Q. Wang, L. Zhang and T. Wu, *J. Phys. Chem. B*, 2008, **112**, 7515.
163. D. Luo, Z. Zhao, L. Zhang, Q. Wang and J. Wang, *Mol. Simul.*, 2013, **40**, 431.
164. A. Molinelli, J. O'Mahony, K. Nolan, M. R. Smyth, M. Jakusch and B. Mizaikoff, *Anal. Chem.*, 2005, **77**, 5196.
165. S. Monti, C. Cappelli, S. Bronco, P. Giusti and G. Ciardelli, *Biosens. Bioelectron.*, 2006, **22**, 153.
166. R. Liu, X. Li, Y. Li, P. Jin, W. Qin and J. Qi, *Biosens. Bioelectron.*, 2009, **25**, 629.
167. Y. Li, X. Li, Y. Li, C. Dong, P. Jin and J. Qi, *Biomaterials*, 2009, **30**, 3205.
168. B. Liu, L. Ou, F. Zhang, Z. Zhang, H. Li, M. Zhu and S. Wang, *J. Sep. Sci.*, 2014, **37**, 3579.
169. S. A. Piletsky, E. V. Piletska, K. Karim, K. W. Freebairn, C. H. Legge and A. P. F. Turner, *Macromolecules*, 2002, **35**, 7499.

170. S. A. Piletsky, A. Guerreiro, E. V. Piletska, I. Chianella, K. Karim and A. P. F. Turner, *Macromolecules*, 2004, **37**, 5018.
171. S. A. Piletsky, I. Mijangos, A. Guerreiro, E. V. Piletska, I. Chianella, K. Karim and A. P. F. Turner, *Macromolecules*, 2005, **38**, 1410.
172. M. Yoshida, Y. Hatate, K. Uezu, M. Goto and S. Furusaki, *Colloids Surf., A*, 2000, **169**, 259.
173. E. Toorisaka, K. Uezu, M. Goto and S. Furusaki, *Biochem. Eng. J.*, 2003, **14**, 85.
174. G. D. Olsson, K. Niedergall, M. Bach, B. C. G. Karlsson, G. Tovar and I. A. Nicholls, *Polym. J.*, 2015, **47**, 827.
175. A. Katz and M. E. Davis, *Macromolecules*, 1999, **32**, 4113.
176. Y. Zhang, D. Song, L. M. Lanni and K. D. Shimizu, *Macromolecules*, 2010, **43**, 6284.
177. J. O'Mahony, S. Wei, A. Molinelli and B. Mizaikoff, *Anal. Chem.*, 2006, **78**, 6187.
178. J. O'Mahony, M. Moloney, M. McCormack, I. A. Nicholls, B. Mizaikoff and M. Danaher, *J. Chromatogr. B*, 2013, **931**, 164.
179. H. Henschel, N. Kirsch, J. Hedin-Dahlström, M. J. Whitcombe, S. Wikman and I. A. Nicholls, *J. Mol. Catal. B: Enzym.*, 2011, **72**, 199.
180. G. D. Olsson, B. C. G. Karlsson, E. Schillinger, B. Sellergren and I. A. Nicholls, *Ind. Eng. Chem. Res.*, 2013, **52**, 13965.
181. E. Schillinger, M. Möder, G. D. Olsson, I. A. Nicholls and B. Sellergren, *Chem.–Eur. J.*, 2012, **18**, 14773.
182. S. Shoravi, G. D. Olsson, B. C. G. Karlsson, F. Bexborn, Y. Abghoui, J. Hussain, J. G. Wiklander and I. A. Nicholls, *Org. Biomol. Chem.*, 2016, **14**, 4210.
183. K. Golker, B. C. G. Karlsson, G. D. Olsson, A. M. Rosengren and I. A. Nicholls, *Macromolecules*, 2013, **46**, 1408.
184. K. Golker, B. C. G. Karlsson, A. M. Rosengren and I. A. Nicholls, *Int. J. Mol. Sci.*, 2014, **15**, 20572.
185. D. Cleland, G. D. Olsson, B. C. G. Karlsson, I. A. Nicholls and A. McCluskey, *Org. Biomol. Chem.*, 2014, **12**, 844.
186. M. Azenha, B. Szefczyk, D. Loureiro, P. Kathirvel, M. N. D. S. Cordeiro and A. Fernando-Silva, *Langmuir*, 2013, **29**, 2024.
187. *Multivariate Data Analysis in Practice*, ed. K. H. Esbensen, Camo Process AS, Oslo, 2002.
188. R. Carlsson, *Design and Optimization in Organic Synthesis*, Elsevier, Amsterdam, 1992.
189. L. Eriksson, E. Johansson, N. Kettaneh-Wold and S. Wold, *Multi- and Megavariate Data Analysis. Principles and Applications*, Umetrics Academy, Umeå, 2002.
190. F. Navarro-Villoslada, B. S. Vicente and M. a. C. Moreno-Bondi, *Anal. Chim. Acta*, 2004, **504**, 149.
191. F. Navarro Villoslada and T. Takeuchi, *Bull. Chem. Soc. Jpn.*, 2005, **78**, 1354.
192. C. R. T. Tarley, M. G. Segatelli and L. T. Kubota, *Talanta*, 2006, **69**, 259.

193. W. d. J. R. Santos, P. R. Lima, C. R. T. Tarley and L. T. Kubota, *Anal. Bioanal. Chem.*, 2007, **389**, 1919.
194. A. R. Koohpaei, S. J. Shahtaheri, M. R. Ganjali, A. R. Forushani and F. Golbabaei, *Iran. J. Environ. Health Sci. Eng.*, 2008, **5**, 283.
195. C. R. T. Tarley and L. T. Kubota, *Anal. Chim. Acta*, 2005, **548**, 11.
196. M. P. Davies, V. De Biasi and D. Perrett, *Anal. Chim. Acta*, 2004, **504**, 7.
197. H. Kempe and M. Kempe, *Macromol. Rapid Commun.*, 2004, **25**, 315.
198. G. Ceolin, F. Navarro-Villoslada, M. C. Moreno-Bondi, G. Horvai and V. Horvath, *J. Comb. Chem.*, 2009, **11**, 645.
199. W. d. J. R. Santos, P. R. Lima, C. R. T. Tarley, N. F. Höehr and L. T. Kubota, *Anal. Chim. Acta*, 2009, **631**, 170.
200. A. R. Koohpaei, S. J. Shahtaheri, M. R. Ganjali, A. R. Forushani and F. Golbabaei, *J. Hazard. Mater.*, 2009, **170**, 1247.
201. A. Valero-Navarro, P. C. Damiani, J. F. Fernández-Sánchez, A. Segura-Carretero and A. Fernández-Gutiérrez, *Talanta*, 2009, **78**, 57.
202. T. Alizadeh, M. R. Ganjali, P. Nourozi and M. Zare, *Anal. Chim. Acta*, 2009, **638**, 154.
203. C. Rossi and K. Haupt, *Anal. Bioanal. Chem.*, 2007, **389**, 455.
204. I. Mijangos, F. Navarro-Villoslada, A. Guerreiro, E. Piletska, I. Chianella, K. Karim, A. Turner and S. Piletsky, *Biosens. Bioelectron.*, 2006, **22**, 381.
205. A. R. Koohpaei, S. J. Shahtaheri, M. R. Ganjali, A. R. Forushani and F. Golbabaei, *Talanta*, 2008, **75**, 978.
206. C. Baggiani, L. Anfossi, C. Giovannoli and C. Tozzi, *J. Chromatogr. B*, 2004, **804**, 31.
207. A. M. Rosengren, J. G. Karlsson, P. O. Andersson and I. A. Nicholls, *Anal. Chem.*, 2005, **77**, 5700.
208. C. Nantasenamat, T. Naenna, C. I. N. Ayudhya and V. Prachayasittikul, *J. Comput.-Aided Mol. Des.*, 2005, **19**, 509.
209. A. M. Rosengren, K. Golker, J. G. Karlsson and I. A. Nicholls, *Biosens. Bioelectron.*, 2009, **25**, 553.
210. B. J. Wythoff, *Systems*, 1993, **18**, 115.
211. A. Nezhadali and R. Shadmehri, *Sens. Actuators, B*, 2014, **202**, 240.
212. N. Salimraftar, S. Noee, M. Abdouss, G. Riazi and Z. M. Khoshhesab, *Polym. Bull.*, 2013, **71**, 19.
213. H. Kempe, A. Parareda Pujolràs and M. Kempe, *Pharm. Res.*, 2014, **32**, 375.
214. C. Nantasenamat, C. Isarankura-Na-Ayudhya, T. Naenna and V. Prachayasittikul, *Biosens. Bioelectron.*, 2007, **22**, 3309.

CHAPTER 8

Molecularly Imprinted Polymer-based Optical Chemosensors for Selective Chemical Determinations

M. C. MORENO-BONDI*, E. BENITO-PEÑA, S. CARRASCO AND J. L. URRACA

Complutense University, Department of Analytical Chemistry, Av. Complutense s/n, Madrid, 28040, Spain
*E-mail: mcmbondi@ucm.es

8.1 Introduction

Chemical sensors[1] and biosensors[2–4] are self-integrated devices capable of providing selective quantitative or semi-quantitative analytical information about the target species by using a chemical or biological, respectively, recognition element (a chemical or biochemical receptor) in direct spatial contact with a transducing element. Optical biosensors measure photons. The need for rapid diagnosis and improved sensing characteristics such as selectivity, cost-effectiveness, long-term operational and storage stability, and ease of preparation have promoted the development of intrinsically stable synthetic receptors approaching biorecognition elements (*i.e.*, biomimetic receptors) in affinity and selectivity.[5–10] Over the past decade, molecularly imprinted polymers (MIPs) have raised

Polymer Chemistry Series No. 28
Molecularly Imprinted Polymers for Analytical Chemistry Applications
Edited by Wlodzimierz Kutner and Piyush Sindhu Sharma

Published by the Royal Society of Chemistry, www.rsc.org

increasing expectations towards replacing biocompounds as the selective recognition elements in biomimetic sensors based on various transduction schemes.

Molecular imprinting is a template-directed technique that allows designing and preparing materials with well-defined artificial receptor sites for a wide range of chemical and biochemical compounds.[11,12] MIPs are obtained by having the selected print molecule (namely, the analyte or a surrogate compound) interact through covalent or non-covalent bonding with functional monomers that are polymerized in the presence of a cross-linking monomer to form a three-dimensional structure.[11,13] Upon removal of the template, the polymer will be left with vacated molecularly imprinted cavities bearing recognition sites with the size, geometry and arrangement of these sites complementary to those of the target analyte. As a result, MIPs have the ability to recognize target molecules selectively, thus mimicking, *e.g.*, natural antibodies.

Molecularly imprinted materials are highly robust. They exhibit excellent operational stability under a wide variety of conditions. In principle, they can be used for sensing purposes in both organic and aqueous media. Moreover, they are easier and more inexpensive to prepare than antibodies, require using no laboratory animals and are subject to none of the difficulties associated to the production of antibodies for toxic compounds or immuno-suppressants. MIPs can be engineered to exhibit high sensitivity and selectivity for various analytes of medical, environmental, and industrial interest, and have been used as recognition elements for single analytes or analyte groups in affinity chromatography, molecularly imprinted solid-phase extraction (MISPE), catalysis, binding assays, drug release and sensor development applications.[7,8,10,14–19] However, their use for optical sensing is subject to some obstacles. Thus, they are usually difficult to integrate with the transducer, a problem that is being addressed by using new monomers with responsive functionalities or nanostructured MIPs for sensing. Unlike biological receptors, which possess well-defined recognizing sites, imprinted cavities in MIPs are usually heterogeneously distributed with given stability constants of complexes formed with analytes. Moreover, these constants of MIPs are usually lower and their binding kinetics slower than those of biological receptors. In addition, MIPs typically exhibit limited selective recognition in aqueous solutions, where biomolecules perform extremely well. However, extensive research in recent years has focused on optimizing MIP composition by using monomers that interact more strongly with the template; this has resulted in materials with much desired recognition properties and higher complex stability constants – even for biomolecules – in water.

MIP-based devices can be of two types in terms of detection principle, namely: (1) affinity sensors ("plastic bodies") and (2) catalytic sensors ("plastic-zymes"). Most MIP-based chemosensors are of the former type, however.[15] In any case, the MIP must be prepared in the right format for coupling with the transducer in order to translate the chemical signal resulting from interaction with the target compound into a quantifiable output signal. Various optical techniques including UV-vis, infrared, fluorescence, chemiluminescence, surface plasmon resonance (SPR) and surface-enhanced Raman scattering (SERS) spectroscopy have been used for MIP-based chemosensor development. The present chapter summarizes progress in the development of MIP-based optical sensors over the past five years.

8.2 Fluorescence-based MIP Chemosensors

Fluorescence is the most common signal transduction mechanism used in biosensors.[20] Applications of fluorescence biosensors have extended into the development of MIP-based chemosensors due to the high sensitivity, high spatial or temporal resolution, broad linear dynamic concentration range, simplicity of the method, and the ability to recognize chemical events.[21,22] Fluorescence is a very versatile phenomenon and several parameters including intensity, time, wavelength, polarization, or position in the case of fluorescence microscopy, can be recorded and applied for sensing.[23] Therefore, MIP-based fluorescent chemosensors combine the advantages of the high sensitivity and selectivity of fluorescence detection with the selectivity of MIP recognition.

Over the last five years (2011–2016) a number of books, book chapters, and review articles have been published covering a wide range of applications of fluorescence-based MIP chemosensors in various fields.[6–10,24–26] Although in most cases detection is based on the evaluation of the fluorescence intensity at a single wavelength and the measurement of the fluorescence enhancement (turn-on), or fluorescence quenching (turn-off), in the presence of the target compound, the chemosensor can be designed to carry out ratiometric measurements at two different wavelengths, to avoid signal drifts. Fluorescence lifetime, an intrinsic molecular property within certain constraints independent of the fluorophore concentration, provides an alternative way to avoid changes in concentration caused by photobleaching or fluctuations in excitation source intensity.[27]

Based on the native characteristics of the target compounds, MIP-based fluorescence sensing can be classified into the following categories.[6]

(1) The analyte is intrinsically fluorescent and the binding of the analyte by the MIP can be monitored directly. Most chemical analytes are non-fluorescent and the number of publications related to direct MIP-based fluorescent detection is scarce.[27–30]
(2) The analyte does not display fluorescence properties so, it is necessary to synthesize a fluorescent analogue that can compete with the analyte for the polymer recognizing sites; the measuring principle is based on competitive or displacement assays.[31–33]

(3) The polymer is labeled with a fluorescent reporter that will modify the polymer emission intensity, lifetime, and/or emission wavelength upon analyte binding. Detection can mainly be based on two approaches. One approach involves the polymerization of fluorescent functional monomers that can form covalent or non-covalent bonds with the template molecules and exhibit fluorescence quenching[34–38] or enhancement[39–42] upon analyte binding. In the other approach, a polymerizable fluorescent monomer that does not contain a group interacting with the analyte is incorporated in the polymer network.[43–45] Alternatively, a fluorescent group,[46] a fluorescent developer,[47] or luminescent nanoparticles[48] can be attached to the MIP through chemical modification for signal generation. This classification does not always apply and a sensor can fit in more than one category.

Table 8.1 summarizes some relevant MIP-based assays and sensors based on fluorescent detection.

8.2.1 Direct MIP-based Fluorescence Detection

Some recent examples of direct MIP-based fluorescence detection are associated with devising micro- and nanostructured MIPs in an effort to increase the available recognition surface and improve mass transfer and binding kinetics as well as to facilitate coupling with the transducer.

Moreno-Bondi *et al.*[28] reported on the synthesis of a linear multifunctional copolymer, poly(methacrylic acid-*co*-2-methacrylamidoethylmethacrylate), P(MAA-*co*-MAAEMA), for the preparation of nano-patterned rhodamine 123, (R123)-MIP, film arrays on silicon substrates by direct writing using electron beam lithography (EBL). The P(MAA-*co*-MAAEMA) copolymer showed a positive-tone behavior for both deep UV (DUV) lithography and EBL in the dose range of 0.1 to 8 mC cm^{-2}. The MIP array exhibited excellent selectivity towards R123 against other structurally related fluorescent compounds, as monitored by fluorescence microscopy. This approach paves the way for the preparation of cost-effective miniaturized MIP film-based arrays for multiple target detection. The same groups described a new approach for the fabrication of 2-D MIP arrays, with sub-micrometre lateral resolution, by photoinduced local polymerization within Al nanoholes behaving as isolated metal-clad circular waveguides. The size of the MIP nanostructures selective to R123 is tuned by controlling the exposure time and the incident light power of the laser beam at 532 nm that is used for photopolymerization. Fluorescence lifetime imaging microscopy (FLIM) with single-photon timing measurements was used as the detection technique to prove the recognition ability and the selectivity of the nanostructures for the target analyte, R123, over other related fluorescent dyes.

Xiong *et al.*[29] prepared a microvolume MIP-based optical-fiber evanescent wave sensor for the determination of bisphenol A (BPA) released from plastic products heated at different temperatures. The BPA selective MIP film was

Table 8.1 Analytical characteristics of fluorescence-based molecularly imprinted assays.[a]

Template/analyte	MIP receptor type/composition	Luminescence technique	Measuring principle	Sample	Analytical characteristics	Ref.
Dibutyl phthalate (DBP)/DBP	QDs@MIP: QDs : ZnS : Mn/ MIP shell: APTES + TEOS	FI	Fluorescence quenching of QDs: Photo-induced electron transfer between DBP and QDs	Brand distilled spirit drinks and wine	R% : 98.1% DR: 5.0–50 μmol L^{-1} LOD: 0.27 μmol L^{-1} CR% minimum with DEP, DAP and DMP	49
Tetracycline (Tc)/Tc	QDs@MIP: QDs: CdTe/ MIP shell: 4-VP-Tc complex/EDMA/CuCl/(Me_6TREN) and poly(glyceryl monomethacrylate) brushes.	FI	Fluorescence quenching. Spectrofluorimetric titration studies	Bovine/goat milk and serum	R%: 95–105% DR: 0.5–15 μmol L^{-1} (milk) and 0.5–50 μmol L^{-1} (serum) LOD: 0.14 μmol L^{-1} (milk) and 1.9 μmol L^{-1} (milk)	50
λ-Cyhalothrin (LC)/LC	SiO_2-APTES-FITC@MIP: MIP shell: LC/DVB/AA	FI	Fluorescence dynamic quenching due to a charge transfer mechanism	River water samples	DR: 0–160 nM LOD: 10.26 nM Detection rate: 8 min CR% approx. 10% with cTc. No CR% with Van, Lex, Chl or atenolol.	51
2,4,6-Trinitrophenol (TNP)/TNP	3-D MIP film on ITO slides MIP: TNP/NH_2-S4/CLM/(TBA)ClO_4	FI	Steady state fluorescence spectroscopy λ_{ex} 350 nm; λ_{em} 600 nm	Buffer	DR: 0.2–20.4 ng L^{-1} LOD: 0.2 ng L^{-1} Detection rate: 8 min CR%: TNT (5 times less sensitive), phenol (6 times less sensitive)	52

(*continued*)

Table 8.1 *(continued)*

Template/analyte	MIP receptor type/ composition	Luminescence technique	Measuring principle	Sample	Analytical characteristics	Ref.
Tamoxifen and its metabolites/ clomiphene	Bulk polymer (sieved 25–45 μm) MIP: VCC/MAA/EDMA/ clomiphene	FI	Fluorescence quenching	Buffer	Strong recognition properties of VCC-MIP to clomiphene and tamoxifen. IF (clomiphene): 21.7, IF (tamoxifen): 18.7	53
Cocaine (COC)/ COC	QDs@MIP: QDs: Mn-doped ZnS modified with PEG/ MIP: COC/DVB/ EDMA	FI	Fluorescence quenching	Urine	Direct method: LOD: 0.076 mg L^{-1} SPE-based method: LOD: 0.0042 mg L^{-1} Positive CR% with BZE and EME. No response of other drugs of abuse	54
Papain (Pap)/Pap	SiO_2@PDA@MIP: MIP shell: Pap/DA	Resonance light scattering (RLS)	Direct measurement of ΔI_{RLS} (I_{RLS} (presence of Pap) − I_0 (no Pap)). λ_{max} = 323 nm	Dietary supplement	R%: 97.5–105.3% DR: 2–20 nM LOD: 0.63 nM CR% no significant for elastase, albumin and lysozyme	55
Bisphenol A (BPA)/BPA	CDs/MIP composite: MIP: BPA/CDs/APTES/ TEOS	FI	Fluorescence quenching	River water	R%: 97.5–102.5% DR: 100–4200 nM LOD: 30 nM CR% no significant for phenol, *o*-NP, 4-NP, K^+, Na^+, Ca^{2+}, Zn^{2+}, CO_3^{2-}, SO_4^{2-}, Fe^{3+}, Ag^+, Pb^{2+}, Hg^{2+} and Cu^{2+}	56

Dicyandiamide (DCD)/DCD	MIP decorated with MPTS-QDs MIP: DCD/MAA/EDMA	FI	Fluorescence quenching	Powder milk	R%: 95–106% DR: 5–1600 μmol LOD: 2.7 μmol CR% no significant for CM, MM, CA	57
Bisphenol A (BPA)/BPA	Thin MIP films (5 μm) on a flow cell in contact with an optical fiber. MIP: MAA/EDMA/BPA	EW	Direct measurement of BPA emission ($\lambda_{ex}/\lambda_{em}$ = 276 nm/306 nm)	Plastic products treated at different temperatures	DR: 0.003–5 mg L^{-1} LOD: 1.7 μg L^{-1} CR% no significant for glucose, ascorbic acid, lactose bisphenol F, K^+, Na^+, Ca^{2+}, Zn^{2+}, CO_3^{2-}, SO_4^{2-}, Fe^{3+}, Ag^+, Pb^{2+}, Hg^{2+} and Cu^2	29
2,4,6-Trichlorophenol (2,4,6-TCP)/ 2,4,6-TCP	GO-QDs@SiO_2-MIP composite: QDs: CD-Te MIP shell: APTES/ 2,4,6-TCP/GO–CdTe@ SiO_2/EDMA	FI	(a) Direct fluorescence quantification and (b) MISPE/HPLC	River water	(a) DR: 0.2–40 μM LOD: 0.18 μM (b) DR: 0.02–2.5 μM LOD: 0.25 μM	58

[a]FI, fluorescence intensity; QD, quantum dot; R%, recovery; DR, linear dynamic concentration range; LOD, limit of detection; CR%, cross-reactivity; APTES, 3-aminopropyltriethoxysilane; TEOS, tetraethyl orthosilicate; DEP, diethyl phthalate; DAP, diallyl phthalate; DMP, dimethyl phthalate; Me_6TREN, *tris*(2-(dimethylamino)ethyl)amine; FITC, fluorescein isothiocyanate; cTc, chlortetracycline; Van, vancomycin; Lex, cephalexin; Chl, chloramphenicol; NH2-S4, bis(2,2′-bithienyl)-(4-aminophenyl)methane; CLM, 3,3′-bis[2,2′-bithiophene-5-yl]thianaphthene; TNP, 2,4,6-trinitrophenol; TNT, 2,4,6-trinitrotoluene; VCC, vinylcoumarin-4-carboxylic acid; AA, acrylamide; EDMA, ethylene glycol dimethacrylate; MMA, methacrylic acid; IF, imprinted factor; PEG, poly(ethylene glycol); BZE, benzoylecgonine; EME, ecgonine methyl ester; PDA, polydopamine; DA, dopamine; Pap, papain; RLS, resonance light scattering; BPA, bisphenol A; CD, carbon dot; *o*-NP, *o*-nitrophenol; 4-NP, paranitrophenol; MPTS, 3-methacryloxypropyl trimethoxy silane; CM, cyanamide; MM, melamine; CA, cyanuric acid; EW, evanescence wavelength; GO, graphene oxide.

placed on the fiber core, which was further inserted into a transparent capillary to produce a microchannel (about 2.0 μL) that acted as a flow cell. Fluorescence of BPA selectively preconcentrated on the MIP layer was excited (λ_{ex} = 276 nm, λ_{em} = 306 nm), by the evanescent wave produced on the fiber core surface and the emission intensity was correlated to the concentration of the compound in the range of 3 to 5000 ng mL^{-1}.

Haupt and coworkers[30] reported on the direct determination of enrofloxacin (ENRO), an intrinsically fluorescent analyte, and other structurally related piperazine-based fluoroquinolones, by monitoring the amount of antimicrobials bound to the MIP with fluorescence polarization measurements. The technique allows direct detection of the antimicrobial bound to the MIP nanoparticle in the presence of the free analyte without the need of a previous separation step. The assay was successfully applied to the determination of ENRO in tap water and milk samples with a low limit of detection, LOD = 1 nM.

The combination of MIPs with magnetic nanoparticles (*m*-MIPs) is especially useful for analytical separations as these materials allow fast and easy template removal from complex samples. In principle, *m*-MIPs can also be applied for sensing purposes, as demonstrated by Valero-Navarro *et al.*,[59] who prepared polyurethane-based *m*-MIPs for the selective determination of a fluorescent compound, 1-naphthylamin (1-NA), in drinking water. The polymers were prepared by encapsulating magnetite-coated-oleic acid nanoparticles into a lipophilic polymeric matrix, poly(methacrylic acid-*co*-ethylene glycol dimethacrylate), P(MMA-*co*-EDMA), followed by coating it with the selective MIP by precipitation polymerization. An optical fiber coupled with a magnetic separator was used to monitor the intrinsic fluorescence of the 1-NA compound retained in the *m*-MIP that was concentrated for readout by sensor spot formation. The chemosensor showed a high selectivity for 1-NA in water even in the presence of other related compounds, such as 2-naphthylamine, 1-naphthol, 2-naphthol, or 1-naphthalenemethylamine, with limits of detection and quantification of 18 and 59 ng mL^{-1}, respectively.

8.2.2 Indirect MIP-based Fluorescence Detection Using Labelled Analytes

As discussed previously, optically active materials are currently the preferred option for the detection of non-fluorescent molecules. However, most targets are not fluorescent and competitive or displacement assays, based on the use of fluorescent probes structurally related to the analyte that compete with it for the recognizing sites in the imprinted material, have shown to be an alternative of remarkable utility for chemosensor development.[60] In these devices, detection is usually based on the fluorescence intensity changes of the sensing material upon analyte binding, however, lifetime, fluorescence anisotropy or Förster resonance energy transfer (FRET) based approaches have been reported as well. Several selected examples are discussed below.

For example, Murase *et al.*[31] have developed a fluorescence polarization-based sensing platform for the determination of free cortisol. The measuring principle is based on the displacement of dansyl-cortisol bound to the hormone-imprinted cavities in the MIP-NPs with the corresponding variation in the fluorescence anisotropy of the solution. Cortisol imprinted core–shell MIP-NPs are prepared by emulsifier-free emulsion using monodispersed poly(styrene-*co*-DVB) particles as cores. Reference particles (R-MIP-NPs), prepared without itaconic acid, showed lower fluoresce anisotropy variations in the presence of dansyl-cortisol than the MIP-NPs suggesting that itaconic acid and cortisol-21-monomethacrylate-derived methacrylic acid residues work cooperatively. The fluorescence polarization signal is correlated to the concentration of cortisol in the sample with an LOD of 80 nM. No significant cross-reactivity was observed for progesterone, although some response was observed for testosterone.

Descalzo *et al.*[33] developed a versatile FRET-based competitive biomimetic assay for the determination of ENRO. Silica nanoparticles (200 nm) were doped with *tris*(1,10-phenanthroline)ruthenium(II), $Ru(phen)_3^{2+}$, to act as the FRET donor. They were further coated with an ENRO-selective MIP nanolayer. ENRO was measured by equilibrating the antibiotic with a constant amount of SiO_2@MIP doped NPs and a cyanine-labelled ENRO, as the luminescent FRET acceptor. When the labelled antibiotic enters the MIP receptor, the red emission of the core-encapsulated $Ru(phen)_3^{2+}$ is quenched by the FRET acceptor. This effect decreases in the presence of the ENRO. The novel photochemical approach achieved a LOD of 2 μM. Some cross-reactivity was observed only for structurally related fluoroquinolones.

The fabrication of MIP-based microarrays is of great interest as these platforms allow high throughput screening of multiple analytes in a short time at a low cost.[61] Despite the advantages, applications of MIPs in this field are still in their infancy. Carrasco *et al.*[32] developed an optical fiber array using an MIP and NIP spherical microparticles for the ENRO determination with the future potential for multiplexed sensing. The sensing platform was prepared by chemical etching an optical fiber bundle, of ~50 000 individual fibers, to create a planar array of microwells. Coumarin-30 labelled MIP spherical microparticles and *tris*(4,7-diphenyl-1,10-phenanthroline)ruthenium(II), $Ru(dip)_3^{2+}$, encoded NIPs were then pooled in the microwells to form a randomly ordered high density array. The sensing mechanism is based on a fluorescent competitive assay between ENRO and a novel highly fluorescent 4,4-difluoro-4-bora-3a,4a-diaza-*s*-indacene (BODIPY)-labelled ENRO for recognition by the selective cavities in the MIP spherical microparticles. The LOD was 40 nM (Figure 8.1a) and selectivity to ENRO was excellent in the presence of other related or non-related antibiotics. The device has been used in the presence of organic solvents and demonstrated excellent reproducibility and short response times in comparison to other similar bioassays. It was applied to the ENRO determination in medicated sheep serum samples and the results were validated by high-performance liquid chromatography (HPLC) with fluorescence detection. As suggested by the authors, this

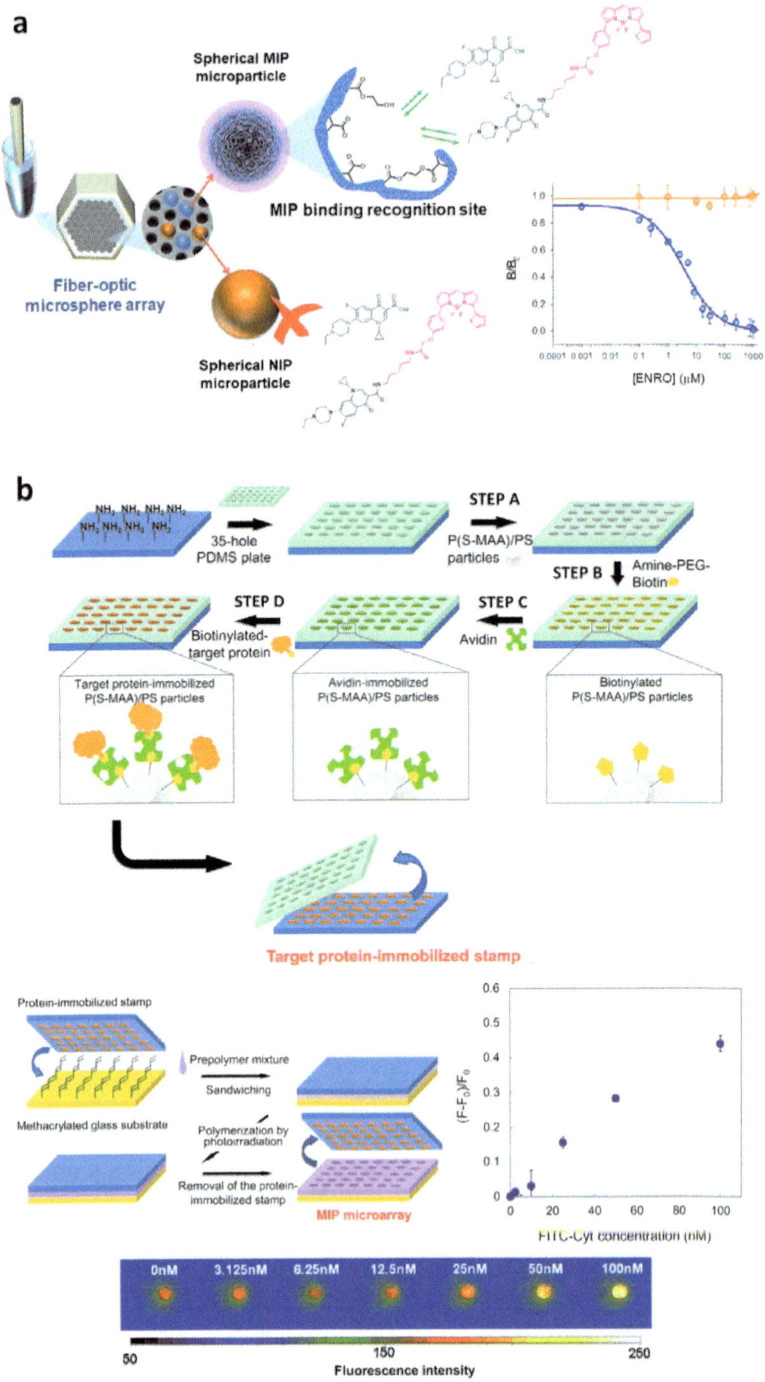

Figure 8.1 (a) Left: Workflow of the assay protocol. ENRO quantification was based on a competitive assay, in which the target analyte competed with the

sensing platform was quite flexible and future applications could be coupled with MIP spherical microparticles to target different molecules selectively.

An interesting approach was introduced by Takeuchi and co-workers,[62] where MIP arrays were prepared using multiple protein-immobilized dots as stamps for the protein determination *via* transcription-type molecular recognition. Biotinylated polystyrene nanoparticles (PS-NPs) were immobilized on aminated glass substrates and were used as scaffolds for immobilizing biotinylated proteins, such as cytochrome *c* (cyt *c*), ribonuclease A, myoglobin, and lysozymes *via* a unified avidin–biotin interaction protocol (Figure 8.1b). For the preparation of the transcription-type protein–MIP array, the prepolymerization mixture, consisting of acrylic acid (AA) as the functional monomer, acrylamide (AM) as the co-monomer, and *N*,*N*′-methylene *bis*(acrylamide) (MBA) as the cross-linking monomer, was sandwiched between the protein-immobilized stamps and a methacrylated glass substrate, resulting in transcribed protein moulds for the corresponding positions of the immobilized proteins. Competitive binding experiments, using the fluorescein-labelled proteins as competitors, confirmed the successful syntheses on the same substrate of different MIPs bearing imprinted cavities selective towards the chosen proteins. As stated by the authors, this approach represented a highly stable, low-cost, and tailor-made alternative for protein determination.

8.2.3 Indirect MIP-based Fluorescence Detection Using Labelled Polymers

8.2.3.1 MIP Chemosensors Based on Fluorescent Monomers

The use of fluorescent dyes, with an intrinsic response to the target compound as monomers for the preparation of MIP-based chemosensors, avoids the need of analyte labelling or the implementation of competitive assays facilitating real-time transduction as well as miniaturization. A thorough review has been published recently on the use of dyes as fluorescent monomers for the development of MIPs.[63] Typically, the fabrication of fluorescent MIPs

---

labelled ENRO (BODIFLOXACIN) to bind to selective recognizing sites on the spherical MIP microparticles. Right: Calibration plots for MIP (blue points, $n = 4$, 14 points) and spherical NIP microparticles (orange points, $n = 2$, 9 points) in 50 : 50 (*v*/*v*) MeCN : HEPES (25 mM pH = 7.5). The B/B_0 is plotted *versus* ENRO concentration, which competes with BODIFLOXACIN in both microparticles. Reproduced from ref. 31 with permission from The Royal Society of Chemistry. (b) Above: Fabrication of patterned multiple protein-immobilized dot stamps for transcribed MIPs and; below: the protocol for preparation of an MIP array, binding behaviour of F-Cyt toward MIP-Cyt array and microscopic images of F-Cyt adsorbed on an MIP array. Reproduced with permission from X. D. Wang and O. S. Wolfbeis, *Anal. Chem.*, 2016, **88**, 203–227, Copyright 2015 American Chemical Society.

can be accomplished in different ways. (i) A fluorescent dye, sensitive to the local environment or to specific functional groups or electronic features of the template, is integrated into the MIP matrix.[34,38,64] In many cases these monomers lack specific groups recognizing the template molecule, which results in low complex stability constants, signal heterogeneity, and high background fluorescence. The use of triadic monomers, including the dye molecules, spacers, and polymerizable groups, may avoid some of these limitations.[65] (ii) The use of polymerizable fluorescent probe monomers usually consisting of four or more subunits, such as fluorophore–(linker)–receptor–(linker)–polymerizable units. As described by Wan *et al.*,[63] a fluorophore–(linker)–receptor is the fluorescent probe in the classical sense. In this case the interaction with the target compound can be very selective and the fluorescent response better defined.[66]

Awino *et al.*[67] synthesized water-compatible molecularly imprinted NPs against naphthyl derivatives using a fluorescent dansyl derivative with two polymerizable methacrylate groups as the fluorescent monomer. The NPs were prepared by surface-core cross-linking of surfactant micelles in water, following the protocol previously reported by the same group.[68,69] The naphthalene-containing template served as the FRET donor for the dansyl acceptor incorporated into the molecularly imprinted NP through co-polymerization. On average, each NP had two dansyl groups to enable the naphthyl template to be within the Förster distance ($R_0 = 2.2$ nm)[70] of the dansyl acceptor after binding to the selective cavities. Non-specific binding of very similar structures, by hydrophobic or electrostatic interactions, did not trigger the FRET response. As confirmed by isothermal titration calorimetry (ITC), the MIP-NPs were highly selective to the target compound with the complex stability constant of $K_a = 0.43 \times 10^6$ M^{-1}; however, none of the anionic analogues tested, regardless of their size, showed any comparable binding. In principle, this approach to combining the recognition properties of molecular imprinting with easy-to-perform FRET-based detection is quite versatile for the development of chemosensors for small fluorescent molecules in water.

Haupt *et al.*[71] reported the development of optical fiber fluorimetric chemosensors based on micrometre-sized MIP tips photopolymerized *in situ*. They explored several approaches for sensor development. (i) A competitive assay format for the determination of carbobenzyloxy-L-phenylalanine (Z-*l*-Phe) using the fluorescent analogue, dansyl-L-phenylalanine (dansyl-L-Phe). (ii) The development of fluorescent MIPs using the fluorescent monomer, *N*-2-{6-(4-methylpiperazin-1-yl)1,3-dioxo-1*H*-benzo[de]isoquinolin-2(3*H*)-yl-ethyl} acrylamide (FIM), for the detection of a non-fluorescent herbicide, 2,4-dichlorophenoxyacetic acid (2,4-D). The resulting MIP showed a light-up response because of the protonation of the piperazinyl residue upon interaction with carboxyl groups. This is the most preferred signalling mode as it is connected to a specific interaction, whereas non-specific quenching may result in false positives.[63] The use of tips prepared by adopting a core–shell approach resulted in faster responses because of the shorter diffusion distances. The incorporation of gold NPs in the MIP formulation resulted in significant enhancement of the fluorescent MIP signal (LOD of 250 pM for 2,4-D), which was partially attributed to scattering by the gold nanoparticles back into the fiber, as the authors did not have a clear evidence of plasmonic enhancement taking place. In subsequent work, the same group[72] applied the FIM monomer for the determination of 2,4-D and citrinin using disposable tapered polystyrene waveguides. The MIP coating was created either by dip-coating with the polymeric particles synthesized beforehand or by direct evanescent-wave polymerization. Analyte binding to the MIP resulted in an increase in the fluorescence emission of FIM upon excitation by the evanescent wave. The chemosensor displayed a certain degree of cross-reactivity for related molecules with LODs in a low nM range.

The use of polymerizable fluorophore–receptor–linker–polymer unit designs has become especially attractive in recent years as the usually smaller size of the probes allows the direct electronic coupling of the receptor and the fluorophore, which results in the direct detection of the target compound.[63] The urea group has been widely used for this approach as this moiety can efficiently interact with oxoanionic templates.[35,42] In a recent work, the groups of Sellergren and Rurack[73] have reported the synthesis of core–shell nanoparticles using a nitrobenzoxadiazole (NBD)–urea-monomer as the binary hydrogen-bond donor targeting the carboxyl group of sialic acid (SA). The NBD fluorophore shows a light-up response upon SA binding, and signal equilibration times were as short as 15 s. The polymer was prepared by a mixed covalent and non-covalent approach based on reversible amine catalyzed boronate esterification, hydrogen bond stabilization through the guest-responsive NBD-appended urea-monomer and electrostatic stabilization. The SA-MIP shell was grafted from 200 nm sized RAFT functionalized silica cores with a thickness of ~10 nm. The chemosensor displayed a strong affinity for SA in methanol/water mixtures ($5.9 \times 10^3\ M^{-1}$). Binding of potential cross-reactants, such as glucuronic acid or other monosaccharides, was not significant. The SA-MIP nanoparticles were applied to the selective staining of the cell surface glycans containing SA, and, as observed for lectins, the authors found a ~1000-fold enhanced affinity for the cell surface compared to the free monosaccharide.

8.2.3.2 MIP Chemosensors Based on Luminescent Nanoparticles

In the last few years, the combination of nanomaterials with MIPs has allowed the implementation of new sensing schemes with unique capabilities not attainable by conventional methods.[61] In particular, the application of semiconductor nanocrystals (quantum dots, QDs), carbon dots or rare earth complexes, among others, has broadened the field of application of these devices. As such, the field now has excellent opportunities for the development of high-sensitivity fluorescence sensors with improved functionality, increased adsorption capacity, decreased nonspecific adsorption, accelerated mass transfer, and unique optical properties.[74,75] QDs are spherically shaped highly luminescent colloidal semiconductor nanocrystals, so called because the quantum confinement takes place in three spatial dimensions, with the size range of 1 to 10 nm, containing typically 10^3 to 10^4 atoms, composed of sulphides, selenides or tellurides of heavy metals (groups II–VI, III–V, or IV–VI of the periodic table) with unique optical properties for sensing organic and inorganic species.[76–78] In comparison to organic dyes and fluorescent proteins, QDs exhibit several advantages including high photostability and photoluminescence quantum yields, size-tunable emission, wide absorption spectra, narrow and symmetrical emission spectrum, and resistance to photobleaching.[79,80] Thus, the emission wavelength can be tuned from the UV to the near-infrared (NIR) region by varying their size and chemical composition and, in principle, the detection of multicolor QDs is possible using a single-light source.[81] The main disadvantage of QDs, however, is their potential toxicity, because of the presence of heavy metals. Nevertheless, this risk can be mitigated by using coatings that may also improve the compatibility of QDs in water.[82] The synthesis of highly crystalline QDs of CdSe is well established.[83] These nanoparticles can be further coated with a layer of a wider bandgap semiconductor, such as ZnS and CdS, to minimize cytotoxicity and improve the photoluminescent yield, thus protecting the QDs from oxidation.[84,85] QDs comprised of CdSe or the CdSe/ZnS core/shell are currently most widely used in biological applications[86] and they have also found applications with MIPs.[48,87–94] MIP-based fluorescent chemosensors using QDs, (abbreviated as QD@MIP chemosensors), are commonly synthesized following two procedures. In one, the QDs were embedded into the bulk or monolithic MIP.[95–98] In the other, the QD solid core was overcoated with an MIP shell.[58,99,100]

Stringer *et al.*[101] prepared a composite combining MIP microparticles with QDs of CdSe, in a postprocessing step, for the detection of nitroaromatic explosives in aqueous samples. The polymer was prepared with methacrylic acid as the functional monomer and the carbodiimide chemistry were utilized to activate the carboxyl groups in the polymer for amide bonding with amine-functionalized QDs.[102] The resulting material was highly robust and the QDs were dispersed throughout the polymer matrix. Fluorescence of the CdSe QDs was effectively quenched by 2,4,6-trinitrotoluene (TNT) in less than one minute, especially when chloroform was used as the porogen. The polymer was highly cross-reactive to 3,5-dinitrotoluene (DNT) and sensitivity was relatively low in comparison to other chemosensors for nitroaromatic explosives, with the LODs of 30.1 μM and 40.7 μM for DNT and TNT, respectively.

Liu and co-workers[103] developed an optosensing chip for the determination of sesamol in sesame oil using MIP-coated CdSe/ZnS QDs. The sensing layer was prepared by mixing the template with the QDs, the functional monomer (MAA) and the cross-linking monomer (EDMA). After polymerization, the polymers were ground, sieved, and suspended in chitosan before being deposited *via* drop-coating onto the surface of a quartz crystal chip. The fluorescence intensity of the chip decreased significantly in the presence of the target sesamol analyte, which was attributed to a charge transfer from the QDs to sesamol. The LOD was 72 nM. The selectivity of the device was evaluated against structural analogues of sesamol, such as piperonal (PN), *tert*-butylhydroquinone and *tert*-butylhydroxyanisole. Only the first showed a slight cross-reactivity although the adsorption capacity for PN was much lower than that for sesamol. The system was validated by HPLC-UV for the determination of sesamol in sesame oil samples. In a different approach, the same group described the synthesis of an optosesing material for the determination of *N*ε-(carboxymethyl)lysine, based on QDs coated with an MIP sensitized with graphene (Gra-QDs@MIP).[104] The presence of graphene improved the stability and kinetic binding properties of the MIP nanoparticles, prepared with (3-aminopropyl)triethoxysilane (APTES) as the functional monomer and tetraethoxysilane (TEOS) as the cross-linking monomer. The chemosensor showed the LOD of 3.0 μg L^{-1} and was applied to sesamol determination in dairy product samples.

Moreover, the detection of biomolecules (proteins, DNA, *etc.*) using MIPs is of great interest for applications in the field of diagnostics. For example, Lee and co-workers[105] prepared composite nanoparticles by embedding QDs into molecularly imprinted poly(ethylene-*co*-vinyl alcohol) NPs for the determination of amylase, lipase, and lysozyme in saliva. Target protein binding was monitored by fluorescence quenching of the QDs-MIP in the presence of the proteins with an LOD of 0.1 μg mL^{-1} for amylase and lipase, and 13 ng mL^{-1} for lysozyme. Physiological concentrations of amylase, lipase, and lysozyme ranged from 0.1 to 10 μg mL^{-1} thus, the devised MIP system was potentially useful for the protein determination in clinical samples. Performance of the new material was compared with that of the Architect ci8200 system, showing an accuracy for both amylase and lipase of *ca.* 96.2 ± 2% (where 100% means a complete agreement with the Architect analysis system).

Different synthetic approaches have been reported to graft MIPs from solid cores with different effects on the properties of the core material and composite performance.[106] The preparation of thin and reproducible coatings for colloidal nanoparticles is complex and sometimes uncontrollable; however, it is a prerequisite for most applications of these materials including the fabrication of MIP-based chemosensors.[107,108] A sol–gel process was developed by Tan and co-workers[109] to fabricate a fluorescent turn-on chemosensor for α-fetoprotein (AFP), a tumour biomarker. Mn-doped ZnS QDs were endcapped with 3-mercaptopropyl triethoxysilane (MPTS), and further activated with APTES and TEOS. A thin layer of MIP was grafted by polymerization of methyl methacrylate (MMA) and 4-vinylphenylboronic acid (VPBA) in the presence of γ-methacryloxypropyl trimethoxysilane (KH570) as the coupling agent. The boronic acid group covalently reacts with the *cis*-diol groups in the template molecule, thus generating highly selective recognizing sites in the resulting composite. In the absence of the target protein, the fluorescence of the Mn-doped ZnS QDs is quenched through photoinduced electron transfer by the boronic acid groups. The AFP binding increases the fluorescence intensity as the boron moieties become negatively charged and the electrostatic interaction with the QDs is hindered. The LOD for this chemosensor was 4.8×10^{-8} g L^{-1} and no cross-reactivity was measured for other

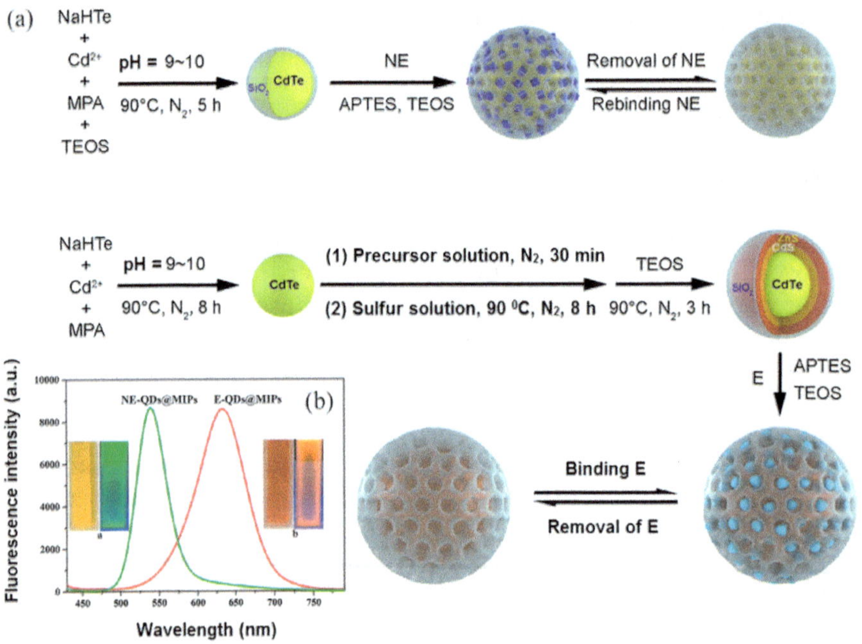

Figure 8.2 (a) Schematic of fabrication of dual-colour QDs@MIPs and (b) fluorescence spectra of QDs@MIPs and photographs of NE-QDs@MIP and E-QDs@MIP under natural light (left) and UV light (right), respectively. Reprinted from *Sensors and Actuators B: Chemical*, **229**, F. Wei, G. Xu, Y. Wu, X. Wang, J. Yang, L. Liu, P. Zhou and Q. Hu, Molecularly imprinted polymers on dual-colour quantum dots for simultaneous detection of norepinephrine and epinephrine, 38–46,[109] Copyright (2016) with permission from Elsevier.

proteins, such as horseradish peroxidase, myoglobin, ovalbumin, transferrin, or ribonuclease B. The chemosensor was applied to the AFP determination in clinical serum samples and the results were in good agreement with those obtained by an ELISA method. Moreover, the core–shell approach was applied by Wei *et al.*[110] for the simultaneous detection of norepinephrine (NE) and epinephrine (E) using two different coloured QDs decorated with a silica-based MIP shell. NE-QDs@MIP and E-QDs@MIP were synthesized using CdTe@SiO_2 and CdTe/CdS/ZnS/SiO_2 QDs, respectively. APTES was used as the functional monomer and TEOS as the cross-linking monomer (Figure 8.2a). The two types of QDs@MIPs can be excited at the same wavelength (365 nm). However, NE-QDs emit at 539 nm and E-QDs@MIP at 632 nm. The fluorescence intensity of the QDs@MIPs decreased with the increase of concentration of the target analyte. The spectral overlap of the fluorescence spectra for both nanoparticles was lower than 5%, thus allowing the simultaneous detection of NE and E (Figure 8.2b). The linear dynamic concentration ranges extended from 0.08 to 20 μM for both analytes and the LOD was 9 nM and 12 nM for NE and E, respectively.

Recently, Haupt and co-workers[111] reported an interesting procedure for the preparation of MIP coated QDs with applications to the simultaneous multiplexed pseudoimmuno labelling and imaging of human keratinocytes. First, stable and robust hydrophilic MIP shells, selective to glucuronic acid (GlcA) or *N*-acetylneuraminic acid (NANA), were generated using the visible fluorescent light emitted from QDs upon excitation by UV light as an internal light source for photopolymerization. Green- and red-emitting InP/ZnS QDs were initially coated with a water-compatible shell using 2-hydroxyethyl methacrylate (HEMA) and *N*,*N*′-ethylenebis(acrylamide) (EbAM) as well as eosin Y (green QDs) or methylene blue (red QDs)/trimethylamine (TEA) as initiators. Next, an MIP shell selective for GlcA (green-QDs) or NANA (red-QDs) was grafted on top of the hydrophilic first shell using UV light (365 nm) excitation to locally prepare a thin polymer film. A stoichiometric functional monomer, (4-acrylamidophenyl)(amino)methaniminium acetate (AB), that interacts electrostatically with the –COOH moiety of GlcA and NANA, was used in combination with methyl methacrylate (MMA) for polymer preparation. The latter was added to decrease MIP hydrophobicity and prevent aggregation in the aqueous cell-imaging medium (Figure 8.3a and 8.3b).

Selectivity was confirmed by competitive binding assays with other monosaccharides (*e.g.* *N*-acetylgalactosamine, *N*-acetylglucosamine, galactose, and glucose) that can be present in the terminal parts of glycolipids or glycoproteins and could potentially interfere during cell imaging. Less than 1% cross-reactivity was observed in all cases with the exception of GlcA, which showed <10% cross-reactivity when MIPNANA was tested. For quantitative cell imaging, a standard immunostaining protocol was adopted for visualizing human keratinocytes (HaCaT cells) (Figure 8.3c) and human KU812 cells (leukemia cells) (Figure 8.3d). Both MIPGlcA-QDs and MIPNANA-QDs showed 42% and 4%, respectively, more binding to the cells than their respective negative controls (NIP-QDs). This work demonstrates for the first time the potential of MIPs conjugated to quantum dots of different emission colors as a versatile imaging tool for multiplexed determination of glycosylations in cells.

Other NPs, such as up-converting nanoparticles (UCNPs) have attracted a great interest in MIP development due to their unique photochemical properties for multicolour imaging applications.[112,113] UCNPs consist of a rare earth element crystalline host (*e.g.* $NaYF_4$, $NaGdF_4$, and Gd_2O_3) doped with different combinations of lanthanide ions (*e.g.* Yb^{3+}, Tm^{3+}, and Er^{3+}) in order to tune the emission wavelength[114] (Figure 8.4). This tuning is particularly attractive for multiplexed imaging. UCNPs emit spectrally narrow anti-Stokes shifted light, display higher sensitivity (by a factor of 10) than conventional Stokes-shifting luminescent reporters, and can be excited by NIR rather than UV light, thus minimizing photo damage of biological species, maximizing

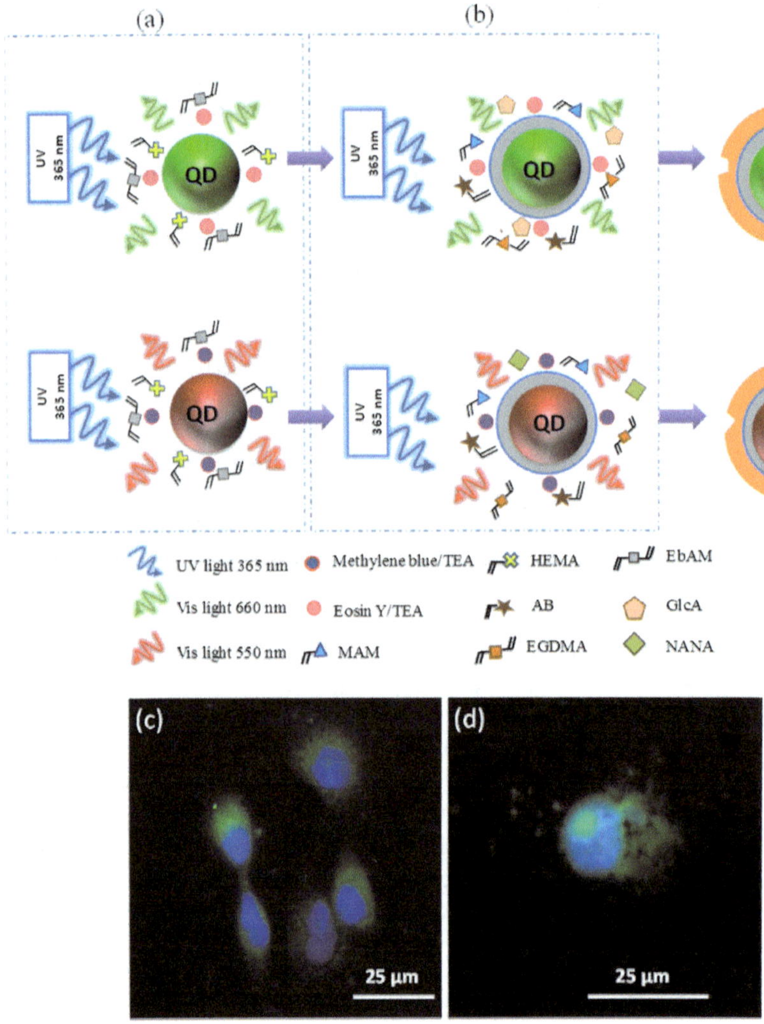

Figure 8.3 (a) Red or green light emitted from InP/ZnS quantum dots excited by UV light is used to synthesize *in situ* a polymeric inner shell around the particles by photopolymerization. (b) An outer shell of an MIP synthesized in the presence of functional and cross-linking monomers, and a molecular template (GlcA or NANA). Staining of (c) keratinocytes and (d) KU812 cells with MIP-QDs (green); nuclear staining with 4′,6-diamidino-2-phenylindole (DAPI; blue). Reproduced/adapted from ref. 111 with permission. Copyright 2016 Wiley-VCH.

the penetration depth, and showing minimal background autofluorescence because blood and tissues are transparent in the NIR spectral range. Moreover, they enable the acquisition of autofluorescence-free data with superb contrast for optical imaging applications.[115,116]

Following the versatile synthetic strategy for the coating of QDs described above,[111] Beyazit *et al.*[117] described the preparation of UCNPs coated with

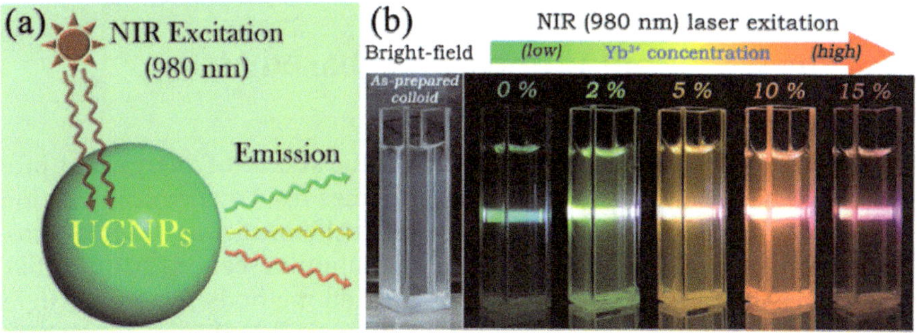

Figure 8.4 (a) Schematic illustration of the UC process for UCNPs under NIR excitation. (b) Digital UC fluorescence images of the $Gd_2O_3:Yb^{3+}$(0–15 mol%)/Er^{3+} UCNP colloids under excitation at 980 nm (no optical filters were used). Adapted from ref. 114 with permission from the PCCP Owner Societies.

MIP shells selective to trypsin. They used visible-light ($NaYF_4:Yb^{3+}$, Er^{3+}) and UV-light ($Na_{0.6}K_{0.4}YF_4:Yb^{3+}$, Tm^{3+}) emitting UCNPs as cores. The light emitted from UCNPs, when excited at 980 nm, was used to create a first water-compatible polymeric shell by photopolymerization, as was done with the QDs.[111] The first shell (UCNP-vis) was prepared from a mixture containing UCPs, HEMA/EbAM, eosin/TEA, and *N*-acryloyl-*p*-aminobenzamidine as the anchoring monomer for trypsin. The recognition properties of the UCNP-MIPs were evaluated by equilibrium binding experiments with a fluorescently labelled trypsin (fluorescein isothiocyanate (FITC)-trypsin). Alternatively, UCNP-UV MIPs were prepared in a similar way using benzophenone/TEA initiation with either 2-carboxyethyl acrylate (CEA) or propargyl acrylamide (PA) as additional monomers to allow for amide coupling or click chemistry, respectively. The UCNP-UV-CEA NPs were conjugated with FITC-streptavidin for imaging applications.

MIP–(carbon dots) (MIP-CDs) composites have been investigated for the fabrication of MIP-based chemosensors in the last few years.[118–120] These composites combine the high selectivity of MIPs with the excellent signal brightness, low toxicity, eco-friendly behaviour, ease of preparation, and excellent photostability of CDs. In most cases, the MIP-CD composites are prepared by encapsulating the CDs in the pores of the MIP network.[121–123] For example, Li *et al.*[124] prepared a new type of thermo-sensitive receptor based on CDs/SiO_2/MIP composites for the recognition of cyt *c*, using the epitope approach, with the AYLKATNE and GDEKKKI peptides as templates. An aqueous solution was used for polymer synthesis, which involved mixing the peptide bound CDs/SiO_2 NPs with the *N*-isopropylacrylamide (NIPAN) and MAA functional monomers, and the MBA cross-linking monomer. Fluorescence was quenched upon selective binding of cyt *c* to the MIP with a linear dynamic concentration range of 0.1 to 40 μM, and the LOD of 89 nM. The authors suggest that the CDs/SiO_2/MIP receptors can be used as temperature gates to control the recognition and release of the target protein by changing the temperature.

8.3 Chemiluminescence and Electrochemiluminescence-based MIP Chemosensors

The different types of luminescence differ in the source of energy that produces the excited state. That is, the energy comes from a chemical reaction in chemiluminescence (CL), and from an electrochemical reaction in electrochemiluminescence or electrogenerated chemiluminescence (ECL). Unlike fluorescence-based methods, CL- and ECL-based methods require no light source; moreover, the latter are not subject to typical problems of photoluminescence, such as light scattering and source instability. These advantages have enabled devising chemosensors with improved characteristics, such as higher sensitivity and lower detection limits for determining trace analytes, especially in combination with a wide variety of recently devised micro- and nanomaterials used as supports and platforms.[125,126]

Unlike CL, ECL affords the electrochemical regeneration of some reactants at the electrode surface. This regeneration results in improved sensitivity and allows electrogeneration of unstable CL reagents and intermediates, and the electrochemical transformation of some compounds into CL active species. In addition, emission in ECL occurs near the electrode surface, which affords finer control.

MIPs have successfully been used with CL and ECL systems (Table 8.2) to devise chemosensors with improved selectivity by combining their properties of the imprinted material with those of the transduction technique. Most MIP-based CL chemosensors use the (hydrogen peroxide)–luminol–(sodium hydroxide) CL reaction for detection. For example, Qiu *et al.*[132] devised a CL array chemosensor for the simultaneous determination of benzenediol isomers, namely, hydroquinone, resorcinol and catechol, in waste waters with graphene–magnetite MIPs. Graphene improved adsorption capabilities, sensitivity, and binding kinetic properties, and Fe_3O_4 nanoparticles facilitated separation and immobilization. MIP-coated magnetic nanoparticles have also been used to devise CL chemosensors for determining dibutyl phthalate in drinks[138] and chrysoidine in foods, using flow-injection analysis CL systems (Table 8.2).[150]

ECL can be produced in various ways most of which use the *tris*(2,2′-bipyridine)ruthenium(II) complex, $[Ru(bpy)_3]^{2+}$, as the luminophore. Thus, Wu *et al.*[151] devised a novel MIP ECL chemosensor based on $[Ru(bpy)_3]^{2+}$–Au nanoparticles (AuNPs) decorated multi-walled carbon nanotubes (MWCNT) composites and a molecularly imprinted polymer to determine isoniazid. The target analyte enhanced the ECL reaction of the metal complex, fixed on an AuNP–MWCNT-Nafion modified glassy carbon electrode, which enabled its determination with a low LOD (0.08 $\mu g\ cm^{-3}$) over a wide linear concentration range (0.1 to 110 $\mu g\ cm^{-3}$) in pharmaceutical formulations and human urine. The chemosensor was made reusable by cyclic voltammetry. Moreover, it exhibited no cross-reactivity in the presence of compounds, such as

Table 8.2 Analytical characteristics of chemiluminescence (CL) and electrochemical luminescence (ECL) MIP-baseds chemosensors.[a]

Template	Functional monomers	Cross-linking monomer/Porogen	Detection technique	Linear dynamic concentration range	Detection limit	Sample	Ref.
Vanillin	AM	EDMA/ethanol	CL	3.3×10^{-7}–1.2×10^{-5} mol L^{-1}	1.1×10^{-7} mol L^{-1}	Vanilla beverages	127
Carbofuran Omethoate	AM	EDMA/acetonitrile	CL	Carbofuran: 4×10^{-8}–9×10^{-6} g mL^{-1} Omethoate: 1×10^{-7}–9×10^{-6} g mL^{-1}	2.5×10^{-8} g mL^{-1} 4.2×10^{-8} g mL^{-1}	Foodstuff	128
Isoniazid	AM	MBAA/acetonitrile	ECL	0.1–110 μg cm^{-3}	0.08 μg cm^{-3}	Urine	129
Doxycicline	MAA	EDMA/acetonitrile	ECL	2×10^{-10}–5×10^{-8} mol L^{-1}	5.17×10^{-11} mol L^{-1}	Urine	130
Sulfadiazine	MAA	EDMA/acetonitrile	CL	4×10^{-7}–1×10^{-4} mol L^{-1}	1.54×10^{-7} mol L^{-1}	Urine	131
Benzenediol isomers (Hydroquinone, resorcinol, catechol)	MAA	EDMA/acetonitrile	CL	Hydroquinone: 2.359×10^{-8}–5.680×10^{-5} mol L^{-1} Resorcinol: 6.981×10^{-7}–3.659×10^{-4} mol L^{-1} Catechol: 1.359×10^{-7}–5.361×10^{-4} mol L^{-1}	1.047×10^{-9} mol L^{-1} 2.047×10^{-8} mol L^{-1} 3.069×10^{-8} mol L^{-1}	Water	132
Lysozyme	MAA	DMAEMA/water	CL	1×10^{-9}–8×10^{-8} mg mL^{-1}	3×10^{-10} mg mL^{-1}	Urine	133
Ochratoxin A	MAA	EDMA/water	ECL	1×10^{-5}–11.13 ng mL^{-1}	3 fg mL^{-1}	Corn/Serum	134

(*continued*)

Table 8.2 (*continued*)

Template	Functional monomers	Cross-linking monomer/ Porogen	Detection technique	Linear dynamic concentration range	Detection limit	Sample	Ref.
Sulfamethoxazole	AM	EDMA/acetone	CL	1×10^{-7}–2×10^{-3} mol L^{-1}	3×10^{-8} mol L^{-1}	Drugs	135
Quercetin	AM	EDMA/acetone	CL	1.4×10^{-6}–1.6×10^{-4} M	9.3×10^{-7} M	Drugs	136
Olaquindox	-	-	CL	2×10^{-8}–1×10^{-6} g mL^{-1}	7×10^{-9} g mL^{-1}	Chick feeds	137
Dibutyl phthalate	MAA	EDMA/ chloroform	CL	3.84×10^{-8}–2.08×10^{-5} mol L^{-1}	2.09×10^{-9} mol L^{-1}	Drink	138
Thifensulfuron-methyl	AM	EDMA/ acetonitrile	CL	1.0×10^{-9}–5.0×10^{-5} mol L^{-1}	8.3×10^{-10} mol L^{-1}	Waters	139
Lysozyme	DMAEMA	EDMA/water	CL	5.04×10^{-9}–4.27×10^{-7} g mL^{-1}	1.90×10^{-9} g mL^{-1}	Egg samples	140
Ribonuclease A	MAA, MA, DMAEMA	EDMA/ethanol	CL	1.0×10^{-9}–1.0×10^{-7} mg mL^{-1}	3.2×10^{-10} mg mL^{-1}	Biological samples	141
Chlorpyrifos	AM	EDMA/ acetonitrile	CL	1.0×10^{-10}–5.0×10^{-7} M	1.0×10^{-11} M	Waters	142
Fenpropathrin	AM	EDMA/ acetonitrile	CL	5.0×10^{-8}–1.5×10^{-5} g mL^{-1}	3.07×10^{-8} g mL^{-1}	Food samples	143
Chrysoidine	MAA	EDMA/ethanol	CL	2.0×10^{-6}–1.0×10^{-4} M	6.19×10^{-7} M	Food samples	150
Haemoglobin	APBA	APBA/water	ECL	0.1–4×10^{4} pg mL^{-1}	0.023 pg mL^{-1}	Serum samples	144
Bisphenol A	MAA	EDMA/ acetonitrile	CL	1.0×10^{-9}–2.0×10^{-8} g mL^{-1}	7.13×10^{-10} g mL^{-1}	Feed and water samples	145

Glyphosate	AM	EDMA/chloroform and methanol (4 : 1, *v* : *v*)	CL	0.5–35 μg mL^{-1}	0.046 μg mL^{-1}	Foodstuffs	146
Dopamine	AM	EDMA/ethanol	CL	8.0–200.0 ng mL^{-1}	1.5 ng mL^{-1}	Urine and dopamine hydrochloride injection samples	147
Dichlorvos	MAA	EDMA/ acetonitrile	CL	3.0 ng mL^{-1}–1.0 μg mL^{-1}	0.8 ng mL^{-1}	Vegetables	148
Chrysoidine	MAA	EDMA/ethanol	CL	1.0×10^{-7}–1.0×10^{-5} M	3.2×10^{-8} M	Paper and fabric cloths	149

[a]AM, acrylamide; 2,4-D, 2,4-dichlorophenoxyacetic acid; MAA, methacrylic acid; 4-VPy, 4-vinylpyridine; EDMA, ethylene glycol dimethacrylate; TRIM, trimethylolpropane trimethacrylate; DVB, *p*-divinylbenzene; MBAA, *N*,*N*′-methylene diacrylamide; DMAEMA, dimethylaminoethyl methacrylate; APBA, 3-aminobenzeneboronic acid; APTES, 3-aminopropyl triethoxysilane; TEOS, tetraethoxysilane.

uric acid, urea, dopamine, glutamic acid, glucose, citric acid, or ascorbic acid, which allowed the direct determination of isoniazid in urine samples.

Li *et al.*[130] constructed ECL MIP membranes for the determination of doxycycline. The polymer was prepared by electropolymerization of Alizarin Red-doped pyrogallol in the presence of the doxycycline template. Poly-pyrogallol MIP was auto-oxidized generating an excited intermediate (PPG*) that reacted with the oxidized derivatives of Alizarin Red produced at the electrode. Binding of doxycycline to selective cavities increased the ECL signal intensity through energy transfer from excited Alizarin Red (AR*) to doxycycline with formation of excited doxycycline. The chemosensor was used to determine doxycycline residues with a low LOD (5.17×10^{-11} mol L^{-1}) in fish muscle. The chemosensor signal decreased by ~10.2% after one month of storage in the dark when not in use.

Moreover, ECL has been used to detect metal ions with MIPs. Towards that, Li *et al.*[152] devised a chemosensor for the selective determination of ultra-traces of Be^{2+}. The complex between the metal ion and 4-(2-pyridylazo)-resorcinol (PAR) was used as the polymer template. A decrease in the ECL signal was observed upon immersion of the MIP chemosensor in the PAR complex solution. This signal was correlated with the Be^{2+} concentration in the samples. The dynamic concentration range extended from 7×10^{-11} to 8.0×10^{-9} mol L^{-1} Be^{2+} and the LOD was 2.35×10^{-11} mol L^{-1}. Other metal ions, such as Bi^{3+}, Ba^{2+}, Cr^{3+}, Cd^{2+}, and Cu^{2+}, exhibited no cross-reactivity. The chemosensor proved reusable with high reproducibility (RSD = 3.14%) for at least 5 times. Furthermore, the ECL intensity changed by less than 5% after 15 days.

ECL transduction in combination with MIP recognition has also proved useful for detecting biocompounds. For example, Zhou *et al.*[153] developed a sandwich ECL immunosensor to detect human immunodeficiency virus type1 antibody (anti-HIV-1) by using magnetic MIPs as the capture probes and HIV-1 antigen conjugated with horseradish peroxidase (HRP-HIV-1) as the label. The polymer was deposited on the surface of silicate-coated magnetic iron oxide nanoparticles (m-MIPs) using 3-aminobenzeneboronic acid (APBA) as the functional monomer and the cross-linking agent. Human immunoglobulin G (HIgG) was selected as the template because it falls in the same Fc region as anti-HIV-1 but in a different Fab region. The labelled derivative, HRP-HIV-1, competed with anti-HIV-1 for specific recognizing sites in the m-MIPs, the amount of label binding to the polymer being monitored through the ECL signal in the presence of the luminol-H_2O_2 system. The response was linear at anti-HIV-1 dilution ratios of 1 : 20 000 to 1 : 50 and the LOD of 1 : 60 000 ($S/N = 3$). The immunosensor was successfully used to monitor anti-HIV-1 in human serum samples.

8.4 Absorption-based MIP Chemosensors

Absorbance spectroscopy, commonly referred to as "spectrophotometry", is based on measuring the amount of light absorbed by a sample at a given wavelength. Spectrophotometry, particularly in the visible (vis) and ultraviolet (UV) regions of the electromagnetic spectrum, is now one of the most flexible and widely used techniques for devising chemosensors. The wavelength range generally used for this purpose spans from 190 nm (UV region) to ~1000 nm (near-infrared region). Measurements in the visible region are very easy to make, even with bare eyes, and hence especially useful for *in situ* determinations. This section describes selected examples of MIP chemosensors based on measurements in the visible spectral region.

Incorporating molecularly imprinted materials into photonic crystals has provided a versatile platform for developing label-free colorimetric sensors. Xue *et al.*[154] designed and synthesized a molecularly imprinted polymer colloidal array (MIPCA) combining photonic crystal colorimetric detection with the molecular imprinting technique to detect *p*-nitrophenol (PNP) in surface waters. The chemosensor was constructed by following a self-assembly approach. Thus, monodisperse PNP imprinted colloidal spherical NPs, ~200 ± 5 nm in diameter, were fully dispersed in de-ionized water and placed in a clean Petri dish into which a glass slide was then immersed vertically. Water evaporation allowed MIPCA films to form on both sides of the glass slide that were then transferred to an adhesive tape. In the presence of PNP, the colloidal spherical NPs swelled and caused a red shift in the diffraction wavelength of the crystalline colloidal array – a shift exceeding 50 nm in the presence of 30 nM PNP – by the effect of a change in lattice spacing, hence, in an effective diffractive index. Peng *et al.*[155] followed a similar procedure to develop MIP photonic hydrogels (MIPH) as colorimetric chemosensors for the rapid, label-free determination of vanillin in buffered solutions. Figure 8.5 depicts the MIPH preparation protocol. The Bragg diffraction peak shifted from 451 to 486 nm in response to a vanillin concentration change from 10^{-12} to 10^{-3} M within 60 s. However, the peak oscillated erratically or remained nearly constant with methyl and ethyl vanillin, which was ascribed to selective recognition of the target analyte by the sensing film. Films were regenerated with methanol, and then exhibited high reproducibility (standard deviation within 5%) after five measuring cycles. Moreover, the chemosensor exhibited long-term stability, with little decay in performance after 3 months of storage at 4 °C. However, no application to real samples was reported.

Instead of silica beads, Zhang *et al.*[156] used monodispersed polystyrene colloids to prepare photonic crystal templates in order to prepare MIP films with a periodic inverse opal structure selective to L-pyroglutamic acid (L-Pga). A linear relationship was found between the shift in the diffraction peak and the concentration of L-Pga over the range of 0.01 to 0.20 mM at pH = 4.0, the resulting colour change being easily detected by the naked eye. The chemosensor exhibited high selectivity for L-Pga in the presence of other amino acids in the L- form and only a slight shift was observed with films immersed in buffer solutions of D-Pga. Moreover, recoverability and reproducibility were high with no increase of the response time or appreciable change in the diffraction wavelength after at least four measurement cycles. However, non-specific binding was under optimal.

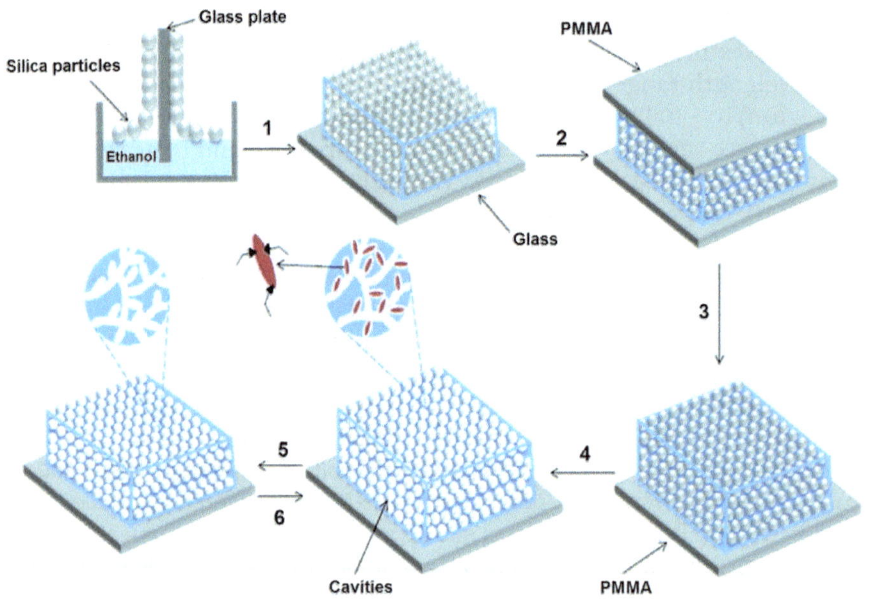

Figure 8.5 Schematic illustration of the procedure used for preparation of molecularly imprinted photonic hydrogels. (1) Preparation of a silica colloid by crystals moulding on a glass substrate, (2) preparation of a "sandwich structure" and infiltration with the precursor solution, (3) photopolymerization, (4) removal of silica particles with hydrofluoric acid, (5) removal of vanillin using methanol and (6) target vanillin binding. Reproduced with permission from F. Xue, Z. Meng, Y. Wang, S. Huang, Q. Wang, W. Lu and M. Xue, *Anal. Methods*, 2014, **6**, 831, Copyright 2012 American Chemical Society.

Behbahani *et al.*[157] constructed a colorimetric chemosensor for the qualitative determination of uranyl ions (UO_2^{2+}) at trace levels. The polymer was synthesized by precipitation polymerization using a mixture of MAA as the functional monomer and 2-(5-bromo-2-pyridylazo)-5-(diethylamino)phenol as the uranyl-complexing ligand. Its selective colour change from yellow to violet in the presence of UO_2^{2+} enabled preconcentration and visual detection of the ion at concentrations down to 0.0005 ppm and spanning a wide linear dynamic concentration range (0.001 to 100 ppm). Although the chemosensor was made reusable by treatment with hydrochloric acid, the UO_2^{2+} recovery decreased after five adsorption–desorption cycles, presumably through leaching of the ligand and a memory effect in the polymer. Other metal ions present in the waste and environmental waters, and potentially interacting with the uranyl-complexing ligand, were investigated as interferences. The chemosensor exhibited tolerance limits above 25 mg L^{-1} in the presence of 0.5 mg L^{-1} UO_2^{2+}, which afforded testing the distilled, tap, and sea water samples from different areas in the Caspian Sea and Persian Gulf.

Sergeyeva *et al.*[158] followed a different approach to devise MIP-based colorimetric sensors. They prepared porous MIP membranes based on semi-interpenetrating polymer networks to detect phenol in natural and waste waters. The target phenol was selectively adsorbed at recognizing sites in the MIP membrane and detected visually upon reacting with 4-aminoantipyrine in an alkaline medium containing potassium hexacyanoferrate to form a pink-coloured product. The LOD was 50 nM and the chemosensor proved highly selective towards phenol relative to its structural analogues. Although the sensing films were not reused, they exhibited high stability for at least 12 months of storage at room temperature. Foguel *et al.*[159] devised evanescent wave-based MIP-based optical fiber chemosensors for the textile dye Basic Red 9 (BR9). An MIP layer was immobilized on a disposable polystyrene optical waveguide by dip-coating the fiber in the polymer suspension. BR9, bound to the MIP coating, strongly absorbed at 545 nm and unabsorbed light was recorded through the same optical fiber probe. Colour intensity was correlated with the BR9 concentration in the low micromolar range. Binding was about 4 to 13 times stronger for BR9 than that for structurally related and unrelated dyes.

Imprinted materials prepared using a sol–gel process instead of organic polymers have also been used to devise colorimetric chemosensors. For that, Li *et al.*[160] fabricated MIP chemosensors based on TiO_2 sol–gel layers prepared by liquid-phase deposition in the presence of a template protein (lysozyme) electrostatically and physically immobilized onto $Cu(OH)_2$ nanoparticles [TiO_2@$Cu(OH)_2$]. The polymers were attached to the bottom of 96-well high binding ELISA plates, and then proteins were recognized based on the oxidative ability of $Cu(OH)_2$ towards *o*-phenylenediamine (OPDA) causing a change in both colour and fluorescence. The colorimetric chemosensor featured the LOD of 0.04 mg mL^{-1} for lysozyme and the MIPs reached adsorption saturation at 0.2 mg mL^{-1}. Slight cross-reactivity was observed in other proteins, such as cyt *c* or bovine haemoglobin (BHb), but not in tryptophan or bovine serum albumin (BSA).

8.5 Infrared and Surface-enhanced Raman Scattering (SERS)-based MIP Chemosensors

Infrared spectroscopy (IR) and, particularly, Fourier transform infrared (FTIR) spectroscopy, is a powerful technique for chemical elucidation of imprinted polymers. Although not extensively, this technique has also been used with MIP-based chemosensors. This combination has the potential for solving problems derived from cross-reactivity or non-specific interactions by facilitating discrimination between analytes. For instance, Mizutani *et al.*[161] fabricated nanometric TiO_2 imprinted sol–gel films for the selective determination of the chiral (*R*)- and (*S*)-propranolol analogues. Imprinted films were deposited onto the surface of quartz plates by using the spin-coating technique, and then template binding was assessed by UV–vis and FTIR spectroscopy techniques. Enantioselectivity was improved by tuning film thickness and by thermally annealing the template–TiO_2 films, which afforded 100% (specific) selectivity for (*R*)-propranolol and 95% for (*S*)-propranolol.

Raman scattering spectroscopy, which is based on the inelastic scattering of photons, is a fast non-destructive optical technique for quantifying and identifying chemicals. Currently, its greatest challenge is the selective determination of target compounds at low concentrations in very small sample volumes. However, the optical signal can substantially be amplified by placing the molecules in the close vicinity of, or physically or chemically attached to, nanostructures of a noble metal, such as gold or silver.[162] The extremely high electromagnetic fields generated on the surface of metal films or nanoparticles in surface-enhanced Raman scattering (SERS) considerably improve the typically faint Raman scattering signals, thereby affording the LOD even at the single molecule level.[163] Experimentally, a synergistic effect in addition to signal enhancement (*viz.*, plasmonic coupling) has been observed depending on the nature, size, and shape of the metal, the type and strength of the bond between the analyte and metal surface, atomic roughness and/or the excitation source wavelength. Amplification mechanisms are described in detail elsewhere.[164]

The use of SERS in combination with MIPs as selective recognition elements for chemosensors has expanded considerably in recent years.[165–171] For instance, Ye *et al.*[172] prepared MIP nanoparticles selective to (*S*)-propranolol that were attached by photo-conjugation to a SERS-active Klarite® substrate. The chemosensor was successfully used to quantify (*S*)-propranolol at concentrations down to 7.7×10^{-4} M in urine samples. In an alternative approach, molecularly imprinted core–shell MIP nanoparticles were covalently linked onto gold substrates through a carbodiimide coupling reaction.[173] Then, surface-exposed amino groups on the MIP nanoparticles were functionalized with tannic acid/citrate-protected gold NPs to obtain suitable substrates for SERS sensing of propranolol. Plasmonic coupling between surface-bonded Au colloids, in combination with selective binding of propranolol by MIP core–shell particles, allowed more effective enhancement of the Raman signal in comparison to a SERS commercial substrate. Clearly visible Raman bands were observed at 1385 cm^{-1} related to the naphthalene moiety in propranolol upon exposure to propranolol at the concentration of 3.4×10^{-4} M. Ye *et al.*[174] also reported a novel synthetic strategy using spherical MIP nanoparticles that were made selective to nicotine by RAFT polymerization. After synthesis, dithioester moieties of the RAFT agent remaining on the surface of MIP beads were reduced to thiol groups to increase their affinity for gold-coated substrates without the need for further metal surface functionalization. Three different approaches to producing SERS active recognition sites were investigated, namely: (i) direct sputtering of gold nanoparticles on MIP-SH covered microspheres; (ii) immobilization of gold colloids through thiol groups in the MIP; and (iii) trapping of MIP spherical beads in a patterned SERS substrate. The first method was the most cost-effective but resulted in relatively weak analytical signals; besides, using colloidal gold, although time-consuming, yielded the strongest signals. The simplest, but most expensive approach was that using the commercial Klarite substrate. Although the equilibration time was relatively long (3 h) in all cases, the chemosensors exhibited high selectivity in the presence of other compounds, such as propranolol or rhodamine B.

Kantarovich *et al.*[175] followed an alternative approach to prepare dot-based MIP microarrays for (*S*)-propranolol determination by depositing droplets of the MIP pre-polymerization mixture onto SERS-active patterned grids with the nano-fountain pen printing method. After polymerization, SERS was used to monitor the uptake and release of this β-blocking drug *via* the band at 1385 cm^{-1}. This approach was also used to monitor the distribution of the printed droplets in line and array; however, quantitative detection of propranolol was limited by variability in thickness within and between droplets. Holthoff *et al.*[176] used xerogel precursors containing 3-mercaptopropyltrimethoxysilane (MPTMS) to facilitate adhesion of the polymer to the SERS substrate. For this purpose, an MIP xerogel film was cast over the substrate by spin coating to obtain micro-thick films selective to the TNT explosive. The SERS intensity of the asymmetric stretching band for nitrate at 1352 cm^{-1} was used as the analytical signal. The sol–gel chemosensor yielded LOD of 3×10^{-6} M for TNT in solution after 24 h exposure. Unfortunately, the sensitivity was inadequate for detecting the explosive at sub-micromolar concentrations. The chemosensor exhibited high selectivity even in the presence of other molecules structurally similar to the target compounds thanks to both the selective recognition of TNT by the MIP and the characteristic molecular "fingerprinting" afforded by SERS measurements. The sensing MIP films were stable for at least 6 months and no statistical differences in Raman intensity were observed after ten sorption–desorption cycles, which testifies to the repeatability of the chemosensor.

The number of molecules near the SERS substrate and/or the distance-dependence of the SERS effect can limit the sensitivity of SERS-based sensors.[163,177] The limitations, however, can be overcome by using metal nanoparticles coated with MIP films to increase the number of analyte recognizing sites near the metal surface.[178] As can be seen from Table 8.3, silver and gold nanoparticles are the most typical cores used as scaffolds for MIP shells by virtue of their having localized surface plasmon resonances (LSPRs) in the visible and near-infrared regions, which make them suitable transducers for sensor fabrication.[179]

In general, plasmonic coupling between silver nanoparticles and MIP-bound target compounds results in LODs at least three orders of magnitude lower than those of gold nanoparticles, consistent with the increased plasmonic intensity of Ag NPs. However, gold colloids are still the most widely used nanostructures for sensing purposes as their stability in buffered and saline media exceeds that of Ag NPs[188] – making them the most suitable candidates for analyses of real samples.

Ahmad *et al.*[186] used Au nanorods instead of isotropic spherical nanoparticles to prepare individually dispersed core–shell hybrid SERS sensors.

Table 8.3 Analytical characteristics of surface-enhanced Raman scattering (SERS) nanosensors based on (Ag or Au core)-MIP shell nanocomposites.

Nanocomposite	Polymerization[a]	Analyte	Analytical characteristics	Ref.
Ag@MIP	FRP-P	Rhodamine B	LAC: 10^{-10} M; EF: 1.6×10^5	180
			LOD: 10^{-12} M; EF: 1.01×10^7; ET: 2 h; Reusability: 3 cycles	181
		4-MBA	LAC: 10^{-15} M; DR: 10^{-15}–10^{-3} M; ET: 24 h	182
		Crystal violet	LAC: 10^{-9} M; DR: 10^{-9}–10^{-4} M; EF: 6.55×10^6	183
		Rhodamine 6G	LOD: 10^{-14} M; DR: 10^{-12}–10^{-6} M; ET: 15 min	184
	Sol–gel	Sialic acid	LAC: 1.25×10^{-5} g mL^{-1}; DR: 1.25–30 $\times 10^{-5}$ g mL^{-1}; ET: 20 min; R: 5 cycles; S: 5 days	185
Au@MIP	RAFT	Folic acid	LAC: 10^{-7} M; DR: 10^{-7}–10^{-4} M	186
	Sol–gel	Bisphenol A	DR: 0.5–22.8 $\times 10^{-6}$ g mL^{-1}; LOD: 1.2×10^{-7} g mL^{-1}; ET: 10 min; R > 60.1%	187

[a]Polymerization techniques: FRP-P, free radical polymerization with polymerizable groups attached to the gold surface; CRP, controlled radical polymerizations, including reversible addition fragmentation chain transfer (RAFT) reactions; 4-MBA, 4-mercaptobenzoic acid; LOD, limit of detection; LAC, lowest assayed or detected concentration; DR, linear dynamic concentration range; ET, exposure time; R, recovery; EF, enhancement factor.

The nanoparticles exhibited strong anisotropic optical properties and intense optical absorption peaks by the effect of the LSPRs along their long and short axes. The magnitude of the local electromagnetic field in the vicinity of gold nanorods increased dramatically upon excitation of the LSPR by an external electromagnetic field – a property that is widely used in SERS. Chemosensors were prepared by coating gold nanorods with MIP shells, thereby combining aryl diazonium salt chemistry with the iniferter method (Figure 8.6). The intensity of the characteristic SERS bands of folic acid (1595 and 1365 cm^{-1}) was used as analytical signals; both linearly increased with the increase of concentration of the target compound over the range of 10^{-8} to 10^{-4} M in water. The ensuing imprinted nanohybrids proved more selective than non-imprinted materials.

In a recent work, Carrasco *et al.*[189] prepared multi-branched gold–silica MIP (bAu@mSiO_2@MIP) core–shell nanoparticles for the selective label-free determination of the antimicrobial ENRO. Nanoparticle branching was done in several steps prior to polymerization in order to control the process and avoid wreckage of recognizing sites. The imprinted recognizing sites had to be close to hot spots generated by the gold branches in order to increase the spectroscopic response of the nanosensors. The chemosensors exhibited high selectivity for ENRO in the presence of potential interfering species; this

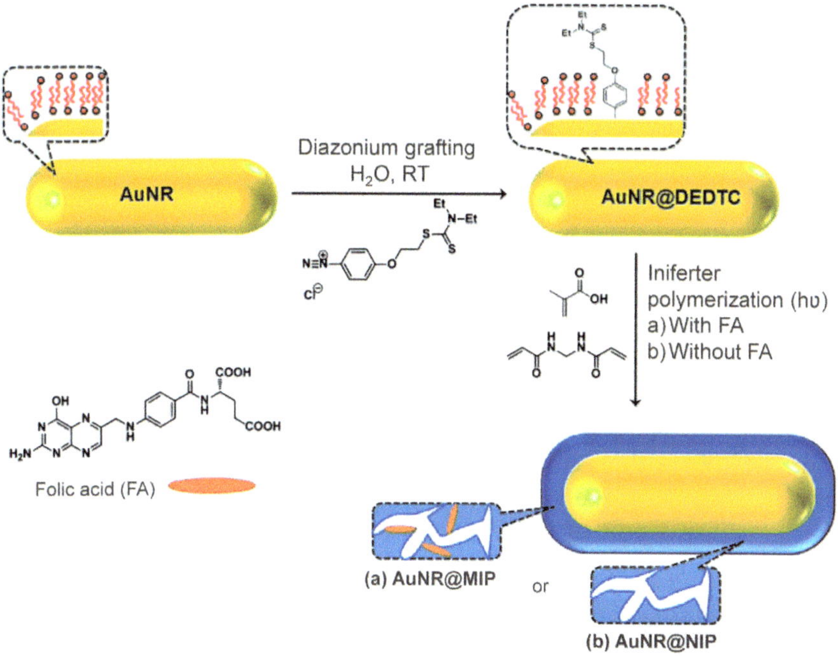

Figure 8.6 Schematic illustration of the surface modification strategy for coating AuNRs with an MIP layer by combining the diazonium salt chemistry and the iniferter photopolymerization method. Reproduced from ref. 186 with permission from The Royal Society of Chemistry.

was a result of the combination of the imprinting process and the SERS technique, as SERS spectra differ among targets. The LOD was very low (1.5 nM) and the response time was 10 min.

Guo *et al.*[190] combined the separation properties of magnetic nanoparticles with the effect of Au nanoparticles on SERS signals. Magnetic Fe_3O_4 NPs were synthesized and coated with an MIP shell selective to ciprofloxacin in order to selectively remove this antibiotic from foetal bovine serum samples. After preconcentration, the composites were dispersed in a silver sol and metal nanoparticles absorbed around the magnetic MIP through coordinate interaction between the nitrogen atom in the antimicrobial and silver atoms. Monitoring the characteristic SERS peak at 1383 cm^{-1} allowed ciprofloxacin detecting at concentrations down to 10^{-9} M in water and 10^{-7} M in serum.

New metal–MIP nanohybrids with improved detection limits and selectivity will likely lead to an increasing use of MIP SERS for chemical and biochemical analysis. Moreover, SERS chemosensors have the potential for multiplex determination as a result of their combining the selective recognition properties of the bio(mimetic)receptor element, with narrow unique Raman vibrational bands, and the large signal enhancements from noble metal nanomaterials.[177]

8.6 Surface Plasmon Resonance (SPR)-based MIP Chemosensors

SPR spectroscopy is a label-free technique of great potential for real-time and cost-effective detection of analyte binding to a receptor element immobilized on a sensing surface.[191] Among other advantages, it allows for determination of the affinity and rate constants, and the quantization of complex formation. Moreover, SPR spectroscopy is one of the main optical techniques for developing affinity biosensors and characterizing thin films. So far, it has been used in combination with MIPs to develop a number of label-free biomimetic sensors for detecting chemical and biochemical targets.

SPR is a charge-density oscillation that may exist at the interface of two media with dielectric constants of opposite signs (*e.g.*, a metal film of Ag or Au and a dielectric medium, such as air).[191,192] The electromagnetic wave associated to the charge density wave, known as the "surface plasma wave" (SPW), is a transverse magnetic (TM) mode; therefore, its magnetic vector lies in the plane of the metal–dielectric interface and normal to the direction of propagation of the wave. The intensity of the magnetic field peaks at the metal–dielectric boundary decays evanescently into both media – most of the field concentrates in the dielectric, however. The penetration depth depends on the wavelength and on the permittivity of the materials. Any change in the refractive index of the dielectric medium will affect the propagation constant of SPWs.[192] One of the requirements for an SPW to be excited is that the wave vector parallel to the metal surface of the incident light should match that of the SPW. Light and the SPW can be coupled through a prism, a waveguide or a grating mainly.[192,193] The prism coupling technique can be used in two different ways, namely: the Kretschmann and Otto geometries, the latter being the most widely used in SPR sensors. In the Kretschmann configuration (Figure 8.7), a light wave is completely reflected at the interface between a prism coupler and a thin metal layer (~50 nm thick) to excite an SPW at the outer boundary of the metal by evanescently tunnelling through the metal

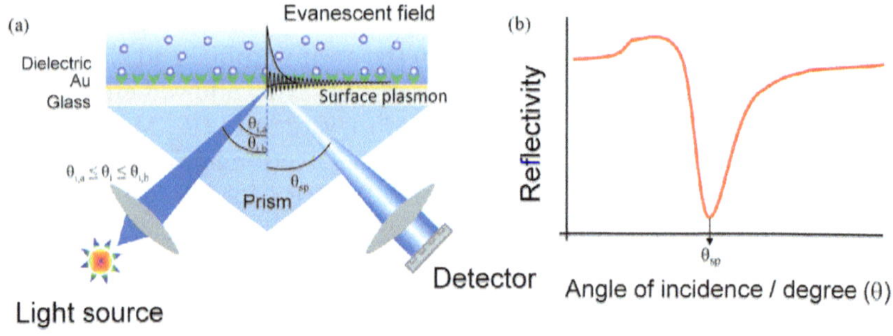

Figure 8.7 (a) Configuration of a prism coupler-based SPR system. (b) A typical SPR reflectivity curve for an Au film.

layer.[191,192] Based on the characteristics of the light wave interacting with the surface plasmon, SPR sensors can be of the angular, wavelength, intensity, phase or polarization modulation type.[191,192,194,195]

Some authors have developed MIP-based SPR chemosensors by attaching MIP nanoparticles onto the surface of gold chips[196,197] and used them for purposes such as the determination of lysozyme in chicken egg white (LOD, 84 pM; response time, 45 min)[196] or chloramphenicol in honey samples (LOD, 40 ng kg^{-1}; response time, 10 min).[197] Moreover, Abdin *et al.*[198] prepared endotoxin films imprinted onto glass beads by using the solid-phase approach and reversible addition fragmentation chain transfer (RAFT)-controlled radical polymerization. The polymer composition was optimized by molecular modelling. Imprinted beads were polymerized with *N*-(3-aminopropyl)-methacrylamide in order to add amine groups to the surface for further attachment to gold chips functionalized with carboxyl groups. The chemosensor allowed determining endotoxins from *Escherichia coli* 0111:B4 at concentrations of 15.6 to 500 ng mL^{-1} with the LOD of 1.56×10^{-8} g mL^{-1}. Only 4 min of analyte exposing and 1 min for chip regeneration was required. No changes in chemosensor performance were observed after 28 binding-regeneration cycles. The same procedure was used to determine diclofenac at concentrations of 1.24 to 80 ng mL^{-1} in aqueous buffer with the LOD of 1.2×10^{-9} g mL^{-1} for gold-conjugated diclofenac.[199]

Wang and Wei[200] proposed a versatile strategy for constructing SPR chemosensors for femtomolar detection of testosterone in aqueous media using a water compatible macroporous MIP film. This film was prepared by co-polymerization of the functional monomers (methacrylic acid and 2-hydroxyethyl methacrylate) with the cross-linking monomer, ethylene glycol dimethacrylate (EDMA), polystyrene nanoparticles (PS), and testosterone as the template, onto gold chips. The macroporous MIP film obtained after removing the PS nanoparticles exhibited high sensitivity and accessibility for binding of the hormone relative to conventional MIP films. Moreover, it proved more selective for the target analyte than for its analogues, such as oestradiol or progesterone. The polymer films showed undetectable non-specific absorption in artificial or human urine, which enabled the determination of testosterone in phosphate buffer saline (PBS) and artificial urine at concentrations down to 10 to 15 g mL^{-1}. The SPR chemosensor exhibited high stability and reproducibility over 8 months of storage at room temperature with no special protection.

Despite the high sensitivity of the previous testosterone sensor, one of the main problems of SPR is its low sensitivity to changes in refractive index caused by compounds of low molecular weight (~1 kDa and smaller), which hinders detection of small organic molecules at very low concentrations. One way to overcome this drawback is by preparing AuNP/MIP composite layers (AuNPs@MIPs) on SPR chips.[74] The SPR spectrum is markedly affected by the coupling of localized plasmons in AuNPs and the surface plasmon wave associated to thin gold films; thus, a small change in dielectric properties by effect of binding of the analyte enhances SPR shifts and facilitates detection of low-molecular-weight targets. Furthermore, the specific surface area of the nanocomposite can be increased by including AuNPs in the MIP matrix to improve accessibility to recognition sites and expedite binding.[74]

Alternatively to AuNPs, some authors have used nanoparticles of graphene, carbon, or titanium oxide to further enhance the physical properties of AuNPs@MIPs. To this end, Yao *et al.*[201] prepared MIP-coated SPR chips with enhanced sensitivity from nano-hybrid films based on MIPs coated with gold nanoparticles and reduced graphene oxide (RGO) for the determination of ractopamine. The films were prepared by using the "grafting-to" spin-coating method to control thickness and decrease brittleness. The 1-butylpyridinium hexafluorophosphate ionic liquid was used as the binder to ensure adequate dispersion between the AuNPs and RGO. The chemosensor allowed the selective determination of ractopamine concentrations down to 5×10^{-9} g mL^{-1} after 5 min of exposure.

As an alternative to prism-based configurations, Taguchi *et al.*[202] fabricated an optical waveguide surface plasmon resonance (OWG-SPR) chemosensor integrated in a microfluidic system to determine BPA. They used a slab-optical fiber coated by successive deposition of parallel gold and silver bands in the line of plasmon flow to simultaneously measure two independent SPR signals due to differences in the resonant reflection spectra for the metals. Core-shell nanoparticles were prepared by using poly(styrene-*co*-divinylbenzene) seeds coated with an MIP layer selective to BPA. The surface of the gold and silver bands in the OWG was functionalized with a BPA thiol derivative to obtain self-assembled monolayers. Moreover, the MIP NPs were further bound to the dense BPA surface. Then, BPA grafted Au nanoparticles (BPA-AuNPs) were immobilized onto the MIP NPs and a red shift (0.7 nm) was observed in the Ag-based SPR spectra as free BPA (0.1 mM) was flowed over the functionalized OWGs. This change reflected an enhancement of the binding signal in the presence of the AuNP-BPA suggesting that their localized surface plasmon resonance affected that on the Ag band. Table 8.4 summarizes the analytical features of selected SPR-based chemosensors using thin MIP films attached to optical fibers.

Furthermore, MIP-capped Fe_3O_4 nanoparticles have been used as recognition elements and signal amplifiers in SPR chemosensors.[209] Magnetic MIPs were synthesized by self-polymerization of dopamine on the surface of Fe_3O_4 nanoparticles in the presence of the chlorpyrifos pesticide as the template. After selective preconcentration of the pesticide and separation of Fe_3O_4@polydopamine nanoparticles (Fe_3O_4@PDA NPs) by using an external magnetic field, the resulting hybrid nanoparticles were integrated into an SPR through interaction between the chlorpyrifos molecules bound to the MIP recognition sites and acetylcholinesterase immobilized on the gold chip. The signal was considerably amplified by the effect of the high molecular weight of the Fe_3O_4@PDA NPs. The SPR chemosensor exhibited the LOD of 7.6×10^{-10} M and a linear dynamic concentration range of 0.001 to 10 μM.

Table 8.4 Surface plasmon resonance (SPR)-based chemosensors using thin MIP films attached to optical fibers.[a]

Analyte	Sample	Analytical characteristics	Ref.
Furfural	Transformer oil	LOD: 10^{-8} g mL^{-1}; DR: up to 10^{-7} g mL^{-1}	225
Furfural		LOD: 9×10^{-9} g mL^{-1}; DR: up to 3×10^{-8} g mL^{-1}	203
TNT		LOD: 5.1×10^{-5} M; ET: 5 min	204
Tetracycline		LOD: 2.2×10^{-9} M; DR: 10^{-8}–10^{-5} M	205
Vitamin B3		LOD: 5×10^{-4} g mL^{-1}; DR: 0.5–10 × 10^{-3} g mL^{-1}	206
L-Nicotine		LOD: 1.86×10^{-4} M; DR: up to 10^{-3} M; ET: 10 min	207
Melamine		LAC: 10^{-7} M; DR: 10^{-7}–10^{-1} M	208

[a]TNT, 2,4,6-trinitrotoluene; LOD, limit of detection; LAC, lowest assayed or detected concentration; DR, linear dynamic concentration range; ET, exposure time.

It was successfully used to analyse apple samples for chlorpyrifos. Binding of chlorpyrifos to the Fe_3O_4@PDA NPs took 12 h. After magnetic separation, the nanoparticles were kept in contact with the enzyme-functionalized gold chip for 30 min.

As can be seen in Table 8.4, most MIP-based SPR chemosensors reported in the last few years were based upon thin 2-D MIP films deposited onto gold chips. The recognition film is prepared by deposition of the pre-polymerization mixture onto the chip surface, which can previously be functionalized with the initiator or a radical controller.[210–213] Film thickness can be improved by casting the pre-polymerization solution between the (functionalized) SPR gold chip and an inert support, followed by application of an external pressure during polymerization.[41,214–221] One advantage of this approach is that the support can be pre-functionalized with the template so that, upon polymerization, recognition sites will be generated on the outer surface of the MIP layer immobilized onto the gold chip in a process known as "micro-contact imprinting".[217–219] This procedure has been used to determine high-molecular-weight targets, such as proteins.[7]

Sharma *et al.*[222] adopted a different approach to construct 3-D composite film architectures capable of accommodating increased amounts of MIP nanoparticles relative to a 2-D monolayer. For this purpose, gold chips were spin-coated with a thin layer (~10 nm) of the SU-8 polymer that was further functionalized with a hydrogel polymeric network. To this end, nanoparticles imprinted with L-Boc-phenylalanine-anilide were embedded in a photo-cross-linkable poly(*N*-isopropylacrylamide)-based polymer with an open structure through which the target molecule had ready access to the imprinted moieties. Moreover, the composite films acted as optical waveguides affording accurate readings of refractive index changes by effect of the target analyte interacting with the MIPs. Measurements were made by using a conventional SPR optical set-up that allowed label-free determination of L-Boc-phenylalanine-anilide with a LOD of 2×10^{-6} M.

Surface electropolymerization is another flexible approach to grafting MIP films of controlled thickness onto Au-coated SPR chips.[223] The resulting layers usually exhibit uniform surface coverage and strong adhesion to the metal surface. However, the number of commercially available conducting polymers and monomer precursors is relatively small, which restricts the scope of this approach relative to others.[224,225] By way of example, Willner *et al.*[226] (Figure 8.8) electropolymerized thioaniline-modified AuNPs capped with phenylboronic acid ligand onto a thioaniline-functionalized Au-coated glass surface to develop sensitive, selective matrices for detecting a series of antibiotics containing vicinal hydroxyl groups.

The resulting bisaniline-cross-linked AuNP composites allowed the determination of neomycin, kanamycin, and streptomycin with very low LODs (2.00 ± 0.21 pM, 1.00 ± 0.10 pM, and 200 ± 30 fM, respectively). Moreover, these authors[227] reported on the stereoselective and chiroselective analysis of amino acids (particularly glutamic acid) by co-functionalizing thioaniline-modified AuNPs with cysteine units to prepare ligands for recognition of the target molecules through zwitterionic and hydrogen-bonding intermolecular interactions.

Irrespective of the particular technique used to prepare an MIP film, the ultimate goal is, as noted above, to ensure that the recognition layer will fit

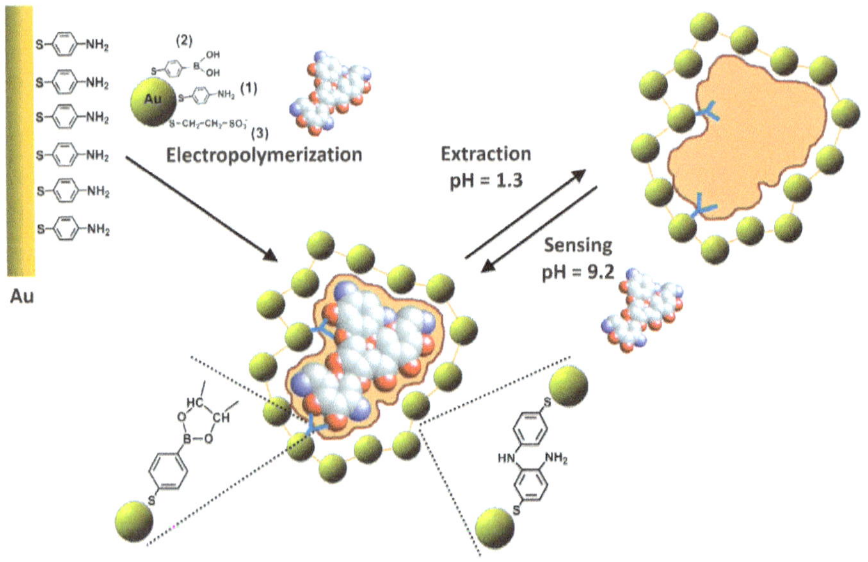

Figure 8.8 Imprinting molecular recognition sites for antibiotic substrates (*e.g.* neomycin) through the electropolymerization of a bisaniline cross-linked AuNPs composite on an Au surface. Reproduced with permission from G. Gupta, A. S. B. Bhaskar, B. K. Tripathi, P. Pandey, M. Boopathi, P. V. L. Rao, B. Singh and R. Vijayaraghavan, *Biosens. Bioelectron.*, 2011, **26**, 2534, Copyright 2010 American Chemical Society.

inside the interaction area of the evanescent wave generated on the surface of the gold chip (~200 nm).[228–233] Table 8.5 shows the analytical characteristics of relevant MIP-based SPR chemosensors published in years 2011 to 2016.

8.7 MIP Chemosensors Using Other Optical Transduction Techniques

Other optical transduction techniques, such as interferometry, have also been used in recent years to fabricate MIP-based chemosensors (especially for implementing label-free detection methods).

Interferometry is an optical technique that allows measurement of small changes along the propagation path of an optical beam. These changes may arise from variations in the path length of the propagating light beam, wavelength, the refractive index of the medium the probe light travels through, or a combination of these factors. Changes in the phase (ϕ) of the light will be directly proportional to the path length (L) and refractive index (n), and inversely proportional to the wavelength (λ, eqn (8.1)):[234]

$$\phi = 2\pi Ln/\lambda \tag{8.1}$$

Changes in refractive index resulting from the interaction between the recognition element and the target compound are commonly monitored in sensing applications.[234] In the simplest configuration, the probe light beam propagates along the selective receptor layer and the change in refractive index is measured by comparison with a stable reference wave. The two beams combine to generate an interference pattern of alternating dark and light fringes that provides a direct way to convert phase into detectable intensity.[235,236] Interferometers can be of (a) the wavefront-splitting type, in which distinct optical paths are deflected so they are no longer parallel and hence cross each other or (b) the amplitude-splitting type, where a single optical path splits to take different paths before crossing or combining. Wavefront-splitting configurations involve Young's double-slit interference and diffraction, whereas amplitude-splitting interferometers include Michelson, Mach–Zehnder, and Fabry–Perot configurations, among many others.[237] Interested readers can find more detailed information elsewhere.[234,235,237]

Diltemiz *et al.*[238] devised a reflectometric interferometric (RIf) nanosensor for sarcosine based on MIP spherical NPs prepared by mini-emulsion polymerization. The NPs were immobilized onto the surface of a glass chip and the sensor was successfully used to determine the prostate cancer marker in urine samples with the LOD of 45 nM. The chemosensor required 20 min for equilibration and was responsive to analyte concentrations of 0.25 to 3 mM. A dual-polarisation interferometry (DPI) sensor was devised by Reddy *et al.*[239] for the determination of BHb in plasma. For that, thin MIP films were prepared by applying pressure to hydrogel solutions over DPI chips. MIPs prepared with BSA, a protein of similar molecular weight to that of

Table 8.5 Analytical characteristics of selected surface plasmon resonance (SPR)-based chemosensors using thin MIP films attached to gold chips and different polymerization techniques.

Analyte	Fabrication method	Polymerization[a]	Sample	Analytical characteristics	Ref.
Lysozyme	Deposition of the pre-polymerization mixture onto the SPR chip	RAFT		LAC: 5×10^{-7} M; ET: 120 s	210
Sudan I		FRP-I		LOD: 3×10^{-8} g mL^{-1}; DR: $(5–40) \times 10^{-8}$ g mL^{-1}; ET: 400 s, Reusability: 20 cycles	211
Histamine			Carp fish	LAC: 2.5×10^{-8} g mL^{-1}; DR: $(2.5–100) \times 10^{-8}$ g mL^{-1}; TT: 1 h; R: > 89%	212
D-Glucose		FRP	Urine	LOD: 2×10^{-5} g mL^{-1} (deionized water), 1.2×10^{-4} g mL^{-1} (urine), DR: $(0.5–5) \times 10^{-3}$ g mL^{-1} (spiked sample); ET: 2 min; TT: 30 min	213
Testosterone	Deposition of the pre-polymerization mixture onto the SPR chip and application of an external pressure	FRP		LAC: 10^{-12} M; DR: $10^{-12}–10^{-6}$ M	214
			Sea water	LOD: 10^{-12} M (pure water); DR: $10^{-12}–10^{-8}$ M; ET: 15 min; TT: 40 min; Reusability: 7 cycles; S: 90 days; R: > 98.5%	215
Methyl parathion				LAC: 10^{-13} M; DR: $10^{-13}–10^{-10}$ M; ET: 20 min; TT: 30 min	216
HAS		FRP-I		LAC: 2.5×10^{-7} M; DR: $2.5–40 \times 10^{-7}$ M; TT: 150 s	41
Myoglobin		FRP-P	Serum	LOD: 8.76×10^{-8} g mL^{-1}; DR: $0.1–10 \times 10^{-6}$ g mL^{-1} (pure water), $(0.3–1) \times 10^{-6}$ g mL^{-1} (serum); ET: 25 min; TT: 55 min	217
PSA			Serum	LOD: 1.8×10^{-13} M (pure water); DR: $0.1–50 \times 10^{-9}$ g mL^{-1} (pure water), $0.1–10 \times 10^{-9}$ g mL^{-1} (serum); ET: 50 min; TT: 60 min; Reusability: 50 cycles; R: > 87%	218
Procalcitonin			Serum	LOD: 9.9×10^{-9} g mL^{-1}; DR: $(2–400) \times 10^{-8}$ g mL^{-1} (pure water), $(2-100) \times 10^{-8}$ g mL^{-1} (serum); TT: 80 min	219
Bisphenol A			Bottled water, Polycarbonate	LOD: 6.15×10^{-9} M (bottled water), 6.84×10^{-9} M (polycarbonate); DR: $1 \times 10^{-8}–1 \times 10^{-6}$ M; S: 3 months	220
			Pure water, Tap water, Waste water	LOD and DR: 2×10^{-11} g mL^{-1}, $(2–8) \times 10^{-8}$ g mL^{-1} (pure water), 6×10^{-11} g mL^{-1}, $(6–20) \times 10^{-8}$ g mL^{-1} (tap water), 8×10^{-11} g mL^{-1}, $(8–30) \times 10^{-8}$ g mL^{-1} (waste water); TT: 75 min; Reusability: 6 cycles; R: > 98.7%	221

Haemoglobin	Electropolymerization	FRP		LOD: 4.35×10^{-7} g mL^{-1}; DR: $(5–500) \times 10^{-5}$ g mL^{-1}; ET: 20 min; TT: 25 min; S: 14 days	224
T-2 toxin				LOD: 10^{-16} M; DR: $2.1–33.6 \times 10^{-15}$ M	225
Citrinin	Spin coating	FRP-P	Red yeast rice	LOD: 1.7×10^{-12} g mL^{-1}; DR: $(5–1000) \times 10^{-12}$ g mL^{-1}; R: > 97.96%	228
Amoxicillin			Egg	LOD: 2.2×10^{-11} g mL^{-1}; DR: $(0.1–2) \times 10^{-9}$ g mL^{-1}; ET: 50 min; Reusability: 5 cycles; R: > 96%	229
Amikacin			Serum	LOD: 4.3×10^{-9} M; DR: $(1–15) \times 10^{-8}$ g mL^{-1}; ET: 50 min; Reusability: 6 cycles; S: 60 days; R: > 97.69%	230
Triclosan			Waste water	LOD: 1.7×10^{-11} g mL^{-1}; DR: $(5–100) \times 10^{-11}$ g mL^{-1}; ET: 50 min; Reusability: 5 cycles; R: > 98%	231
Oxytocin			Milk	LOD: 3×10^{-12} g mL^{-1}; DR: $(1–100) \times 10^{-11}$ g mL^{-1}; ET: 50 min; Reusability: 5 cycles; RC: > 98.08%	232
TNT		Sol–gel		LOD: 2.6×10^{-10} g mL^{-1}; ET: 2 h	233

[a]Polymerization techniques: FRP, free radical polymerization with (-P) polymerizable groups or initiator (-I) attached to the gold surface; CRP, controlled radical polymerizations, including reversible addition fragmentation chain transfer (RAFT) reactions; HSA, human serum albumin; PSA, prostate specific antigen; TNT, 2,4,6-trinitrotoluene; LOD, limit of detection; LAC, lowest assayed or detected concentration; DR, linear dynamic concentration range; ET, exposure time; TT, total analysis time; R, recovery; S, stability without affinity losses.

haemoglobin, were used as control polymers. Upon injection over the chips, BHb bound selectively and strongly to them, thereby resulting in a sustained increase in film thickness and mass; however, binding to MIP-BSA chips was only transient and reversible. The chemosensor exhibited a linear response up to 2×10^{-4} g mL^{-1} haemoglobin with a LOD of 2×10^{-6} g mL^{-1}. Moreover, the BSA-MIPs retained their selectivity after exposure to real biological samples only if used in 1:100 diluted serum.

Diffraction-based sensors are one class of interferometric sensors. As stated above, diffraction arises in a wavefront-splitting interference configuration in which one propagating set of wavefronts interacts with the sensing film whereas another does not. This causes a change in their mutual diffraction pattern that can be detected and correlated with the amount of analyte present in the recognition layer.[237] As with biosensors, various types of diffraction-based MIP chemosensors can be designed including polymer gratings on planar substrates, MIPs immobilized on micro-structured surfaces, cantilevers, and MIP-bead based chemosensors capturing beads in different spatial patterns diffracting a probe beam.[237]

Barrios *et al.*[240] reported the first MIP-based 2-D diffraction grating (DG) for fabrication of label-free optical chemosensors selective towards the antimicrobial enrofloxacin. They prepared micro-patterned polymeric gratings by micro-transfer moulding based on SiO_2/Si moulds. The films were removed from the mould after polymerization and placed in a home-made cell so that the incident light would be normal to, and pass through, the films. The first-order diffraction efficiency was used as the chemosensor response because it was greater than those of the higher-order diffraction beams. Assayed analyte concentrations ranged from 30 to 60 μM and the LOD of 18 μM was achieved. In a subsequent work, Barrios *et al.*[241] reported the characterization of an antibiotic-selective MIP as the photonic material for integrated optics. They prepared MIP and NIP DGs (Figure 8.9) on borosilicate microscope slides by using SU-8/Si moulds and microtransfer moulding. Figure 8.9 shows the far-field transmission diffraction pattern obtained with the 2-D MIP DG. The diffracted beams were much weaker than the undiffracted beam, thereby confirming "thin grating". Molecular recognition increased the grating optical thickness, which was ascribed to a change in the polymer refractive index, an increase in grating thickness or both. The main limitation of these 2-D MIP DGs were the long equilibration times needed with the instrumental set-up used for measurements (up to 30 h).

Haupt and co-workers[242,243] have reported an interesting work on the preparation of micro- and nano-patterned MIP structures on various types of substrates. For example, they were the first to prepare holographic diffraction gratings directly by structuring the polymer matrix during the synthetic process for the label-free determination of testosterone.[244] The diffraction gratings were prepared by interference photolithography using two coherent laser beams that were combined to create a stable pattern of light intensity through interference. The 3-D interfering pattern was transferred to the polymer precursor mixture, confined between two glass slides. They used the

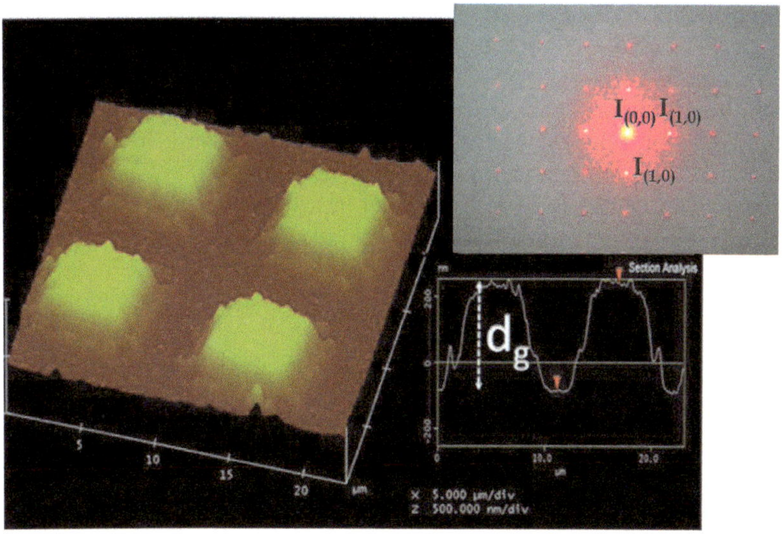

Figure 8.9 The AFM image and cross-section scan of a representative portion of an MIP 2-D grating. Inset: far-field diffraction pattern of the MIP (2-D-DG). Reprinted from *Biosensors and Bioelectronics*, **26**(5), C. A. Barrios, C. Zhenhe, F. Navarro-Villoslada, D. López-Romero and M. C. Moreno-Bondi, Molecularly imprinted polymer diffraction gratis as label-free optical bio(mimetic)sensor, 2801–2804,[239] Copyright (2011) with permission from Elsevier.

organic dye *bis*(cyclopentadienyl)titanium dichloride, which absorbs light in the green region, as the photoinitiator. The diffraction efficiency, measured from the intensities of transmitted and first-order diffracted beams, was related to the concentration of testosterone over the range of 1 to 100 µM.

Ahmadi *et al.*[245] devised a nanosensor for the determination of mefenamic acid based on elastic resonance light scattering (RLS). Light scattered by magnetite nanoparticles coated with MIP shells was monitored with a luminescence spectrometer and the magnitude of the RLS intensity measured by synchronous scanning at $\lambda_{ex} = \lambda_{em}$ ($\Delta\lambda = 0$ nm). The efficiency of scattered light was strongly affected by various factors including pH, agglomeration, and nanoparticle concentration. The response was linear over the concentration range of 100 to 2000 ng L^{-1} and the LOD was 50 ng L^{-1}. Avoiding agglomeration required sonication for 20 min before a stable signal was obtained.

An original optical probe was developed by Lépinay *et al.*[246] for detecting low-molecular-weight compounds, such as maltol (3-hydroxy-2-methyl-4*H*-pyran-4-one), by coupling an MIP and Bragg grating refractometry in an optical fiber. The chemosensor was fabricated by inscribing tilted grating planes in the fiber core and immobilizing the maltol imprinted polymer film by photopolymerization on the fiber cladding surface over the Bragg grating. The presence of the target compound altered the transmission spectrum for the tilted fiber Bragg grating coated with the polymer by effect of the MIP layer capturing molecules. The cladding resonance wavelength (~1538 nm) was especially sensitive to coupling interactions and changes in the transmission signal. Those were monitored by tracking the intensity of the transmission loss peak and the shifting of its wavelength away from its original position. A linear relationship was found up to 60 nM and the LOD of 8.1 nM was reached. The chemosensor was capable of detecting maltol concentrations as low as 10 nM in food jelly preparations. Reusability was limited to 20 detection–regeneration cycles, a restriction that was ascribed to damage and loss of the mirror downstream the fiber or to a decline in recognition capacity of the polymer.

8.8 Conclusions and Outlook

MIP-based optical biomimetic sensors (particularly those based on fluorescence detection) have attracted much attention as promising alternatives to biosensors by virtue of the ease of preparation, low cost, and long-term stability relative to their biological counterparts. The number of applications of MIP-based chemosensors using optical detection has grown steadily over the past decade. Moreover, applications have expanded to the detection of a wide range of analytes including compounds of both small molecules and large biomacromolecules in different fields (*e.g.* diagnostics, nutrition, and environmental monitoring). There have been substantial advances in the search for flexible surface polymerization methods and controlled polymerization techniques to facilitate integration of the MIP recognition element with the transducer which have led to the fabrication of chemosensors with improved sensitivity, selectivity, binding capacity and response to the analyte.

The combination of MIPs with luminescent nanoparticles (DQs, UCNPs) has enabled the preparation of chemosensors with very low detection limits, fast response to the analyte, and the potential for multi-analyte detection. Furthermore, using metal (Au, Ag) NPs in combination with the SPR or SERS spectroscopy has extended the applicability of MIP-based chemosensors to the label free-detection of various analytes.

Despite their obvious potential, MIPs have not yet replaced biological recognition elements for sensor fabrication. In fact, there remain some challenges including controlled synthesis of MIP materials (especially as regards reproducibility), and improving the sensitivity and selectivity of MIP-based devices, which have so far restricted use in ultra-trace analyses of complex matrices in relation to that of well-established biosensors. In fact, improved procedures for preparing MIPs with greater affinity and more homogeneous distribution of recognition sites are required in order to extend their use to areas where low detection limits are typically needed (*e.g.*, clinical diagnostics and biological applications). Although substantial progress has been made in preparing MIPs with high recognition abilities in aqueous samples, applications to the analysis of complex samples are still in its infancy and, therefore, further research is required to prepare commercially available devices.

MIPs can be easily integrated in standard microfabrication processes to obtain micro- and nano-patterned functional films coupled to optical transducers with excellent performance relative to conventionally sized counterparts as regards removal of the template, response times, and ease of combination with other functionalities. In fact, preparing MIPs for multi-functional chemosensors integrating additional functionalities without adversely affecting selective molecular recognition is now one of the priority research areas. The additional functions include the ability to respond to magnetic fields, electric fields, mechanical shear, or temperature and can be highly interesting for theranostic applications, for example. Post-imprinting modification of the MIP surface and nanoparticle conjugation are the preferred synthetic strategies to prepare multi-functional MIPs because only the polymer surface is affected, without compromising the selective recognition cavities inside the MIP material. Other important trends in this field are miniaturization, high-throughput analysis, and multi-sensing. Advanced microfabrication and signal processing techniques, and innovative platforms with multiplexing capabilities, are already being explored. Research in this field will be of paramount interest for improved optical sensing applications in the coming years.

With these considerations in mind, we are confident that MIP-based devices will be used in the not too distant future as sensing tools in many analytical areas including food analysis, environmental monitoring, safety detection and healthcare, and also that the new sensing devices will reach commercial status.

List of Abbreviations

1-NA	1-Naphthylamine
2,4-D	2,4-Dichlorophenoxyacetic acid
4-VPy	4-Vinylpyridine
AA	Acrylic acid
AB	4-Acrylamidophenylaminomethaniminium acetate
AFP	α-Fetoprotein
AM	Acrylamide
APBA	3-Aminobenzeneboronic acid
APTES	3-Aminopropyltriethoxysilane
AR	Alizarin Red
AuNPs	Au nanoparticles
bAu@mSiO_2	Multi-branched gold–silica particles
BHb	Bovine haemoglobin
BODIPY	4,4-Difluoro-4-bora-3a,4a-diaza-*s*-indacene
BPA	Bisphenol A
BR9	Basic Red 9
BSA	Bovine serum albumin
CEA	2-Carboxyethyl acrylate

CL	Chemiluminescence
cyt *c*	Cytochrome *c*
dansyl-L-Phe	Dansyl-L-phenylalanine
DG	Diffraction grating
DMAEMA	Dimethylaminoethyl acrylate
DNT	3,5-Dinitrotoluene
DPI	Dual polarisation interferometry
DUV	Deep UV
DVB	*p*-Divinylbenzene
E	Epinephrine
EbAM	*N*,*N*′-Ethylenebisacrylamide
EBL	Electron beam lithography
ECL	Electrogenerated chemiluminescence
EDMA	Ethylene glycol dimethacrylate
ENRO	Enrofloxacin
FIM	*N*-2-(6-(4-Methyl piperidin-1-yl)1,3-dioxo-1*H*-benzo[de] isoquinolin-2(3*H*)-yl-ethyl)acrylamide
FITC	Fluorescein isothiocyanate
FLIM	Fluorescence lifetime imaging microscopy
FRET	Förster resonance energy transfer
FTIR	Fourier transform infrared (FTIR) spectroscopy
GlcA	Glucuronic acid
Gold nanoparticle	AuNP
Gra	Graphene
HEMA	2-Hydroxyethyl methacrylate
HIgG	Human immunoglobulin G
HPLC	High-performance liquid chromatography
HRP	Horseradish peroxidase
IR	Infrared spectroscopy
ITC	Isothermal titration calorimetry
KH570	γ-Methacryloxypropyl trimethoxy silane
LOD	Limit of detection
L-Pga	L-Pyroglutamic acid
LSPR	Localized surface plasmon resonance
MAA	Methacrylic acid
MBA	*N*,*N*′-Methylene bisacrylamide
MBAA	*N*,*N*′-Methylene diacrylamide
MIP	Molecularly imprinted polymer
MIPCA	Molecularly imprinted polymer colloidal array
MIPH	Molecularly imprinted polymer photonic hydrogels
MISPE	Molecularly imprinted solid-phase extraction
MMA	Methyl methacrylate
m-MIP	Magnetic molecularly imprinted polymer
MPTMS	3-Mercaptopropyl trimethoxysilane
MPTS	3-Mercaptopropyl triethoxysilane
NANA	*N*-Acetylneuraminic acid
NBD	Nitrobenzoxadiazole

NE	Norepinephrine
NIPAN	*N*-Isopropylacrylamide
NIR	Near infrared
NP	Nanoparticle
OPDA	*o*-Phenylenediamine
OWG	Optical waveguide
PA	Propargyl acrylamide
PAR	4-(2-Pyridylazo)-resorcinol
PBS	Phosphate buffer saline
PDA	Polydopamine
P(MAA-*co*-MAAEMA)	Poly(methacrylic acid-*co*-2-methacrylamidoethylmethacrylate)
P(MMA-*co*-EDMA)	Poly(methyl methacrylate-*co*-ethylene glycol dimethacrylate)
PN	Piperonal
PNP	*p*-Nitrophenol
PS	Polystyrene
QD	Quantum dot
R123	Rhodamine 123
RAFT	Reversible addition-fragmentation chain transfer
RGO	Reduced graphene oxide
RIf	Reflectometric interferometry
RLS	Elastic resonance light scattering
$Ru(dip)_3^{2+}$	*Tris*(4,7-diphenyl-1,10-phenanthroline) ruthenium(II)
$Ru(phen)_3^{2+}$	*Tris*(1,10-phenanthroline)ruthenium(II)
SA	Sialic acid
SERS	Surface-enhanced Raman scattering
SH	Thiol group
SPR	Surface plasmon resonance
SPW	Surface plasma wave
TEA	Trimethylamine
TEOS	Tetraethoxysilane
TM	Transverse magnetic mode
TNT	2,4,6-Trinitrotoluene
TRIM	Trimethylolpropane trimethacrylate
UCNPs	Up-converting nanoparticles
UV-vis	Ultraviolet-visible spectroscopy
VPBA	4-Vinylphenylboronic acid
Z-*l*-Phe	Carbobenzyloxy-L-phenylalanine

Acknowledgements

The authors are thankful for financial support from the Spanish Ministry of Economy and Competitiveness (grant CTQ2015-69278-C2-1R/AIE), and the European Funds for Regional Development (FEDER).

References

1. A. Hulanicki, S. Glab, F. Ingman, K. Toth, R. A. Durst and G. S. Wilson, *Pure Appl. Chem.*, 1991, **63**, 1247.
2. A. Hulanicki, S. Glab, F. Ingman, K. Toth, R. A. Durst and G. S. Wilson, *Pure Appl. Chem.*, 1991, **63**, 1247.
3. D. R. Thevenot, K. Toth, R. A. Durst and G. S. Wilson, *Anal. Lett.*, 2001, **34**, 635.
4. D. R. Thevenot, K. Toth, R. A. Durst and G. S. Wilson, *Biosens. Bioelectron.*, 2001, **16**, 121.
5. C. I. L. Justino, A. C. Freitas, R. Pereira, A. C. Duarte and T. A. P. Rocha Santos, *TrAC, Trends Anal. Chem.*, 2015, **68**, 2.
6. *Molecularly Imprinted Sensors: Overview and Applications*, ed. S. Li, Y. Ge, S. A. Piletsky and J. Lunec, Elsevier, The Netherlands, 2012.
7. *Molecular Imprinting*, ed. K. Haupt, Springer, Berlin, 2012.
8. *Artificial Receptors for Chemical Sensors*, ed. V. Mirsky and A. Yatsimirsky, Wiley–VCH, Weinheim, 2011.
9. *Designing Receptors for the Next Generation of Biosensors*, ed. S. A. Pieltsky and M. J. Whitcombe, Springer, Berlin, 2013.
10. *Handbook of Molecular Imprinting. Advanced Sensor Applications*, ed. S. W. Lee and T. Kunitake, Pan Stanford Publishing, Singapore, 2013.
11. *Molecularly Imprinted Polymers: Man-made Mimics of Antibodies and Their Applications in Analytical Chemistry*, ed. B. Sellergren, Elsevier, Amsterdam, 2001.
12. C. Alexander, H. S. Andersson, L. I. Andersson, R. J. Ansell, N. Kirsch, I. A. Nicholls, J. O'Mahony and M. J. Whitcombe, *J. Mol. Recognit.*, 2006, **19**, 106.
13. G. Wulff, *Angew. Chem., Int. Ed.*, 1995, **34**, 1812.
14. S. A. Piletsky and A. P. F. Turner, *Optical Biosensors: Present and Future*, Elsevier, The Netherlands, 2002, pp. 397–426.
15. M. C. Moreno-Bondi, F. Navarro-Villoslada, E. Benito-Peña and J. L. Urraca, *Curr. Anal. Chem.*, 2008, **4**, 316.
16. L. Ye and K. Mosbach, *Chem. Mater.*, 2008, **20**, 859.
17. Y. Hoshino and K. J. Shea, *J. Mater. Chem.*, 2011, **21**, 3517.
18. M. J. Whitcombe, *Nat. Chem.*, 2011, **3**, 657.
19. G. Wulff and J. Liu, *Acc. Chem. Res.*, 2012, **45**, 239.
20. S. M. Borisov and O. S. Wolfbeis, *Chem. Rev.*, 2008, **108**, 423.
21. W. J. Akers, M. Y. Berezin, H. Lee, K. Guo, A. Almutairi, J. M. J. Frechet, G. M. Fischer, E. Daltrozzo and S. Achilefu, *Proc. SPIE*, 2010, **7576**, 757612.
22. J. Yao, M. Yang and Y. X. Duan, *Chem. Rev.*, 2014, **114**, 6130.
23. A. P. Demchenko, *Introduction to Fluorescence Sensing*, Springer, The Netherlands, 2009.
24. T. S. Bedwell and M. J. Whitcombe, *Anal. Bioanal. Chem.*, 2015, **408**, 1735.
25. S. Wang, X. Zhu and M. Zhao, in *Smart Nanomaterials for Sensor Application*, ed. S. Li, Y. Le and H. Li, Bentham Science Publishers, Oak Park, IL, 2012, pp. 60–75.

26. *Biomedical Nanomaterials*, ed. Y. Zhao and Y. Shen, Wiley-VCH, Weinheim, 2016.
27. J. L. Urraca, C. A. Barrios, V. Canalejas-Tejero, G. Orellana and M. C. Moreno-Bondi, *Nanoscale*, 2014, **6**, 8656.
28. S. Carrasco, V. Canalejas-Tejero, F. Navarro-Villoslada, C. A. Barrios and M. C. Moreno-Bondi, *J. Mater. Chem. C*, 2014, **2**, 1400.
29. Y. Xiong, Z. Ye, J. Xu, Y. Liu and H. Zhang, *Anal. Bioanal. Chem.*, 2014, **406**, 2411.
30. X. A. Ton, V. Acha, K. Haupt and B. T. S. Bui, *Biosens. Bioelectron.*, 2012, **36**, 22.
31. N. Murase, S. Taniguchi, E. Takano, Y. Kitayama and T. Takeuchi, *J. Mater. Chem. B*, 2016, **4**, 1770.
32. S. Carrasco, E. Benito-Pena, D. R. Walt and M. C. Moreno-Bondi, *Chem. Sci.*, 2015, **6**, 3139.
33. A. B. Descalzo, C. Somoza, M. C. Moreno-Bondi and G. Orellana, *Anal. Chem.*, 2013, **85**, 5316.
34. S. C. L. Pinheiro, A. B. Descalzo, I. M. Raimundo Jr, G. Orellana and M. C. Moreno-Bondi, *Anal. Bioanal. Chem.*, 2012, **402**, 3253.
35. R. Wagner, W. Wan, M. Biyikal, E. Benito-Peña, M. C. Moreno-Bondi, I. Lazraq, K. Rurack and B. Sellergren, *J. Org. Chem.*, 2013, **78**, 1377.
36. L. Gao, J. Wang, X. Li, Y. Yan, C. Li and J. Pan, *Anal. Bioanal. Chem.*, 2014, **406**, 7213.
37. S. P. Wren, T. H. Nguyen, P. Gascoine, R. Lacey, T. Sun and K. T. V. Grattan, *Sens. Actuators, B*, 2014, **193**, 35.
38. Z. Chen, M. Alvarez-Perez, F. Navarro-Villoslada and M. C. Moreno-Bondi, *Sens. Actuators, B*, 2014, **191**, 137.
39. T. H. Nguyen, S. A. Hardwick, T. Sun and K. T. V. Grattan, *IEEE Sens. J.*, 2012, **12**, 255.
40. J. Zhao, L. Nie, L. Zhang, Y. Jin, Y. Peng, S. Du and N. Jiang, *Anal. Methods*, 2013, **5**, 3009.
41. Y. Inoue, A. Kuwahara, K. Ohmori, H. Sunayama, T. Ooya and T. Takeuchi, *Biosens. Bioelectron.*, 2013, **48**, 113.
42. W. Wan, M. Biyikal, R. Wagner, B. Sellergren and K. Rurack, *Angew. Chem., Int. Ed.*, 2013, **52**, 7023.
43. Y. Li, C. Dong, J. Chu, J. Qi and X. Li, *Nanoscale*, 2011, **3**, 280.
44. J. Huang, H. Liu, H. Men, Y. Zhai, Q. Xi, Z. Zhang, J. Zhang, Z. Yin and L. Li, *Macromol. Res.*, 2013, **21**, 1021.
45. L. Gao, X. Li, Q. Zhang, J. Dai, X. Wei, Z. Song, Y. Yan and C. Li, *Food Chem.*, 2014, **156**, 1.
46. P. K. Ivanova-Mitseva, A. Guerreiro, E. V. Piletska, M. J. Whitcombe, Z. Zhou, P. A. Mitsev, F. Davis and S. A. Piletsky, *Angew. Chem., Int. Ed.*, 2012, **51**, 5196.
47. J. Zdunek, E. Benito-Peña, A. Linares, A. Falcimaigne-Cordin, G. Orellana, K. Haupt and M. C. Moreno-Bondi, *Chem.–Eur. J.*, 2013, **19**, 10209.
48. S. Xu, H. Lu, J. Li, X. Song, A. Wang, L. Chen and S. Han, *ACS Appl. Mater. Interfaces*, 2013, **5**, 8146.

49. T. Li, Z. Gao, N. Wang, Z. Zhou, W. Xu, J. Zheng and W. Yang, *RSC Adv.*, 2016, **6**, 54615.
50. Y. Yang, H. Niu and H. Zhang, *ACS Appl. Mater. Interfaces*, 2016, **8**, 15741.
51. J. Wang, H. Qiu, H. Shen, J. Pan, X. Dai, Y. Yan, G. Pan and B. Sellergren, *Biosens. Bioelectron.*, 2016, **85**, 387.
52. T. P. Huynh, A. Wojnarowicz, A. Kelm, P. Woznicki, P. Borowicz, A. Majka, F. D'Souza and W. Kutner, *ACS Sens.*, 2016, **1**, 636.
53. J. V. Ray, F. Mirata, C. Perollier, M. Arotcarena, S. Bayoudh and M. Resmini, *Anal. Bioanal. Chem.*, 2016, **408**, 1855.
54. M. P. Chantada-Vazquez, J. Sanchez-Gonzalez, E. Peña-Vazquez, M. J. Tabernero, A. M. Bermejo, P. Bermejo-Barrera and A. Moreda-Pineiro, *Anal. Chem.*, 2016, **88**, 2734.
55. B. Yang, S. Lv, F. Chen, C. Liu, C. Cai, C. Chen and X. Chen, *Anal. Chim. Acta*, 2016, **912**, 125.
56. G. Liu, Z. Chen Zhuo, X. Jiang, D. Q. Feng, J. Zhao, D. Fan and W. Wang, *Sens. Actuators, B*, 2016, **228**, 302.
57. H. Liu, K. Zhou, D. Wu, J. Wang and B. Sun, *Biosens. Bioelectron.*, 2016, **77**, 512.
58. S. Han, X. Li, Y. Wang and S. Chen, *RSC Adv.*, 2015, **5**, 2129.
59. A. Valero-Navarro, A. L. Medina-Castillo, J. F. Fernandez-Sanchez and A. Fernández-Gutiérrez, *Biosens. Bioelectron.*, 2011, **26**, 4520.
60. M. C. Moreno-Bondi, M. E. Benito-Peña, J. L. Urraca and G. Orellana, *Top. Curr. Chem.*, 2012, **325**, 111.
61. X. D. Wang and O. S. Wolfbeis, *Anal. Chem.*, 2016, **88**, 203–227.
62. T. Kuwata, A. Uchida, E. Takano, Y. Kitayama and T. Takeuchi, *Anal. Chem.*, 2015, **87**, 11784.
63. W. Wan, S. Wagner and K. Rurack, *Anal. Bioanal. Chem.*, 2016, **408**, 1753.
64. Y. Wu, Y. Liu, X. Gao, K. Gao, H. Xia, M. Luo, X. Wang, L. Ye, Y. Shi and B. Lu, *Chemosphere*, 2015, **119**, 515.
65. D. Batra and K. J. Shea, *Org. Lett.*, 2013, **5**, 3895.
66. S. Rouhani and F. Nahavandifard, *Sens. Actuators, B*, 2014, **197**, 185.
67. J. Awino and Y. Zhao, *Chem. Commun.*, 2014, **50**, 5752.
68. S. Zhang and Y. Zhao, *Macromolecules*, 2010, **43**, 4020.
69. S. Zhang and Y. Zhao, *J. Am. Chem. Soc.*, 2010, **132**, 10642.
70. L. Stryer, *Annu. Rev. Biochem.*, 1978, **47**, 819.
71. X.-A. Ton, B. Tse Sum Bui, M. Resmini, P. Bonomi, I. Dika, O. Soppera and K. Haupt, *Angew. Chem., Int. Ed.*, 2013, **52**, 8317.
72. X. A. Ton, V. Acha, P. Bonomi, B. Tse Sum Bui and K. Haupt, *Biosens. Bioelectron.*, 2015, **64**, 359.
73. S. Shinde, Z. El-Schich, A. Malakpour, W. Wan, N. Dizeyi, R. Mohammadi, K. Rurack, A. Gjörloff Wingren and B. Sellergren, *J. Am. Chem. Soc.*, 2015, **137**, 13908.
74. R. Ahmad, N. Griffete, A. Lamouri, N. Felidj, M. M. Chehimi and C. Mangeney, *Chem. Mater.*, 2015, **27**, 5464.
75. M. Niu, C. Pham-Huy and H. He, *Microchim. Acta*, 2016, **183**, 2677.
76. C. R. Kagan and C. B. Murray, *Nat. Nanotechnol.*, 2015, **10**, 1013.

77. V. Kumar Khanna, *Nanosensors Physical Chemical and Biological*, CRC Press, Boca Raton, FL, 2016, pp. 116–124.
78. *Nanocrystal Quantum Dots*, ed. V. I. Klimov, CRC Press, Boca Raton, FL, 2nd edn, 2009.
79. C. Wang, X. Gao and X. Su, *Anal. Bioanal. Chem.*, 2010, **397**, 1397.
80. J. Liu, S. K. Lau, V. A. Varma, B. A. Kairdolf and S. Nie, *Anal. Chem.*, 2010, **82**, 6237.
81. S. K. Sahoo, S. Parveen and J. J. Panda, *Nanomed. Nanotechnol. Biol. Med.*, 2007, **3**, 20.
82. A. F. E. Hezinger, J. Teßmar and A. Göpferich, *Eur. J. Pharm. Biopharm.*, 2008, **68**, 138.
83. C. B. Murray, D. J. Norris and M. G. Bawendi, *J. Am. Chem. Soc.*, 1993, **115**, 8706.
84. I. Medintz, H. T. Uyeda, E. R. Goldman and H. Mattoussi, *Nat. Mat.*, 2005, **4**, 435.
85. W. Zhou, D. T. Schwartz and F. O. Baneyx, *J. Am. Chem. Soc.*, 2010, **132**, 4731.
86. K. Sapsford, T. Pons, I. Medintz and H. Mattoussi, *Sensors*, 2006, **6**, 925.
87. D. Y. Li, X. W. He, Y. Chen, W. Y. Li and Y. K. Zhang, *ACS Appl. Mater. Interfaces*, 2013, **5**, 12609.
88. F. Wei, Y. Wu, G. Xu, Y. Gao, J. Yang, L. Liu, P. Zhou and Q. Hu, *Analyst*, 2014, **139**, 5785.
89. Y. Q. Yang, X. W. He, Y. Z. Wang, W. Y. Li and Y. K. Zhang, *Biosens. Bioelectron.*, 2014, **54**, 266.
90. B. T. Huy, M. H. Seo, X. Zhang and Y. Lee III, *Biosens. Bioelectron.*, 2014, **57**, 310.
91. W. Zhang, X. W. He, W. Y. Li and Y. K. Zhang, *Chem. Commun.*, 2012, **48**, 1757.
92. H. Liu, D. Liu, G. Fang, F. Liu, C. Liu, Y. Yang and S. Wang, *Anal. Chim. Acta*, 2013, **762**, 76.
93. C. I. Lin, A. K. Joseph, C. K. Chang and Y. D. Lee, *Biosens. Bioelectron.*, 2014, **55**, 127.
94. H. Liu, G. Fang, C. Li, M. Pan, C. Liu, C. Fan and S. Wang, *J. Mater. Chem.*, 2012, **22**, 19882.
95. H. Liu, D. Wu, K. Zhou, J. Wang and B. Sun, *Sens. Actuators, B*, 2016, **211**, 34.
96. X. Wei, T. Hao, Y. Xu, K. Lu, H. Li, Y. Yan and Z. Zhou, *RSC Adv.*, 2015, **5**, 79511.
97. S. Xu and H. Lu, *Biosens. Bioelectron.*, 2015, **73**, 160.
98. Z. Zhang, J. Li, X. Wang, D. Shen and L. Chen, *ACS Appl. Mater. Interfaces*, 2015, **7**, 9118.
99. Y. P. Chen, D. N. Wang, Y. M. Yin, L. Y. Wang, X. F. Wang and M. X. Xie, *J. Agric. Food Chem.*, 2012, **60**, 10472.
100. S. Gam-Derrouich, C. Bourdillon, W. Daney De Marcillac, L. Coolen, A. Maître, C. Mangeney and C. Schwob, *Chem. Commun.*, 2015, **51**, 14933.
101. R. C. Stringer, S. Gangopadhyay and S. A. Grant, *Anal. Chem.*, 2010, **82**, 4015.

102. *Bioconjugate Techniques*, ed. G. T. Hermanson, Elsevier, Oxford, UK, 2nd edn, 2008.
103. H. Liu, D. Wu, Y. Liu, H. Zhang, T. Ma, A. Aidaerhan, J. Wang and B. Sun, *J. Agric. Food Chem.*, 2015, **63**, 2545.
104. H. Liu, X. Chen, L. Mu, J. Wang and B. Sun, *J. Agric. Food Chem.*, 2016, **64**, 4801.
105. M. H. Lee, Y. C. Chen, M. H. Ho and H. Y. Lin, *Anal. Bional. Chem.*, 2010, **397**, 1457.
106. J. Wackerlig and R. Schirhagl, *Anal. Chem.*, 2016, **88**, 250.
107. X. Michalet, F. F. Pinaud, L. A. Bentolila, J. M. Tsay, S. Doose, J. J. Li, G. Sundaresan, A. M. Wu, S. S. Gambhir and S. Weiss, *Science*, 2005, **307**, 538.
108. P. D. McNaughter, J. C. Bear, D. C. Steytler, A. G. Mayes and T. Nann, *Angew. Chem., Int. Ed.*, 2011, **50**, 10384.
109. L. Tan, K. Chen, C. Huang, R. Peng, X. Luo, R. Yang, Y. Cheng and Y. Tang, *Microchim. Acta*, 2015, **182**, 2615.
110. F. Wei, G. Xu, Y. Wu, X. Wang, J. Yang, L. Liu, P. Zhou and Q. Hu, *Sens. Actuators, B*, 2016, **229**, 38.
111. M. Panagiotopoulou, Y. Salinas, S. Beyazit, S. Kunath, L. Duma, E. Prost, A. G. Mayes, M. Resmini, B. Tse Sum Bui and K. Haupt, *Angew. Chem., Int. Ed.*, 2016, **55**, 8244.
112. F. Wang, D. Banerjee, Y. S. Liu, X. Y. Chen and X. G. Liu, *Analyst*, 2010, **135**, 1839.
113. M. Wang, G. Abbineni, A. Clevenger, C. B. Mao and S. K. Xu, *Nanomedicine*, 2011, **7**, 710.
114. J. Liu, H. Deng, Z. Huang, Y. Zhang, D. Chen and Y. Shao, *Phys. Chem. Chem. Phys.*, 2015, **17**, 15412.
115. A. B. Chinen, C. M. Guan, J. R. Ferrer, S. N. Barnaby, T. J. Merkel and C. A. Mirkin, *Chem. Rev.*, 2015, **115**, 10530.
116. C. T. Xu, Q. Zhan, H. Liu, G. Somesfalean, J. Qian, S. He and S. Andersson-Engels, *Laser Photonics Rev.*, 2013, **7**, 663.
117. S. Beyazit, S. Ambrosini, N. Marchyk, E. Palo, V. Kale, T. Soukka, B. Tse Sum Bui and K. Haupt, *Angew. Chem., Int. Ed.*, 2014, **53**, 8919.
118. P. Karfa, E. Roy, S. Patra, D. Kumar, R. Madhuri and P. K. Sharma, *Biosens. Bioelectron.*, 2016, **78**, 454.
119. J. Hou, H. Li, L. Wang, P. Zhang, T. Zhou, H. Ding and L. Ding, *Talanta*, 2016, **146**, 34.
120. S. Li, J. Luo, G. Yin, Z. Xu, Y. Le, X. Wu, N. Wu and Q. Zhang, *Sens. Actuators, B*, 2015, **206**, 14.
121. P. Zuo, J. Gao, J. Peng, J. Liu, M. Zhao, J. Zhao, P. Zuo and H. He, *Microchim. Acta*, 2016, **183**, 329.
122. G. Liu, Z. Chen, X. Jiang, D. Q. Feng, J. Zhao, D. Fan and W. Wang, *Sens. Actuators, B*, 2016, **228**, 302.
123. H. Nui, Y. Yang and H. Zhang, *Biosens. Bioelectron.*, 2015, **74**, 440.
124. D. Y. Li, X. M. Zhang, Y. J. Yan, X. W. He, W. Y. Li and Y. K. Zhang, *Biosens. Bioelectron.*, 2016, **79**, 187.

125. C. Dodeigne, L. Thunus and R. Lejeune, *Talanta*, 2000, **51**, 415.
126. L. Huab and G. Xu, *Chem. Soc. Rev.*, 2010, **39**, 3275.
127. H. Duan, X. Li, L. Li, X. Wang, J. Feng, M. Sun and C. Luo, *Anal. Methods*, 2014, **6**, 8706.
128. S. Ge, P. Zhao, M. Yan, D. Zang and J. Yu, *Anal. Methods*, 2012, **4**, 3150.
129. B. Wu, Z. Wang, Z. Xue, X. Zhou, J. Du, X. Liu and X. Lu, *Analyst*, 2012, **137**, 3644.
130. S. Li, J. Li, Q. Lin and X. Wei, *Analyst*, 2015, **140**, 4702.
131. F. Lu, H. Li, M. Sun, L. Fan, H. Qiu, X. Li and C. Luo, *Anal. Chim. Acta*, 2012, **718**, 84.
132. H. Qiu, C. Luo, M. Sun, F. Lu, L. Fan and X. Li, *Anal. Chim. Acta*, 2012, **744**, 75.
133. H. Duan, X. Wang, Y. Wang, Y. Sun, J. Li and C. Luo, *Anal. Chim. Acta*, 2016, **918**, 89.
134. Q. Wang, M. Chen, H. Zhang, W. Wen, X. Zhang and S. Wang, *Biosens. Bioelectron.*, 2016, **79**, 561.
135. Q. Huamina, F. Lulua, X. Li, L. Li, S. Mina and L. Chuannan, *Carbohydr. Polym.*, 2013, **92**, 394.
136. H. Qiu, C. Luo, M. Sun, F. Lu, L. Fan and X. Li, *Food Chem.*, 2012, **134**, 469.
137. Z. Xu, J. Song, L. Li, X. Qiao and H. Chen, *J. Sci. Food Agric.*, 2012, **92**, 2696.
138. H. Qiu, L. Fan, X. Li, L. Li, M. Sun and C. Luo, *J. Pharm. Biomed. Anal.*, 2013, **75**, 123.
139. C. Xie, S. Gao, H. Zhou and H. Li, *Luminescence*, 2011, **26**, 271.
140. Y. Wang, H. Duan, L. Li, X. Wang, J. Li, Y. Bu and C. Luo, *RSC Adv.*, 2016, **6**, 12391.
141. H. Duan, L. Li, X. Wang, Y. Wang, J. Li and C. Luo, *RSC Adv.*, 2015, **5**, 18850.
142. J. Wu, X. Fu, C. Xie, M. Yang, W. Fang and S. Gao, *Sens. Actuators, B*, 2011, **160**, 511.
143. P. Zhao, J. Yu, S. Liu, M. Yan, D. Zang and L. Gao, *Sens. Actuators, B*, 2012, **162**, 166.
144. J. Zhou, N. Gan, F. Hu, T. Li, H. Zhou, X. Li and L. Zheng, *Sens. Actuators, B*, 2013, **186**, 300.
145. W. Cao, Y. Chao, L. Liu, Q. Liu and M. Pei, *Sens. Actuators, B*, 2014, **204**, 704.
146. P. Zhao, M. Yan, C. Zhang, R. Peng, D. Ma and J. Yu, *Spectrochim. Acta, Part A*, 2011, **78**, 1482.
147. H. Duan, L. Li, X. Wang, Y. Wang, J. Lia and C. Luo, *Spectrochim. Acta, Part A*, 2015, **139**, 374.
148. W. Liu, Y. Guo, J. Luo, J. Kou, H. Zheng, B. Li and Z. Zhang, *Spectrochim. Acta, Part A*, 2015, **141**, 51.
149. H. Duan, L. Li, X. Wang, Y. Wang, J. Li and C. Luo, *Spectrochim. Acta, Part A*, 2016, **153**, 535.
150. F. Lu, M. Sun, L. Fan, H. Qiu, X. Li and C. Luo, *Sens. Actuators, B*, 2012, **173**, 591.

151. B. Wu, Z. Wang, Z. Xue, X. Zhou, J. Du, X. Liu and X. Lu, *Analyst*, 2012, **137**, 3644.
152. J. Li, F. Ma, X. Wei, C. Fu and H. Pan, *Anal. Chim. Acta*, 2015, **871**, 51.
153. J. Zhou, N. Gan, T. Li, F. Hu, X. Li, L. Wang and L. Zheng, *Biosens. Bioelectron.*, 2014, **54**, 199.
154. F. Xue, Z. Meng, Y. Wang, S. Huang, Q. Wang, W. Lu and M. Xue, *Anal. Methods*, 2014, **6**, 831.
155. H. Peng, S. Wang, Z. Zhang, H. Xiong, J. Li, L. Chen and Y. Li, *J. Agric. Food Chem.*, 2012, **60**, 1921.
156. Y. Zhang, P. Zhao and L. Yu, *Sens. Actuators, B*, 2013, **181**, 850.
157. M. Behbahani, S. Salimi, H. S. Abandansari, F. Omidi, M. Salarian and A. Esrafili, *RSC Adv.*, 2015, **5**, 59912.
158. T. A. Sergeyeva, D. S. Chelyadina, L. A. Gorbach, O. O. Brovko, E. V. Piletska, S. A. Piletsky, L. M. Sergeeva and A. V. El'skaya, *Biopolym. Cell*, 2014, **30**, 209–215.
159. M. V. Foguel, X.-A. Ton, M. V. B. Zanoni, M. D. Sotomayor, K. Haupt and B. Tse Sum Bui, *Sens. Actuators, B*, 2015, **218**, 222.
160. Y. Li, Y. Li, L. Huang, Q. Bin, Z. Lin, H. Yang, Z. Cai and G. Chen, *J. Mat. Chem. B*, 2013, **1**, 1256.
161. N. Mizutani, D. H. Yang, R. Selyanchyn, S. Korposh, S. W. Lee and T. Kunitake, *Anal. Chim. Acta*, 2011, **694**, 142.
162. L. Jensen, C. M. Aikens and G. C. Schatz, *Chem. Soc. Rev.*, 2008, **37**, 1061.
163. G. C. Schatz, M. A. Young and R. P. Duyne, in *Surface-enhanced Raman Scattering: Physics and Applications*, ed. K. Kneipp, M. Moskovits and H. Kneipp, Springer Berlin Heidelberg, Berlin, Heidelberg, 2006, vol. 2, pp. 19–45.
164. S. Zou and G. C. Schatz, in *Surface-enhanced Raman Scattering: Physics and Applications*, ed. K. Kneipp, M. Moskovits and H. Kneipp, Springer Berlin Heidelberg, Berlin, Heidelberg, 2006, vol. 3, pp. 67–85.
165. M. Bompart, L. A. Gheber, Y. De Wilde and K. Haupt, *Biosens. Bioelectron.*, 2009, **25**, 568.
166. K. Kantarovich, A. S. Belmont, K. Haupt, I. Bar and L. A. Gheber, *Appl. Phys. Lett.*, 2009, **94**, 194103.
167. D. McStay, A. H. Al-Obaidi, R. Hoskins and P. J. Quinn, *J. Opt. A: Pure Appl. Opt.*, 2005, **7**, S340.
168. K. Sreenivasan, *Anal. Chim. Acta*, 2007, **583**, 284.
169. A. F. Che, L. S. Wan, J. Ling, Z. M. Liu and Z.-K. Xu, *J. Phys. Chem. B*, 2009, **113**, 7053.
170. J. Y. Wang, Z. L. Xu, P. Wu and S. J. Yin, *J. Membr. Sci.*, 2009, **331**, 84.
171. R. Voicu, K. Faid, A. A. Farah, F. Bensebaa, R. Barjovanu, C. Py and Y. Tao, *Langmuir*, 2007, **23**, 5452.
172. T. Kamra, C. Xu, L. Montelius, J. Schnadt, S. A. Wijesundera, M. Yan and L. Ye, *ACS Appl. Mater. Interfaces*, 2015, **7**, 27479.
173. T. Kamra, S. Chaudhary, C. Xu, L. Montelius, J. Schnadt and L. Ye, *J. Colloid Interface Sci.*, 2016, **461**, 1.

174. T. Kamra, T. Zhou, L. Montelius, J. Schnadt and L. Ye, *Anal. Chem.*, 2015, **87**, 5056.
175. K. Kantarovich, I. Tsarfati-BarAd, L. Gheber, K. Haupt and I. Bar, *Plasmonics*, 2013, **8**, 3.
176. E. L. Holthoff, D. N. Stratis-Cullum and M. E. Hankus, *Sensors*, 2011, **11**, 2700.
177. A. A. Volkert and A. J. Haes, *Analyst*, 2014, **139**, 21.
178. M. Bompart, Y. De Wilde and K. Haupt, *Adv. Mater.*, 2010, **22**, 2343.
179. J. Langer, S. M. Novikov and L. M. Liz-Marzán, *Nanotechnology*, 2016, **26**, 322001.
180. S. Chen, X. Li, Y. Guo and J. Qi, *Analyst*, 2015, **140**, 3239.
181. S. N. Chen, X. Li, S. Han, J. H. Liu and Y. Y. Zhao, *RSC Adv.*, 2015, **5**, 99914.
182. L. Chang, Y. Ding and X. Li, *Biosens. Bioelectron.*, 2013, **50**, 106.
183. S. Chen, X. Li, Y. Zhao, L. Chang and J. Qi, *Chem. Commun.*, 2014, **50**, 14331.
184. Y. Guo, L. Kang, S. Chen and X. Li, *Phys. Chem. Chem. Phys.*, 2015, **17**, 21343.
185. D. Yin, S. Wang, Y. He, J. Liu, M. Zhou, J. Ouyang, B. Liu, H. Y. Chen and Z. Liu, *Chem. Commun.*, 2015, **51**, 17696.
186. R. Ahmad, N. Felidi, L. Boubekeur-Lecaque, S. Lau-Truong, S. Gam-Derouich, P. Decorse, A. Lamouri and C. Mangeney, *Chem. Commun.*, 2015, **51**, 9678.
187. J. Q. Xue, D. W. Li, L. L. Qu and Y. T. Long, *Anal. Chim. Acta*, 2013, **777**, 57.
188. P. Zijlstra, M. Orrit and A. Koenderink, in *Nanoparticles: Workhorses of Nanoscience*, ed. C. de Mello Donegá, Springer, Berlin, Heidelberg, 2014, vol. 3, pp. 53–98.
189. S. Carrasco, E. Benito-Peña, F. Navarro-Villoslada, J. Langer, M. N. Sanz-Ortiz, J. Reguera, L. M. Liz-Marzán and M. C. Moreno-Bondi, *Chem. Mater.*, 2016, **28**, 7947.
190. Z. Guo, L. Chen, H. Lu, Z. Yu and B. Zhao, *Anal. Methods*, 2014, **6**, 1627.
191. J. Homola, S. Yee and G. Gauglitz, *Sens. Actuators, B*, 1999, **54**, 3.
192. J. Homola, in *Surface Plasmon Resonance Based Sensors*, ed. J. Homola, Springer, Berlin Heidelberg, 2006, vol. 1, pp. 3–69.
193. E. K. Akowuah, T. Gorman and S. Haxha, *Opt. Express*, 2009, **17**, 23511.
194. S. A. Maier, in *Plasmonics: Fundamentals and Applications*, ed. S. A. Maier, Springer, New York, 2007, vol. 3, pp. 39–52.
195. H. S. Leong, J. Guo, R. G. Lindquist and Q. H. Liu, *J. Appl. Phys.*, 2009, **106**, 124314.
196. G. Sener, L. Uzun, R. Say and A. Denizli, *Sens. Actuators, B*, 2011, **160**, 791.
197. M. Kara, L. Uzun, S. Kolayli and A. Denizli, *J. Appl. Polym. Sci.*, 2013, **129**, 2273.
198. M. J. Abdin, Z. Altintas and I. E. Tothill, *Biosens. Bioelectron.*, 2015, **67**, 177.

199. Z. Altintas, A. Guerreiro, S. A. Piletsky and I. E. Tothill, *Sens. Actuators, B*, 2015, **213**, 305.
200. Q. Zhang, L. Jing, J. Zhang, Y. Ren, Y. Wang, T. Wei and B. Liedberg, *Anal. Biochem.*, 2014, **463**, 7.
201. T. Yao, X. Gu, T. Li, J. Li, J. Li, Z. Zhao, J. Wang, Y. Qin and Y. She, *Biosens. Bioelectron.*, 2016, **75**, 96.
202. Y. Taguchi, E. Takano and T. Takeuchi, *Langmuir*, 2012, **28**, 7083.
203. N. Cennamo, L. De Maria, G. Agostino, L. Zeni and M. Pesavento, *Sensors*, 2015, **15**, 8499.
204. N. Cennamo, G. D'Agostino, R. Galatus, L. Bibbò, M. Pesavento and L. Zeni, *Sens. Actuators, B*, 2013, **188**, 221.
205. M. S. Anand, K. M. Satyendra and D. G. Banshi, *Mater. Res. Express*, 2015, **2**, 035007.
206. R. Verma and B. D. Gupta, *Sens. Actuators, B*, 2013, **177**, 279.
207. N. Cennamo, G. D'Agostino, M. Pesavento and L. Zeni, *Sens. Actuators, B*, 2014, **191**, 529.
208. A. M. Shrivastav, S. K. Mishra and B. D. Gupta, *Sens. Actuators, B*, 2015, **212**, 404.
209. G. H. Yao, R. P. Liang, C. F. Huang, Y. Wang and J. D. Qiu, *Anal. Chem.*, 2013, **85**, 11944.
210. H. Sunayama, T. Ooya and T. Takeuchi, *Chem. Commun.*, 2014, **50**, 1347.
211. X. Y. Xu, X. G. Tian, L. G. Cai, Z. L. Xu, H. T. Lei, H. Wang and Y. M. Sun, *Anal. Methods*, 2014, **6**, 3751.
212. S. Jiang, Y. Peng, B. Ning, J. Bai, Y. Liu, N. Zhang and Z. Gao, *Sens. Actuators, B*, 2015, **221**, 15.
213. J. Wang, S. Banerji, N. Menegazzo, W. Peng, Q. Zou and K. S. Booksh, *Talanta*, 2011, **86**, 133.
214. Q. Q. Wei and T. X. Wei, *Chin. Chem. Lett.*, 2011, **22**, 721.
215. Y. Tan, L. Jing, Y. Ding and T. Wei, *Appl. Surf. Sci.*, 2015, **342**, 84.
216. Y. Tan, I. Ahmad and T. X. Wei, *Chin. Chem. Lett.*, 2015, **26**, 797.
217. B. Osman, L. Uzun, N. Beşirli and A. Denizli, *Mater. Sci. Eng., C*, 2013, **33**, 3609.
218. G. Ertürk, H. Özen, M. A. Tümer, B. Mattiasson and A. Denizli, *Sens. Actuators, B*, 2016, **224**, 823.
219. G. Sener, E. Ozgur, A. Y. Rad, L. Uzun, R. Say and A. Denizli, *Analyst*, 2013, **138**, 6422–6428.
220. C. Zhu, L. Zhang, C. Chen and J. Zhou, *Anal. Lett.*, 2015, **48**, 1537.
221. H. Shaikh, G. Sener, N. Memon, M. I. Bhanger, S. M. Nizamani, R. Uzek and A. Denizli, *Anal. Methods*, 2015, **7**, 4661.
222. N. Sharma, C. Petri, U. Jonas, M. Bach, G. Tovar, K. Mrkvová, M. Vala, J. Homola, W. Knoll and J. Dostálek, *Macromol. Chem. Phys.*, 2014, **215**, 2295.
223. G. Bidan, in *Electropolymerization*, ed. S. Cosnier and A. Karyakin, Wiley-VCH Verlag GmbH & Co. KGaA, Weinheim, 2010, vol. 1, pp. 1–26.
224. Y. Wang, Q. Zhang, Y. Ren, L. Jing and T. Wei, *Chem. Res. Chin. Univ.*, 2014, **30**, 42.

225. G. Gupta, A. S. B. Bhaskar, B. K. Tripathi, P. Pandey, M. Boopathi, P. V. L. Rao, B. Singh and R. Vijayaraghavan, *Biosens. Bioelectron.*, 2011, **26**, 2534.
226. M. Frasconi, R. Tel-Vered, M. Riskin and I. Willner, *Anal. Chem.*, 2010, **82**, 2512.
227. M. Riskin, R. Tel-Vered, M. Frasconi, N. Yayo and I. Willner, *Chem.–Eur. J.*, 2010, **16**, 7114.
228. N. Atar, T. Eren and M. L. Yola, *Food Chem.*, 2015, **184**, 7.
229. M. L. Yola, T. Eren and N. Atar, *Sens. Actuators, B*, 2014, **195**, 28.
230. M. L. Yola, N. Atar and T. Eren, *Sens. Actuators, B*, 2014, **198**, 70.
231. N. Atar, T. Eren, M. L. Yola and S. Wang, *Sens. Actuators, B*, 2015, **216**, 638.
232. M. L. Yola, N. Atar and A. Erdem, *Sens. Actuators, B*, 2015, **221**, 842.
233. G. Della Giustina, A. Sonato, E. Gazzola, G. Ruffato, S. Brusa and F. Romanato, *Mater. Lett.*, 2016, **162**, 44.
234. D. P. Campbell and C. J. McCloskey, in *Optical Biosensors: Present and Future*, ed. F. S. Ligler and C. A. Rowe Taitt, Elsevier, Amsterdam, 1st edn, 2002, vol. 9, pp. 277–304.
235. M. Abolbashari, F. Farahi and J. L. Santos, in *Handbook of Optical Sensors*, ed. J. L. Santos and F. Farahi, CRC Press, New York, 2014, vol. 5, pp. 111–144.
236. A. Brandenburg, V. Hinkov and W. Konz, in *Sensors: Optical Sensors*, ed. E. Wagner, R. Dändliker and K. Spenner, Wiley-VCH Verlag GmbH, Weinheim, 2008, vol. 6, 16, pp. 399–420.
237. D. D. Nolte, in *Optical Interferometry for Biology and Medicine*, ed. D. D. Nolte, Springer, New York, 2012, vol. 1, pp. 3–48.
238. S. E. Diltemiz and O. Uslu, *Biotechnol. Prog.*, 2015, **31**, 55.
239. S. M. Reddy, D. M. Hawkins, Q. T. Phan, D. Stevenson and K. Warriner, *Sens. Actuators, B*, 2013, **176**, 190–197.
240. C. A. Barrios, C. Zhenhe, F. Navarro-Villoslada, D. López-Romero and M. C. Moreno-Bondi, *Biosens. Bioelectron.*, 2011, **26**, 2801.
241. C. A. Barrios, S. Carrasco, P. Yurrita, F. Navarro-Villoslada and M. C. Moreno-Bondi, *Sens. Actuators, B*, 2012, **161**, 607.
242. Y. Fuchs, O. Soppera and K. Haupt, *Anal. Chim. Acta*, 2012, **717**, 7.
243. C. Ayela, G. Dubourg, C. Pellet and K. Haupt, *Adv. Mater.*, 2014, **17**, 5876.
244. Y. Fuchs, O. Soppera, A. G. Mayes and K. Haupt, *Adv. Mater.*, 2013, **25**, 566.
245. M. Ahmadi, T. Madrakian and A. Afkhami, *Anal. Chim. Acta*, 2014, **852**, 250.
246. S. Lépinay, A. Ianoul and J. Albert, *Talanta*, 2014, **128**, 401.

CHAPTER 9

Protein Determination Using Molecularly Imprinted Polymer (MIP) Chemosensors

MACIEJ CIEPLAK*[a] AND WLODZIMIERZ KUTNER*[a,b]

[a]Institute of Physical Chemistry, Polish Academy of Sciences, Kasprzaka 44/52, 01-224 Warsaw, Poland; [b]Faculty of Mathematics and Natural Sciences. School of Sciences Cardinal Stefan Wyszynski University in Warsaw, Wóycickiego 1/3, 01-815 Warsaw, Poland
*E-mail: mcieplak@ichf.edu.pl, wkutner@ichf.edu.pl

9.1 Introduction

Molecularly imprinted polymers (MIPs), synthesized with the use of macromolecular templates ($M_w > 1500$ Da), and proteins in particular, have received significant attention over much of the last two decades. Despite a rapidly growing number of publications, there was only a limited progress made in this field till the beginning of the last decade. Most applications of these MIPs are limited to support materials for solid-phase extraction (SPE) and liquid chromatography. These applications were extensively reviewed elsewhere.[1–8] In the present chapter, we critically survey papers involving application of MIPs as recognition units in chemosensors selective to chosen proteins. In the recent few years, several new strategies and ideas have been proposed with that respect.

Polymer Chemistry Series No. 28
Molecularly Imprinted Polymers for Analytical Chemistry Applications
Edited by Wlodzimierz Kutner and Piyush Sindhu Sharma

Published by the Royal Society of Chemistry, www.rsc.org

9.2 Historical Background of Molecular Imprinting

Artificial recognition materials based on the molecular imprinting concept date back to the pioneering work of Dickey in 1955.[9] He prepared organic dye (methyl orange and its homologs) selective silica adsorbents by acidifying silica solutions of these dyes. However, low repeatability in preparation of these absorbents and their low selectivity with respect to target analytes – the dyes – resulted in low interest of the analytical chemistry community in this concept, until Wulff *et al.*[10] and Klotz *et al.*[11] in the early seventies of the last century reported on imprinting templates in organic polymers. Illustrated using organic polymers, this concept opened up new horizons for designing molecular cavities in a very precise way. From that time, many different strategies of molecular imprinting were proposed (Scheme 9.1). However, a real breakthrough came in 1993, when Mosbach *et al.*[12] introduced a radiolabeled ligand-binding assay, proving that MIPs can replace antibodies as recognition units in biosensors.

In the earliest attempts of protein imprinting, a copper(II) complex of *N*-(4-vinyl)benzyliminodiacetic acid was used as the functional monomer. This complex was capable of binding the protein (RNase A) template by copper ions.[14] The polymer was deposited on the surface of silica beads modified with vinyl groups. After protein template removal, these beads were subsequently used as a column packing material for HPLC. Most applications of protein–MIPs, especially bulk polymers, involve use as chromatography and SPE column and cartridge, respectively, packing materials. But these materials are out of scope of the present chapter.

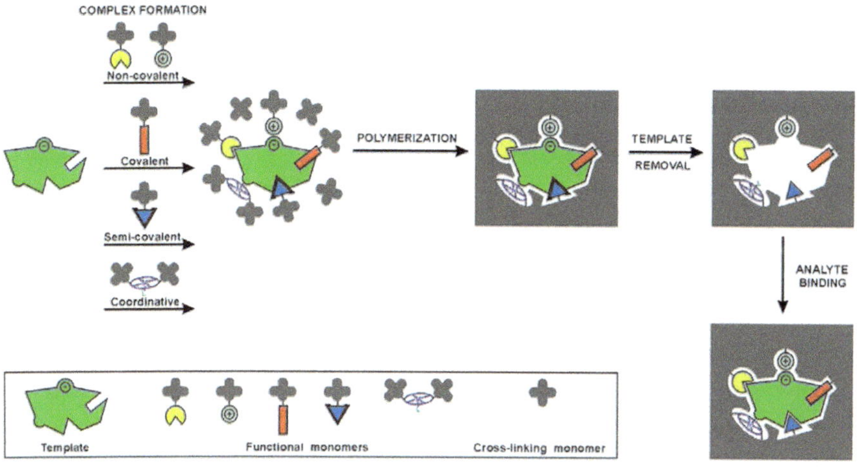

Scheme 9.1 Different routes to molecular cavity imprinting in polymers. Adapted from *TrAC Trends in Analytical Chemistry*, **51**, Piyush Sindhu Sharma, Marcin Dabrowski, Francis D'Souza and Wlodzimierz Kutner, Surface development of molecularly imprinted polymer films to enhance sensing signals, 146–157. Copyright (2013) with permission from Elsevier.

9.3 Methods of Protein Imprinting

Methods developed for protein imprinting involve both whole protein molecule imprinting and epitope imprinting, as discussed below.

9.3.1 Whole Protein Imprinting

For chemosensor fabrication, an MIP should preferably be deposited as a thin film on the transducer surface (Scheme 9.2A). That way, recognition units of the chemosensor are prepared. There are many methods for deposition of these films; however, electropolymerization seems to be most elegant. Therefore, there is ever increasing interest in engaging electropolymerization in preparation of both conducting and non-conducting MIP films.[15,16] MIP recognition units have found their application in many different chemosensors.

9.3.1.1 Piezoelectric Microgravimetry for Protein Sensing

In the first attempt of fabrication an MIP chemosensor selective with respect to proteins, an Au film-coated quartz crystal resonator (Au-QCR) was spin coated with a film of a human serum albumin (HSA) imprinted polyacrylamide.[17] For that, 3-dimethylaminopropyl methacrylamide (DMAPMA) was used as the functional monomer. The Au-QCR modified that way was used to determine HSA by piezoelectric microgravimetry (PM) with a quartz crystal microbalance (QCM). However, the dynamic linear concentration range appeared to be very narrow (60 to 160 μg mL^{-1} HSA) and linearity of the response was low. The selectivity of the chemosensor was magnificent but a low imprinting factor (IF), equal to 1.43, suggested that this selectivity originated from strong unspecific adsorption of HSA molecules on the polymer surface rather than from their binding in the imprinted cavities.

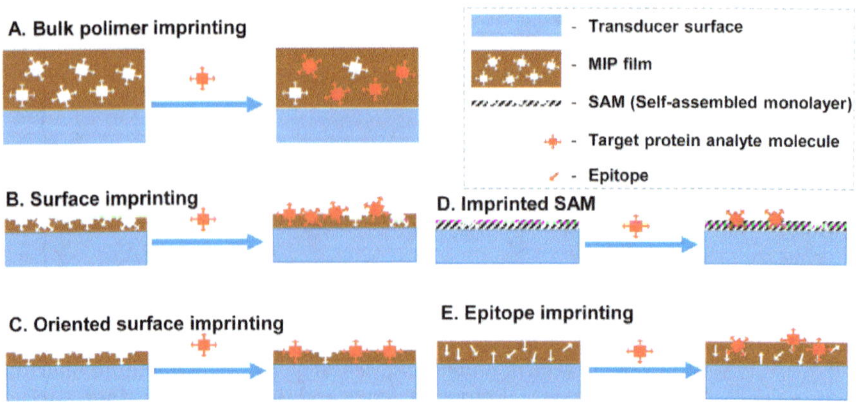

Scheme 9.2 Five (A–E) different strategies of protein imprinting.

Preparation of a protein imprinted MIP film is not trivial. This is because many factors can influence this process. To examine these factors, three different methods of preparation of a trypsin imprinted acrylamide polymer were compared, by way of example, including (i) polymerization from solution, (ii) pressing a protein crystal stamp against a thin layer of the pre-polymerization solution, and (iii) deposition of lyophilized protein on the Au-QCR surface with subsequent polymerization.[18] It appeared that (Au-QCR)-MIP chips prepared by chemical polymerization from solution, method (i), revealed superior sensitivity. That is, the resonance frequency change caused by the trypsin presence in solution was ~4 times higher than that for methods (ii) and (iii). The PM chemosensor fabricated was so selective that it even allowed for detecting changes in trypsin conformation under denaturing conditions.

Both electrostatic interactions and also the hydrophilicity of functional monomers greatly influenced performance of an MIP hydrogel.[19] In an illustrative example of the hydrophilicity effect, bovine hemoglobin (BHb) imprinted hydrogels prepared with acrylamide (AM), *N*-hydroxymethylacrylamide, or *N*-isopropylacrylamide were compared. Despite the fact that the binding capacity, determined in spectroscopic studies, of the hydrophobic *N*-isopropylacrylamide based MIP was the highest, it appeared that the hydrophilic *N*-hydroxymethylacrylamide based MIP-BHb was the most selective with respect to five protein interferences in the PM at QCM determinations.

In a subsequent attempt of fabrication of PM chemosensors for protein determination, Au-QCRs were coated with several different polymer films including silica sol–gel,[20] acrylamide gel,[21,22] or poly(ethylene-*co*-vinyl alcohol) (EVAL),[23] imprinted with HSA,[20] a rubber latex allergen (Hev b1 G),[21,22] α-amylase EC 3.2.1.1, lipase EC 3.1.1.3 (both from hog pancreas) and lysozyme (Lyz, EC 3.2.1.17) from a hen egg.[23] Moreover, electrochemical quartz crystal microbalance (EQCM) allowed for easy combining of PM determinations with electrochemical experiments, *e.g.*, electrochemical impedance spectroscopy (EIS).[20]

9.3.1.2 *Electrochemical MIP Chemosensing of Proteins*

Moreover, an MIP film can be deposited on an electrode surface. In early experiments on determination of BHb, a target protein imprinted polyacrylamide film was deposited on a glassy carbon electrode (GCE). For that, acrylamide was indirectly electropolymerized in the presence of BHb with potassium peroxydisulfate serving as the polymerization initiator.[24] Unfortunately, the resulting MIP film coated GCE allowed determining the BHb concentration in the quite narrow concentration range of 0.1 to 1 mg mL^{-1} by differential pulse voltammetry (DPV). Importantly, however, the chemosensor response to BHb was higher than that to bovine serum albumin (BSA), cytochrome *c* (Cyt *c*), glucose oxidase (GOx), and ovalbumin (Ova).

Electropolymerization is the most convenient method of MIP deposition on the electrode surface. In one example, poly(*o*-phenylenediamine) was deposited in the presence of the cardiac troponin T (TnT), a biomarker of myocardial infarction, on an Au electrode surface.[25] Then, this electrode was used for TnT determination under DPV conditions in the presence of a signaling redox probe. In another example, caffeic acid was oxidatively electropolymerized under potentiostatic conditions in the presence of the microseminoprotein beta (MSMB) template and dopamine serving as a positively charged functional monomer.[26]

Atom transfer radical polymerization (ATRP) is another efficient method for deposition of a thin polymer film. In one study, a reduced Hb served as both the template and catalyst of electrochemically induced ATRP.[27] The Au electrode surface was subsequently modified with *p*-hydroxythiophenol and 2-bromoisobutyryl bromide to provide an ATRP initiator. When a potential of −0.51 V was applied to the working electrode, the Fe(III) ion in the Hb molecule was reduced to Fe(II), and then this molecule catalyzed living polymerization of acrylamide and *N*,*N*′-methylene bisacrylamide resulting in an MIP film.

Peptide or protein imprinting is quite challenging. This is mainly because of their large molecule size and conformation susceptibility to experimental conditions. In the case of non-covalent imprinting,[28] it is very difficult to estimate which and how many functional groups on the surface of the template molecule are accessible for binding. For that, proper selection of a suitable molar ratio of functional monomers to peptide in the pre-polymerization solution seems to be crucial. This is because a too high relative concentration of the functional monomer would lead to appearance of recognizing sites unbound to the template molecules and randomly distributed inside MIP. This site randomization would result in a low IF and low selectivity. However, a too low ratio would lead to a low number of recognition sites in the molecular cavities and, hence, low affinity to the peptide analyte. Therefore, optimization of the composition of the pre-polymerization solution for non-covalent imprinting is time and labor consuming. To overcome these drawbacks, a semi-covalent imprinting was introduced (Scheme 9.3).[29] The usefulness of this type of imprinting was illustrated with imprinting of HSA. This imprinting involved covalent binding of functional monomers to appropriate functional groups accessible on the surface of the HSA molecule. These monomers contained the electropolymerizable *bis*(2,2′-bithienyl)methane moiety and either the amino or carboxyl group responsible for recognition of complementary functional groups on the target HSA analyte. The monomer excess was then removed by size-exclusion chromatography. Subsequently, the obtained adduct was deposited by potentiodynamic electropolymerization on the surface of a gold disk electrode in the presence of a cross-linking monomer. Both high selectivity and imprinting factor (IF > 20) of the chemosensor proved that the imprinting was successful.

Some attempts were undertaken to amplify the signal generated by the electrodes coated with the MIP-protein films. Towards that, first, a layer of

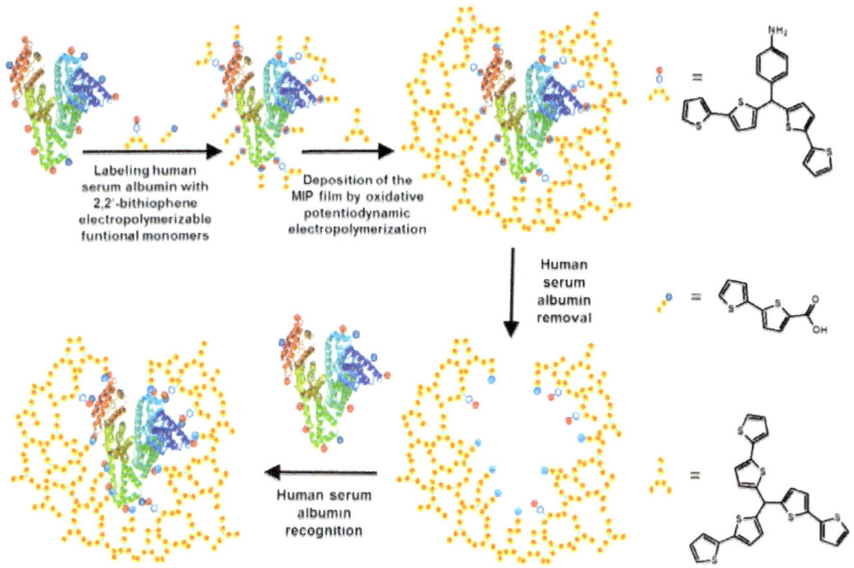

Scheme 9.3 Semi-covalent imprinting of HSA.

hierarchical dendritic gold microstructures (HDGMs), and then a Prussian Blue (PB) film was deposited by potentiodynamic electropolymerization on the GCE surface.[30] Subsequently, the BHb imprinted acrylamide film was deposited on top of the PB surface by indirect electropolymerization in the presence of the ammonium persulfate initiator. Moreover, an EIS detection signal was amplified with dopants, such as the Ponceau S, Coomassie BB R250, and I-Carrageenan organic dyes capable of strong interaction with target proteins.[31] To this end, the ricin toxin chain A (RTA) protein was physically absorbed on the surface of an array of gold electrodes. Then, the dopants were allowed to form complexes with the RTA and, finally, a polypyrrole film was deposited on top of this construct by potentiodynamic electropolymerization. The MIP-polypyrrole film, prepared in the absence of dopants, was not selective with respect to BSA. The highest selectivity was obtained with Coomassie BB R250.

In almost all DPV and EIS electrochemical determinations, the authors used a redox probe, *e.g.*, $[Fe(CN)_6]^{3-/4-}$, to generate an electrochemical signal. A decrease of this signal is governed by the so called "gate effect", *i.e.*, blocking of redox probe diffusion through the MIP film to the electrode substrate because of the analyte filling the MIP cavities.[32] However, some proteins contain electroactive groups in their structure. Therefore, it is possible to determine directly their binding to the MIP. For example, a multi-wall carbon nanotube (MWCNT) ceramic electrode, modified with the BSA imprinted polymer, was prepared.[33] Towards that, tetraethyleneglycol-3-morpholine propionate acrylate (TEGMPA) and diacryloyl urea (DAU) was used as the functional and cross-linking monomer, respectively. After electroreduction, the DPV peak current corresponding to electro-oxidation of the BSA disulfide bonds was recorded in the −0.70 to +0.70 V *vs.* (Ag/AgCl) potential range. Furthermore, a label-free MIP electrochemical sensor based on hydrogel imprinted with bovine oxygenated hemoglobin (BoHb) was devised.[34] For that, a BoHb imprinted hydrogel was prepared. Small pieces of this hydrogel were mechanically fastened to the GCE surface with a nylon net and a dialysis membrane. The CV cathodic peak corresponding to electrochemical reduction of BoHb was used to characterize electrochemical properties of both the MIP and the non-imprinted polymer (NIP) hydrogel.

In another study, a photosensitive copolymer, composed of poly(γ-glutamic acid) grafted with the 7-amino-4-methylcoumarin fluorophore was mixed with Hb and allowed to arrange around this protein.[35] Then, the resulting aggregates were immobilized on a GCE surface *via* electrophoretic deposition, and finally crosslinked under UV light illumination. An Hb electro-oxidation DPV peak served as the analytical signal. Moreover, a Cyt *c* imprinted non-conducting polyscopoletin film was deposited by electropolymerization, from an aqueous solution of the Cyt *c* template, on an Au electrode surface modified with mercaptoundecanoic acid.[36] The presence of carboxyl groups on the electrode surface was crucial for this imprinting because the amount of Cyt *c* incorporated in the MIP during electropolymerization dropped by 70% in their absence. Moreover, electrostatic interactions between carboxyl groups and the Cyt *c* template molecules forced the Cyt *c* molecule to assume an orientation enabling direct electron transfer between the Cyt *c* heme group and the electrode. Therefore, the electrode process of the MIP-entrapped Cyt *c* was manifested with a pair of CV redox peaks with the formal potential of 0.04 ± 0.01 V *vs.* Ag/AgCl. This value was close to that of Cyt *c* adsorbed by mercaptoundecanoic acid, thus suggesting that the Cyt *c* structure and conformation were not much affected by the imprinting. In another approach, the amount of acetylcholinesterase (AChE) molecules bound to an MIP film was determined by *in situ* measuring activity of this enzyme.[37] For that purpose, thiocholine, a product of acetylthiocholine enzymatic hydrolysis was determined with amperometry using the MIP film coated gold wire. The current of oxidation of the released thiocholine was dependent on the AChE concentration.

9.3.1.3 *Optical Sensing of Proteins*

The most convenient methods of protein sensing, especially for "in field determinations", are optical methods. In one example, microscope glass slides with mica wafers placed on top were coated with 100 μm flat films of MIP hydrogels selective with respect to maltose binding protein (MBP) labeled with the sulfoindocyanine *N*-hydroxysuccinimidyl ester (Cy3-NHS) organic dye as the template.[38] In some determinations, however, protein labeling with organic dyes is not even needed. For instance, binding BSA and Ova to MIP was demonstrated with their tryptophan moiety fluorescence measured using a deep-UV fluorescence image microscope (UVFLIM).[39] For this measurement, proteins immobilized on 15 μm diameter silica beads were used as molecular stamps to form micrometre size holes with imprinted walls. The UVFLIM images indicated that target proteins adsorbed only on walls of those holes and the flat surface of the polymer around the holes showed no trace of fluorescence.

Instead of labeling proteins, molecular cavities of an MIP were labeled in another approach with molecules of an organic dye.[40] For that, a special functional monomer was devised. It bore three functional groups, *i.e.*, a polymerizable methacryloyl group, a secondary amine group for fluorescent dye conjugation, and a benzoic acid moiety capable of interacting with the target protein. The Lyz imprinted polymer films were prepared by free-radical polymerization on the initiator-immobilized glass slide substrates. Another study introduced a cleavable functional monomer bearing a disulfide bond and a terminal maleimide moiety capable of binding cys-17 located on the Cyt *c* surface.[41] After deposition of the acrylamide (Cyt *c*)-imprinted polymer film on a glass substrate, the Cyt *c* template was removed by chemical reduction of the disulfide bond to thiol. Then, the imprinted molecular cavities were labeled with the thiol reacting dye, 4-(*N*,*N*-dimethylaminosulfonyl)-7-fluoro-2,1,3-benzoxadiazole (DBD-F). However, well-defined molecular cavities were presumably not formed during the polymerization. Therefore, the selectivity determined with the use of the fluorescence protocol might originate from interactions of the DBD-F dye with the cys-17 of Cyt *c* adsorbed on the MIP surface rather than with the imprinted molecular cavities. Moreover, these doubts were strengthened for another Lyz imprinted polymer where 4-[2-(*N*-methacrylamido)ethylaminomethyl] benzoic acid (MABA) was used as the functional monomer electrostatically interacting with Lyz. For this polymer, the response to interfering Cyt *c*, caused by interaction of Cyt *c* molecules with dye molecules located outside the MIP cavities, was stronger than to the Lyz analyte.[42] To eliminate this lack of selectivity, after MIP deposition but before extraction of the protein template, all functional monomers located outside the molecularly imprinted cavities were capped with *p*-isothiocyanatophenyl α-D-mannopyranoside. Subsequently, *i.e.*, after template removal, the secondary amine residues of MABA located inside the cavities were reacted with fluorescein isothiocyanate (FITC). A dramatic increase of the MIP selectivity in Lyz determination according to the fluorescence protocol strongly supported the usefulness of the capping concept. Moreover, introduction of the capping step in the MIP synthesis did not alter selectivity of surface plasmon resonance (SPR) spectroscopy determinations. This result proved that the fluorescence change measured after capping originated from binding proteins only inside the cavities. Furthermore, {[[2-(2-methacrylamido) ethyldithio]ethylcarbamoyl]methoxy} acetic acid (MDTA) was introduced as the functional monomer.[43] Its cleavable disulfide bonds were used for exchanging carboxyl functional groups for amine groups, and subsequent labeling of them with fluorescein isothiocyanate inside the molecular cavities. Besides, an acrylic MIP-HSA bearing *O*-acryl-L-4-hydroxyproline amide of the dansylethylenediamine organic dye inside the polymer matrix as the functional monomer appeared to be selective to HSA.[44] The proline amide moiety provided very high affinity of this functional monomer to the protein template.

Another transduction method commonly used for determination of macromolecular biocompounds is SPR spectroscopy. For that purpose, SPR chips with the olefin modified surface were subsequently coated with MIP films by free-radical co-polymerization of acrylic acid (AA) and *N*,*N'*-methylenebisacrylamide in the presence of Lyz. Selectivity of the resulting chemosensor with respect to interfering proteins including Cyt *c*, Myo, RNase A, and lactoalbumin was very high.[45] Moreover, the SPR chips coated with the methacrylic polymer imprinted with hepatitis B surface antibody (HBsAb) were used to detect infection in plasma samples donated by HBs positive patients.[46] Comparison of the above results with those of the commercial ELISA indicated that the results obtained with the use of the MIP-HBsAb chemosensors were reliable in the 0 to 120 mU Ml^{-1} HBs concentration range. Moreover, the SPR chips, coated with films of Myo imprinted polyhydroxyethyl methacrylate-*N*-methacryloyl-L-tryptophan methyl ester were used for determination of Myo in serum samples donated by a patient suffering from acute myocardial infarction.[47] Furthermore, MIP chemosensors for glycoproteins were devised. The MIPs involved were based on covalent imprinting of glycoproteins with 3-aminophenylboronic acid used as the functional monomer. These MIPs were deposited by electropolymerization on SPR chips. However, the resulting SPR chemosensor using an MIP-BSA recognition unit was not very sensitive and selective with respect to BSA.[48]

Moreover, some attempts of the SPR signal improvement were undertaken. For instance, micropatterned surface-imprinted polymers were fabricated photolithographically.[49] For that purpose, 5 µm wide and 2 µm deep microchannels were formed with the AZ® photoresist on the Au surface of SPR chips, and polycarbonate was then deposited in these microchannels by spin coating. After removal of the photoresist, avidin (Av) was adsorbed on the surface of the polycarbonate, and the poly(3,4-ethylenedioxythiophene)/poly(styrenesulfonate) (PEDOT/PSS) conducting copolymer was then deposited by potentiostatic electropolymerization in free spaces left after photoresist removal. Finally, the polycarbonate was dissolved in dichloromethane resulting in formation of a micropatterned surface-imprinted MIP film. These film coated SPR chips were used for the determination of Av.

The combination of photonic crystals with MIPs seems promising. These crystals are fabricated by templating polymers or hydrogels with colloidal crystals that leads to formation of porous films with a highly ordered 3-D structure. Periodic variation in the refractive index of these films strongly enhanced their interesting optical properties. For instance, binding analyte molecules inside molecularly imprinted cavities caused a pronounced swelling or shrinking of these crystals and, therefore, significantly changing their properties. Moreover, formation of thin polymer walls between pores substantially increased polymer surface area and enhanced rapid response to the analyte presence. For example, a photonic hydrogel was prepared that way.[50] That is, it was imprinted with BSA by immobilization of the BSA template on surface of the silica colloidal particles by spin-casting. Then, empty spaces between these nanoparticles were filled with the acrylamide hydrogel by free-radical polymerization. After subsequent removal of the nanoparticles with surface-immobilized BSA template molecules, a hierarchical hydrogel film with the photonic crystal structure was left. Binding the BSA target analyte in this film caused a shift in the Bragg diffraction. This shift allowed determining the BSA in an impressively wide linear dynamic concentration range (Table 9.1).

9.3.1.4 Imprinting of Immobilized Proteins

Deposition of a thin MIP film from bulk solution is not very efficient in terms of the protein rational use. This is because most of the protein molecules stay in solution, thus not participating in imprinting. Proteins that are less available and, therefore, more expensive should be imprinted more efficiently. The most adequate procedure serving that purpose seems to be protein immobilization on the substrate surface before polymer deposition. This procedure ensures that all protein molecules participate in formation of imprinted molecular cavities. Moreover, it allows fabricating films as thin as the protein molecules themselves, with only one layer of molecular cavities placed in close proximity of the chemosensor surface. Therefore, the response signal of an MIP chemosensor prepared that way is high. Desirably, a precise control is needed over the deposited film thickness. This is to enable depositing of a film just slightly thinner than the diameter of the imprinted protein. That way the tops of the imprinted cavities are left open (Scheme 9.2B). For example, the surface of a gold screen printed electrode (Au-SPE) was modified with the succinimydyl ester groups.[51] Then, the electrode was immersed in the Myo template solution for template immobilization. After that, the acrylamide polymer was deposited on the electrode surface free of Myo molecules by free-radical polymerization. For this polymerization, 2-acrylamido-2-methyl-1-propane sulfonic acid sodium salt (AMPSA) and 2-aminoethyl methacrylate hydrochloride (AEMA) served as charged functional monomers. The AFM images, taken after template extraction, showed small, ~5 nm diameter, holes in the polymer film. These holes were absent in the NIP film, thus directly indicating formation of the imprinted molecular cavities in the MIP. Moreover, the Au-SPE chip modified with cysteamine[52] or a film of PVC-COOH[53] was used for Myo template immobilization with such activating agents as glutaraldehyde[52] and 1-ethyl-3-(3-dimethylaminopropyl) carbodiimide combined with *N*-hydroxysuccinimide.[53] In other studies, Myo molecules were physically adsorbed on the Au-SPE surface[54] and immunoglobulin (IgG) molecules were covalently immobilized with a linker containing the cleavable

Table 9.1 Comparison of MIP-protein chemosensors using different transduction methods.

No.	Imprinted protein	Functional monomer	Cross-linking monomer	Transduction method	Dynamic linear concentration range	Limit of detection	Selectivity with respect to other proteins	Imprinting factor	Ref.
1.	HSA	DMAPMA	AM	PM (QCM)	60 to 160 μg mL^{-1}	—	Cyt *c*,160; Lyz, 1900; Myo, 30	1.43	17
2.	HSA	Phenyltrimethoxysilane, methyltrimethoxysilane	Tetraethyl orthosilicate (TEOS)	PM (EQCM)	1 ng mL^{-1} to 500 ng mL^{-1}	500 pg mL^{-1}	Hb, 3.5; HRP, 4.9; trypsin, 5; egg yolk, 9.3; egg white, 8.8	—	20
				EIS (EQCM)	2 ng mL^{-1} to 1 μg mL^{-1}	1 ng mL^{-1}	Hb, 3.8; HRP, 5; trypsin, 5.6; egg yolk, 9.8; egg white, 9.2	—	
3.	Trypsin	AM, AA, 1-vinyl-2-pyrrolidone	Ethylene glycol dimethacrylate, divinylbenzene, *N*,*N*-9-(1,2-dihydroxyethylene) bisacrylamide	PM (QCM)	100 ng mL^{-1} to 2 mg mL^{-1}	—	Lyz, Pepsin	—	18
4.	Hev 1b G	*N*,*N*-Methylene bisacrylamide, *N*,*N*-(1,2-dihydroxyethylene) bisacrylamide	AM, MAA, 1-vinyl-2-pyrrolidone	PM (QCM)	10 ng mL^{-1} to 1.5 μg mL^{-1}	1 ng mL^{-1}	Lyz, BSA, Ova, Hev b1 L, Hev b2 L, Hev b3 L	—	21
5.	Hev b1 G	*N*,*N*′-1,2-(Dihydroxy ethylene) bisacrylamide	MAA, vinylpyrrolidone	PM (QCM)	10 to 900 ng mL^{-1}	—	Lyz, BSA, Ova, Hev b1 L, Hev b2 L, Hev b3 L	—	22
6.	BHb	AA		PM (QCM)	—	—	Trypsin, Lyz, BSA	—	19
		N-hydroxymethylacrylamide			—	—	BSA	—	
		N-isopropylacrylamide			—	—	BSA	—	

7.	α-Amylase				1 ng mL^{-1} to 1 μg mL^{-1}	—	No selectivity	2.28	
	Lipase	EVAL		PM (QCM)	100 pg mL^{-1} to 2 μg mL^{-1}	—	No selectivity	2.13	23
	Lyz				100 pg mL^{-1} to 10 ng mL^{-1}	—	Only MIP-Lyz showed moderate selectivity to other two proteins	2.47	
8.	BHb	AM	*N*,*N*-Methylene-bisacrylamide	DPV (GCE)	0.1 to 1 mg mL^{-1}	—	BSA, Cyt *c*, GOx, Ova	—	24
9.	BSA	3-Morpholine propionate acrylate	1,3-Diacryloyl urea	DPV (MWCNT ceramic electrode)	1.99 to 30.91 ng mL^{-1}	0.42 ng mL^{-1}	Insulin, Hb, HSA, Ova, Lyz, glutamic acid, cysteine, tryptophan, ascorbic acid, histidine, dopamine	—	33
10.	Hb	Photosensitive copolymer composed of poly (γ-glutamic acid) grafted with 7-amino-4-methylcoumrin		DPV (GCE)	5 to 100 μg mL^{-1}	—	Lyz, HRP, BSA, Ova, papain	—	35
11.	TnT	*o*-Phenylenediamine		DPV (Au disk electrode)	9 to 800 pg mL^{-1}	9 pg mL^{-1}	—	—	25
12.	MSME	Caffeic acid		SWV	0.5 to 100 ng mL^{-1}	0.12 ng mL^{-1}	—	—	26
13.	Cyt *c*	Scopoletin		CV (Au electrode modified with mercaptoundecanoic acid)	1 to 4 μg mL^{-1}	—	BSA, Lyz, Myo	2	36

(continued)

Table 9.1 *(continued)*

No.	Imprinted protein	Functional monomer	Cross-linking monomer	Transduction method	Dynamic linear concentration range	Limit of detection	Selectivity with respect to other proteins	Imprinting factor	Ref.
14.	Hb	*N*,*N*′-Methylene bisacrylamide	AM	DPV (*p*-hydroxy-thiophenol and 2-bromoisobutyryl bromide modified Au electrode)	100 pg mL^{-1} to 1 mg mL^{-1}	78 fg mL^{-1}	Lyz, 5.40; BSA, 6.88; HSA, 32.4; IgG, 34.92	—	27
15.	HSA	2,2′-Bithiophene-5-carboxylic acid, *p*-bis(2,2′-bithien-5-yl) methyl alanine	5,5′,5″-Methane-triyltris(2,2′-bithiophene)	DPV (Au disk electrode)	0.8 to 20 µg mL^{-1}	16.6 ng mL^{-1}	—	20.4	29
				EIS (Au disk electrode)	4 to 80 µg mL^{-1}	800 ng mL^{-1}	Creatinine, glucose, urea, uric acid, Cyt *c*, Myo - no response; Lyz, 3.6	26.8	
16.	BHb	PB		DPV (hierarchical dendritic gold microstructures modified GCE)	0.1 to 500 µg mL^{-1}	0.05 µg mL^{-1}	Cyt *c*, Ova, BSA, Lyz	7.6	30
17.	Cy3-NHS labeled MBP	*N*-Isopropyl-acrylamide, AM, MAA, *N*-[3-(dimethyl-amino) propyl] methacrylamide	*N*,*N*-Methylenebis (acrylamide)	Fluorescence	20 to 1000 µg mL^{-1}	—	BSA, 3; Ova, 3	3.3	38

18.	Lyz	AM, 4-[2-(*N*-methacrylamido) ethylaminomethyl]benzoic acid	*N*,*N*′-Methylene-bisacrylamide	Fluorescence (fluorescein isothiocyanate labeled MIP film)	0.1 to 0.5 μM	—	Myo, lactoalbumin - no response, Cyt *c* - opposite response	4.4	40
				SPR	0.1 to 0.5 μM	—	Myo, lactoalbumin - no response; Cyt *c*, 2.5	2.4	
19.	Cyt *c*	AM, tailor designed cleavable monomer	*N*,*N*′-Methylene-bisacrylamide	Fluorescence (4-(*N*,*N*-dimethylaminosulfonyl)-7-fluoro-2,1,3-benzoxadiazole labeled MIP film)	0.2 to 0.8 μM	—	Hb, HSA, IgG, RNase A - opposite response	—	41
				SPR	0.1 to 0.6 μM	—	HSA, RNase A - no response, Hb - moderate, IgG - no selectivity	—	
20.	HSA	*O*-Acryl-L-4-hydroxyproline amide of dansylethylendiamine, AM	*N*,*N*′-Methylene-bisacrylamide	Fluorescence (MIP film labeled with dansyl-ethylene = diamine)	0.5 to 5 μM	—	Lyz, Av, chymotrypsin, BSA - high selectivity	—	44
				SPR	0.5 to 2 μM	—	Lyz, chymotrypsin - high; BSA, Av - moderate	Very low	
21.	Lyz	AA	*N*,*N*′-Methylene-bisacrylamide	SPR	—	—	Cyt *c*, Myo, RNase A, lactoalbumin, 9.8	—	45
22.	HBsAb	MA		SPR	0 to 120 mU mL^{-1}	—	—	—	46
23.	Myo	Hydroxyethyl methacrylate-*N*-methacryloyl-L-tryptophan methyl ester		SPR	0.1 to 1.0 μg mL^{-1}	was 26 ng mL^{-1}	Lyz, 6.7; Cyt *c*, 10.7; BSA, 39.1	Not determined	47

(*continued*)

Table 9.1 (*continued*)

No.	Imprinted protein	Functional monomer	Cross-linking monomer	Transduction method	Dynamic linear concentration range	Limit of detection	Selectivity with respect to other proteins	Imprinting factor	Ref.
24.	BSA	3-Aminophenyl boronic acid	Aniline	SPR	20 to 800 µg mL^{-1}	20 µg mL^{-1}	Lyz, 1.6; Hb, 1.8	2.2	48
25.	Av	Styrene sulfonate	3,4-Ethylenedioxythiophene	SPR (micro-patterned surface-imprinted MIP film fabricated photolithographically)	1 µg mL^{-1} to 10 µg mL^{-1}	—	Streptavidin, neutravidin, extravidin, BSA, and Lyz	4	49
26.	BSA	AM		Shift in Bragg diffraction of the MIP photonic crystal	1 ng mL^{-1} to 10 mg mL^{-1}	—	EA, and Lyz – no response	—	50

disulfide bond on the Au-QCR surface.[55] Then, coating these proteins with dozen-nanometre-thick polymer films and subsequent protein removal resulted in fabrication of selective MIP chemosensors.

Modification of a transducer surface with low-molecular-weight ligands capable of specific interaction with only one part of the target protein, *e.g.*, enzyme substrates or receptor agonists, is an important step forward in development of surface imprinting *via* protein immobilization. This immobilization not only enforces location of molecules of the protein template close to the surface but also the same predefined arrangement of all these molecules. In effect, the shape of the resulting cavities is, advantageously, identical (Scheme 9.2C). Moreover, these ligands may equally well serve as recognition sites located inside imprinted cavities, thus enhancing the cavity affinity to molecules of the target protein. An illustrative example of this approach is the glutathione *S*-transferase (GTS) immobilization with glutathione on an SPR chip.[56] Towards that, a self-assembled monolayer (SAM) comprising maleimide groups and bromoisobutyryl groups was anchored to a gold surface of the chip. Then, the maleimide groups were reacted with thiol groups of glutathione. Finally, MIP thin films of different thickness were deposited by controlled/living free-radical polymerization. Selectivity of the resulting chemosensor with respect to HSA and fibrinogen was highest for the 14 nm thick film. In another study, a thiolated DNA aptamer was used for site specific immobilization of the PSA antigen, a cancer biomarker.[57] Next, a 10 nm thick MIP film was deposited *via* potentiodynamic electropolymerization of dopamine. After subsequent PSA template removal, this chemosensor was applied for PSA determination using EIS within the linear dynamic concentration range of 100 pg mL^{-1} to 100 ng mL^{-1} with LOD ($S/N = 3$) equal to 1 pg mL^{-1}. Moreover, the propidium lipolic acid amide[37] or thiolated oligoethyleneglycol mannose conjugate[58] self-assembled monolayers were used for directed immobilization of AChE and concanavalin V, respectively. After MIP film deposition and template removal, propidium and mannose groups served as additional recognizing sites present in imprinted molecular cavities, that way providing very high selectivity of the resulting MIP based chemosensors.

When enzyme substrates are immobilized on the transducer surface, enzymes are then attached on top with their active sites. MIPs prepared that way demonstrate one additional property, namely, they can operate as enzyme inhibitors.[59]

In some instances, however, immobilization of a protein template on the transducer surface before polymerization is disadvantageous because a very precise thickness control of the deposited MIP film is then necessary. That is, deposition of a too thick film would result in entrapping molecules of the protein template deeply inside the MIP matrix with no possibility of their subsequent removal. Back-side surface imprinting may easily solve this problem (Scheme 9.4).

For example, oxidized low-density lipoprotein (oxLDL), a protein template, was covalently immobilized on the drop-cast carboxylated poly(vinyl

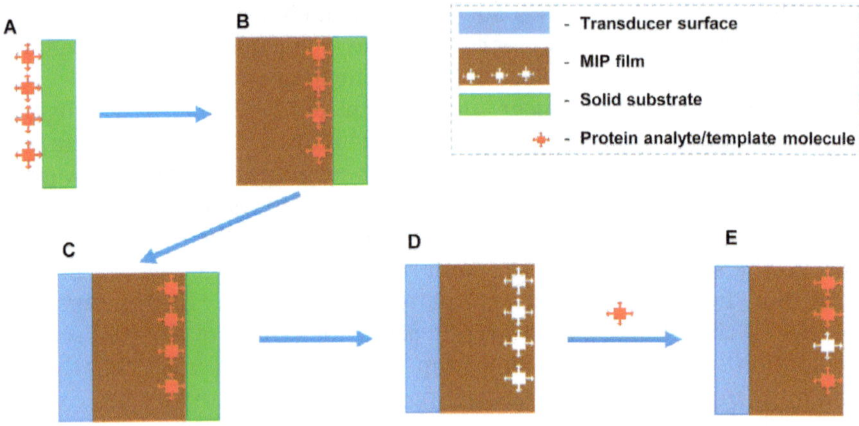

Scheme 9.4 Consecutive steps of back-side imprinting: (A) immobilization of template molecules on the solid support surface, (B) deposition of the thick polymer film, (C) attachment of the MIP film to the transducer surface, (D) detachment of the solid support combined with template extraction, and (E) selective determination of the target protein.

chloride), PCV-COOH, film coated Au-SPE.[60] Then, this immobilized oxLDL was fully coated with a thick film of polymethacrylate using methacrylic acid (MAA) and ethylene glycol dimethacrylate (EGDMA) as the functional and cross-linking monomer, respectively. Afterwards, the top side of the polymethacrylate film was covalently attached to another Au-SPE. Next, the bottom Au-SPE was detached, and then the protein was extracted from the film, thus emptying molecular cavities located exactly on the surface of the prepared MIP film. The chemosensor fabricated that way determined LDL in fetal calf serum (FCS). In another example of a back-side surface imprinting, BSA,[61,62] PSA,[63] and procalcitonin (PCT)[64] were immobilized on glass slides to form "protein stamps". Then, these stamps were pressed to microcontact Au electrodes[61–63] and an SPR[64] chip with a drop of solution for polymerization between them. After polymerization, these "stamps" were stripped off resulting in formation of MIP films with all cavities located on the surface.

Recently, back-side surface imprinting was used for MIP deposition on methacrylated glass slides in a form of 1.25 mm diameter dots.[65] For that, a polydimethylsiloxane (PDMS) plate with 35 drilled holes of 1.25 mm in diameter was pressed against a glass slide. In the next step, biotinylated polystyrene@poly(styrene-methacrylic acid) NPs were immobilized on the surface of a glass slide that was not coated with PDMS. Afterwards, biotinylated Cyt *c*, the protein template, was attached to the immobilized NPs, and PDMS was then removed to result in a "protein stamp". Finally, the pre-polymerization solution was placed between the methacrylated glass substrate and the "protein stamp". Subsequent photopolymerization, and then removal of the "stamp" resulted in MIP-(Cyt *c*) preparation in a form of

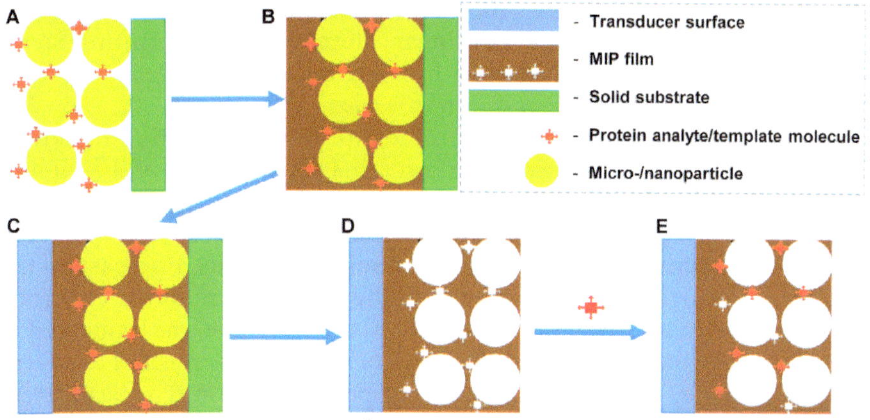

Scheme 9.5 Consecutive steps of back-side imprinting combined with surface development: (A) deposition of a micro-/nanoparticles on the solid support surface combined with template protein immobilization, (B) deposition of a thick polymer film, (C) attachment of the MIP film to the transducer surface, (D) detachment of the solid support combined with extraction of the template and micro-/nanoparticle, and (E) selective determination of the target protein.

35 dots on the glass slide surface. Then, this slide was used for determination of fluorescently labeled Cyt *c*. Devising this protein chemosensor opens up a general route toward high-throughput screening of multiprotein samples because each MIP dot can easily be imprinted with a different protein template.

Moreover, the back-side surface imprinting was combined with surface development techniques (Scheme 9.5). For instance, a macroporous poly(-methacrylate) film imprinted with Lyz was deposited on the Au-QCR surface.[66] For that purpose, a clean round glass cover slide was coated with a thick optically visible layer of NPs of $CaCO_3$ by vertical dipping it in and taking it out several times from the solution of NPs. Next, it was spin coated with Lyz, and afterwards a drop of the solution for polymerization was dispensed on the surface of the $CaCO_3$ NPs-Lyz coated glass slide. Once the layer was soaked with this suspension, a bare Au-QCR was placed on top. Then, the UV light initiated polymerization was performed, and the bottom glass slide was subsequently detached. Finally, the NPs of $CaCO_3$ and protein templates were removed with 2% HCl. Comparison of the PM at QCM response to Lyz and Myo of the MIP-Lyz film coated Au-QCRs with that of the control NIP film coated Au-QCR indicated that well defined selective molecular cavities were formed. Moreover, comparison of a macroporous MIP film with that nonporous showed that the PM signal increased two to four times depending on the Lyz concentration. The same strategy was adopted for ribonuclease A (RNase A) imprinting.[67] For that purpose, *N*-methacryloyl-histidine was used as the functional monomer. Unfortunately, the devised PM MIP-chemosensor responded to the target protein in a microgram per milliliter concentration range only. However, the calibration plot in this range was not linear. Moreover, the imprinting factor (IF < 2) and selectivity with respect to Lyz were rather moderate. Apparently, imprinting was not very successful in this case.

A similar way to circumvent the problem of the polymer film thickness control during surface imprinting of proteins involved immobilization of the BHb template on $-NH_2$ group modified silica beads of several micrometres in diameter.[68] The $-NH_2$ groups on the bead surface were then activated with glutaraldehyde. Next, the bead suspension was drop-cast on the Au electrode surface, and subsequently the resulting modified electrode was immersed in a BHb solution. After that, the electrode was immersed in an aqueous pyrrole solution, and polypyrrole was then deposited in spaces between beads by potentiodynamic electropolymerization. The beads were then dissolved in 4% HF. Finally, the BHb template was extracted with an oxalic acid solution from the MIP film prepared. This extraction resulted in a macroporous MIP film with empty imprinted cavities located exactly on the surface of pore walls. The DPV response to BHb of this MIP chemosensor was linear in a very broad linear dynamic concentration range of 100 fg mL^{-1} to 1 μg mL^{-1}. The IF exceeded 5. Moreover, selectivity with respect to BSA and egg albumin (EA) was high. However, selectivity to Lyz was lower, equaling 2.5, because its molecule is much smaller than that of BSA. Although electrode surface enlargement should dramatically improve analytical properties of a chemosensor, these properties were only slightly improved in this case compared to those of the previously devised similar chemosensor featuring a nonporous MIP film.[69]

9.3.1.5 *Protein Imprinting in 2-D MIP Films*

Template imprinting in 2-D SAMs is the most extreme example of surface imprinting (Scheme 9.2D). Advantages of imprinted SAMs include fast response, high concentration of molecular cavities in the resulting MIP film, and location of these cavities directly on the transducer surface. For this method of preparation of MIPs, SAMs of hydroxyl-terminated alkane thiol were deposited on gold electrodes to imprint several proteins and viruses.[70,71] The alkane thiols are insoluble in water, a natural environment of proteins. Therefore, these thiols were first dissolved in acetic acid, and the resulting mixture was then mixed with a water solution of the template protein at the 1-to-19 volume ratio. Subsequently, gold-coated silicon chips were immersed in the final solution to allow the hydroxyl-terminated alkane thiol and protein molecules to assemble on the Au surface. The resulting SAM coated Au electrodes were applied for potentiometric titration of the target proteins. Moreover, hyaluronan-linked protein 1 (HAPLN1), a biomarker of malignant pleural mesotheliomas, was imprinted in SAM containing hydroxyl-terminated alkane thiols on the gold electrode surface.[72] Because of the high cost of the target HAPLN1 protein, all imprinting conditions were tested with much cheaper BSA, and then those resulting in the best performance were used for HAPLN1 imprinting. In potentiometric titration, this SAM chemosensor responded to HAPLN1 in a wide linear dynamic concentration range (Table 9.2).

Another example brings a SAM chemosensor for the PSA glycoprotein.[140] For preparation of this chemosensor, the surface of an SPR chip was coated with an acrylamide–alkyne cysteine derivative. Then, a covalent adduct of PSA and 3-acrylamidophenylboronic acid was deposited on this modified surface by polymerization of the acrylamide moiety. Finally, the chip surface left uncoated with PSA was blocked. For that, azide-terminated heptaethylene glycol was used forming a SAM *via* Cu catalyzed [2 + 3] cycloaddition. The linear dynamic concentration range of the resulting chemosensor extended from 65 to 650 nM PSA.

9.3.1.6 MIP Films Deposited on Nanoparticle and Nanoelectrode Surfaces and MIP Nanoparticles

Nanomaterials, such as metal NPs, graphene, or MWCNTs, allow for enhancing of the generated chemosensor signal, thus significantly improving sensitivity and detectability.[107,108]

The easiest way of triggering this enhancement is to immobilize these nanostructures on the transducer surface before depositing an MIP film (Scheme 9.6B). One such example involves incorporation of MWCNTs in a C-reactive protein (CRP) imprinted polymer.[87] For that purpose, 2-acryl amidoethyldihydrogen phosphate, a monomer that behaves as a phosphotidyl choline natural binder analogue, was synthetized and used as the functional monomer. The acrylamide MIP-CRP was deposited by thermally inducted polymerization on the SPE material surface, which was first spin-coated with a film of MWCNTs. The engineered functional monomer allowed for a very high selectivity (over 15) with respect to chosen proteins including BSA, insulin, Hb, and Lyz. This chemosensor was used for determination of CRP in blood serum samples of four different patients. Moreover, in another study, QDs were covalently attached with an L-cystein linker to COCl-modified MWCNTs, which were spin-coated on the pencil graphite electrode (PGE).[88] Then, the diacryl urea and *N*-methacryloyl glutamic acid were co-polymerized, in the presence of Hb, on the surface of these modified electrodes. The heme iron cation of Hb allowed detection of Hb binding in the differential pulse cathodic stripping voltammetry (DPCSV) study. This chemosensor did not respond to such interferences as BSA, tryptophan, CRP, glutamic acid, dopamine, cysteine, ascorbic acid, and insulin, each in the concentration of 155 ng mL^{-1}. It was also used for determination of Hb in blood serum samples of four different patients. In another study, MWCNTs were decorated with sodium *N*,*N*-diethyldithiocarbamate and MnO_2 nanoparticles, and then deposited on a PGE.[97] A nanocomposite prepared that way served as an iniferter, *i.e.*, a chemical compound acting as the initiator, transfer agent, and terminator in controlled free-radical polymerization. Moreover, it served as a support for controlled radical

Table 9.2 Chosen examples of new and more sophisticated methods of MIP-protein chemosensor devising.

No.	Imprinted protein	Monomers	Details of imprinting	Transduction method	Dynamic linear concentration range	Limit of detection	Selectivity	Imprinting factor	Ref.
1.	Myo	AM, AMPSA, AEMA	Protein immobilized on the SPE material surface before MIP deposition	EIS	Down to 3.5 μg mL^{-1}	1.5 μg mL^{-1}	—	—	51
				SWV	Down to 0.28 μg mL^{-1}	0.28 μg mL^{-1}	TnT, BSA, urea	—	
2.	Myo	AM and *N*,*N*′-methylene = bisacrylamide	Protein immobilized on the SPE material surface before MIP deposition	EIS	9 to 36 μg mL^{-1}	2.25 μg mL^{-1}		—	
				SWV	9 to 36 μg mL^{-1}	8.5 μg mL^{-1}	Hb, BSA, sodium glutamate, urea	—	52
				Potentiometry	350 ng mL^{-1} to 24 μg mL^{-1}	130 ng mL^{-1}		—	
3.	Myo	2-Amino phenol	Protein immobilized on the SPE material surface before MIP deposition	EIS	4 to 53 μmg mL^{-1}	3.5 μg mL^{-1}	—	—	54
				SWV	2.2 to 53 μmg mL^{-1}	0.8 μg mL^{-1}	TnT, creatine kinase	—	
4.	Myo	AM, *N*,*N*-methylenebis = acrylamide	Protein immobilized on the coated with PVC-COOH film SPE material surface before MIP deposition	EIS	0.852 to 4.26 μg mL^{-1}	2.25 μg mL^{-1}	TnT, 14.3; BSA, 9; urea, 50	—	53
				SWV	1.1 to 2.98 μg mL^{-1}	—		—	
5.	PSA	Dopamine	Thiolated aptamer DNA used for protein immobilization on the electrode surface	EIS	100 pg mL^{-1} to 100 ng mL^{-1}	1 pg mL^{-1}	Human kallikrein 2, HSA	—	57
6.	oxLDL	MAA, ethylene glycol dimethacrylate	Backside-surface imprinting	EIS	2.5 to 12.5 mg mL^{-1}	—	Myo, Hb	—	60

7.	PCT	HEMA, EGDMA	Backside-surface imprinting	SPR	20 to 1000 ng mL^{-1}	3 ng mL^{-1}	Cyt *c*, 11.3; Myo, 2.17; HSA, 3.96	6.5	64
8.	Lyz	AM	Surface imprinted macroporous MIP film	PM (QCM)	10 to 500 µg mL^{-1}	—	Myo	—	66
9.	BHb	Pyrrole	Surface imprinted macroporous MIP film	DPV	100 fg mL^{-1} to 1 µg mL^{-1}	—	BSA and EA - high, Lyz - moderate	5	68
10.	HAPLN1	Hydroxyl-terminated alkane thiols	SAM	Potentiometric titration	1 pM to 1 nM	—	—	—	72
11.	Amylase			Fluorescence quenching	0.5 to 3 mg mL^{-1}	—	Lipase, Lyz	3.79	
12.	Lipase	EVAL	MIP film coated QDs		0.5 to 3 mg mL^{-1}	—	Amylase, Lyz	2.57	73
13.	Lyz				0.1 to 10 µg mL^{-1}	—	Amylase, lipase	2.14	
14.	—	—	MIP film coated QDs	Fluorescence quenching	10 nM to 10 µM	—	—	—	74–77
15.	BHb	3-Amino propyltriethoxysilicate (APTES), tetraethyl orthosilicate (TEOS)	Imprinted silica coated QDs	Fluorescence quenching	20 nM to 2.1 µM	9.4 nM	Ova, BSA, Lyz	4	78
16.	BHb	AM, MAA, γ-methacryl oxypropyl trimethoxy silane.	QDs coated with an imprinted hybrid of organic/inorganic film	Fluorescence quenching	0.1 to 100 µM	32 nM	BSA, Lyz, Ova - very high, BHb - moderate	3.1	79
17.	Cyt *c*	TEOS, APTES	Upconversion NPs coated with an MIP film	Fluorescence quenching	1 to 24 µM	0.73 µM	BSA, Lyz	3.19	80

(*continued*)

Table 9.2 (*continued*)

No.	Imprinted protein	Monomers	Details of imprinting	Transduction method	Dynamic linear concentration range	Limit of detection	Selectivity	Imprinting factor	Ref.
18.	BHb	*N*-Isopropyl acrylamide, *N*,*N*-methylenebisacrylamide	Upconversion NPs coated with a MOF and a thermoresponsive MIP film	Fluorescence quenching	100 to 600 μg mL^{-1}	62 μg mL^{-1}	BSA, Cyt *c*	Very low	81
19.	Lyz	AM, *N*,*N*′-dimethyl acrylamide	MWCNTs coated with an MIP film	Quenching of fluorescence of an MIP film labeled with fluorescein isothiocyanate	3 to 30 μM	—	Hb, BSA, Cyt *c*	3.4	82
20.	HPR	*m*-Aminophenyl – boronic acid, dopamine	Macroporous MIP film photolithographically deposited on the 96- and 384-spot plate; after binding, the target glycoprotein molecules were labeled with AgNPs	SERS	Of 1 ng mL^{-1} to 10 μg mL^{-1}	—	BSA, transferritin, RNase B, glucose	11.5	83
	AFP				Of 1 ng mL^{-1} to 10 μg mL^{-1}	—	—	—	

21.	HSA	AM, *N*,*N*-methylenebis = acrylamide	Cryogel prepared by graft copolymerization of monomers in the presence of chitosan, graphene, and ferrocene	DPV	100 pg mL^{-1} to 1 µg mL^{-1}	50 pg mL^{-1} HSA	Ascorbic acid, uric acid, urea, creatinine, BSA, 6.25	14	84
22.	cTnT	Pyrrole, pyrrole-3-carboxylic acid 3-pyrrol carboxylic acid	An SPE material coated with reduced graphene oxide, and subsequently with an MIP film	DPV	10 to 100 pg mL^{-1}	6 pg mL^{-1}	—	—	85
23.	BHb	Dopamine	Magnetic NPs coated with an MIP film	DPV (magnetic glassy carbon electrode)	0.5 to10 µg mL^{-1}	11.8 ng mL^{-1}	BSA, Lyz, HRP, Cyt *c* - very high	Very high	86
24.	CRP	AM, 2-acryl amidoethyl-dihydrogen phosphate	MWCNTs incorporated in an MIP film	DPV - SPE	0.18 to 8.51 µg mL^{-1}	0.04 µg mL^{-1}	BSA, insulin, Hb, Lyz > 15	Very high	87
25.	Hb	Diacryl urea, *N*-methacryloyl glutamic acid	QDs, covalently attached to MWCNTs, deposited on the PGE, and then coated with an MIP film	DPCSV	27.8 to 444 ng mL^{-1}	6.7 ng mL^{-1}	No response to BSA, tryptophan, CRP, glutamic acid, dopamine, cysteine, ascorbic acid, and insulin	Very high	88

(*continued*)

Table 9.2 (*continued*)

No.	Imprinted protein	Monomers	Details of imprinting	Transduction method	Dynamic linear concentration range	Limit of detection	Selectivity	Imprinting factor	Ref.
26.	BSA	Aniline	Carbon nanotubes, covalently attached to graphene, deposited on the carbon electrode, and then coated with an MIP film	DPV	100 pg mL^{-1} to 100 μg mL^{-1}	62 pg mL^{-1}	HSA, 10; BHb, 13	—	89
27.	BSA	Pyrrole	Chitosan coated magnetic NPs, covalently attached to MWCNTs, drop-cast on the carbon electrode, and then coated with an MIP film	DPV	100 pg mL^{-1} to 100 μg mL^{-1}	28 pg mL^{-1}	HSA, 10; BHb, 7.5	—	90
28.	BSA	Pyrrole	Graphene/ionic liquid/chitosan deposited on the GCE, and then coated with an MIP film	DPV	100 pg mL^{-1} to 100 μg mL^{-1}	20 pg mL^{-1}	HSA, 10.6; BHb, 14.2	7	91
29.	BHb	Dopamine	AuNPs grafted on the surface of a gold disk electrode by electroreduction of hydrogen tetrachloroaurate, and then coated with an MIP film	DPV	10 fg mL^{-1} to 10 μg mL^{-1}	—	BSA, EA Lyz	—	92

30.	Human ferritin	Phenol	Array of carbon nanotube tips serving as nano-electrodes coated with an MIP film	DPV and EIS	1 fg mL^{-1} to 100 pg mL^{-1}	—	No response to BSA, ahsFtn, neither to hsFtn	—	93
	Human papilloma virus E7 protein (type-16)				—	10 ag mL^{-1}	Human papillomavirus E6 protein (type-16) was not detected	—	
31.	CA 125	Phenol	Gold nanoelectrode ensembles coated with an MIP film	DPV	1 to 400 U mL^{-1}	0.5 U mL^{-1}	HSA	—	94
32.	BSA	*o*-Phenyl diamine, 3-aminophenylboronic acid	Gold-graphene hybrid NPs, deposited on the GCE surface, and then coated with an MIP film. Protein molecules, bound in an MIP film, were reacted with AuNPs labeled with 4-mercaptophenyl boronic acid and 6-(ferrocenyl) hexanetiol	DPV	10 fg mL^{-1} to 10 ng mL^{-1}	7.5 fg mL^{-1}	IgA, 7.8; IgB, 8.5	2.65	95
33.	BSA	TEOS, APTES	MWCNT-QD composite coated with an MIP film	Fluorescence quenching	0.5 to 3.5 μM	80 nM	BHb, Lyz	4.17	96

(*continued*)

Table 9.2 (*continued*)

No.	Imprinted protein	Monomers	Details of imprinting	Transduction method	Dynamic linear concentration range	Limit of detection	Selectivity	Imprinting factor	Ref.
34.	PSA	EGDMA	MWCNTs-QDs composite, deposited on a PGE, and then coated with an MIP film	SWASV	1 pg mL^{-1} to 35 μg mL^{-1}	25 pg mL^{-1}	NaCl, insulin, ferritin, urea, uric acid, globulin, albumin, lysine, histidine, glutamic acid, arginine, tyrosine, citric acid - very high	12.1	97
				DPASV	10 pg mL^{-1} to 60 μg mL^{-1}	3.04 pg mL^{-1}	—	—	
35.	TnT	AM, *N*,*N*′-methylenebisacrylamide	MWCNTs, coated with an MIP film, and then incorporated in a PVC membrane	Potentiometry	Down to 4.2 nmol mL^{-1}	157 ng mL^{-1}	Creatinine, sucrose, fructose, Myo, sodium glutamate, thiamine, urea	—	98
36.	Myo	4-Styrenesulfonic acid, 2-aminoethyl methacrylate hydrochloride, EGDMA	Graphite powder beads, coated with an MIP film, incorporated in a PVC membrane deposited on the SPE material surface	SWV	1 to 21 μg mL^{-1}	—	Hb, BSA, creatinine, NaCl	—	99

37.	Hb	Styrene, divinylbenzene, 2-aminoethyl methacrylate hydrochloride, and styrene sulfonic acid	Silica beads coated with an MIP film incorporated in a PVC membrane	Potentiometric titration	Down to 83.8 μg mL^{-1}	43.8 μg mL^{-1}	Myo, glucose, creatine kinase, ascorbic acid, creatinine, BSA	—	100
38.	PSA	AM, vinylbenzyl trimethylammonium chloride, vinyl benzoate	Graphene sheets, coated with an MIP film, incorporated in a PVC membrane	Potentiometric titration	2 to 89 ng mL^{-1}	2 ng mL^{-1}	Creatinine, urea, glucose, Hb, BSA	—	101
39.	BHb	Dopamine	Graphene sheets, coated with an MIP film, drop-cast on the GCE surface	DPV	1 pg mL^{-1} to 100 μg mL^{-1}	200 fg mL^{-1}	BSA, 10; EA, 8.5; papain, 4.5; Lyz, 2.5	7.2	102
40.	HRP	Aniline, APBA, AA	Fe_3O_4@Au NFs, coated with an MIP film, immobilized on GCE with chitosan	DPV	10 to 300 μg mL^{-1}	5 μg mL^{-1}	OVA, 2.74; phytohaemagglutinin E, 6.63; BHb, 4.36; BSA, 42.06; Cyt *c* - 1.59	7.8	103
41.	Lyz	Ethylene glycol dimethacrylate-*N*-methacryloyl-L-histidine methyl ester	MIP NPs attached to the 3-mercaptopropene modified surface of the SPR chip and Au-QCR	SPR	21 nM to 1.4 μM	—	Cyt *c*, 10.94; albumin – no response	4	104
				PM at QCM	14 nM to 105 μM	—	—	—	105
42.	IgG	*N*-Methacryl-L-histidine methyl ester	An MIP imprinted with the F_{ab} fragment of IgG	SPR	20 μg mL^{-1} to 1 mg mL^{-1}	56 ng mL^{-1}	BSA, 21	Not tested	106

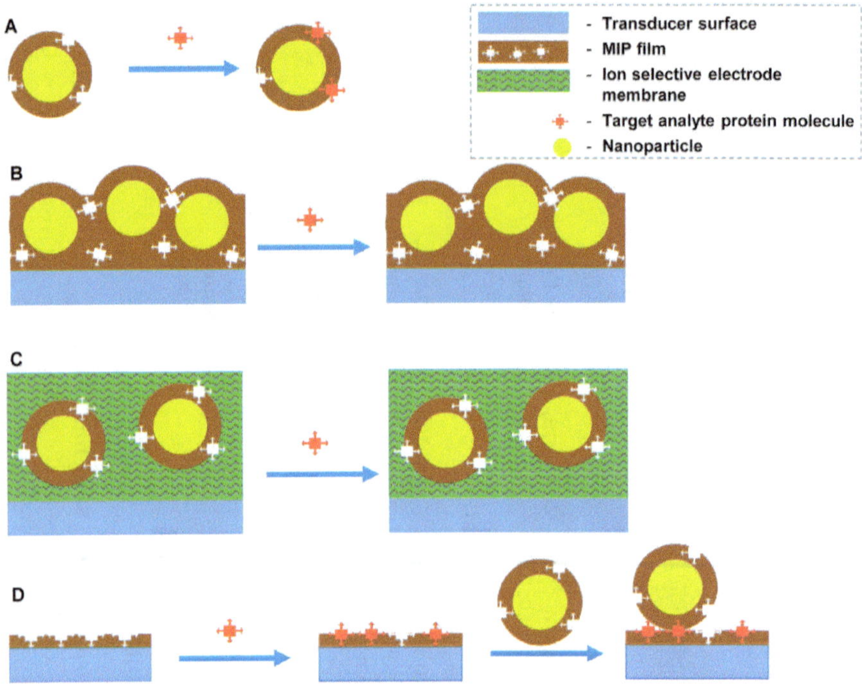

Scheme 9.6 Four different strategies of combining MIPs with nanotechnology and nanoparticles: (A) deposition of MIP films on the surface of NPs, (B) embedding NPs in the MIP film, (C) embedding MIP coated NPs in ion selective electrode membranes, and (D) enhancing an MIP-chemosensor signal with the MIP film coated NPs with boronic acid moieties present on their surface or NPs capable of selective protein binding.

polymerization of the PSA antigen resulting in an MIP-PSA. The electrochemical sensor devised that way was used for determination of PSA with a LOD of 25 pg L^{-1} and 3.04 pg L^{-1} in the square wave anodic stripping voltammetry (SWASV) and differential pulse anodic stripping voltammetric (DPASV) determinations, respectively.

Moreover, MWCNTs modified with vinyl groups were coated with an insulin imprinted MIP film.[109] For that, the (*p*-acryloylaminophenyl)-[(4-aminophenyl)-diethylammonium]-ethylphosphate functional monomer was copolymerized with EGDMA *via* free-radical polymerization. Finally, the PGEs were spin-coated with the MWCNTs-(MIP-insulin) adduct films. Subsequently, the insulin template was extracted by immersing of these modified PGEs in the AcOH–ethanol (9 : 1, *v* : *v*) solution. The resulting electrodes were used for DPASV determination of insulin in real human blood serum samples. Moreover, this electroanalytical technique was combined with molecularly imprinted polymer micro-solid phase extraction (MIP-MSPE) on silica nanofibers coated with a similar MIP film.[110] This combination appeared suitable for achieving sufficient sensitivity for ultra-trace insulin determinations in human blood serum samples without any non-specific (false-positive) contributions.

Graphene is another support nanomaterial used for fabrication of MIP chemosensors. In one example, the carbon electrode surface was drop-coated with graphene.[89] Then, CNTs were covalently attached to the graphene surface by diazonium salt coupling. Finally, the BSA imprinted polyaniline film was deposited by potentiodynamic electropolymerization on the carbon electrode surface modified that way. The dynamic linear concentration range of this chemosensor was very broad, covering 100 pg mL^{-1} to 100 μg mL^{-1} BSA, with a LOD of 62 pg mL^{-1}. The response of the chemosensor to similar proteins, *e.g.*, HSA and BHb, was 10 and 13 times, respectively, lower than that to BSA at the same concentration. The protein response signal was much higher when a mesoporous MIP film was combined with graphene. For that, the MIP-HSA cryogel was prepared by graft copolymerization of acrylamide with *N*,*N*-methylenebisacrylamide in the presence of chitosan, graphene, and ferrocene.[84] The sub-zero temperature polymerization caused water forming ice crystals, which resulted in formation of a macroporous structure when the crystals were thawed. The ferrocene entrapped in the cryogel matrix served as an internal redox probe in the DPV determination of HSA. Moreover, the DPV current signal was enhanced because of the presence of graphene, which was also entrapped in the cryogel. Sensitivity of the chemosensor was 1.3 times higher if graphene was evenly spread over the whole cryogel volume rather than it formed a layer at the polymer-(gold substrate) interface. Sensitivity of this chemosensor was as high as 1.36 μA for the logarithm unit of the HSA concentration. Furthermore, a polydopamine film imprinted with BHb was deposited on the graphene surface.[102] In this case, graphene oxide served as both the solid support and the oxidant. The GCE surface was drop-coated with NPs of the MIP-graphene composite prepared that way. The response to BHb of the resulting modified electrode was linear in the range of 100 μg mL^{-1} to 1 pg mL^{-1} and the IF was 7.2.

In another study, an SPE material was coated with reduced graphene oxide, and subsequently with the cardiac troponin T (cTnT) imprinted polypyrrole MIP film.[85] 3-Pyrrolcarboxylic acid was used as the functional monomer in this MIP preparation. The usefulness of this chemosensor was tested with BSA determination in the BSA spiked human serum samples. In other research, chitosan coated magnetic NPs were covalently attached to carboxyl-functionalized MWCNTs.[90] The resulting composite NPs were drop-cast on the carbon electrode and, finally, a BSA imprinted polypyrrole film was deposited by electropolymerization on top of the NP film deposited. The analytical properties of this chemosensor were quite similar to those of the former. The dynamic linear concentration range and the limit of BSA detection was 100 pg mL^{-1} to 100 μg mL^{-1} BSA and 28 pg mL^{-1}, respectively. The response of the chemosensor to HSA and BHb was 10 and 7.5 times, respectively, lower than that to BSA at the same concentration. Other research explored deposition of a graphene/(ionic liquid)/chitosan composite on the GCE surface to improve analytical performance and biocompatibility of the resulting chemosensor.[91] Finally, the electrode surface was coated with polypyrrole by potentiodynamic electropolymerization in the BSA presence. This chemosensor allowed selective BSA determining in a very broad concentration range of 100 pg mL^{-1} to 100 μg mL^{-1} with a LOD of 20 pg mL^{-1} BSA.

Moreover, noble metal NPs appeared useful for signal enhancement in electrochemical MIP chemosensors. In one example, AuNPs were grafted on the surface of a gold disk electrode by electroreduction of hydrogen tetrachloroaurate.[92] Then, this electrode was coated with a 10 nm thick MIP film by self-polymerization of dopamine in the presence of BHb under mild basic conditions (pH = 8.5). The response to BHb of the resulting chemosensor was linear in a very wide concentration range of 10 fg mL^{-1} to 10 μg mL^{-1}.

The most impressive example of signal amplification involves carbon nanotube (CNT) nanoelectrodes.[93] These nanotubes were embedded in a supporting polymer to form an array, and their tips were then coated with 13 nm thick polyphenol films imprinted with human ferritin. This array allowed for DPV and EIS determination of human ferritin in an extraordinarily wide concentration range of 1 fg mL^{-1} to 100 pg mL^{-1}. Moreover, this array responded neither to BSA, horse apoferritin (ahsFtn), nor horse ferritin (hsFtn) at concentrations reaching 100 ng mL^{-1}. To demonstrate other capabilities, a similar array of CNTs nanoelectrodes was coated with polyphenol films imprinted with human papillomavirus E7 protein (type-16). The protein LOD of this array was 10 ag mL^{-1} and a very similar interference, human papillomavirus E6 protein (type-16) was not detected. Moreover, a nanoelectrode CNTs array, prepared in the same way, and then coated with Ca^{2+}–calmodulin complex imprinted polyphenol film, was used to study conformational changes in the calmodulin protein. Ca^{2+} coplexation with calmodulin caused significant conformational changes in this protein. Therefore, the complex could be detected by the MIP-film coated nanoelectrode array even in a large excess of free calmodulin present in the test solution.

In another study, an array of gold nanoelectrode ensembles (GNEEs) was coated with molecularly imprinted polyphenol films.[94] This array was prepared by electroless gold deposition in the 50 nm diameter pores of polycarbonate particle track-etched membranes. Next, the polycarbonate was dissolved, and the resulting gold nanoelectrodes were then coated with an MIP film by potentiodynamic electropolymerization of phenol in the presence of CA 125 protein, a cancer biomarker. This GNEE array was successfully applied for determination of CA 125 in human plasma samples.

Apparently, hybrid graphene–gold nanoparticles, Gr-AuNPs, are among the state of the art in the use of NPs to enhance the MIP electrode sensitivity.[95] These Gr-AuNPs were prepared by reduction of $HAuCl_4$ in the presence of graphene oxide. These nanoparticles were then drop cast on the GCE surface. The Gr-AuNPs operated as "electron antennas" efficiently channeling electrons between the electrode substrate and the electroactive species. The GCE modified that way was then coated with the MIP-BSA film by potentiodynamic electropolymerization of *o*-phenyldiamine in the presence of BSA and the 3-aminophenyl boronic acid as the functional monomer. This monomer reversibly formed strong covalent bonds with the carbohydrate moieties of the BSA glycoprotein. This preparation procedure resulted in an MIP of strong affinity of the BSA molecule to the molecular cavity and, hence, high selectivity. Moreover, the 6-ferrocenylhexanethiol modified AuNPs were used to enhance the DPV signal (Scheme 9.6D). The GCE-(Gr-AuNPs)-MIP was consecutively immersed in solutions of BSA, 4-mercaptophenylboronic acid, AuNPs, 6-ferrocenylhexanethiol, and finally in 0.1 M $NaClO_4$ for DPV determination. A complete determination procedure was rather complicated and time consuming. However, the results were astonishingly positive. That is, this electrode linearly responded to the logarithm of the BSA concentration in the range of 10 fg mL^{-1} to 10 ng mL^{-1} with a LOD of 7.5 fg mL^{-1}. To evaluate applicability of the chemosensor devised, samples of diluted human plasma, spiked with BSA, were used to determine the BSA analyte.

MIP film coated micro- and nanoparticles may serve as selective ionophores in PVC-plasticizer membranes of ion selective electrodes in potentiometric titrations of highly charged proteins (Scheme 9.6C). In this potentiometric transduction, the dependence of the signal of potential on the target analyte concentration allowed for determination of differently charged protein analytes in a very broad concentration range. For this purpose, MWCNTs,[98] graphite particles,[99] graphene sheets,[101] and aminopropylated silica particles[100,111] were coated with MIP films selective with respect to TnT,[98] Myo,[99,111] Hb,[100] and PSA.[101] In some examples, protein templates were immobilized on the nanosubstrate surface with glutaraldehyde[99,111] or *N*-hydroxysuccinimide,[100,101] which resulted in a higher control over the imprinting process.

Deposition of MIPs on the surface of magnetic nanoparticles (MNPs) appeared very promising. MNPs coated that way may serve for transporting analytes from bulk solution to the chemosensor surface. For that, they can be dispensed in a solution to bind target proteins, and then collected on the chemosensor surface by the external magnetic field. This concept of analyte accumulation was recently explored with the DPV determination using a magnetic glassy carbon electrode (MGCE) to determine BHb that was transported to the MGCE proximity from bulk solution by surface imprinted MNPs.[86] In other research, magnetic Fe_3O_4@Au nanofibers (NFs) were prepared, and their surface was then modified with aniline as well as 3-aminophenylboronic acid (APBA) and acrylic acid used as functional monomers.[103] Then, the horseradish peroxidase (HRP) glycoprotein was imprinted *via* radical induced graft copolymerization of the monomers on the surface of these NFs. The resulting NPs of MIP-HPR were mostly used for absorption studies. In one experiment, however, they were immobilized on the GCE surface using chitosan, and then applied for HPR determination.

Moreover, MIP film coated NPs may serve as transducers in optical chemosensors (Scheme 9.6A). For this application, MIPs are predominantly deposited on the surface of quantum dots (QDs) and the target protein concentration is then determined by fluorescence quenching. For instance, QDs of CdTe were coated using a sol–gel procedure with the imprinted silica or thermo-sensitive imprinted polymer films.[74–77] The fluorescence quenching linearly depended on the target protein concentration in the 10^{-8} to 10^{-5} M concentration range. In another example, QDs were coated with an RNase A template imprinted TiO_2 film using a liquid-phase deposition (LPD) procedure.[112] This procedure consists of immobilization of a template on the QD surface followed by direct deposition of TiO_2 from a water solution of ammonium hexafluorotitanate and boronic acid. Moreover, QDs of CdTe were coated with the BHb imprinted silica.[78] The resulting MIP coated QDs were used for BHb determination in 100-fold diluted urine and 6000-fold diluted bovine blood samples. Moreover, QDs of CdTe were coated with thermo- and pH-responsive copolymer of *N*-isopropylacrylamide and 4-vinylphenylboronic acid, the latter serving as the functional monomer, in the presence of HRP.[113] Fluorescence quenching in the presence of HPR could reversibly be "turned off" by temperature controlled polymer swelling and/or shrinking, or by lowering affinity of boronic acid moieties to glycoprotein by decreasing solution pH.

In another study, the surface of QDs of amine modified Mn-doped ZnS was coated with an inorganic/acrylamide hybrid grafted MIP for determination of BHb.[79] In view of there being no commercial siloxanes bearing functional groups necessary to complex the target BHb protein, organic functional monomers, *viz.* acrylamide and MAA, were copolymerized with γ-methacryloxypropyl trimethoxy silane. Moreover, NPs of the QD-MIP and EVAL composite were introduced.[73] Fluorescence quenching of these NPs allowed for selective determination of three different proteins, *viz.*, amylase, lipase, and Lyz. Preparation of different composites with three different quantum dots, each coated with different MIP, enabled simultaneous determination of all three proteins in one sample just by measuring fluorescence quenching at three different wavelengths.

Furthermore, upconversion NPs were coated with MIP-(Cyt *c*) using a sol–gel procedure.[80] These NPs absorbed two photons, one at 543.5 and the other at 980 nm, and then emitted light at 660.5 nm. Recently, upconversion NPs were consecutively coated with metal–organic frameworks (MOFs), and then with the BHb imprinted thermoresponsive MIP film.[81] The resulting nanomaterial provided two functionalities. One was the temperature dependence of adsorption capacity while the other involved quenching fluorescence of these NPs in the 100 to 600 μg mL^{-1} BHb linear dynamic concentration range.

Moreover, MWCNTs were used to improve optosensing properties of the QD-MIP composites.[96] These composites were used to circumvent such QD deficiencies as the long response time and low stability of the fluorescence signal. Moreover, they served to avoid enhanced photocatalytic effects and, hence, photocurrents unnecessarily contributing to electrochemical sensing. These composites were coated with MIP-BSA films using the sol–gel procedure. Because of the developed surface area of the resulting sensing material, more molecular cavities were generated, thus leading to a much higher imprinting factor (IF = 4.17) than that for the MIP-QD in the absence of MWCNTs (IF = 2.70). Moreover, the following mechanism of fluorescence quenching in the presence of BSA was proposed. The UV-vis absorbance band for BSA was close to the energy gap band of QDs, being different to the emission band of the MIP-BSA film coated composites. Therefore, the negative charge of the conduction band of the MIP-BSA coated chemosensor could be transferred to the lowest unoccupied molecular orbital (LUMO) of BSA. Hence, it was concluded that the electron transfer from the transducer to the BSA was responsible for the fluorescence quenching.

Moreover, Ag@CdS core–shell NPs were labeled with a double bond containing cysteine derivative.[114] Then, these NPs were coated with a ferritin imprinted polymer by (activator regenerated electron transfer)–(atom transfer radical polymerization), ARGET-ATRP. The energy of the conduction band of the CdS semiconductor is −1.0 V. As the energy of the Fermi level of silver is 0.15 V, the large energy difference occurring facilitates the electron transfer from the semiconductor to the metal resulting in enhanced optical properties of NPs of Ag@CdS and improved electrochemical performance of sensor electrodes. The fluorescence intensity was enhanced 1.3 times and the currents were three- and tenfold higher in the CV and DPASV determination, respectively.

Moreover, NPs may serve only as supports in fluorescent protein determination. However, MIP labeling with dyes is needed for that. For instance, MWCNTs were coated with a Lyz imprinted *N*,*N*′-dimethylacrylamide polymer film using acrylamide as the functional monomer.[82] After Lyz template extraction, the MIP cavities were labeled with fluorescein isothiocyanate. The fluorescence quenching of MIP-Lyz coated NPs was linearly dependent on the Lyz concentration.

An MIP film of the macroporous poly(*m*-aminophenylboronic acid-*co*-dopamine), selective with respect to the HPR and α-fetoprotein (AFP) glycoproteins was photolithographically deposited on the 96- and 384-spot plate. After binding in the MIP cavities, the target glycoprotein molecules were labeled with silver nanoparticles (AgNPs) modified with 4-mercaptophenylboronic acid.[83] This modification enabled SERS determination of HPR or AFP. The Raman signal measured at 1072 cm^{-1} linearly increased with the logarithm of the target protein concentration in the range of 1 ng mL^{-1} to 10 μg mL^{-1}. The same 96-spot plate loaded with the MIP-AFP film was used for fabrication of an ELISA-like analytical system as well.[115] In this system, AFP molecules entrapped in the MIP cavities were reacted with the HPR labeled anti-AFP antibodies, and 3,3,5′,5′-tetramethylbenzidine dihydrochloride was subsequently added. The AFP concentration was then determined as the increase of the solution absorbance at λ = 650 nm. The chemosensor response to AFP was linear in the 5 to 50 ng mL^{-1} concentration range.

Furthermore, the use of NPs may affect the SPR signal. For instance, gold nanocages (AuNCs) were coated for that purpose by polymerization of organo-siloxane monomers in the presence of surface-immobilized neutrophil gelatinase-associated lipocalin (NGAL).[116] Unfortunately, the detection signal was not linearly dependent on the NGAL concentration. In another study, a surface imprinted silica (SIS) colloid served for BHb determination.[117] That is, the SIS nanoparticles were cast on the surface of glass slides, and the red shift in the reflectance spectra was measured. As in the above case, however, the response was not linear with the BHb concentration.

Similarly to antibodies in immunosensors, NPs coated with MIP films may be immobilized on the surface of electrodes. Then, they can serve as recognition units in the same way as the MIP thin films described above. Moreover, NPs of MIPs can play the role of recognition units of chemosensors. One of many examples of that kind involves Lyz imprinted poly(ethylene glycol dimethacrylate-*N*-methacryloyl-L-histidine methyl ester) NPs. These NPs were attached to the 3-mercaptopropene modified surface of the SPR chip and the QCM resonator.[104,105] The dynamic linear concentration ranges of these two chemosensors for Lyz were 21 nM to 1.4 μM and 14 nM to 105 μM, respectively. These chemosensors were capable of detecting Lyz even in complex natural matrices, such as chicken egg white. Moreover, the MIP technology was combined with ELISA protocols. In this case, some antibodies used in typical ELISA tests appeared to be replaceable with MIPs. For this, NPs of MIP were labeled with NPs of electrochemiluminescent magnetic Ru@SiO_2@Au,[118] QDs,[119] and biotinylated HRP[120] or by deposition of a Lyz imprinted TiO_2 film on the surface of catalytic NPs of $Cu(OH)_2$ NPs.[121] Then, the target proteins were determined in two ways. In one, suspensions of NPs of MIP and target proteins were mixed in wells of the Nunc-Immuno™ MicroWell™ 96-well microplate and changes in the solution color[121] or electrochemiluminescence of NPs of Ru@SiO_2@Au entrapped by external magnetic field[118] were measured. In another study, target proteins were adsorbed on walls of the microplate wells and, subsequently, suspensions of NPs of the MIP were added.[119,120] Then, target protein concentrations were indirectly determined by quantifying the amount of NPs of MIP that were adsorbed on walls of the wells.

9.3.2 Epitope Imprinting

Proteins are demanding templates for imprinting. They are not easy to handle because of conformational instability, sensitivity to such external conditions as an elevated temperature, or the presence of acids, bases, or aggressive organic solvents. Moreover, they are always obtained by extraction from living organisms, and then tediously purified. Therefore, obtaining proteins in a pure form is expensive, laborious, and time consuming. The strategy of molecular imprinting of epitopes seems to be the most promising way to overcome these drawbacks (Scheme 9.2E). This strategy involves devising MIPs with only a small characteristic part of a protein used as the template. This strategy was first tested for imprinting the oxytocin nonapeptide.[122] For that purpose, a selective MIP was prepared using as a template only an YPLG tetrapeptide with the same three amino acid sequence at the C-terminal as that of the target oxytocin.

One of the most impressive achievements of the epitope strategy of protein imprinting involves preparation of selective MIPs imprinted with three different 9-(amino acid) peptide epitopes, namely AYLKKATNE, GRYVVDTSK, and VVSTQTALA of Cyt *c*, alcohol dehydrogenase, and BSA protein, respectively.[123] Selectivity of these MIPs was very high. Moreover, these MIPs recognized target proteins in a five-protein mixture. The epitopes were immobilized with their *N*-termini on planar Si/SiO_2 supports. Subsequently, thin MIP films were deposited on top of these epitopes, and the films were then peeled off the supports. In effect, MIPs with recognition cavities placed on the surface were prepared, thus enabling rapid detection of the target proteins.

So far, epitope imprinting found only limited application in MIP chemosensor fabrication. One example involves a QCM chemosensor for determination of the human immunodeficiency virus type 1 (HIV-1) related glycoprotein 41 (gp41). This chemosensor used a 35-(amino acid) peptide imprinted in polydopamine.[124] Moreover, a series of acrylamide polymers, imprinted with 9-to-14-(amino-acid)-long epitopes were tested to devise a QCM chemosensor for the anthrax protective antigen (PA_{83}), a biomarker of the anthrax infection.[125] Moreover, an F_{ab} fragment of IgG was imprinted with the use of *N*-methacryloyl-L-histidine methyl ester (MAH) as the functional monomer.[106] Then, the resulting MIP-F_{ab} was deposited on the surface of an SPR chip for the real-time determination of IgG. The LOD of IgG was 56 ng mL^{-1} and the chemosensor response to IgG was 21 times higher than that to BSA.

Presumably, the epitope imprinting allows the introducing into an MIP more and better defined molecular cavities than the whole protein imprinting does. The fluorescence quenching response of QDs of CdTe, coated with an MIP film imprinted with a 6-(amino acid) peptide of a surface-exposed BSA C-terminus, was higher and, moreover, the IF was higher than that for the MIP film imprinted with the whole BSA.[126]

Although the epitope imprinting strategy is very attractive, it suffers from several serious limitations. That is, the fragment chosen for imprinting should be specific for the target protein only and, moreover, it should readily be accessible for binding, *e.g.*, it should occur on the surface of the target protein molecule. Furthermore, this epitope should easily be available commercially at a reasonable price. A rational and systematic approach to selection of epitopes for imprinting is rather demanding.[127] For that, a list of hundreds short peptides of the NT-proBNP protein (a marker of the risk of cardiovascular events), which could be prepared by cleavage of this protein with different hydrolytic enzymes, was generated *in silico*. Then, potential epitopes were evaluated with respect to parameters, such as the peptide length, hydrophobicity, and location in the 3-D peptide structure. The sequences of the listed epitopes were compared with all protein sequences gathered in the

UniProtKB database to check specificity of these sequences. Then, the two most promising epitopes were synthetized, and then successfully used in preparing MIP beads in application as an SPE column packing material.

Original publications focused only on designing polymers molecularly imprinted with potential epitopes of proteins without experimental verification involving detection of a complete protein are not discussed herein.

9.4 Miscellaneous

One of the most unusual applications of MIP-proteins involves facilitation of protein crystallization.[128] X-ray crystallography is the most convenient method for determining protein structure. This method provides a complete 3-D atom arrangement of the investigated compound, if monocrystals are grown. With the rapidly increasing number of proteins of biological and clinical interest, an efficient method of protein crystallization becomes more and more desired. For that purpose, one hydrogel was imprinted with BHb and the other with trypsin. Then, small (0.2 μL) drops of each hydrogel were immersed in five times lager drops of the protein solutions by micropipetting. Next, crystallization was allowed to proceed at room temperature. Then, in five days, the MIP-BHb produced crystals of only their target BHb protein. Importantly, no crystals were obtained of any other protein tested. In the case of MIP-trypsin, trypsin crystals were obtained in seven days. However, selectivity of the MIP-trypsin hydrogel was low. That is, crystals of three other, similar in size proteins, *viz.*, human macrophage migration inhibitor factor (MIF), lysozyme, and thaumatin, were grown as well. Moreover, optimization of crystallization conditions with MIP nucleating agents was automated. For that, five different MIPs, each imprinted with a different protein, were prepared in the same way as that used in previous studies.[129] These MIPs were administered to automatic crystallization trials by a robotic system, without blocking robotic dispensing tips, and this system was then used for automatic harvesting of protein crystals. Usually, addition of zwitterions to the solution for crystallization improves the protein crystals growth. Therefore, using the zwitterionic [3-(methacryloylamino)propyl]-dimethyl(3-sulfopropyl) ammonium hydroxide functional monomer, instead of methacrylamide, enabled substantial improving yield of growth of protein crystals (Figure 9.1).[130]

Selectivity of the MIP based chemosensors can be so high that even protein conformations can be distinguished.[131] In one example, poly(3-aminophenylboronic acid) MIPs, imprinted with different β-lactoglobulin conformational isomers (BGLs), were deposited on Au-QCRs. The BGL conformational isomers were prepared either by heating or treating solutions of this protein with 1,1,1,3,3,3-hexafluoroisopropanol. In both cases, affinity to denatured BGL of the MIPs was higher than that to that native in PM determinations. Moreover, MIPs imprinted with the native BGL were selective with respect to denatured BGL. Considering that the studied proteins differ in conformation only, one has to admit that this result is impressive,

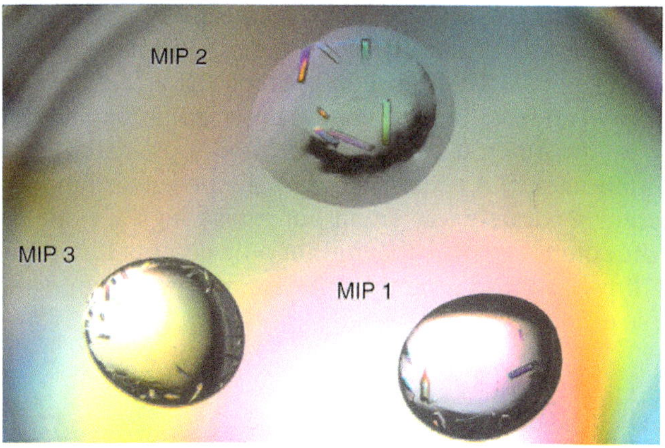

Figure 9.1 The microphotograph of protein crystals grown with MIP supports. Adapted with permission from Reddy *et al.*, *Biomacromolecules*, 2012, **13**, 3959–3965.[128] Copyright (2012) American Chemical Society.

even though the selectivity reached is moderate. These MIPs may serve as recognition units in chemosensors for diagnosing different diseases caused by improper protein folding. Importantly, the issue of MIP sensitivity to protein conformation may be the reason for a low progress in protein imprinting. Apparently, addition of functional monomers commonly used for imprinting caused significant changes in the secondary structure of lysozyme and BHb in concentrations far below those reported in literature for MIP fabrication.[132]

Interactions between the target protein and the MIP film may be followed not only with analyte binding experiments. When the target protein, Cyt *c*[133] or BHb,[134] was immobilized on the tip of an AFM cantilever, it was possible to measure strong force changes caused by the high affinity of the protein immobilized on the tip to molecular cavities located on the surface of an MIP film deposited on the AFM support. The AFM imaging allowed not only determining of the interactions but also mapping localization and distribution of imprinted cavities on the MIP surface at the nanometre scale.[135] Mapping the MIP–streptavidin film with a regular tip enabled determination of the topology of the film. If the tip with immobilized streptavidin was used instead, black spots appeared in places where the imprinted cavities were located. Moreover, when images were taken after immersing the MIP in a streptavidin solution, there was no visible change in polymer topology but most of these black spots disappeared.

Interactions of NPs of an MIP with template molecules may conveniently be monitored by QCM. In one example, the precipitation polymerization method was employed for synthesis of the NPs of the MIP imprinted with the melittin 26-(amino acid) peptide[136] and the 9-(amino acid) epitope of the green fluorescent protein (GFP-9).[137] In another example, NPs of an MIP were synthetized in glass columns filled with solid supports (glass beads) with surface immobilized trypsin, kallikrein, or RNase A.[138] Then, the Au-QCRs with the target protein immobilized on the Au surface were used to detect binding of NPs of the MIP with the Au-QCM. Moreover, Lyz and Cyt *c* form a 1:1 (mole:mole) adduct. This adduct was used as a template for depositing an MIP film on the Au-QCR surface, with 3-aminophenylboronic acid as the functional monomer.[139] A response of the chemosensor fabricated that way to each protein separately was small but it was high if both proteins were present in the solution. This behavior not only demonstrated formation of a Lyz–(Cyt *c*) dimer but also a significant conformational change of both proteins because of dimerization.

9.5 Conclusions

Molecularly imprinted polymers synthesized with the use of macromolecular templates ($M_w > 1.5$ kDa), and proteins in particular, have received significant attention for nearly the last two decades. Despite an extensive research effort invested, the progress made in this field unitl the beginning of the present decade was rather limited. New ideas and strategies proposed in the last few years have raised protein imprinting to a completely new level. Now, it seems that commercial application of MIP-protein chemosensors are just a matter of time.

List of Abbreviations

AA	Acrylic acid
AChE	Acetylcholinesterase
AEMA	2-Aminoethyl methacrylate hydrochloride
AFM	Atomic force microscopy
AFP	α-Fetoprotein
ahsFtn	Horse apoferritin
AIBN	Azoisobutyronitrile
AM	Acrylamide
AMPSA	2-Acryl-amido-2-methyl-1-propane sulfonic acid sodium salt
APBA	3-Aminophenylboronic acid
APTES	3-Amino propyltriethoxysilane
ARGET-ATRP	Activator regenerated by electron transfer–atom transfer radical polymerization
ATRP	Atom transfer radical polymerization
AuNC	Gold nanocage
Au-QCR	Gold film-coated quartz crystal resonator
Au-SPE	Gold screen printed electrode
Av	avidin
BGL	β-Lactoglobulin conformational isomer
BHb	Bovine hemoglobin
BoHb	Bovine oxygenated hemoglobin

BSA	Bovine serum albumin
CNT	Carbon nanotube
CRP	C-reactive protein
cTnT	Cardiac troponin T
Cy3-NHS	Sulfoindocyanine *N*-hydroxysuccinimidyl ester
Cyt *c*	Cytochrome *c*
DAU	Diacryloyl urea
DBD-F	4-(*N*,*N*-Dimethylaminosulfonyl)-7-fluoro-2,1,3-benzoxadiazole
DMAPMA	3-Dimethylaminopropyl methacrylamide
DPASV	Differential pulse anodic stripping voltammetry
DPCSV	Differential pulse cathodic stripping voltammetry
DPV	Differential pulse voltammetry
EA	Egg albumin
EAv	Extravidin
EGDMA	Ethylene glycol dimethacrylate
EIS	Electrochemical impedance spectroscopy
ELISA	Enzyme-linked immunosorbent assay
EQCM	Electrochemical quartz crustal microbalance
EVAL	Poly(ethylene-*co*-vinyl alcohol)
FCS	Fetal calf serum
FITC	Fluorescein isothiocyanate
GCE	Glassy carbon electrode
GFP-9	Green fluorescent protein
GNEE	Gold nanoelectrodes ensemble
GOx	Glucose oxidase
Gr-AuNP	Hybrid gold-graphene nanoparticle
GTS	Glutathione S-transferase
HAPLN1	Hyaluronan-linked protein 1
Hb	Hemoglobin
HBs	Hepatitis B
HBsAb	Hepatitis B surface antibody
HDGM	Hierarchical dendritic gold microstructure
HEMA	2-Hydroxyethyl methacrylate
HIV-1	Human immunodeficiency virus type 1
HPLC	High-performance liquid chromatography
HRP	Horseradish peroxidase
HSA	Human serum albumin
hsFtn	Horse ferritin
IF	Imprinting factor
IgG	Immunoglobulin G
LDL	Low-density lipoprotein
LOD	Limit of detection
LPD	Liquid-phase deposition
LUMO	Lowest unoccupied molecular orbital
Lyz	Lysozyme
MAA	Methacrylic acid

MABA	4-[2-(*N*-Methacrylamido)ethylaminomethyl] benzoic acid
MAH	*N*-Methacryloyl-L-histidine methyl ester
MALDI-TOF	Matrix-assisted laser desorption/ionization time-of-flight (mass spectrometry)
MBP	Maltose binding protein
MDTA	{[[2-(2-Methacrylamido)ethyldithio]ethylcarbamoyl]methoxy} acetic acid
MGCE	Magnetic glassy carbon electrode
MIF	Migration inhibitor factor
MIP	Molecularly imprinted polymer
MNP	Magnetic nanoparticle
MOF	Metal-organic framework
MSMB	Microseminoprotein beta
MSPE	Micro-solid phase extraction
MWCNT	Multi-wall carbon nanotube
Myo	Myoglobin
NA	Neutravidin
NF	Nanofiber
NGAL	Neutrophil gelatinase-associated lipocalin
NIP	Non-imprinted polymer
NP	Nanoparticle
oNPOE	*o*-Nitrophenyloctyl ether
Ova	Ovalbumin
oxLDL	Oxidized low-density lipoprotein
PB	Prussian Blue
PCT	Procalcitonin
PDMS	Polydimethylsiloxane
PEDOT	Poly(3,4-ethylenedioxythiophene)
PEGDMA	Poly(ethylene glycol) dimethacrylate
PGE	Pencil graphite electrode
PM	Piezoelectric microgravimetry
PSA	Prostate-specific antigen
PSS	Polystyrene sulfonate
PVC	Poly(vinyl chloride)
PCV-COOH	Carboxylated poly(vinyl chloride)
QCM	Quartz crystal microbalance
QD	Quantum dot
RTA	Ricin toxin chain A
RNase A	Ribonuclease A
SAM	Self-assembled monolayer
SEB	Staphylococcal enterotoxin B
SERS	Surface-enhanced Raman spectroscopy
SIS	Surface imprinted silica
SPE	Solid-phase extraction
SPR	Surface plasmon resonance

ST	Streptavidin
SWASV	Square wave anodic stripping voltammetry
SWV	Square wave voltammetry
TEGMPA	Tetraethyleneglycol-3-morpholine propionate acrylate
TEOS	Tetraethyl orthosilicate
TnT	Cardiac troponin T
UVFLIM	Deep-UV fluorescence image microscope

Acknowledgements

We thank the National Science Center of Poland for financial support (Grant No. NCN 2014/15/B/NZ/01011 to W. K.).

References

1. A. Bossi, F. Bonini, A. P. F. Turner and S. A. Piletsky, *Biosens. Bioelectron.*, 2007, **22**, 1131–1137.
2. D. S. Janiak and P. Kofinas, *Anal. Bioanal. Chem.*, 2007, **389**, 399–404.
3. N. M. Bergmann and N. A. Peppas, *Prog. Polym. Sci.*, 2008, **33**, 271–288.
4. E. Verheyen, J. P. Schillemans, M. van Wijk, M. A. Demeniex, W. E. Hennink and C. F. van Nostrum, *Biomaterials*, 2011, **32**, 3008–3020.
5. K. G. Yang, L. H. Zhang, Z. Liang and Y. K. Zhang, *Anal. Bioanal. Chem.*, 2012, **403**, 2173–2183.
6. D. R. Kryscio and N. A. Peppas, *Acta Biomater.*, 2012, **8**, 461–473.
7. Y. Q. Lv, T. W. Tan and F. Svec, *Biotechnol. Adv.*, 2013, **31**, 1172–1186.
8. S. J. Li, S. S. Cao, M. J. Whitcombe and S. A. Piletsky, *Prog. Polym. Sci.*, 2014, **39**, 145–163.
9. F. H. Dickey, *J. Phys. Chem.*, 1955, **59**, 695–707.
10. G. Wulff and A. Sarhan, *Angew. Chem., Int. Ed.*, 1972, **11**, 341–344.
11. T. Takagish and I. M. Klotz, *Biopolymers*, 1972, **11**, 483–491.
12. G. Vlatakis, L. I. Andersson, R. Muller and K. Mosbach, *Nature*, 1993, **361**, 645–647.
13. P. S. Sharma, M. Dabrowski, F. D'Souza and W. Kutner, *TrAC, Trends Anal. Chem.*, 2013, **51**, 146–157.
14. M. Kempe, M. Glad and K. Mosbach, *J. Mol. Recognit.*, 1995, **8**, 35–39.
15. P. S. Sharma, A. Pietrzyk-Le, F. D'Souza and W. Kutner, *Anal. Bioanal. Chem.*, 2012, **402**, 3177–3204.
16. C. Malitesta, E. Mazzotta, R. A. Picca, A. Poma, I. Chianella and S. A. Piletsky, *Anal. Bioanal. Chem.*, 2012, **402**, 1827–1846.
17. T. Y. Lin, C. H. Hu and T. C. Chou, *Biosens. Bioelectron.*, 2004, **20**, 75–81.
18. O. Hayden, C. Haderspock, S. Krassnig, X. H. Chen and F. L. Dickert, *Analyst*, 2006, **131**, 1044–1050.
19. H. F. El-Sharif, H. Aizawa and S. M. Reddy, *Sens. Actuators, B*, 2015, **206**, 239–245.

20. Z. Zhang, Y. Long, L. Nie and S. Yao, *Biosens. Bioelectron.*, 2006, **21**, 1244–1251.
21. C. Sontimuang, R. Suedee, B. Canyuk, N. Phadoongsombut and F. L. Dickert, *Anal. Chim. Acta*, 2011, **687**, 184–192.
22. C. Sontimuang, R. Suedee and F. Dickert, *Anal. Biochem.*, 2011, **410**, 224–233.
23. M. H. Lee, J. L. Thomas, H. Y. Tseng, W. C. Lin, B. D. Liu and H. Y. Lin, *ACS Appl. Mater. Interfaces*, 2011, **3**, 3064–3071.
24. S. G. Wu, W. G. Tan and H. H. Xu, *Analyst*, 2010, **135**, 2523–2527.
25. N. Karimian, M. Vagin, M. H. Zavar, M. Chamsaz, A. P. Turner and A. Tiwari, *Biosens. Bioelectron.*, 2013, **50**, 492–498.
26. T. S. C. R. Rebelo, C. M. Pereira, M. G. F. Sales, J. P. Noronha and F. Silva, *Sens. Actuators, B*, 2016, **223**, 846–852.
27. Y. Sun, H. Y. Du, Y. T. Lan, W. J. Wang, Y. J. Liang, C. L. Feng and M. Yang, *Biosens. Bioelectron.*, 2016, **77**, 894–900.
28. B. Sellergren, M. Lepisto and K. Mosbach, *J. Am. Chem. Soc.*, 1988, **110**, 5853–5860.
29. M. Cieplak, K. Szwabinska, M. Sosnowska, B. K. Chandra, P. Borowicz, K. Noworyta, F. D'Souza and W. Kutner, *Biosens. Bioelectron.*, 2015, **74**, 960–966.
30. Y. X. Li, Y. J. Li, M. Hong, Q. Bin, Z. Y. Lin, Z. Lin, Z. W. Cai and G. N. Chen, *Biosens. Bioelectron.*, 2013, **42**, 612–617.
31. E. Komarova, M. Aldissi and A. Bogomolova, *Analyst*, 2015, **140**, 1099–1106.
32. Y. Yoshimi, R. Ohdaira, C. Iiyama and K. Sakai, *Sens. Actuators, B*, 2001, **73**, 49–53.
33. B. B. Prasad, A. Prasad and M. P. Tiwari, *Biosens. Bioelectron.*, 2013, **39**, 236–243.
34. S. M. Reddy, G. Sette and Q. Phan, *Electrochim. Acta*, 2011, **56**, 9203–9208.
35. R. L. Zhang, S. Xu, J. Luo and X. Y. Liu, *Microchim. Acta*, 2015, **182**, 175–183.
36. M. Bosserdt, N. Gajovic-Eichelman and F. W. Scheller, *Anal. Bioanal. Chem.*, 2013, **405**, 6437–6444.
37. K. J. Jetzschmann, G. Jagerszki, D. Dechtrirat, A. Yarman, N. Gajovic-Eichelmann, H. D. Gilsing, B. Schulz, R. E. Gyurcsanyi and F. W. Scheller, *Adv. Funct. Mater.*, 2015, **25**, 5178–5183.
38. M. Zayats, M. Kanwar, M. Ostermeier and P. C. Searson, *Macromolecules*, 2011, **44**, 3966–3972.
39. K. Fukazawa, Q. Li, S. Seeger and K. Ishihara, *Biosens. Bioelectron.*, 2013, **40**, 96–101.
40. H. Sunayama, T. Ooya and T. Takeuchi, *Biosens. Bioelectron.*, 2010, **26**, 458–462.
41. Y. Suga, H. Sunayama, T. Ooya and T. Takeuchi, *Chem. Commun.*, 2013, **49**, 8450–8452.

42. H. Sunayama, T. Ooya and T. Takeuchi, *Chem. Commun.*, 2014, **50**, 1347–1349.
43. H. Sunayama and T. Takeuchi, *ACS Appl. Mater. Interfaces*, 2014, **6**, 20003–20009.
44. Y. Inoue, A. Kuwahara, K. Ohmori, H. Sunayama, T. Ooya and T. Takeuchi, *Biosens. Bioelectron.*, 2013, **48**, 113–119.
45. T. Matsunaga, T. Hishiya and T. Takeuchi, *Anal. Chim. Acta*, 2007, **591**, 63–67.
46. L. Uzun, R. Say, S. Unal and A. Denizli, *Biosens. Bioelectron.*, 2009, **24**, 2878–2884.
47. B. Osman, L. Uzun, N. Besirli and A. Denizli, *Mater. Sci. Eng., C*, 2013, **33**, 3609–3614.
48. Y. Wang and T. X. Wei, *Chin. Chem. Lett.*, 2013, **24**, 813–816.
49. G. Lautner, J. Kaev, J. Reut, A. Opik, J. Rappich, V. Syritski and R. E. Gyurcsanyi, *Adv. Funct. Mater.*, 2011, **21**, 591–597.
50. X. B. Hu, G. T. Li, J. Huang, D. Zhang and Y. Qiu, *Adv. Mater.*, 2007, **19**, 4327–4332.
51. F. T. C. Moreira, S. Sharma, R. A. F. Dutra, J. P. C. Noronha, A. E. G. Cass and M. G. F. Sales, *Biosens. Bioelectron.*, 2013, **45**, 237–244.
52. F. T. C. Moreira, R. A. F. Dutra, J. P. C. Noronha, J. C. S. Fernandes and M. G. F. Sales, *Sens. Actuators, B*, 2013, **182**, 733–740.
53. F. T. C. Moreira, R. A. F. Dutra, J. P. C. Noronha and M. G. F. Sales, *Electrochim. Acta*, 2013, **107**, 481–487.
54. F. T. C. Moreira, S. Sharma, R. A. F. Dutra, J. P. C. Noronha, A. E. G. Cass and M. G. F. Sales, *Sens. Actuators, B*, 2014, **196**, 123–132.
55. A. Tretjakov, V. Syritski, J. Reut, R. Boroznjak, O. Volobujeva and A. Opik, *Microchim. Acta*, 2013, **180**, 1433–1442.
56. Y. Kamon, R. Matsuura, Y. Kitayama, T. Ooya and T. Takeuchi, *Polym. Chem.*, 2014, **5**, 4764–4771.
57. P. Jolly, V. Tamboli, R. L. Harniman, P. Estrela, C. J. Allender and J. L. Bowen, *Biosens. Bioelectron.*, 2016, **75**, 188–195.
58. D. Dechtrirat, N. Gajovic-Eichelmann, F. F. Bier and F. W. Scheller, *Adv. Funct. Mater.*, 2014, **24**, 2233–2239.
59. H. Q. Zhang, J. S. Jiang, H. T. Zhang, Y. Zhang and P. C. Sun, *ACS Macro Lett.*, 2013, **2**, 566–570.
60. G. Cabral-Miranda, M. Gidlund and M. G. F. Sales, *J. Mater. Chem. B*, 2014, **2**, 3087–3095.
61. D. B. G. Ertürk, M. Hedström and B. Mattiasson, *Biotechnol. Rep.*, 2014, **3**, 65–72.
62. G. Erturk, H. Ozen, M. A. Tumer, B. Mattiasson and A. Denizli, *Sens. Actuators, B*, 2016, **224**, 823–832.
63. G. Erturk, M. Hedstrom, M. A. Tumer, A. Denizli and B. Mattiasson, *Anal. Chim. Acta*, 2015, **891**, 120–129.
64. G. Sener, E. Ozgur, A. Y. Rad, L. Uzun, R. Say and A. Denizli, *Analyst*, 2013, **138**, 6422–6428.

65. T. Kuwata, A. Uchida, E. Takano, Y. Kitayama and T. Takeuchi, *Anal. Chem.*, 2015, **87**, 11784–11791.
66. D. Z. Zhou, T. Y. Guo, Y. Yang and Z. P. Zhang, *Sens. Actuators, B*, 2011, **153**, 96–102.
67. S. Liu, D. Z. Zhou and T. Y. Guo, *Biosens. Bioelectron.*, 2013, **42**, 80–86.
68. L. Li, L. L. Yang, Z. L. Xing, X. J. Lu and X. W. Kan, *Analyst*, 2013, **138**, 6962–6968.
69. X. W. Kan, Z. L. Xing, A. H. Zhu, Z. Zhao, G. L. Xu, C. Li and H. Zhou, *Sens. Actuators, B*, 2012, **168**, 395–401.
70. Y. T. Wang, Y. X. Zhou, J. Sokolov, B. Rigas, K. Levon and M. Rafailovich, *Biosens. Bioelectron.*, 2008, **24**, 162–166.
71. Y. T. Wang, Z. Q. Zhang, V. Jain, J. J. Yi, S. Mueller, J. Sokolov, Z. X. Liu, K. Levon, B. Rigas and M. H. Rafailovich, *Sens. Actuators, B*, 2010, **146**, 381–387.
72. A. Mathur, S. Blais, C. M. V. Goparaju, T. Neubert, H. Pass and K. Levon, *PLoS One*, 2013, **8**, e57681.
73. M. H. Lee, Y. C. Chen, M. H. Ho and H. Y. Lin, *Anal. Bioanal. Chem.*, 2010, **397**, 1457–1466.
74. W. Zhang, X. W. He, Y. Chen, W. Y. Li and Y. K. Zhang, *Biosens. Bioelectron.*, 2011, **26**, 2553–2558.
75. W. Zhang, X. W. He, Y. Chen, W. Y. Li and Y. K. Zhang, *Biosens. Bioelectron.*, 2012, **31**, 84–89.
76. W. Zhang, X. W. He, W. Y. Li and Y. K. Zhang, *Chem. Commun.*, 2012, **48**, 1757–1759.
77. Y. Z. Wang, D. Y. Li, X. W. He, W. Y. Li and Y. K. Zhang, *Microchim. Acta*, 2015, **182**, 1465–1472.
78. D. Y. Li, X. W. He, Y. Chen, W. Y. Li and Y. K. Zhang, *ACS Appl. Mater. Interfaces*, 2013, **5**, 12609–12616.
79. L. Tan, C. Huang, R. F. Peng, Y. W. Tang and W. M. Li, *Biosens. Bioelectron.*, 2014, **61**, 506–511.
80. T. Guo, Q. L. Deng, G. Z. Fang, C. C. Liu, X. Huang and S. Wang, *Biosens. Bioelectron.*, 2015, **74**, 498–503.
81. T. Guo, Q. Deng, G. Fang, D. Gu, Y. Yang and S. Wang, *Biosens. Bioelectron.*, 2016, **79**, 341–346.
82. Q. L. Deng, J. H. Wu, X. R. Zhai, G. Z. Fang and S. Wang, *Sensors (Basel)*, 2013, **13**, 12994–13004.
83. J. Ye, Y. Chen and Z. Liu, *Angew. Chem., Int. Ed.*, 2014, **53**, 10386–10389.
84. A. Fatoni, A. Numnuam, P. Kanatharana, W. Limbut and P. Thavarungkul, *Analyst*, 2014, **139**, 6160–6167.
85. B. V. M. Silva, B. A. G. Rodriguez, G. F. Sales, M. D. T. Sotomayor and R. F. Dutra, *Biosens. Bioelectron.*, 2016, **77**, 978–985.
86. S. Sun, L. Z. Chen, H. J. Shi, Y. J. Li and X. W. He, *J. Electroanal. Chem.*, 2014, **734**, 18–24.
87. D. Kumar and B. B. Prasad, *Sens. Actuators, B*, 2012, **171**, 1141–1150.
88. B. B. Prasad, A. Prasad and M. P. Tiwari, *Talanta*, 2013, **109**, 52–60.

89. H. J. Chen, Z. H. Zhang, D. Xie, R. Cai, X. Chen, Y. N. Liu and S. Z. Yao, *Electroanalysis*, 2012, **24**, 2109–2116.
90. H. J. Chen, Z. H. Zhang, L. J. Luo and S. Z. Yao, *Sens. Actuators, B*, 2012, **163**, 76–83.
91. J. F. Xia, X. Y. Cao, Z. H. Wang, M. Yang, F. F. Zhang, B. Lu, F. Li, L. Xia, Y. H. Li and Y. Z. Xia, *Sens. Actuators, B*, 2016, **225**, 305–311.
92. L. Li, L. M. Fan, Y. L. Dai and X. W. Kan, *Microchim. Acta*, 2015, **182**, 2477–2483.
93. D. Cai, L. Ren, H. Z. Zhao, C. J. Xu, L. Zhang, Y. Yu, H. Z. Wang, Y. C. Lan, M. F. Roberts, J. H. Chuang, M. J. Naughton, Z. F. Ren and T. C. Chiles, *Nat. Nanotechnol.*, 2010, **5**, 597–601.
94. S. Viswanathan, C. Rani, S. Ribeiro and C. Delerue-Matos, *Biosens. Bioelectron.*, 2012, **33**, 179–183.
95. X. D. Wang, J. Dong, H. M. Ming and S. Y. Ai, *Analyst*, 2013, **138**, 1219–1225.
96. Z. Q. Ding, S. W. A. Bligh, L. Tao, J. Quan, H. L. Nie, L. M. Zhu and X. Gong, *Mater. Sci. Eng., C*, 2015, **48**, 469–479.
97. S. Patra, E. Roy, R. Madhuri and P. K. Sharma, *Biosens. Bioelectron.*, 2015, **66**, 1–10.
98. F. T. C. Moreira, R. A. F. Dutra, J. P. C. Noronha, A. L. Cunha and M. G. F. Sales, *Biosens. Bioelectron.*, 2011, **28**, 243–250.
99. F. T. C. Moreira, S. Sharma, R. A. F. Dutra, J. P. C. Noronha, A. E. G. Cass and M. G. F. Sales, *Microchim. Acta*, 2015, **182**, 975–983.
100. A. P. M. Tavares, F. T. C. Moreira and M. G. F. Sales, *RSC Adv.*, 2013, **3**, 26210–26219.
101. M. S. C. R. Rebelo, C. Santos, J. Costa-Rodrigues, M. H. Fernandes, J. P. Noronha and M. G. F. Sales, *Electrochim. Acta*, 2014, **132**, 142–150.
102. J. Luo, S. S. Jiang and X. Y. Liu, *Sens. Actuators, B*, 2014, **203**, 782–789.
103. Y. Li, M. Hong, Miaomiao, Q. Bin, Z. Lin, Z. Cai and G. Chen, *J. Mater. Chem. B*, 2013, **1**, 1044–1051.
104. G. Sener, L. Uzun, R. Say and A. Denizli, *Sens. Actuators, B*, 2011, **160**, 791–799.
105. G. Sener, E. Ozgur, E. Yilmaz, L. Uzun, R. Say and A. Denizli, *Biosens. Bioelectron.*, 2010, **26**, 815–821.
106. G. Erturk, L. Uzun, M. A. Tumer, R. Say and A. Denizli, *Biosens. Bioelectron.*, 2011, **28**, 97–104.
107. J. Q. Liu, Z. Liu, C. J. Barrow and W. R. Yang, *Anal. Chim. Acta*, 2015, **859**, 1–19.
108. K. Saha, S. S. Agasti, C. Kim, X. N. Li and V. M. Rotello, *Chem. Rev.*, 2012, **112**, 2739–2779.
109. B. B. Prasad, R. Madhuri, M. P. Tiwari and P. S. Sharma, *Electrochim. Acta*, 2010, **55**, 9146–9156.
110. M. P. Tiwari and B. B. Prasad, *J. Chromatogr., A*, 2014, **1337**, 22–31.
111. F. T. C. Moreira, R. A. F. Dutra, J. P. C. Noronha and M. G. F. Sales, *Biosens. Bioelectron.*, 2011, **26**, 4760–4766.

112. J. Inoue, T. Ooya and T. Takeuchi, *Soft Matter*, 2011, **7**, 9681–9684.
113. W. Zhang, W. Liu, P. Li, H. B. Xiao, H. Wang and B. Tang, *Angew. Chem., Int. Ed.*, 2014, **53**, 12489–12493.
114. S. Patra, E. Roy, R. Madhuri and P. K. Sharma, *Biosens. Bioelectron.*, 2015, **63**, 301–310.
115. X. D. Bi and Z. Liu, *Anal. Chem.*, 2014, **86**, 959–966.
116. L. M. Tian, K. K. Liu, J. J. Morrissey, N. Gandra, E. D. Kharasch and S. Singamaneni, *J. Mater. Chem. B*, 2014, **2**, 167–170.
117. W. Chen, W. Lei, M. Xue, F. Xue, Z. H. Meng, W. B. Zhang, F. Qu and K. J. Shea, *J. Mater. Chem. A*, 2014, **2**, 7165–7169.
118. J. Zhou, N. Gan, F. T. Hu, T. H. Li, H. K. Zhou, X. Li and L. Zheng, *Sens. Actuators, B*, 2013, **186**, 300–307.
119. M. H. Lee, J. Thomas, Y. C. Chen, W. T. Chin and H. Y. Lin, *Microchim. Acta*, 2013, **180**, 1393–1399.
120. Y. Yonamine, Y. Hoshino and K. J. Shea, *Biomacromolecules*, 2012, **13**, 2952–2957.
121. Y. X. Li, Y. J. Li, L. H. Huang, Q. Bin, Z. Y. Lin, H. H. Yang, Z. W. Cai and G. N. Chen, *J. Mater. Chem. B*, 2013, **1**, 1256–1262.
122. A. Rachkov and N. Minoura, *J. Chromatogr., A*, 2000, **889**, 111–118.
123. H. Nishino, C. S. Huang and K. J. Shea, *Angew. Chem., Int. Ed.*, 2006, **45**, 2392–2396.
124. C. H. Lu, Y. Zhang, S. F. Tang, Z. B. Fang, H. H. Yang, X. Chen and G. N. Chen, *Biosens. Bioelectron.*, 2012, **31**, 439–444.
125. D. F. Tai, M. H. Jhang, G. Y. Chen, S. C. Wang, K. H. Lu, Y. D. Lee and H. T. Liu, *Anal. Chem.*, 2010, **82**, 2290–2293.
126. Y. Q. Yang, X. W. He, Y. Z. Wang, W. Y. Li and Y. K. Zhang, *Biosens. Bioelectron.*, 2014, **54**, 266–272.
127. A. M. Bossi, P. S. Sharma, L. Montana, G. Zoccatelli, O. Laub and R. Levi, *Anal. Chem.*, 2012, **84**, 4036–4041.
128. S. M. Reddy, Q. T. Phan, H. El-Sharif, L. Govada, D. Stevenson and N. E. Chayen, *Biomacromolecules*, 2012, **13**, 3959–3965.
129. S. Khurshid, L. Govada, H. F. El-Sharif, S. M. Reddy and N. E. Chayen, *Acta Crystallogr., Sect. D: Biol. Crystallogr.*, 2015, **71**, 534–540.
130. Y. Xing, Y. F. Hu, L. Jiang, Z. D. Gao, Z. H. Chen, Z. Z. Chen and X. Q. Ren, *Cryst. Growth Des.*, 2015, **15**, 4932–4937.
131. N. W. Turner, X. Liu, S. A. Piletsky, V. Hlady and D. W. Britt, *Biomacromolecules*, 2007, **8**, 2781–2787.
132. D. R. Kryscio, M. Q. Fleming and N. A. Peppas, *Biomed. Microdevices*, 2012, **14**, 679–687.
133. K. El Kirat, M. Bartkowski and K. Haupt, *Biosens. Bioelectron.*, 2009, **24**, 2618–2624.
134. E. Saridakis, S. Khurshid, L. Govada, Q. Phan, D. Hawkins, G. V. Crichlow, E. Lolis, S. M. Reddy and N. E. Chayen, *Proc. Natl. Acad. Sci. U. S. A.*, 2011, **108**, 11081–11086.
135. M. Duman, *Mater. Sci. Eng., C*, 2014, **43**, 214–220.

136. Y. Hoshino, T. Kodama, Y. Okahata and K. J. Shea, *J. Am. Chem. Soc.*, 2008, **130**, 15242–15243.
137. Z. Y. Zeng, Y. Hoshino, A. Rodriguez, H. S. Yoo and K. J. Shea, *ACS Nano*, 2010, **4**, 199–204.
138. J. J. Xu, S. Ambrosini, E. Tamahkar, C. Rossi, K. Haupt and B. T. S. Bui, *Biomacromolecules*, 2016, **17**, 345–353.
139. J. Rick and T. C. Chou, *Anal. Chim. Acta*, 2005, **542**, 26–31.
140. A. Stephenson-Brown, A. L. Acton, J. A. Preece, J. S. Fossey and P. M. Mendes, *Chem. Sci.*, 2015, **6**, 5114–5119.

CHAPTER 10

Water-compatible Molecularly Imprinted Polymers

HUIQI ZHANG

State Key Laboratory of Medicinal Chemical Biology, Key Laboratory of Functional Polymer Materials (Ministry of Education), Collaborative Innovation Center of Chemical Science and Engineering (Tianjin), College of Chemistry, Nankai University, Tianjin 300071, P.R. China
*E-mail: zhanghuiqi@nankai.edu.cn

10.1 Introduction

Recent years have witnessed considerable interest in synthetic receptors with high affinity and selectivity toward target analytes because of their great potential in many applications including highly selective separation, catalytic processes, and sensitive chemical assays.[1–4] In comparison with biological receptors, synthetic receptors have some distinct advantages, including high physical and chemical stability, easy preparation, and low cost. Therefore, much effort has been devoted to developing synthetic receptors in order to replace their biological counterparts in practical applications. So far, a large number of synthetic receptors have been prepared, which can be divided into two main classes, *i.e.*, small organic synthetic receptors and macromolecular ones. The synthesis of small organic receptors, however, typically involves significant synthetic efforts, which dramatically limits their scale-up and practical applications.[5] It is thus highly desirable to develop other synthetically more accessible receptors. As a new

Polymer Chemistry Series No. 28
Molecularly Imprinted Polymers for Analytical Chemistry Applications
Edited by Wlodzimierz Kutner and Piyush Sindhu Sharma

Published by the Royal Society of Chemistry, www.rsc.org

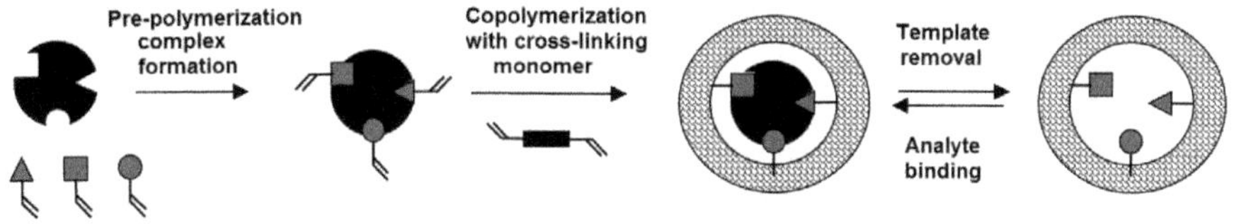

Scheme 10.1 Illustration of the molecular imprinting process. Reprinted from *Polymer*, **55**(3), H. Zhang, Water-compatible molecularly imprinted polymers: Promising synthetic substitutes for biological receptors, 699–714, Copyright 2014 with permission from Elsevier.

class of artificial macromolecular receptors that can be prepared in a facile and efficient way, molecularly imprinted polymers (MIPs) have become the hot research targets in this field.[6–24]

Molecular imprinting techniques have been developed for the preparation of MIPs, which can be simply defined as a template-induced polymerization strategy for the generation of synthetic receptors with predetermined molecular recognition ability.[6–24] It typically involves the copolymerization of a functional monomer and a cross-linking monomer in the presence of a target analyte (*i.e.*, the template) in a porogenic solvent. The subsequent removal of the template from the resulting cross-linked polymer networks leads to MIPs with molecular cavities complementary to the template in shape, size and distribution of recognizing sites, thus revealing selective template recognizing ability (Scheme 10.1). So far, three main types of molecular imprinting approaches (*i.e.*, the covalent, noncovalent, and semi-covalent approaches) have been developed on the basis of different interactions used between the templates and functional monomers (or recognizing sites in imprinted cavities) during the template imprinting and analyte binding steps. Among them, the noncovalent molecular imprinting approach has been mostly used nowadays because it can easily be accomplished without the need of complicated organic syntheses. This imprinting method is applicable to a wide range of template molecules.

So far, many different polymerization methods have been applied into molecular imprinting. Free-radical polymerization is most widely used because of its compatibility with a broad range of functional monomers and templates. In particular, controlled/"living" radical polymerization (CRP) techniques have proven highly versatile in the preparation of MIPs with well-defined structures and improved properties. Therefore, they have received rapidly increasing interest in recent years.[25–29] Moreover, some other polymerization approaches such as sol–gel polymerization[30] and electropolymerization[31] have been widely used to prepare MIPs for different purposes.

The key to the success of molecular imprinting is to choose the appropriate functional monomer, cross-linking monomer, solvent, and their composition. The most widely used functional monomers include (meth)acrylic acid (MAA or AA), 4-vinylpyridine (4-VP), and acrylamide (Am), and the cross-linking monomers typically used are ethylene glycol dimethacrylate (EGDMA), trimethylolpropane trimethacrylate (TMPTMA), and *N*,*N*′-methylene bisacrylamide (MBA).

In general, the use of an apolar solvent and an excessive amount of the functional monomer to the template is preferred in most of the noncovalent molecular imprinting systems because it is helpful to form a stronger supramolecular complex between the template and the functional monomer, which leads to better imprinting effects. In addition, some specially designed functional monomers bearing recognition sites strongly interacting with the template binding sites have been developed, which could form rather stable complexes with chosen templates even in polar solvents through stoichiometric noncovalent interactions, thus leading to MIPs with high affinity toward the templates.[32]

A large number of templates have been successfully imprinted using the noncovalent molecular imprinting approach, including small molecules with molecular weights (M_W) < 1500 Da (*e.g.*, metal ions, amino acids, and sugars, as well as pesticides, herbicides, or drugs), biomacromolecules (M_W > 1500 Da, *e.g.*, proteins or polypeptides), and even living organisms (*e.g.*, bacteria). In particular, significant progress has been made in the molecular imprinting of small templates and the commercial application of such MIPs in the solid-phase extraction area has been realized recently. In comparison, many issues still remain to be addressed for the molecular imprinting of biomacromolecular compounds, mainly because of their large sizes, complex structures, and conformational variability.[15,17,21,22] Nevertheless, some new and useful imprinting strategies have been put forward for the imprinting of biomacromolecules in the past few years, such as the epitope approach[33,34] and surface molecular imprinting method.[35]

To date, the molecular imprinting technique has become a straightforward and versatile approach for the generation of biomimetic macromolecular receptors. General applicability is one of its most distinct characteristics, which allows the preparation of synthetic receptors for a large variety of templates by using similar synthetic protocols. This characteristic makes this technique outstanding amongst the other non-biological approaches. The generated molecular cavities with recognizing sites in the MIPs can have an affinity and a selectivity approaching those of antibody–antigen systems. MIPs are thus also dubbed "antibody mimics".

Although the study on the MIPs has a rather long history and a great number of MIPs have been developed for different purposes since the first MIP was developed over 40 years ago,[12] it is still a significant challenge to develop water-compatible MIPs for small organic analytes. This challenge greatly limits practical applications of the MIPs in many bioanalytical areas involving aqueous media such as food safety control, environmental monitoring, and clinical diagnostics.[20,36]

10.2 Previous Strategies for the Preparation of MIPs Compatible with Simple Aqueous Samples

In 1993, Mosbach and coworkers[37] published a milestone work in Nature, which demonstrated that the MIPs were promising substitutes for biological receptors (*i.e.*, antibodies) in immunoassays. It is also in this paper that the authors pointed out the importance of developing water-compatible MIPs for the first time. Over the past two decades, much effort has been devoted to achieving this goal, and some useful knowledge has been obtained to shed light on the possible reasons for the incompatibility of MIPs in aqueous solutions. (1) The addition of water into the sample solutions can considerably weaken the hydrogen bonding and electrostatic interactions between template molecules and recognizing cavities of the MIPs, thus decreasing their affinity and lowering recognition ability toward the target analytes.[38–41] (2) The hydrophobic interaction between the typical MIPs and small organic molecules is largely enhanced in aqueous solutions in comparison with that in the organic solvents, which will result in large hydrophobicity-induced nonspecific template bindings.[18,42–44] Since the total binding of an MIP toward the analyte can be divided into the specific analyte binding to the recognizing cavities of the MIP and the nonspecific analyte binding to the MIP surfaces, the selectivity of the MIP will be obscured if the nonspecific binding dominates.[45] So far, some approaches have been developed on the basis of the above guidelines for the preparation of MIPs with certain selective molecular recognition ability in simple aqueous samples. These include pure water, surfactant-containing water, aqueous buffer solutions (mostly containing an organic solvent), and beer or solutions that mimic alcoholic beverages. The previously developed approaches include the utilization of the conventional molecular imprinting approach and optimization of binding conditions, the use of special polymerizable monomers (*e.g.*, tailor-designed functional monomers that can form stoichiometric noncovalent interactions with the templates and other hydrophilic functional monomers, cross-linking monomers, or co-monomers), surface post-modification, and interfacial Pickering emulsion polymerization. A detailed overview of these advances made in the preparation of water-compatible MIPs can be found in a recent feature article from our group.[20]

In addition to the aforementioned methods, several recently developed new approaches for the preparation of water-compatible MIPs are worth mentioning here. Zhao and coworkers developed an interesting method for the synthesis of protein-like and water-soluble MIP nanoparticles (NPs) for selective binding of several fluorescent organic templates including bile salts and naphthyl derivatives (salts) in water.[46–49] This method involves, first, generation of micelles in aqueous solutions by the self-assembly of a tripropargylammonium surfactant with a methacrylate-terminated hydrophobic tail, a template with a hydrophobic tail and an anionic head (*i.e.*, a sodium carboxylate or sodium sulfonate group), a hydrophobic cross-linking monomer (divinyl benzene or a fluorescent dansyl derivative with two methacrylate groups), and a photo- or thermal initiator for the free-radical polymerization. Then, these micelles are surface cross-linked *via* click chemistry by using a diazide compound. Then, the resulting surface-cross-linked micelles are further surface functionalized with a sugar-derived azide. Finally, the cores of these micelles are cross-linked by free-radical polymerization. The resulting MIP particles highly resembled protein receptors in terms of their nanodimension (with their diameters being several nanometres), complete water-solubility, functionalizable and hydrophilic exterior, hydrophobic core, and easily accessible and tailor-made hydrophobic binding pockets (typically one binding pocket per MIP NPs). In particular, they exhibited good selective binding toward the templates in aqueous buffer solutions. Lu and coworkers described a "ship-in-a-bottle" strategy for the preparation of hydrophilic spherical MIP microparticles by encapsulating MIPs inside hollow mesoporous spherical silica microparticles. The resulting MIP@SiO_2 composites revealed strong selective binding toward small organic templates, *e.g.*, 2,4-dichlorophenyloxyacetic acid (2,4-D) in aqueous solutions.[50] Liu and coworkers[51,52] reported on the facile preparation of electrochemical sensors through the direct electrodeposition of micelles from colloidal systems formed by self-assembly of the template (glucose or paracetamol) and an amphiphilic copolymer with photocross-linkable groups and some functional groups (that can interact with the templates non-covalently). Subsequently, the electrodeposited polymer film was photocross-linked on the electrode surface. The resulting chemosensors showed a high response and selectivity toward the target analytes in aqueous solutions because of the presence of hydrophilic surfaces.

Despite some progress made in the development of water-compatible MIPs over the past two decades, the design of MIPs directly capable of selective recognition of the targeted small organic analytes in real biological samples remains a formidable challenge because of the complex nature of the sample matrices.[20,42] Because trace analyses of biological samples will become more difficult or even unachievable after their high dilution, it is of the utmost importance to devise MIPs that can be applied directly in the undiluted biological samples.

10.3 Our Approaches to Preparing MIP Micro- or Nanoparticles Compatible with Aqueous Samples and Real Undiluted Biological Samples

Over the past ten years, our research interest in the field of molecular imprinting has mainly been focused on the development of water-compatible MIPs, in particular those that are capable of selective recognition of small-molecule organic analytes in real, undiluted biological samples.[20,36] Our strategy to achieving this goal is to largely improve the surface hydrophilicity of the MIP particles by their facile grafting of an ultrathin hydrophilic shell *via* the CRP techniques, such as iniferter-induced "living" radical polymerization (ILRP),[53] atom transfer radical polymerization (ATRP),[54,55] and reversible addition–fragmentation chain transfer (RAFT) polymerization.[56] The versatility of CRPs allows MIP particles to be readily grafted with tailor-made hydrophilic polymer layers with easily adjustable structural parameters, such as chemical structures, molecular weights, and grafting densities of the hydrophilic polymer brushes. Moreover, it allows controlling chemical structures, layer thickness, and cross-linking degrees of the hydrogel layers. That way, MIP particles are prepared with greatly enhanced surface hydrophilicity and largely decreased hydrophobicity-induced nonspecific template bindings. In particular, the grafted hydrophilic polymer shell could also act as a protective layer to prevent proteins in the biological samples from accumulating on the MIP particle surfaces and blocking the imprinted cavities, thus leading to MIPs fully compatible with complex biological samples.

For the above purpose, we have developed a series of new controlled/"living" radical precipitation polymerization (CRPP) techniques simply by introducing the CRP protocol into the precipitation polymerization system, such as atom transfer radical precipitation polymerization (ATRPP), RAFT precipitation polymerization (RAFTPP), and iniferter-induced "living" radical precipitation polymerization (ILRPP).[28,57–64] These techniques have proven highly versatile in direct preparation of uniform "living" polymer particles with surface-bound CRP-initiating or chain transfer groups, which allow their further grafting of well-defined hydrophilic polymer shells *via* the surface-initiated CRPs. In addition, uniform micro- or nanosized polymer particles with surface-grafted hydrophilic polymer brushes can be directly generated by using hydrophilic macromolecular CRP-initiating or chain transfer agents in the CRPP systems. By introducing CRPP techniques into the molecular imprinting field, we have been able to establish a series of facile, general, and highly efficient approaches to obtaining fully water-compatible MIP micro- or nanoparticles.[20] So far, two main types of approaches have been developed in our laboratory for the preparation of water-compatible MIPs, which are the "two-step approach"[65–73] and the "one-step approach".[74–78] While the "two-step approach" involves, first, the synthesis of "living" MIP particles with surface-bound CRP-initiating or chain transfer groups (either directly by CRPP techniques or by the grafting of MIP layers onto preformed "living" polymer/silica particles *via* surface-initiated CRPs), and their subsequent grafting of hydrophilic polymer shells (either *via* surface-initiated CRPs of hydrophilic monomers[65–72] or *via* such approaches as RAFT coupling chemistry[73]), the "one-step approach" can directly provide hydrophilic MIP micro- or nanoparticles with surface-grafted hydrophilic polymer brushes by utilizing hydrophilic macromolecular chain transfer agents (macro-CTAs) or macromolecular initiating agents in the CRPP systems.[74–78] Note, the "two-step approach" might also involve some other extra steps in order to obtain water-compatible MIPs with more advanced properties.

10.3.1 Preparation of Water-compatible MIPs *via* the "Two-step Approach"

In 2010, we presented the first report on the preparation of pure water-compatible MIP microspheres *via* the above-mentioned "two-step approach".[65] "Living" 2,4-D-imprinted polymer, (2,4-D)-MIP, spherical microparticles with surface-bound dithioester groups were firstly prepared *via* RAFTPP, and they were subsequently grafted with poly(*N*-isopropylacrylamide) (poly(NIPAAm) or PNIPAAm) brushes *via* surface-initiated RAFT polymerization (Scheme 10.2). 4-VP, EGDMA, cumyl dithiobenzoate (CDB), 2,2′-azobisisobutyronitrile (AIBN), and a mixture of methanol and water (4 : 1, (*v*/*v*)) were used as the functional monomer, cross-linking monomer, chain transfer agent, initiator, and porogenic solvent, respectively, in RAFTPP where 4-VP could hydrophobically interact and form ionic bonds with the 2,4-D template. RAFTPP was performed at 60 °C for 24 h with a reactant composition of 2,4-D/4-VP/EGDMA/AIBN/CDB being 1/4/20/0.88/1.76 (molar ratio) and the volume percentage of the porogenic solvent being ~99%. After being thoroughly purified by Soxhlet extraction, the light pink "living" (2,4-D)-MIP particles were obtained. The corresponding non-imprinted polymer or control polymer of (2,4-D)-MIP (*i.e.*, (2,4-D)-CP) was prepared similarly but in the absence of 2,4-D. The (2,4-D)-MIP/(2,4-D)-CP particles had number-average

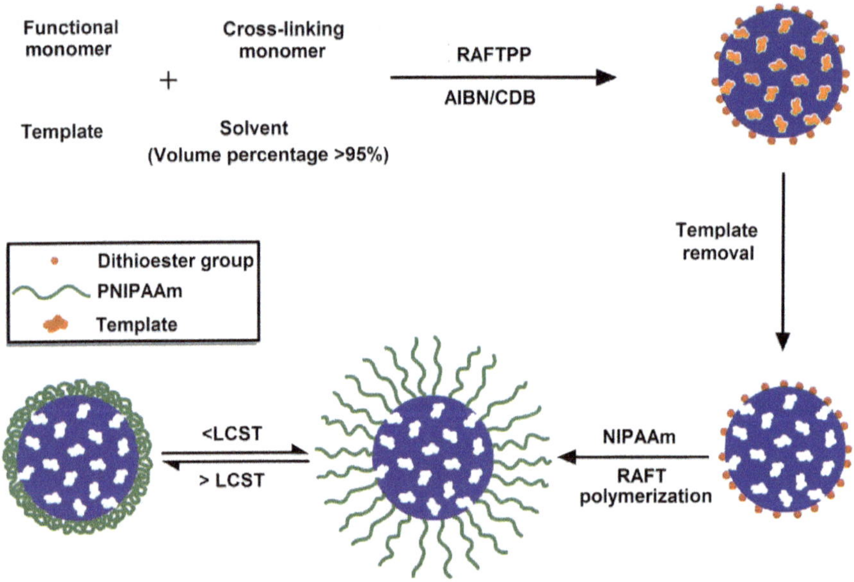

Scheme 10.2 Protocol for the preparation of water-compatible and stimuli-responsive spherical MIP microparticles with surface-grafted PNIPAAm brushes. AIBN – 2,2′-azobisisobutyronitrile, CDB – cumyl dithiobenzoate, LCST – lower critical solution temperature, NIPAAm – *N*-isopropylacrylamide, PNIPAAm – poly(*N*-isopropylacrylamide), RAFT – reversible addition–fragmentation chain transfer, RAFTPP – reversible addition fragmentation chain transfer precipitation polymerization. Reprinted from *Biosensors and Bioelectronics*, **26**(3), G. Pan *et al.*, An efficient approach to obtaining water-compatible and stimuli-responsive molecularly imprinted polymers by the facile surface-grafting of functional polymer brushes *via* RAFT polymerization, 976–982, Copyright 2010 with permission from Elsevier.

diameters (D_n) being 2.77 μM and polydispersity indices (U) being 1.12 and 1.70, respectively. Surface-initiated RAFT polymerization of NIPAAm was then performed at 70 °C to prepare (2,4-D)-MIP/(2,4-D)-CP microspheres bearing PNIPAAm brushes with the above-obtained "living" (2,4-D)-MIP/(2,4-D)-CP microspheres as the immobilized RAFT agent, AIBN as the initiator, and *N*,*N*-dimethylformamide (DMF) as the solvent. Grafting density of the grafted (2,4-D)-MIP and (2,4-D)-CP spherical microparticles was high equaling to ~1.2 and ~1.1, respectively, PNIPAAm chains per nm^2 on their surfaces. The introduction of PNIPAAm brushes onto the (2,4-D)-MIP microspheres significantly improved their surface hydrophilicity at ambient temperature and suppress the hydrophobically driven nonspecific interaction between the MIP surfaces and template molecules, thus leading to MIP particles with excellent specific template bindings (*i.e.*, the template binding differences between the MIP and its CP) in pure water. In addition, the (2,4-D)-MIP/(2,4-D)-CP microspheres with PNIPAAm brushes also showed thermo-responsive properties in the aqueous solution and exhibited thermo-responsive template binding properties because PNIPAAm is a thermo-responsive polymer that can undergo a conformation change between a hydrated (coiled and soluble) and a dehydrated (collapsed and insoluble) state in water around its lower critical solution temperature (LCST).[79,80] In this context, it is worth mentioning that our above findings represent, to our knowledge, the first successful example of achieving thermo-responsive MIPs by the facile surface modification of MIP particles with PNIPAAm brushes, although many thermo-responsive MIPs have been reported up to now.[81–87]

Following the above work, we have also prepared many different water-compatible MIPs with various hydrophilic polymer shells.[66–73] For example, a series of water-compatible (2,4-D)-MIP spherical microparticles with ultrathin hydrophilic poly(2-hydroxyethyl methacrylate) (*i.e.*, poly(HEMA) or PHEMA) shells were also prepared *via* the combined use of RAFTPP and surface-initiated RAFT polymerization.[66] The "living" (2,4-D)-MIP spherical microparticles were first prepared *via* RAFTPP, and they were subsequently grafted with either PHEMA brushes or PHEMA hydrogel layers by the surface-initiated RAFT polymerization of HEMA or a mixture of HEMA and a water-soluble cross-linking monomer, MBA (1 mole % relative to HEMA). Grafting density of the resulting hydrophilic (2,4-D)-MIP spherical microparticles was ~1 chain per nm^2 for PHEMA brushes. The equilibrium template binding capacity of the grafted (2,4-D)-MIP/(2,4-D)-CP microspheres obviously decreased with the increase of the grafting time in pure water (Figure 10.1), which could be attributed to the increase in the surface hydrophilicity of the modified MIP/CP spherical microparticles with the extended grafting time (or with an increase in the chain length of polymer brushes), leading to their reduced hydrophobically driven nonspecific template bindings. In particular, the specific template binding of the grafted (2,4-D)-MIP dramatically increased with the increase of the grafting time in the beginning and then reached their maximum values at a grafting time of 12 h. Apparently, the chain length of PHEMA brushes significantly influenced the

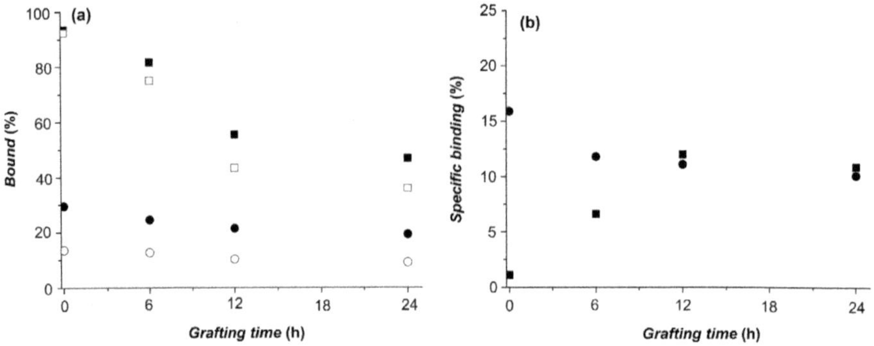

Figure 10.1 (a) Equilibrium bindings of 2,4-D on the ungrafted ($t = 0$ h) and grafted ($t = 6$, 12, and 24 h) spherical (2,4-D)-MIP (filled symbols)/(2,4-D)-CP (open symbols) microparticles in a methanol–water (4 : 1, *v*/*v*) solution (circles) and pure water (squares), respectively. (b) Selective binding of 2,4-D on the ungrafted ($t = 0$ h) and grafted ($t = 6$, 12, and 24 h) (2,4-D)-MIP microspheres in a methanol–water (4/1, *v*/*v*) solution (filled circles) and pure water (filled squares), respectively ($C_{2,4\text{-D}} = 0.02$ mM, $C_{MIP/CP} = 16$ mg mL^{-1}). Reproduced from ref. 66 with permission from The Royal Society of Chemistry.

water-compatibility of the modified MIPs and only those polymer brushes with high enough molecular weights could act as an efficient hydrophilic protective shield for the MIP spherical microparticles under a constant grafting density. These results are of significant importance for the rational design of water-compatible MIPs by using this controlled hydrophilic polymer brush-grafting approach.

Recent years have witnessed considerable interest in stimuli-responsive MIPs due to their great potential in many applications including controlled drug delivery/release, intelligent catalysis, smart assay, and controlled separation.[84–90] In addition to the preparation of the aforementioned water-compatible and thermoresponsive MIPs,[65] our "two-step approach" was also highly versatile in preparing water-compatible and dual or multiple stimuli-responsive MIPs. By taking advantage of our expertise in the fields of both molecular imprinting[64–78,91–95] and stimuli-responsive polymers,[65,96–101] we succeeded in obtaining MIP spherical microparticles with selective template recognition ability and both photo- and thermoresponsive template binding properties in pure aqueous media through the combined use of ATRPP and surface-initiated ATRP (Scheme 10.3).[67] "Living" azobenzene (azo) containing (2,4-D)-MIP spherical microparticles ($D_n = 2.31$ μm, $U = 1.29$) with surface-bound ATRP-initiating groups (*i.e.*, alkyl halide groups) and photoresponsive template binding properties were first prepared *via* ATRPP by using a methacrylate-type azo monomer bearing a pyridine group as the functional monomer, EGDMA as the cross-linking monomer, ethyl 2-chloropropionate/CuCl/*tris*[2-(dimethylamino)ethyl]amine, Me_6TREN, as the initiating system, and acetonitrile as the solvent. Thermoresponsive PNIPAAm brushes were

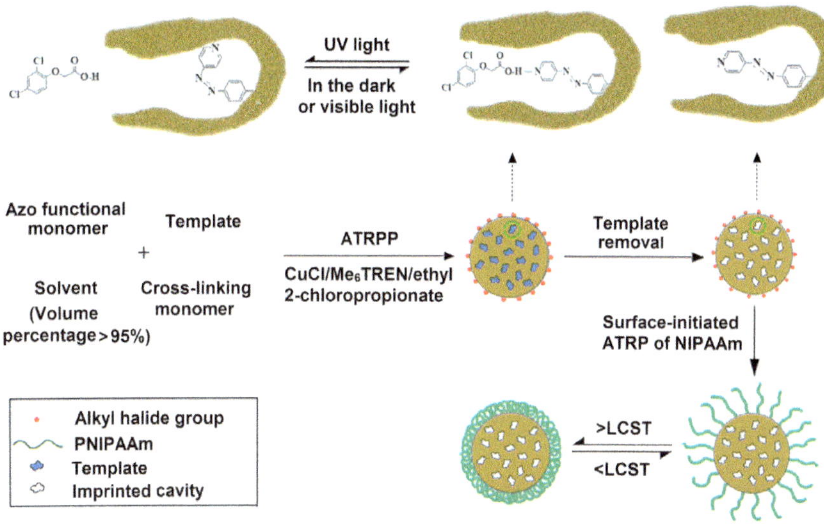

Scheme 10.3 Protocol for the synthesis of azo-containing spherical MIP microparticles with both photo- and thermoresponsive template binding properties in aqueous media by the combined use of ATRPP and surface-initiated ATRP. ATRP – atom transfer radical polymerization, ATRPP – atom transfer radical precipitation polymerization, Me_6TREN – *tris*[2-(dimethylamino)ethyl]amine, LCST - lower critical solution temperature, NIPAAm - *N*-isopropylacrylamide, PNIPAAm – poly(*N*-isopropylacrylamide). Reprinted with permission from L. Fang, S. Chen, X. Guo, *et al.*, *Langmuir*, 2012, **28**, 9767.[67] Copyright 2012 American Chemical Society.

then grafted onto these (2,4-D)-MIP microspheres *via* the surface-initiated ATRP of NIPAAm. The introduction of PNIPAAm brushes onto the photoresponsive MIP spherical microparticles significantly improved their surface hydrophilicity at ambient temperature and imparted thermoresponsive properties to them, thus leading to their pure water-compatible and both photo- and thermoresponsive template binding properties.

Moreover, a facile, general, and highly efficient approach has been developed for the preparation of uniform azo-containing MIP spherical microparticles with dual or multiple stimuli-responsive template binding properties in aqueous media.[68,69] It involves, first, the synthesis of uniform "living" core/shell-structured photoresponsive MIP spherical microparticles by simple grafting of an azo-containing MIP layer onto the uniform "living" polymer spherical microparticles, prepared *via* CRPP techniques, and their subsequent grafting of stimuli-responsive hydrophilic polymer brushes. In these cases, uniform "living" core polymer spherical microparticles with surface-bound dithioester groups were readily prepared *via* the RAFTPP of 4-VP and EGDMA, and then they were successively grafted with an azo-containing propranolol-MIP layer and thermoresponsive (PNIPAAm)[68] or both thermo- and pH-responsive (poly(NIPAAm-*co*-DMAEMA): DMAEMA refers to 2-(dimethylamino)ethyl methacrylate)[69] hydrophilic polymer brushes *via* the two-step surface-initiated RAFT polymerization (Scheme 10.4). The resulting azo-containing MIP spherical microparticles with poly(NIPAAm-*co*-DMAEMA) brushes strongly bound the template selectively revealing desired photo-, thermo-, and pH-responsive template binding properties in pure aqueous solutions. This is the first report on the successful synthesis of intelligent MIPs with multiple stimuli-responsive template binding properties.[85] The versatility of RAFT polymerization for the direct and easy synthesis of uniform "living" core polymer microspheres and growth of uniform MIP/polymer layers with adjustable thickness as well as the easy availability of many (dual or multiple) stimuli-responsive hydrophilic polymer brushes makes the present method promising to develop water-compatible MIP spherical microparticles with various multiple stimuli-responsive template binding properties.

Even though the above-obtained water-compatible and stimuli-responsive MIP particles are highly promising in such applications as controlled and sustained chemical or drug delivery/release systems, the use of the aforementioned polymer brushes with stimuli-responsive collapse and dissolution behavior as the responsive layers has led to hydrophobic MIP particles upon their collapse at elevated temperature or pH values of the MIP aqueous solutions. Therefore, such stimuli-responsive MIP particles are inappropriate for the drug delivery application in aqueous media because their resulting hydrophobic surfaces will inevitably be susceptible to high nonselective bindings of organic compounds and proteins present in biological samples. In addition, the template loading capacities of these MIPs still need to be improved to achieve their optimal drug delivery effects.

To address the above issues, we have recently developed a versatile approach to preparing advanced water-compatible and stimuli-responsive MIP spherical microparticles that can not only retain their hydrophilicity during the stimuli-responsive process, but also have largely enhanced template loading capacities (Scheme 10.5).[70] This approach involves, first, the synthesis of uniform "living" spherical silica submicroparticles with surface-bound ATRP-initiating groups *via* a one-pot sol–gel method, their subsequent surface-grafting of an azo-containing MIP layer and hydrophilic stimuli-responsive block copolymer PNIPAAm-*b*-PHEMA brushes *via* the successive surface-initiated ATRP and, finally, the removal of the silica core. The resulting hydrophilic hollow MIP microparticles showed template binding capacities higher than those of the corresponding solid ones and obvious photo- and thermoresponsive template binding properties in aqueous solutions. In addition, they exhibited pronounced light- and temperature-controlled template release in aqueous milieu. More importantly, high surface hydrophilicity was observed for the hollow MIP spherical microparticles both below and above the LCST of PNIPAAm, mainly because of the presence of external hydrophilic PHEMA layers in the grafted polymer brushes, which paves the way for their applications in such areas as controlled drug/chemical delivery and smart bioanalysis.

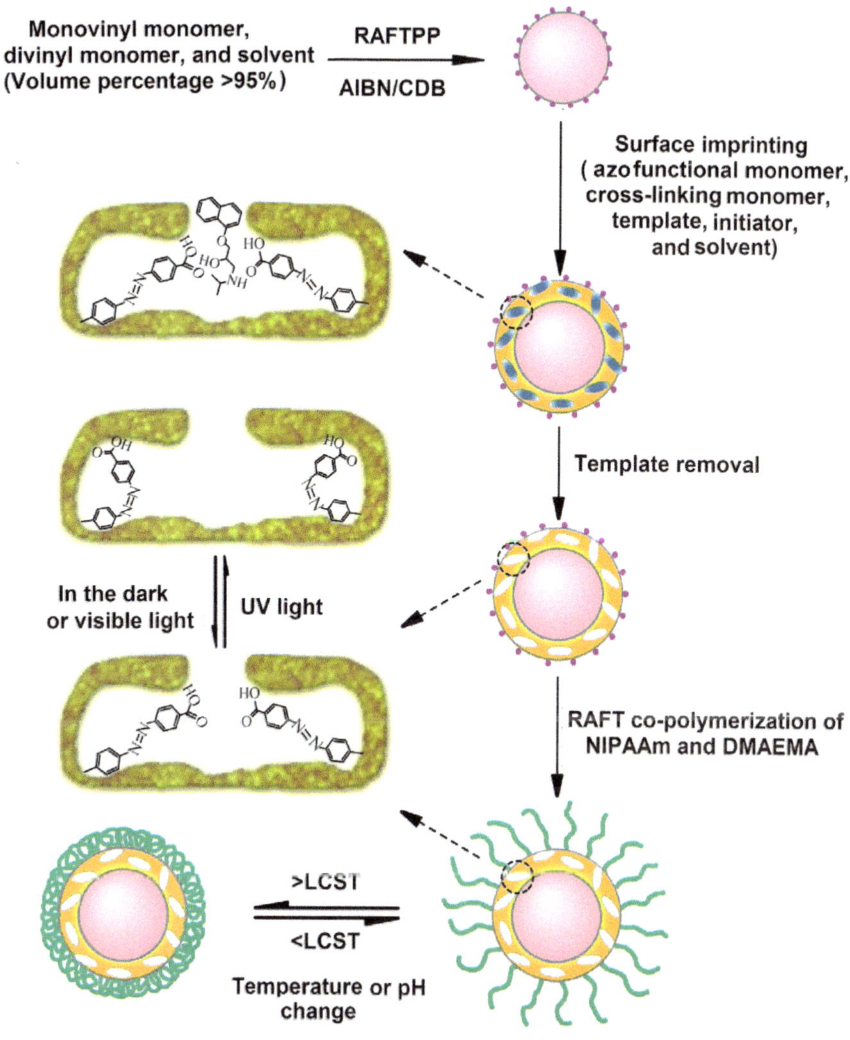

Scheme 10.4 Protocol for the preparation of uniform water-compatible spherical MIP microparticles with photo-, thermo-, and pH-responsive template binding properties by the combined use of RAFTPP and successive surface-initiated RAFT polymerization. AIBN – 2,2′-azobisisobutyronitrile, CDB – cumyl dithiobenzoate, DMAEMA – 2-(dimethylamino)ethyl methacrylate, LCST – lower critical solution temperature, NIPAAm – *N*-isopropylacrylamide, RAFT – reversible addition–fragmentation chain transfer, RAFTPP – reversible addition fragmentation chain transfer precipitation polymerization. Reproduced from ref. 69 with permission from The Royal Society of Chemistry.

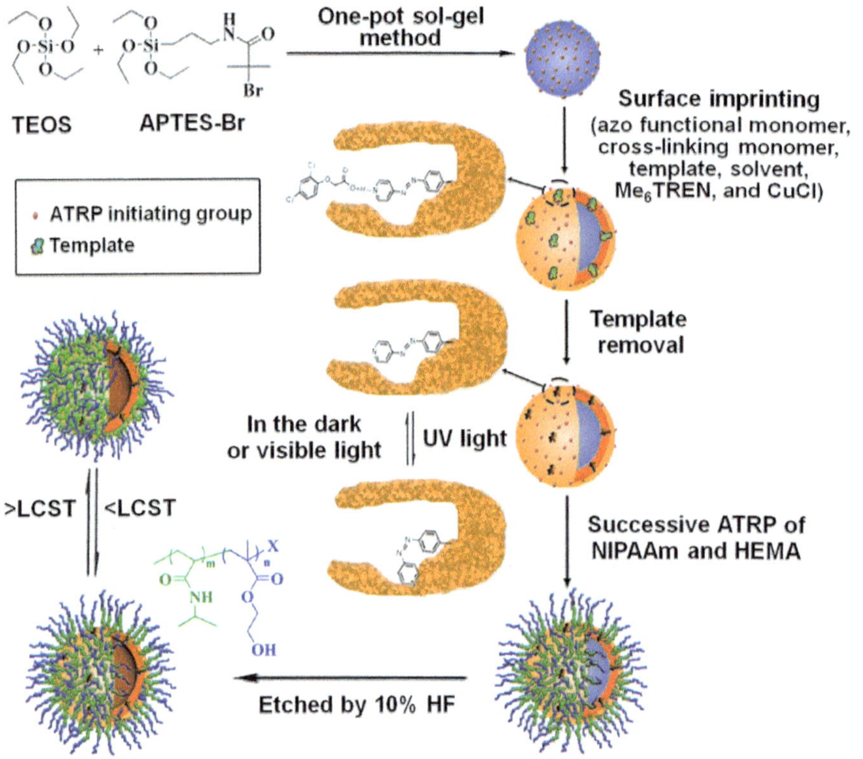

Scheme 10.5 Illustration for the preparation of photo- and thermoresponsive hydrophilic hollow azo-containing MIP microparticles bearing PNIPAAm-*b*-PHEMA brushes. ATRP – atom transfer radical polymerization, HEMA – 2-hydroxyethyl methacrylate, NIPAAm – *N*-isopropylacrylamide, LCST – lower critical solution temperature, Me_6TREN – *tris*[2-(dimethylamino)ethyl]amine, TEOS – tetraethyl orthosilicate. Reproduced with permission from C. Li, Y. Ma, H. Nui, *et al.*, *ACS Applied Materials and Interfaces*, 2015, 7, 27340.[70] Copyright 2015 American Chemical Society.

Furthermore, a facile and highly efficient new approach (namely RAFT coupling chemistry) has been developed in our laboratory for the preparation of well-defined water-compatible MIP spherical microparticles with densely grafted hydrophilic polymer brushes (~1.8 chains per nm^2) of desired chemical structures and molecular weights, and high recognition selectivity toward small-molecule organic analytes in real, undiluted biological samples (Scheme 10.6).[73] Uniform "living" MIP microspheres with surface-bound vinyl and dithioester groups were first prepared *via* RAFTPP, and they were subsequently grafted with hydrophilic polymer brushes by the simple coupling reaction of hydrophilic macro-RAFT agents, *i.e.*, PNIPAAm and poly(ethylene glycol) (PEG) with a dithioester end group, with the vinyl groups on the "living" MIP particles in the presence of a free-radical initiator.

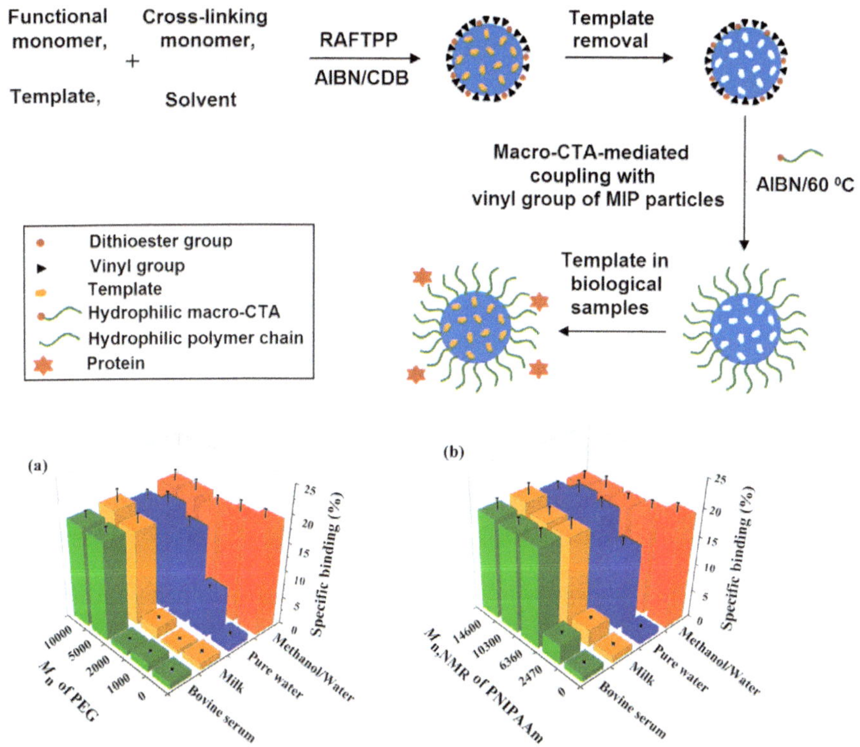

Scheme 10.6 (top) Protocol for the synthesis of well-defined spherical MIP microparticles with densely grafted hydrophilic polymer brushes by the facile RAFT coupling chemistry and their selective recognition of small organic analytes in the real, undiluted biological samples; (bottom) Effects of the molecular weight (M_n) and chemical structure of the grafted hydrophilic polymer: (a) PEG, (b) PNIPAAm, brushes on the selective template binding properties of the spherical (2,4-D)-MIP microparticles. AIBN – 2,2′-azobisisobutyronitrile, CDB – cumyl dithiobenzoate, Macro-CTA – macromolecular chain transfer agent, $M_{n,NMR}$ – number-average M_w determined by H^1 NMR spectroscopy, PEG – poly(ethylene glycol), PNIPAAm – poly(*N*-isopropylacrylamide), RAFTPP – reversible addition–fragmentation chain transfer precipitation polymerization. Reprinted with permission from M. Zhao, X. Chen, H. Zhang, *et al.*, *Biomacromolecules*, 2014, **15**, 1663.[73] Copyright 2014 American Chemical Society.

The well-defined characteristics of the resulting hydrophilic MIP particles allowed the first systematic study on the effects of various structural parameters (including chemical structures, molecular weights, and grafting densities) of the grafted hydrophilic polymer brushes on their water-compatible properties. The water-compatibility of the resulting grafted MIP particles increased with the increase of the hydrophilicity, molecular weights, or grafting densities of the grafted polymer brushes, and only those polymer brushes with sufficiently high molecular weights could act as an efficient hydrophilic protective shield for the MIP particles under a constant grafting density. In addition, much longer polymer brushes were required for the grafted MIP particles to show high compatibility with aqueous samples when the grafting densities of the hydrophilic polymer brushes were lower. Furthermore, PEG brushes were somewhat more efficient than PNIPAAm brushes in improving the pure water compatibility of the grafted MIPs. These results provided important and useful information for rational designing more advanced real biological sample-compatible MIPs.

Although our water-compatible MIPs obtained in the above cases revealed almost the same specific template bindings in aqueous media (or real biological milieu) as those in the organic solvent-based media, their nonspecific template bindings in aqueous media are typically much higher than those in the organic solvent-based systems. This drawback will be very detrimental to the assay and chemosensor applications of the water-compatible MIPs. The design of MIPs with both their specific and nonspecific template bindings in aqueous solutions (especially in real biological samples) being the same as those obtained in the organic solvent-based media is thus highly important.

To this end, we have recently developed an effective approach to obtaining uniform hydrophilic and magnetic MIP spherical microparticles with a molecular recognition ability in a real biological sample (*i.e.*, undiluted bovine serum) as good as that found in organic solvent-based media.[71] Uniform (2,4-D)-MIP spherical microparticles with surface-grafted hydrophilic poly(glyceryl monomethacrylate) (poly(GMMA) or PGMMA) brushes were first prepared by the combined use of RAFTPP and surface-initiated RAFT polymerization of GMMA. Subsequently, magnetic Fe_3O_4 NPs were attached to the PGMMA brushes of the MIP particles *via* simple co-precipitation. The key to the success of this strategy is the introduction of PGMMA brushes onto the MIP spherical microparticles, which allows the easy and stable immobilization of Fe_3O_4 NPs by their strong multidentate interactions with the 1,2-diol groups of the polymer brushes.[102] The presence of Fe_3O_4 NPs on the polymer brushes of the MIP spherical microparticles not only imparted them with appropriate magnetic properties easily, but also largely improved their surface hydrophilicity, thus leading to the water-compatible and magnetic MIPs. Those MIPs showed the same specific and nonspecific template binding in a real biological sample as those obtained in the organic solvent-based medium.

Biosensors and chemical sensors (or chemosensors) have attracted enormous attention in recent years because they are of great importance in a wide range of bioanalytical applications, particularly in controlling the food quality, monitoring the environmental conditions, and performing medical diagnostics.[103,104] Among various studied sensing systems, MIP-based chemosensors have become a particularly hot research targets because of the obvious advantages of the MIPs (including their high stability, easy preparation, and low cost) over the biological recognition elements (*e.g.*, antibody and enzyme) and their ease of integration with different tranducers.[6,17,23,105]

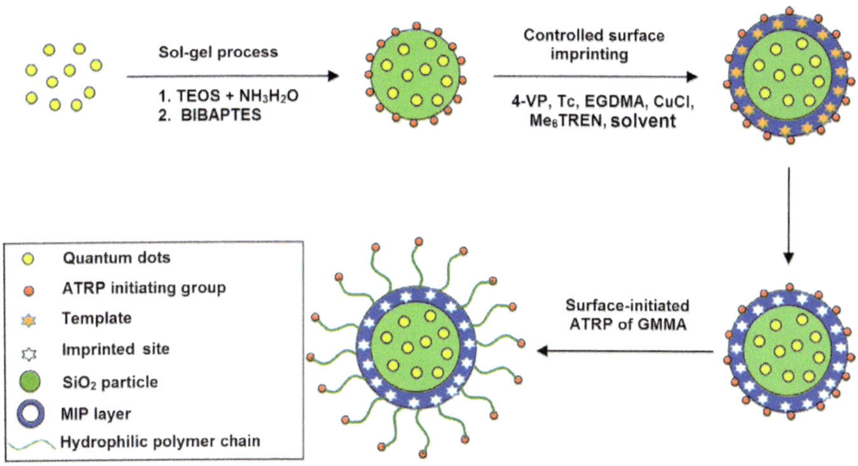

Scheme 10.7 Protocol for the preparation of QD-labeled hydrophilic spherical MIP microparticles. ATRP – atom transfer radical polymrization, BIBAPTES – 3-(*N*-propyl)triethoxysilane 2-bromo-2-methylpropanamide, EGDMA – ethylene glycol dimethacrylate, GMMA – glyceryl monomethacylate, Me_6TREN – *tris*[2-(dimethylamino)ethyl]amine, 4-VP – 4-vinylpyridine, Tc – tetracycline, TEOS – tetraethyl orthosilicate. Reprinted with permission from Y. Yang, H. Niu and H. Zhang, *ACS Applied Materials and Interfaces*, 2016, **8**, 15741.[72] Copyright 2016 American Chemical Society.

Very recently, we have also successfully utilized our "two-step approach" to preparing quantum-dot (QD)-labeled hydrophilic fluorescent MIP microparticles for direct and highly selective optosensing of a tetracycline (Tc) broad-spectrum antibiotic drug in the real, undiluted biological samples (including samples of pure bovine/goat milk and bovine/porcine serum) (Scheme 10.7).[72] "Living" CdTe QD-SiO_2 composite microparticles with surface-bound alkyl halide substituents were readily prepared by introducing CdTe QDs and a 3-(*N*-propyl)triethoxysilane 2-bromo-2-methylpropanamide (BIBAPTES) alkyl halide group-containing silicate monomer into the sol–gel reaction system. Those "living" QD-SiO_2 composite microparticles were subsequently grafted with a Tc-MIP layer and PGMMA brushes *via* the successive surface-initiated ATRP. The introduction of hydrophilic PGMMA brushes and fluorescence labeling onto/into the MIP particles not only significantly improved their surface hydrophilicity and led to their compatibility in the undiluted biological media, but also imparted them with strong fluorescent properties. In particular, significant fluorescence quenching was observed upon their binding with Tc in such complex biological milieu, which makes this MIP-Tc a useful optical chemosensor with its limits of detection down to 0.14 μM in complex biological media. Moreover, a facile and effective approach was developed based on a newly derived equation (from Stern–Volmer equation) to eliminate the false positive problem of the fluorescent chemosensor and provide it with wider linear dynamic concentration ranges in comparison with those obtained using the generally adopted Stern–Volmer equation. The fluorescent MIP chemosensor was also successfully applied for direct, sensitive, selective, and accurate quantification of Tc in different biological media, and the average recovery values were in the range of 95–105%, even when several other drugs and the fluorescently interfering chlortetracycline were present in the samples.

Our "two-step approach" has some obvious advantages over the previously reported methods (*e.g.*, adding hydrophilic monomers into the molecular imprinting systems) in obtaining water-compatible MIPs. These include: (1) all the hydrophilic monomers are grafted on the surface of the MIP particles by using our method, which can thus more efficiently enhance the surface hydrophilicity of the MIPs; (2) the controlled characteristics of CRPs allow for easy tailoring of the structural parameters of the grafted hydrophilic polymer shells, including the chemical structure, molecular weights, and grafting densities of the polymer brushes, as well as the chemical structures, layer thickness, and cross-linking degree of the hydrogel layers, thus readily leading to MIPs with high surface hydrophilicity and water-compatibility; (3) the most distinct advantage of our approach is its synthetic simplicity and flexibility, because CRPP techniques could readily provide "living" MIP particles with high molecular recognition ability and adjustable sizes for a wide range of templates under the normal molecular imprinting conditions (no hydrophilic co-monomer is needed for imprinting, which can avoid much time-consuming effort for optimizing MIP formulations), and the subsequent facile surface-grafting of hydrophilic polymer brushes can in principle make all of them compatible with real biological samples easily; (4) our strategy is extendable to many other supporting substrates, such as flat silica wafers and functional NPs (*e.g.*, magnetic or gold NPs), because their surfaces can easily be modified with "living" CRP-initiating or chain transfer groups, which makes it possible to efficiently prepare more advanced functional and smart MIPs with great potential in various bioanalytical applications such as MIP-based chemosensors, stimuli-responsive drug delivery, bioimaging, and theranostics.

In this context, it is worth mentioning here that many hydrophilic/water-compatible MIP particles have also been prepared by others following our "two-step approach" recently.[106–117] In addition, some MIP spherical microparticles with highly selective template binding properties have also been successfully synthesized *via* CRPP techniques by others.[118–121]

10.3.2 Preparation of Water-compatible MIPs *via* "One-step Approach"

In 2011, we developed a "one-step approach" in order to prepare water-compatible MIPs in a more facile and efficient way.[74] A series of uniform (2,4-D)-MIP spherical microparticles with surface-grafted hydrophilic polymer brushes were directly generated *via* the facile RAFTPP mediated by hydrophilic macro-CTAs (*i.e.*, PNIPAAm and PEG with the dithioester end-group and different

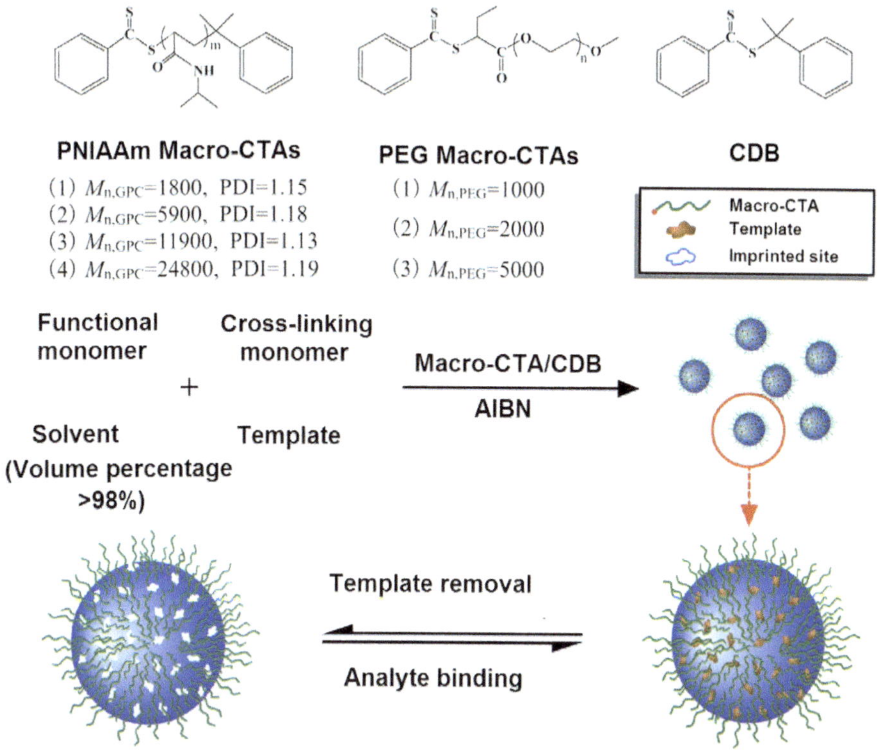

Scheme 10.8 Structural formulas of the utilized RAFT agents (including hydrophilic macro-CTAs and CDB) and the schematic protocol for the one-pot preparation of water-compatible spherical MIP microparticles by RAFTPP. AIBN – 2,2′-azobisisobutyronitrile, CDB – cumyl dithiobenzoate, Macro-CTA – macromolecular chain transfer agent, PEG – poly(ethylene glycol), PNIPAAm – poly(*N*-isopropylacrylamide). Reproduced from G. Pan *et al.*, Efficient One-Pot Synthesis of Water-Compatible Molecularly Imprinted Polymer Microspheres by Facile RAFT Precipitation Polymerization, *Angewandte Chemie*, *International Edition*, John Wiley and Sons, © 2011 WILEY-VCH Verlag GmbH & Co. KGaA, Weinheim.

molecular weights) (Scheme 10.8). The influence of the utilized hydrophilic macro-CTAs on the formation of selective recognizing sites was negligible, as revealed by almost the same specific template binding by the MIP spherical microparticles with different polymer brushes in a methanol–water (4 : 1, *v*/*v*) solution. However, the chain length of the grafted hydrophilic polymer brushes significantly influenced the water-compatibility of the grafted MIP spherical microparticles and only those polymer brushes with sufficiently high molecular weights could act as an efficient hydrophilic protective shield for the MIP spherical microparticles. This effect was demonstrated by a dramatic increase of the specific template bindings by the resulting MIPs in the pure aqueous media when the chain length of the macro-CTAs was increased in the low-molecular-weight range, and then leveled off at $M_w \geq 2000$ Da for PEG macro-CTAs and $M_w \geq 5900$ Da for PNIPAAm macro-CTAs.

MIP NPs have drawn a rapidly increasing interest over the past few years because of their great potential in selective separation, biomimetic immunoassays, chemosensors, drug delivery, bioimaging, and biomedicine.[122–127] The development of water-compatible MIP NPs is thus of the utmost importance. By utilizing the hydrophilic macro-CTA-mediated RAFTPP, but simply changing the above-used hydrophilic PNIPAAm and PEG macro-CTAs to PHEMA macro-CTAs (prepared *via* the RAFT polymerization of HEMA), we have succeeded in preparing uniform hydrophilic MIP NPs with highly selective molecular-recognition ability in real aqueous solutions, including river water and real, undiluted biological samples (*i.e.*, pure milk and bovine serum) (Scheme 10.9).[75] These MIP NPs showed considerably enhanced surface hydrophilicity and high dispersion stability in aqueous solutions. The compatibility of the MIP NPs with biological samples was significantly affected by the molecular weights of their grafted

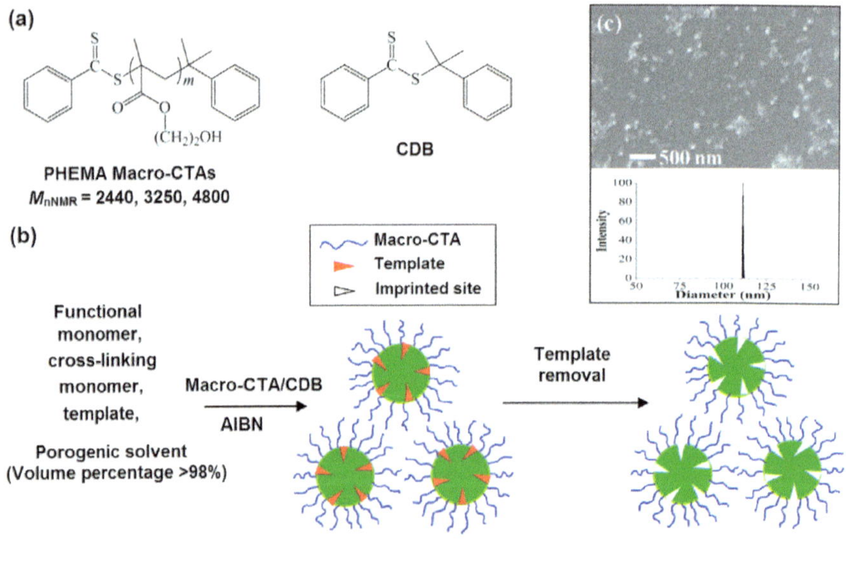

Scheme 10.9 (a) Structural formulas of the RAFT agents used including PHEMA macro-CTAs and CDB. (b) Protocol for the one-pot preparation of MIP NPs compatible with real aqueous solutions *via* RAFTPP. (c) Scanning electron microscopy (SEM) image (top) and dynamic laser scattering (DLS) characterization (bottom) of MIP NPs ($M_{n,NMR}$ of PHEMA brushes: 4800 Da). AIBN – 2,2′-azobisisobutyronitrile, CDB – cumyl dithiobenzoate, Macro-CTA – macromolecular chain transfer agent, PHEMA – poly(2-hydroxyethyl methacrylate). Reproduced from Y. Ma *et al.*, Narrowly Dispersed Hydrophilic Molecularly Imprinted Polymer Nanoparticles for Efficient Molecular Recognition in Real Aqueous Samples Including River Water, Milk and Bovine Serum, *Angewandte Chemie International Edition*, John Wiley and Sons, © 2013 WILEY-VCH Verlag GmbH & Co. KGaA, Weinheim.

hydrophilic polymer brushes, and only sufficiently long polymer brushes could form an effective hydrophilic protective layer to prevent the accumulation of proteins on the surfaces of the MIP NPs and allow them to function properly in complex biological samples. In particular, the general applicability of our strategy was also verified by preparing biological sample-compatible MIP NPs for different template molecules (*i.e.*, 2,4-D and propranolol).[75] To our knowledge, this report describes the first successful preparation of MIP NPs that are capable of selective recognizing small organic compounds directly in undiluted biological samples. This finding is a major breakthrough for molecular imprinting technology, because it paves the way for the facile preparation of biological sample-compatible MIP NPs that are very attractive synthetic substitutes for biological receptors in various bioanalytical applications.

Moreover, the hydrophilic macro-CTA-mediated RAFTPP has been applied to prepare water-compatible MIP NPs with advanced properties for different purposes in our laboratory. For example, water-compatible and photoresponsive MIP NPs have been prepared in a one-pot process by the combined use of PHEMA macro-CTA-mediated RAFTPP and easily available water-insoluble azo functional monomers (Scheme 10.10).[76] The resulting azo-containing MIPs were nanosized particles (with their diameters of ~104–397 nm, as determined by DLS in methanol) with surface-grafted hydrophilic polymer brushes. They exhibited excellent properties of template binding in a pure water system. In addition, they also showed obvious photoregulated template binding behavior and largely accelerated template release

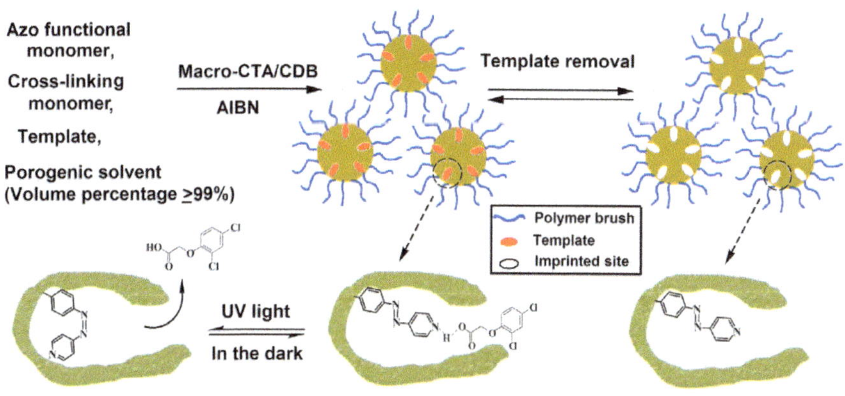

Scheme 10.10 Illustration for the one-pot synthesis of water-compatible and photoresponsive azo-containing MIP NPs *via* hydrophilic macro-CTA-mediated RAFTPP. AIBN – 2,2′-azobisisobutyronitrile, CDB – cumyl dithiobenzoate, Macro-CTA – macromolecular chain transfer agent. Reproduced from Y. Ma *et al.*, Efficient one-pot synthesis of water-compatible and photoresponsive molecularly imprinted polymer nanoparticles by facile RAFT precipitation polymerization, *Journal of Polymer Science Part A: Polymer Chemistry*, John Wiley and Sons, © 2014 Wiley Periodicals, Inc.

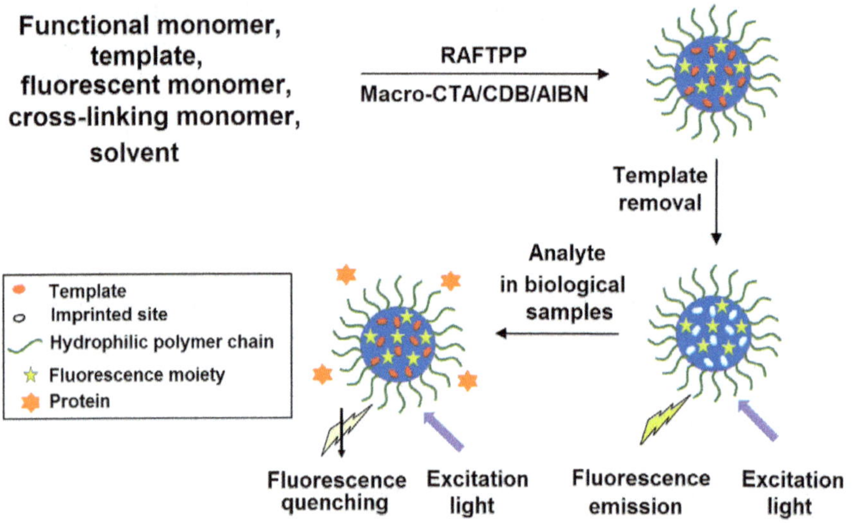

Scheme 10.11 Protocol for the synthesis of hydrophilic and fluorescent MIP NPs for direct and rapid drug optosensing in real, undiluted biological samples. AIBN – 2,2′-azobisisobutyronitrile, CDB – cumyl dithiobenzoate, Macro-CTA – macromolecular chain transfer agent, RAFTPP – reversible addition–fragmentation chain transfer precipitation polymerization. Reprinted from *Biosensors and Bioelectronics*, **74**, H. Niu *et al.*, Efficient one-pot synthesis of hydrophilic and fluorescent molecularly imprinted polymer nanoparticles for direct drug quantification in real biological samples, 440–446, Copyright 2015 with permission from Elsevier.

in aqueous media under UV light irradiation. Furthermore, the general applicability of the strategy was demonstrated by preparing water-compatible and photoresponsive MIP NPs for different templates (*i.e.*, 2,4-D and propranolol).

Recently, we have reported the one-pot synthesis of hydrophilic and fluorescent MIP NPs *via* hydrophilic macro-CTA-mediated RAFTPP and their application for direct drug quantification in the real, undiluted biological samples (*i.e.*, bovine and porcine serums).[77] The general principle was demonstrated by preparing fluorescent Tc-MIP NPs bearing hydrophilic polymer brushes *via* PHEMA CTA-mediated RAFTPP in the presence of a (2-hydroxyethyl anthrancene-9-carboxylate) methacrylate polymerizable fluorescent monomer (Scheme 10.11). The hydrodynamic diameter of the resulting MIP NPs was 49 nm in water (determined by DLS) and their number-average diameter was ~15 nm in the dry state [determined by atomic force microscopy (AFM) height profiling]. The introduction of fluorescent labeling and hydrophilic polymer brushes into/onto MIP NPs allowed them to function properly in biological samples and show obvious template-binding-induced fluorescence quenching, which make them a useful fluorescent chemosensor with a LOD of 0.26 μM in complex biological milieu (*i.e.*, undiluted bovine serum).

Moreover, the application of such an advanced functional MIP NP-based chemosensor for direct, sensitive, and accurate determination of Tc in the undiluted porcine serum has been demonstrated, with an average recovery ranging from 98% to 102%, even in the presence of several interfering drugs.

Very recently, we have also prepared QD-labeled hydrophilic MIP NPs *via* PGMMA macro-CTA-mediated RAFTPP for direct optosensing of folic acid (FA) in undiluted bovine and porcine serums.[78] The resulting FA-MIP NPs with surface-grafted hydrophilic PGMMA brushes and QDs labeling (D_n = 172 nm in DMF by DLS and ~50 nm in the dry state by AFM height profiling) not only showed outstanding selective molecular recognition toward FA in biological samples, but also exhibited high photostability, rapid binding kinetics, and obvious template binding-induced fluorescence quenching. These characteristics make them a useful fluorescent chemosensor for direct and selective optosensing of FA in undiluted bovine and porcine serums, with its LOD being 0.025 μM and average recovery ranging from 98% to 102%, even in the presence of several interfering compounds.

The above results have demonstrated that our "one-pot approach" is a facile, general, and highly efficient strategy for the preparation of various water-compatible MIPs. The versatility of RAFTPP for the controlled preparation of MIP micro- or nanoparticles, together with the easy availability of many different hydrophilic macro-CTAs by either RAFT polymerization of hydrophilic monomers or hydrophilic polymer end group modification, makes this method highly applicable. This has also been clearly demonstrated by the adoption of this method for the synthesis of water-compatible MIPs or other functional polymer spherical microparticles by others.[128–130]

10.4 Summary and Outlook

This book chapter presents a detailed overview of the research progress made in the recent development of water-compatible MIPs, mainly in our group. Evidently, significant advances have been achieved in this field and some facile, general, and highly efficient approaches have been successfully developed for this purpose. Our strategy has proven highly versatile in the preparation of water-compatible MIPs with highly selective molecular-recognition ability in even rather complex biological samples, because it allows easy tailoring of surface hydrophilicity of the MIP particles by the facile and controlled grafting of an ultrathin hydrophilic protective shell with easily adjustable structural parameters (including chemical structures, molecular weights, and grafting densities of the hydrophilic polymer brushes as well as the chemical structures, layer thickness, and cross-linking degrees of the hydrogel layers). The experimental results have clearly demonstrated that the MIP micro- or nanoparticles with surface-grafted hydrophilic polymer brushes are highly promising in many applications including selective analyte separation, drug delivery, chemosensors, bioimaging, and theranostics. Further effort should be devoted to the fabrication of water-compatible MIPs with improved binding properties (*e.g.*, with more homogeneous binding sites) in order to make them more suitable substitutes for their biological counterparts. Moreover, the development of more advanced water-compatible MIP micro- or nanoparticles by incorporating various other functionalities is highly desirable for their practical applications. We believe that these water-compatible MIPs with high molecular recognition ability, stability, and low cost hold great promise in a wide range of bioanalytical and biomedical applications (*e.g.*, food safety control, environmental monitoring, and clinical diagnostics and biomedicine), and their routine use in some of these areas may not lie in the too distant future.

List of Abbreviations and Symbols

AA	Acrylic acid
AFM	Atomic force microscopy
AIBN	2,2′-Azobisisobutyronitrile
Am	Acrylamide
ATRP	Atom transfer radical polymerization
ATRPP	Atom transfer radical precipitation polymerization
Azo	Azobenzene
BIBAPTES	3-(*N*-propyl)triethoxysilane 2-bromo-2-methylpropanamide
CP	Control polymer or non-imprinted polymer
CRP	Controlled/"living" radical polymerization
CRPP	Controlled/"living" radical precipitation polymerization
CDB	Cumyl dithiobenzoate
D_n	Number-average diameter
2,4-D	2,4-Dichlorophenyloxyacetic acid
(2,4-D)-CP	Control polymer of 2,4-D-MIP
(2,4-D)-MIP	2,4-D-imprinted polymer
DLS	Dynamic laser scattering
DMAEMA	2-(Dimethylamino)ethyl methacrylate
DMF	*N,N*-Dimethylformamide
EGDMA	Ethylene glycol dimethacrylate
FA	Folic acid
FA-MIP	FA-imprinted polymer
GMMA	Glyceryl monomethacrylate
HEMA	2-Hydroxyethyl methacrylate
ILRP	Iniferter-induced "living" radical polymerization
ILRPP	Iniferter-induced "living" radical precipitation polymerization
NIPAAm	*N*-Isopropylacrylamide
LOD	Limit of detection
LCST	Lower critical solution temperature
macro-CTA	Macromolecular chain transfer agent
MAA	Methacrylic acid
MBA	*N,N*′-Methylene bisacrylamide
Me_6TREN	*tris*[2-(Dimethylamino)ethyl]amine

MIPs	Molecularly imprinted polymers
M_W	Molecular weight
$M_{n,NMR}$	Number-average M_W determined by ^1H NMR spectroscopy
NP	Nanoparticle
PEG	Poly(ethylene glycol)
PGMMA	Poly(glyceryl monomethacrylate)
PHEMA	Poly(2-hydroxyethyl methacrylate)
PNIPAAm	Poly(*N*-isopropylacrylamide)
PNIPAAm-*b*-PHEMA	Block copolymer of NIPAAm and HEMA
Poly(NIPAAm-*co*-DMAEMA)	Random copolymer of NIPAAm and DMAEMA
QD	Quantum dot
RAFT	Reversible addition–fragmentation chain transfer
RAFTPP	RAFT precipitation polymerization
SEM	Scanning electron microscopy
Tc	Tetracycline
Tc-MIP	Tc-imprinted polymer
TMPTMA	Trimethylolpropane trimethacrylate
U	Polydispersity index
UV	Ultraviolet
4-VP	4-Vinylpyridine

Acknowledgements

The author thanks his coworkers for their contribution to the works presented in this book chapter. The financial supports from National Natural Science Foundation of China (Nos. 20744003, 20774044, 21174067, and 21574070), Natural Science Foundation of Tianjin (Nos. 11JCYBJC01500 and 16JCZDJC36800), Doctoral Fund of Ministry of Education of China (No. 20130031110018), a supporting program for New Century Excellent Talents (Ministry of Education) (NCET-07-0462), and PCSIRT (IRT1257) are also acknowledged.

References

1. D. M. Vriezema, M. C. Aragonès, J. A. A. Elemans, J. J. L. M. Cornelissen, A. E. Rowan and R. J. M. Nolte, *Chem. Rev.*, 2005, **105**, 1445.
2. S. Dutt, C. Wilch and T. Schrader, *Chem. Commun.*, 2011, **47**, 5376.
3. M. C. Moreno-Bondi, M. E. Benito-Peña, J. L. Urraca and G. Orellana, *Top. Curr. Chem.*, 2012, **325**, 111.
4. C. Baggiani, L. Anfossi and C. Giovannoli, *Mol. Imprinting*, 2013, 41.
5. F. Hof, S. L. Craig, C. Nuckolls and J. Rebek Jr, *Angew. Chem., Int. Ed.*, 2002, **41**, 1488.
6. K. Haupt and K. Mosbach, *Chem. Rev.*, 2000, **100**, 2495.
7. B. Sellergren, *Angew. Chem., Int. Ed.*, 2000, **39**, 1031.

8. M. J. Whitcombe and E. N. Vulfson, *Adv. Mater.*, 2001, **13**, 467.
9. G. Wulff, *Chem. Rev.*, 2002, **102**, 1.
10. K. Haupt, *Chem. Commun.*, 2003, 171.
11. J. Z. Hilt and M. E. Byrne, *Adv. Drug Delivery Rev.*, 2004, **56**, 1599.
12. C. Alexander, H. S. Andersson, L. I. Andersson, R. J. Ansell, N. Kirsch, I. A. Nicholls, J. O'Mahony and M. J. Whitcombe, *J. Mol. Recognit.*, 2006, **19**, 106.
13. H. Zhang, L. Ye and K. Mosbach, *J. Mol. Recognit.*, 2006, **19**, 248.
14. L. Ye and K. Mosbach, *Chem. Mater.*, 2008, **20**, 859.
15. T. Takeuchi and T. Hishiya, *Org. Biomol. Chem.*, 2008, **6**, 2459.
16. Y. Hoshino and K. J. Shea, *J. Mater. Chem.*, 2011, **21**, 3517.
17. M. J. Whitcombe, I. Chianella, L. Larcombe, S. A. Piletsky, J. Noble, R. Porter and A. Horgan, *Chem. Soc. Rev.*, 2011, **40**, 1547.
18. K. Haupt, A. V. Linares, M. Bompart and B. Tse Sum Bui, *Top. Curr. Chem.*, 2012, **325**, 1.
19. G. Wulff and J. Liu, *Acc. Chem. Res.*, 2012, **45**, 239.
20. H. Zhang, *Polymer*, 2014, **55**, 699.
21. S. J. Li, S. Cao, M. J. Whitcombe and S. A. Piletsky, *Prog. Polym. Sci.*, 2014, **39**, 145.
22. H. Zhang, *Mol. Imprinting*, 2015, **2**, 63.
23. J. Wackerlig and P. A. Lieberzeit, *Sens. Actuators, B*, 2015, **207**, 144.
24. L. Chen, X. Wang, W. Lu, X. Wu and J. Li, *Chem. Soc. Rev.*, 2016, **45**, 2137.
25. H. Zhang, *Molecular Imprinting: Principles and Applications of Micro- and Nano-structured Polymers*, ed. L. Ye, Pan Stanford Publishing, Singapore, 2013, ch. 4, pp. 105–160.
26. Y. Zhang and H. Zhang, *Chin. J. React. Polym.*, 2008, **17**, 1.
27. M. Bompart and K. Haupt, *Aust. J. Chem.*, 2009, **62**, 751.
28. H. Zhang, *Eur. Polym. J.*, 2013, **49**, 579.
29. S. Beyazit, B. Tse Sum Bui, K. Haupt and C. Gonzato, *Prog. Polym. Sci.*, 2016, **62**, 1.
30. J. E. Lofgreen and A. O. Geoffrey, *Chem. Soc. Rev.*, 2014, **43**, 911.
31. C. Malitesta, E. Mazzotta, R. A. Picca, A. Poma, I. Chianella and S. A. Piletsky, *Anal. Bioanal. Chem.*, 2012, **402**, 1827.
32. G. Wulff and K. Knorr, *Bioseparation*, 2001, **10**, 257.
33. A. Rachkov and N. Minoura, *J. Chromatogr. A*, 2000, **889**, 111.
34. H. Nishino, C. S. Huang and K. J. Shea, *Angew. Chem., Int. Ed.*, 2006, **45**, 2392.
35. X. Ding and P. A. Heiden, *Macromol. Mater. Eng.*, 2014, **299**, 268.
36. H. Zhang, *Chin. Polym. Bull.*, 2013, **1**, 13–25.
37. G. Vlatakis, L. I. Andersson, R. Müller and K. Mosbach, *Nature*, 1993, **361**, 645.
38. I. A. Nicholls, O. Ramström and K. Mosbach, *J. Chromatogr. A*, 1995, **691**, 349.
39. I. A. Nicholls, *J. Mol. Recognit.*, 1998, **11**, 79.
40. S. A. Piletsky, H. S. Andersson and I. A. Nicholls, *Macromolecules*, 1999, **32**, 633.

41. J. G. Karlsson, L. I. Andersson and I. A. Nicholls, *Anal. Chim. Acta*, 2001, **435**, 57.
42. J. L. Bowen, P. Manesiotis and C. J. Allender, *Mol. Imprinting*, 2012, 35.
43. X. Shen, C. Xu and L. Ye, *Ind. Eng. Chem. Res.*, 2013, **52**, 13890.
44. C. Baggiani, L. Anfossi and C. Giovannoli, *Mol. Imprinting*, 2013, 41.
45. L. I. Anderson, R. Müller, G. Vlatakis and K. Mosbach, *Proc. Natl. Acad. Sci. U. S. A.*, 1995, **92**, 4788.
46. J. K. Awino and Y. Zhao, *J. Am. Chem. Soc.*, 2013, **135**, 12552.
47. J. K. Awino and Y. Zhao, *Chem. Commun.*, 2014, **50**, 5752.
48. J. K. Awino and Y. Zhao, *Chem.–Eur. J.*, 2015, **21**, 655.
49. J. K. Awino, L. Hu and Y. Zhao, *Org. Lett.*, 2016, **18**, 1650.
50. S. Yang, N. Li, Q. Xing, D. Chen, Q. Xu, H. Li and J. Lu, *Chem.–Asian J.*, 2015, **10**, 722.
51. Y. Yang, C. Yi, J. Luo, R. Liu, J. Liu, J. Jiang and X. Liu, *Biosens. Bioelectron.*, 2011, **26**, 2607.
52. J. Luo, C. Fan, X. Wang, R. Liu and X. Liu, *Sens. Actuators, B*, 2013, **188**, 909.
53. T. Otsu, *J. Polym. Sci., Part A: Polym. Chem.*, 2000, **38**, 2121.
54. K. Matyjaszewski and J. Xia, *Chem. Rev.*, 2001, **101**, 2921.
55. M. Kamigaito, T. Ando and M. Sawamoto, *Chem. Rev.*, 2001, **101**, 3689.
56. G. Moad, E. Rizzardo and S. H. Thang, *Polymer*, 2008, **49**, 1079.
57. B. Zu, G. Pan, X. Guo, Y. Zhang and H. Zhang, *J. Polym. Sci., Part A: Polym. Chem.*, 2009, **47**, 3257.
58. G. Pan, B. Zu, X. Guo, Y. Zhang, C. Li and H. Zhang, *Polymer*, 2009, **50**, 2819.
59. J. Li, B. Zu, Y. Zhang, X. Guo and H. Zhang, *J. Polym. Sci., Part A: Polym. Chem.*, 2010, **48**, 3217.
60. J. Jiang, Y. Zhang, X. Guo and H. Zhang, *Macromolecules*, 2011, **44**, 5893.
61. J. Jiang, Y. Zhang, X. Guo and H. Zhang, *RSC Adv.*, 2012, **2**, 5651.
62. M. Zhao, H. T. Zhang, F. Ma, Y. Zhang, X. Guo and H. Zhang, *J. Polym. Sci., Part A: Polym. Chem.*, 2013, **51**, 1983.
63. Y. Jiao, J. Jiang, Y. Zhang, H. T. Zhang, K. Shi and H. Zhang, *Eur. Polym. J.*, 2014, **54**, 95.
64. H. Zhang, J. Jiang, H. T. Zhang, Y. Zhang and P. Sun, *ACS Macro Lett.*, 2013, **2**, 566.
65. G. Pan, Y. Zhang, X. Guo, C. Li and H. Zhang, *Biosens. Bioelectron.*, 2010, **26**, 976.
66. G. Pan, Y. Ma, Y. Zhang, X. Guo, C. Li and H. Zhang, *Soft Matter*, 2011, **7**, 8428.
67. L. Fang, S. Chen, X. Guo, Y. Zhang and H. Zhang, *Langmuir*, 2012, **28**, 9767.
68. Y. Ma, Y. Zhang, M. Zhao, X. Guo and H. Zhang, *Mol. Imprinting*, 2012, 3.
69. Y. Ma, Y. Zhang, M. Zhao, X. Guo and H. Zhang, *Chem. Commun.*, 2012, **48**, 6217.
70. C. Li, Y. Ma, H. Niu and H. Zhang, *ACS Appl. Mater. Interfaces*, 2015, **7**, 27340.

71. M. Zhao, C. Zhang, Y. Zhang, X. Guo, H. Yan and H. Zhang, *Chem. Commun.*, 2014, **50**, 2208.
72. Y. Yang, H. Niu and H. Zhang, *ACS Appl. Mater. Interfaces*, 2016, **8**, 15741.
73. M. Zhao, X. Chen, H. T. Zhang, H. Yan and H. Zhang, *Biomacromolecules*, 2014, **15**, 1663.
74. G. Pan, Y. Zhang, Y. Ma, C. Li and H. Zhang, *Angew. Chem., Int. Ed.*, 2011, **50**, 11731.
75. Y. Ma, G. Pan, Y. Zhang, X. Guo and H. Zhang, *Angew. Chem., Int. Ed.*, 2013, **52**, 1511.
76. Y. Ma, H. Li, S. He, H. T. Zhang and H. Zhang, *J. Polym. Sci., Part A: Polym. Chem.*, 2014, **52**, 1941.
77. H. Niu, Y. Yang and H. Zhang, *Biosens. Bioelectron.*, 2015, **74**, 440.
78. Y. Yang, Z. Wang, H. Niu and H. Zhang, *Biosens. Bioelectron.*, 2016, **86**, 580.
79. H. G. Schild, *Prog. Polym. Sci.*, 1992, **17**, 163.
80. H. Kanazawa, *J. Sep. Sci.*, 2007, **30**, 1646.
81. T. Oya, T. Enoki, A. Y. Grosberg, S. Masamune, T. Sakiyama, Y. Takeoka, K. Tanaka, G. Wang, Y. Yilmaz, M. S. Feld, R. Dasari and T. Tanaka, *Science*, 1999, **286**, 1543.
82. D. Y. Ko, H. J. Lee and B. Jeong, *Macromol. Rapid Commun.*, 2006, **27**, 1367.
83. Z. Hua, Z. Chen, Y. Li and M. Zhao, *Langmuir*, 2008, **24**, 5773.
84. F. Puoci, F. Iemma and N. Picci, *Curr. Drug Delivery*, 2008, **5**, 85.
85. S. Xu, H. Lu, X. Zheng and L. Chen, *J. Mater. Chem. C*, 2013, **1**, 4406.
86. Y. Ge, B. Butler, F. Mirza, S. Habib-Ullah and D. Fei, *Macromol. Rapid Commun.*, 2013, **34**, 903.
87. W. Chen, Y. Ma, J. Pan, Z. Meng, G. Pan and B. Sellergren, *Polymers*, 2015, **7**, 1689.
88. A. Goulet-Hanssens and C. J. Barrett, *J. Polym. Sci., Part A: Polym. Chem.*, 2013, **51**, 3508.
89. C. Alvarez-Lorenzo and A. Concheiro, *Chem. Commun.*, 2014, **50**, 7743.
90. Y. B. Wei, Q. Tang, C. B. Gong and M. H. W. Lam, *Anal. Chim. Acta*, 2015, **900**, 10.
91. H. Zhang, W. Verboom and D. N. Reinhoudt, *Tetrahedron Lett.*, 2001, **42**, 4413.
92. H. Zhang, T. Piacham, M. Drew, M. Patek, K. Mosbach and L. Ye, *J. Am. Chem. Soc.*, 2006, **128**, 4178.
93. B. Zu, Y. Zhang, X. Guo and H. Zhang, *J. Polym. Sci., Part A: Polym. Chem.*, 2010, **48**, 532.
94. L. Fang, S. Chen, Y. Zhang and H. Zhang, *J. Mater. Chem.*, 2011, **21**, 2320.
95. Y. Ma, G. Pan, Y. Zhang, X. Guo and H. Zhang, *J. Mol. Recognit.*, 2013, **26**, 240.
96. X. Li, R. Wen, Y. Zhang, L. Zhu, B. Zhang and H. Zhang, *J. Mater. Chem.*, 2009, **19**, 236.
97. Z. Li, Y. Zhang, L. Zhu, T. Shen and H. Zhang, *Polym. Chem.*, 2010, **1**, 1501.

98. X. Li, L. Fang, L. Hou, L. Zhu, Y. Zhang, B. Zhang and H. Zhang, *Soft Matter*, 2012, **8**, 5532.
99. L. Fang, H. T. Zhang, Z. Li, Y. Zhang, Y. Y. Zhang and H. Zhang, *Macromolecules*, 2013, **46**, 7650.
100. G. Han, H. T. Zhang, J. Chen, Q. Sun, Y. Y. Zhang and H. Zhang, *New J. Chem.*, 2015, **39**, 1410.
101. G. Han, J. Nie and H. Zhang, *Polym. Chem.*, 2016, **7**, 5088.
102. S. Wang, Y. Zheng, Y. Liu, H. Yan and K. Liu, *J. Mater. Chem.*, 2005, **15**, 3424.
103. F.-G. Bănică, *Chemical Sensors and Biosensors: Fundamentals and Applications*, John Wiley & Sons, Ltd, West Sussex (UK), 2012.
104. A. P. Turner, *Chem. Soc. Rev.*, 2013, **42**, 3183.
105. H. Zhang, M. Zhao and Y. Yang, *Biomedical Nanomaterials, First Edition*, ed. Y. Zhao and Y. Shen, Wiley-VCH Verlag GmbH & Co. KGaA, 2016, ch. 11, pp. 283–326.
106. L. Yang, Z. Fan, T. Wang, W. Cai, M. Yang, P. Jiang, M. Zhang and X. Dong, *Anal. Lett.*, 2011, **44**, 2617.
107. M. Yang, Y. Zhang, S. Lin, X. Yang, Z. Fan, L. Yang and X. Dong, *Talanta*, 2013, **114**, 143.
108. X. Li, M. Zhou, M. Turson, S. Lin, P. Jiang and X. Dong, *Analyst*, 2013, **138**, 3066.
109. S. Xu, H. Lu and L. Chen, *J. Chromatogr. A*, 2014, **1350**, 23.
110. J. Dai, Y. Zou, Z. Zhou, X. Dai, J. Pan, P. Yu, T. Zou, Y. Yan and C. Li, *RSC Adv.*, 2014, **4**, 1965.
111. J. Dai, Z. Zhou, Y. Zou, X. Wei, X. Dai, C. Li and Y. Yan, *J. Appl. Polym. Sci.*, 2014, **131**, 5829.
112. L. Zhao, F. Zhao and B. Zeng, *Biosens. Bioelectron.*, 2014, **60**, 71.
113. R. Song, X. Hu, P. Guan, J. Li, C. Du, L. Qian and C. Wang, *Mater. Sci. Eng., C*, 2016, **60**, 1.
114. X. Wu, X. Wang, W. Lu, X. Wang, J. Li, H. You, H. Xiong and L. Chen, *J. Chromatogr. A*, 2016, **1435**, 30.
115. C. Gonzato, M. Courty, P. Pasetto and K. Haupt, *Adv. Funct. Mater.*, 2011, **21**, 3947.
116. N. Marchyk, J. Maximilien, S. Beyazit, K. Haupt and B. Tse Sum Bui, *Nanoscale*, 2014, **6**, 2872.
117. Z. Adali-Kaya, B. Tse Sum Bui, A. Falcimaigne-Cordin and K. Haupt, *Angew. Chem., Int. Ed.*, 2015, **54**, 5192.
118. S. Xu, J. Li and L. Chen, *Talanta*, 2011, **85**, 282.
119. Y. Zou, C. Zhao, J. Dai, Z. Zhou, J. Pan, P. Yu, Y. Yan and C. Li, *Colloid Polym. Sci.*, 2014, **292**, 333.
120. T. Zhou, L. Jørgensen, M. A. Mattebjerg, I. S. Chronakisc and L. Ye, *RSC Adv.*, 2014, **4**, 30292.
121. L. Su, X. Guo and S. Han, *Anal. Methods*, 2014, **6**, 2512.
122. F. Canfarotta, M. J. Whitcombe and S. A. Piletsky, *Biotechnol. Adv.*, 2013, **31**, 1585.
123. A. A. Volkert and A. J. Haes, *Analyst*, 2014, **139**, 21.

124. J. Wackerlig and R. Schirhagl, *Anal. Chem.*, 2016, **88**, 250.
125. S. Shinde, Z. El-Schich, A. Malakpour, W. Wan, N. Dizeyi, R. Mohammadi, K. Rurack, A. G. Wingren and B. Sellergren, *J. Am. Chem. Soc.*, 2015, **137**, 13908.
126. M. Panagiotopoulou, Y. Salinas, S. Beyazit, S. Kunath, L. Duma, E. Prost, A. G. Mayes, M. Resmini, B. Tse Sum Bui and K. Haupt, *Angew. Chem., Int. Ed.*, 2016, **55**, 8244.
127. Y. Hoshino, H. Koide, T. Urakami, H. Kanazawa, T. Kodama, N. Oku and K. J. Shea, *J. Am. Chem. Soc.*, 2010, **132**, 6644.
128. W. J. Song, J. P. Wei, S. Y. Wang and H. S. Wang, *Anal. Chim. Acta*, 2014, **832**, 58.
129. L. Zhao, F. Zhao and B. Zeng, *Biosens. Bioelectron.*, 2014, **62**, 19.
130. Y. Liu, S. Fang, J. Zhai and M. Zhao, *Nanoscale*, 2015, **7**, 7162.

CHAPTER 11

Designing of Biomimetic Molecularly Imprinted Catalysts

Z. Y. DONG* AND J. Q. LIU

Jilin University, College of Chemistry, State Key Laboratory of Supramolecular Structure and Materials, 2699 Qianjin Street, 130012, Changchun, China
*E-mail: zdong@jlu.edu.cn

11.1 Introduction of Biomimetic Catalysts

Enzymes are natural macromolecular receptors with three-dimensional structures consisting of linear polymeric peptide sequences. Owing to their nanometre-sized macromolecular structures that possess hydrophobic specific regions on the surface, natural enzymes are able to catalyze many chemical biotransformations with extremely high efficiency and substrate selectivity in a physiological environment. To address the origin of functions of natural enzymes, scientists worldwide have made many efforts to elucidate the reaction mechanism of enzyme catalysis. Owing to various modern analytical techniques, some considerable progresses in exploring the catalytic mechanism of enzymes have recently been achieved.[1–4] Particularly, the role of transition state stabilization has widely been recognized in enzyme catalysis.[5] The research on the design of enzyme special inhibitors strongly underpinned this recognition. Moreover, the use of transition state theory had evoked the development of catalytic antibodies that act

Polymer Chemistry Series No. 28
Molecularly Imprinted Polymers for Analytical Chemistry Applications
Edited by Wlodzimierz Kutner and Piyush Sindhu Sharma

Published by the Royal Society of Chemistry, www.rsc.org

as a kind of enzyme-like biological receptors, which can be expressed by the immune response on the cellular surface. Notably, catalytic antibodies exhibit the ability to generate a site recognizing virtually any substrate of interest.[6] However, natural enzymes and catalytic antibodies are relatively unstable and very sensitive to the pH and temperature. They suffer from a high cost and are difficult to produce. All these disadvantages have seriously restricted further applications of natural enzymes. Therefore, one of the chemical strategies for designing biomimetic catalysts is to properly combine the catalytic site and the recognition site into one synthetic macromolecule.[7,8]

In the past several decades, biomimetic catalysts have been widely studied, not only for understanding the reaction mechanism of enzyme-like catalysis but also for their potential uses in some chemical research areas.[9,10] By taking advantage of the cooperation between the catalytic site and the recognition site, a number of biomimetic catalysts have been designed and synthesized based on various different scaffolds, including macrocycle and acceptor molecules, polymers, supramolecular assemblies, nanometre-sized materials, catalytic antibodies, natural proteins, and so on.[11–14] Although these biomimetic catalysts reported previously have showed strong catalytic capability and high selectivity to some extent, several drawbacks still occur in these artificial systems including instability, incompatibility, tenuous synthesis, poor co-operativity, and low productivity. Benefitting from the macromolecular structures endowed with multiple catalytic sites as well as the possible proper cooperation between catalysis and recognition, however, biomimetic catalysts based on synthetic macromolecular scaffolds are able, promisingly, to overcome these issues.[8] Actually, it remains a challenge for chemists to simulate the catalytic nature of enzymes by using tailor-made polymer materials. The synthetic strategy is too complicated to obtain the desired macromolecular biomimetic catalysts with high efficiency and selectivity. However, by using molecular imprinting strategy, chemists have made a great success in the design of macromolecular biomimetic catalysts. In this regard, therefore, recent remarkable developments focus on the construction of synthetic macromolecules and supramolecules imprinted by selected template molecules that hence represent significant catalytic functions. For example, several seminal works in the area of molecularly imprinted polymer (MIP) catalysts have been reported by Damen and Neckers,[15] and Shea and Wulff *et al.*[16,17] on the stereoselective synthesis of chemical reactions. Since the first example of an MIP using a transition state analogue as the imprinting template was reported by Robinson and Mosbach,[18] various approaches to imprinting for the preparation of MIP artificial enzymes have been explored, including the covalent approach[19] and the stoichiometric noncovalent imprinting method.[20] At the same time, a large variety of chemical reactions, such as dehydrofluorination,[21] β-elimination,[22] the aldolase type II reaction,[23] the Diels–Alder reaction,[24] phosphate hydrolysis,[25] and Suzuki cross-coupling,[26] have been investigated by using different MIP catalysts.

In the present chapter, we summarize recent advances in development of biomimetic molecularly imprinted catalysts. First, the concept of operation of biomimetic catalysts is briefly introduced to provide a basic comprehension on the development of artificial enzymes. Then, the biomimetic molecularly imprinted catalysts are highlighted with respect to molecular catalysis and even supramolecular catalysis. Finally, the application of biomimetic molecularly imprinted polymers in sensing is emphasized as well.

11.2 Development of Biomimetic Molecularly Imprinted Polymers for Catalysis

It remains a challenge for chemists to design artificial enzymes synthetically with the integrated functions of both strong affinity for substrates and high catalytic activity. Numerous small molecule artificial enzymes have been constructed in the past.[14] However, a relatively simplified structure is unable to install the specific microenvironment for substrate binding. To realize the integrated functions, a prerequisite is that the artificial enzymes possess both the substrate binding and catalytically active groups in spatially arranged proper positions, particularly showing strong ability for selective stabilization of a transition state. Molecular imprinting technology has proved to be a powerful tool for constructing artificial enzymes with rationally designed active sites. During molecular imprinting, a template is fixed in space by the copolymerization of excess cross-linking monomer and functional monomer. After eliminating the template, the imprinted cavities are thus generated with special size, shape, and spatial microenvironment matching that of the template molecule (Scheme 11.1). MIPs showed significant

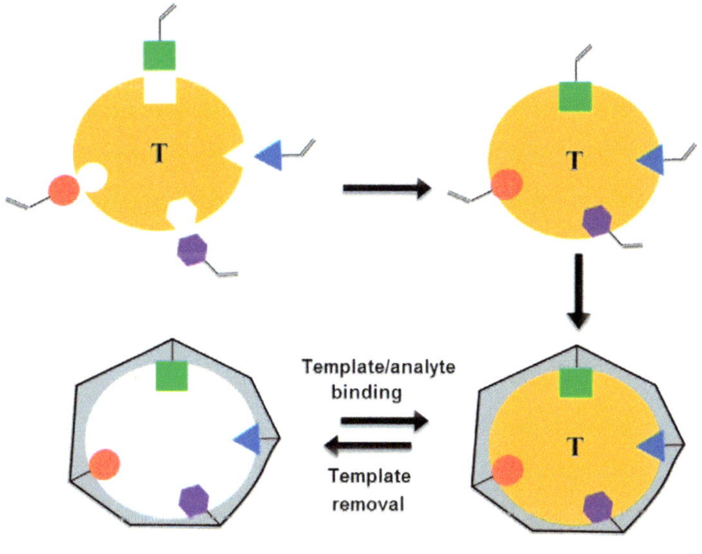

Scheme 11.1 A model of the molecular cavity imprinted in the cross-linked polymer network by a template (T) molecule with three different functional groups.

merits, including high stability and strong affinity to target molecules. As for the chemoimmunization production, MIPs are similar to catalytic antibodies owing to their selectivity, affinity, and catalytic function.[27] In recent years, MIPs have been used to construct artificial enzymes. A significant rate enhancement in a number of chemical reactions has been achieved by MIP-based artificial enzymes, such as the oxidation, reduction, elimination, carbon–carbon bond formation reactions and so on.[28–30]

11.2.1 Imprinting with a Transition State Analogue Template

The role of transition state stabilization has proved crucial in enzyme catalysis. One of successful examples involves catalytic activity of antibodies generated against transition state analogs. Approaches analogous to that used for catalytic antibodies have been utilized to prepare MIPs exhibiting substrate-selective binding and enhanced activity. During molecularly imprinted copolymerization, the active sites were formed on the surface of highly cross-linked network polymers. Therefore, the active centers in MIPs incorporated an array of functional recognition sites with a special microenvironment shape and size complementary to that of the transition state analog.

Numerous esterase mimics have been established by molecular imprinting with transition state analogues. As shown in Scheme 11.2, 4-nitrophenol methylphosphonate **1** is structurally similar to the substrate molecule 4-nitrophenyl acetate **2**, but contains a tetrahedral phosphoryl group in place of the carboxyl group. Therefore, with phosphonate ester **1** as the template molecule that is expected to serve as a transition state analogue for the hydrolysis of substrate **2**, a functionally active polymer, namely, poly[4(5)-vinylimidazole], was formed in the presence of metal binding sites. This MIP hydrolyzed

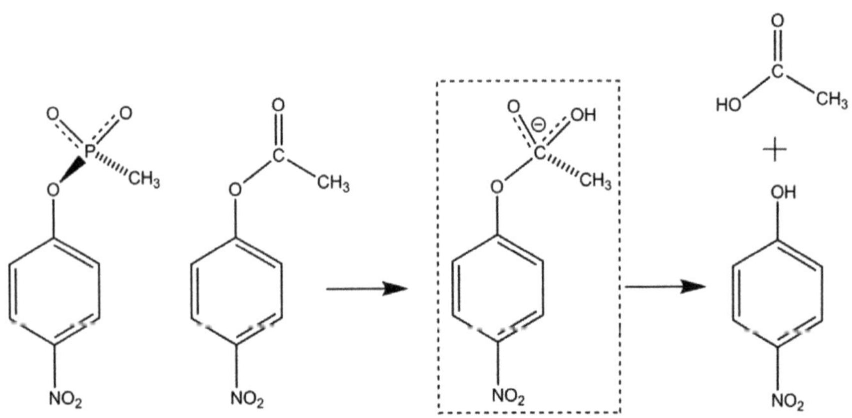

1 2

Scheme 11.2 4-Nitrophenol methylphosphonate **1** is expected to serve as a transition state analogue for the hydrolysis of 4-nitrophenyl acetate **2**.

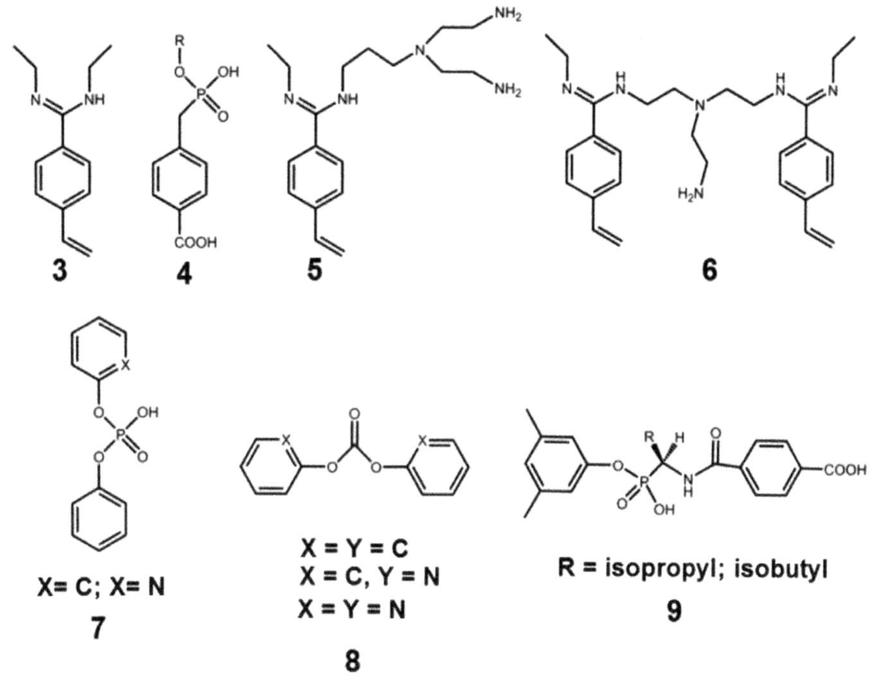

Scheme 11.3 Structural formulas of designed functional templates and monomers for molecularly imprinted polystyrene.

4-nitrophenol acetate **2** at an enhanced rate, and then this hydrolysis was inhibited by addition of the 4-nitrophenol methylphosphonate **1** template.[31] The Wulff group has studied catalytic behavior of esterase mimics by molecular imprinting with transition state analogues as molecular templates.[32,33] The polymerizable amidine **3** (Scheme 11.3) was employed to provide anchoring and catalytic groups for capturing substrates and increasing the esterase activity.[34] The molecularly imprinted artificial enzyme, incorporated with **3** using **4** as the transition state template, showed enhanced catalytic activity towards ester hydrolysis, remarkable selectivity, and a typical enzyme nature. To mimic the well-studied carboxypeptidase A enzyme, in which the active sites consisted of the amino acid residues of His 69, Glu 72, and His 196 chelating a zinc ion, Wulff and coworkers designed the amidinium groups (**5** and **6**) to build the artificial enzymes by using **7** as the transition state analogue template by the molecular imprinting method. In the resulting MIPs, an amidinium group and the Zn^{2+} center were oriented, in a template-imprinted cavity, in a manner similar to that of the active site in carboxypeptidase A. Significantly, the resulting MIP artificial enzyme remarkably enhanced the reaction rate constant, k_{cat}/k_{soln} = 6900, thus proving to be a highly efficient MIP artificial enzyme, comparable to the catalytic antibodies.[35] Subsequently, they developed another highly efficient artificial enzyme to simulate catalytic behaviors of carboxypeptidase A by using a similar molecularly imprinted protocol except with Cu^{2+} instead of Zn^{2+} and different substrates **8** used as the templates.[36,37] The enzyme mimics were synthesized by a molecularly imprinted approach, as shown in Scheme 11.4. The MIP catalysts exhibited typical enzyme-like properties, such as substrate selectivity, saturation kinetics, and competitive inhibition. More interestingly, one of these model enzymes exhibited extremely high catalytic activity (k_{cat}/k_{soln} = 110 000). In addition, Wulff and coworkers constructed by molecular imprinting a highly efficient enzyme mimic with enantioselective ability and esterase-like activity.[38] In this case, the enantiomerically pure phosphonic monoesters **9** were prepared and used as transition state analog templates to produce the MIPs *via* cross-linking by stoichiometric noncovalent interactions to two equivalents of the amidinium binding group **3**. Consequently, recognition for the substrates (K_{ML}/K_{MD} = 0.82) and for the transition state analogs (k_{catL}/k_{catD} = 1.36) was selective. Moreover, remarkable template inhibition was found in catalysis.

During the molecular imprinting process, the transition state analogue template can be either in its free form or in an immobilized form,

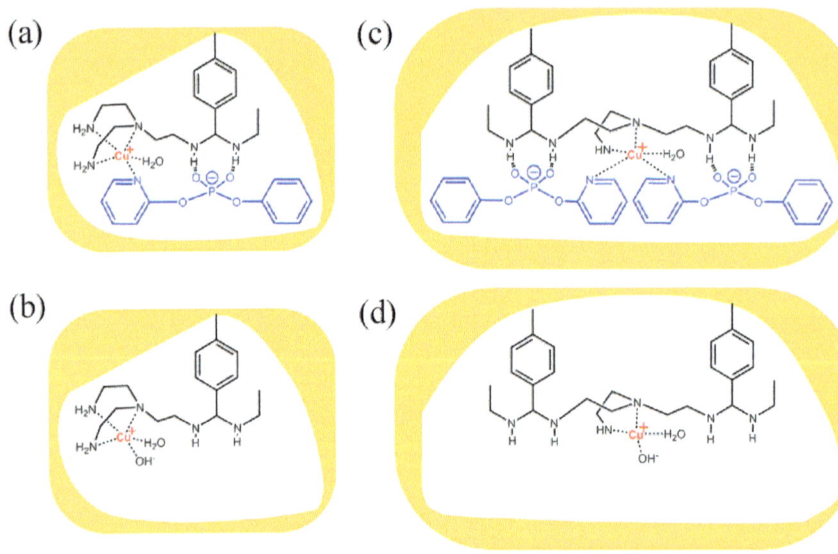

Scheme 11.4 Representation of the catalytic centers of MIP catalysts. (a) Molecular imprinting with functional monomer **5** and substrate analogue template **7** in the presence of Cu^{2+} for the preparation of the imprinted polymer catalyst. (b) The enzyme-like active site of the imprinted polymer catalyst prepared after removal of the template **7**. (c) Molecular imprinting with functional monomer **6** and two molecules of the template **7** in presence of Cu^{2+} for the preparation of the imprinted polymer catalyst. (d) The enzyme-like active site produced after removal of the template **7**. Reprinted with permission from J. Q. Liu, and G. Wulff, *J. Am. Chem. Soc.*, 2008, **130**, 8044.[37] Copyright 2008 American Chemical Society.

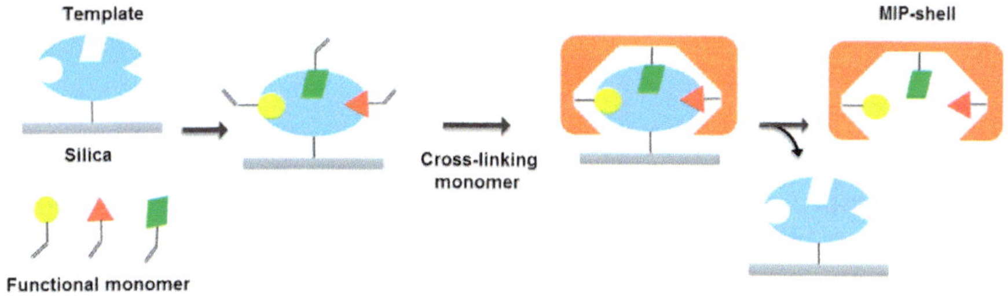

Scheme 11.5 Consecutive steps of the procedures of preparation of MIP shells based on an immobilized template. Reproduced from *React. Funct. Polym.*, **66**, A. Volkmann and O. Brüggemann, Catalysis of an ester hydrolysis applying molecularly imprinted polymer shells based on an immobilized chiral template, 1725.[41] Copyright 2006, with permission from Elsevier.

as appropriate functional monomers which are cross-linked with other monomers *via* copolymerization. When the copolymerization is completed, the immobilized template leaves the molecularly imprinting matrix after extraction from the polymer network. Importantly, immobilization of the transition state analogue template can result in relatively more efficient MIP catalysts.[39,40] Recently, Volkmann and Brüggemann reported,[41] for the first time, the use of an immobilized chiral transition state analog template for the preparation of an active MIP catalyst (Scheme 11.5). In this approach, the molecular imprinting was accomplished by copolymerizing a functional monomer and a cross-linking monomer in the presence of template-imprinted silica particles. Control polymers were generated by a similar polymerization, however, around silica particles which did not contain any transition state analog templates on their surfaces. Both the MIP and the control polymer shells were then studied in terms of their catalytic features on the selected hydrolysis reaction. Consequently, catalytic efficiency in the hydrolysis reaction of the MIP shells was higher than that of the control polymers, whereas substrate conversion was the lowest in the polymer-free solution. With a chiral transition state analog template, the MIP shells equipped with a spatial microenvironment on their inner surfaces provided selectively the catalytic capacity.

To simulate the function of chymotrypsin, the MIP enzyme mimics were prepared for the amidolysis of phenylalanine *p*-nitroanilide, in which the phenyl-1-(*N*-benzyloxycarbonylamino)-2-(phenyl)ethyl phosphonate transition state analog was used as the template.[42] The amidase-like activity of the MIP artificial enzymes meets the pseudo-first-order kinetics. The transition state analog provides a tetrahedral geometry complementary to the transition state intermediate, which is crucial for the catalytic capacity of the MIP artificial enzymes. The MIP artificial enzymes showed stereoselectivity and substrate selectivity in the amidolysis of phenylalanine *p*-nitroanilide. The proper position of the reactive functionalities in the highly cross-linked macroporous matrix for selective substrate binding through hydrogen bonding interactions was responsible for both high catalytic efficiency and substrate selectivity of the MIP catalysts.

11.2.2 Imprinting with Active Sites

Molecular imprinting is a powerful implement for the synthesis of polymeric catalysts with enzyme-like active centers.[43] In designing biomimetic MIP materials, a complex template, formed in different ways including covalent bonds and metal coordination, was used to imprint the specific microenvironment similar to that of the active centers of natural enzymes.[44] Beside the use of the transition state analog template, the reactive functionalities inside the catalytic microenvironment of enzymes can be generated to act as the template for constructing the MIP catalysts. An MIP catalyst can be inactivated because of the competitive inhibition in catalysis when templates of the transition state or product analogs were used. Therefore, during catalytic reaction it is crucial to facilitate the substrate recognition and product diffusion on the surface of the imprinted polymeric matrix. This issue has been addressed by a surface imprinting method, in which water–oil emulsion on an organic–aqueous interface was used as a binding site for template molecules to assure proper orientation of the functional groups toward target molecules fixed at this interface.[45,46] This method was very effective for creating MIP catalysts by using water-soluble or hydrophilic templates, such as metal coordination complexes. Actually, molecular imprinting of a metal complex with a molecule shape matching that of to a template molecule represents a promising way to form a polymer matrix with an artificially active site.

Recently, Muratsugu and Tada[28] designed and prepared molecularly imprinted metal complex catalysts with active sites for shape-selective catalysis. The molecularly imprinted Ru complexes were synthesized by using surface-attached Ru complexes with template ligands and an inorganic–organic surface matrix to control the chemical microenvironment around the active metal complex catalysts on oxide surfaces.[47,48] The (oxide surface)-attached Ru complex catalysts not only improved thermal stability and catalyst dispersion but also provided unique catalytic performance not observed in the homogeneous precursors. Importantly, the use of molecular imprinting can facilitate artificial integration of catalytic functions at surfaces.

To mimic the active sites of metalloenzymes, Czulak *et al.*[49,50] synthesized a kind of MIP catalysts by suspension polymerization of functional monomers in the presence of transition metal ions and 4-methoxybenzylalcohol as the template. Four metal ions, namely, Cu^{2+}, Co^{2+}, Mn^{2+}, and Zn^{2+}, were used for imprinting the polymeric catalysts with the microelements that are most essential in the native enzymes. The Cu^{2+}-, Co^{2+}-, and Zn^{2+}-imprinted catalysts were capable of hydroquinone oxidation in the presence of hydrogen peroxide. However, the Mn^{2+}-imprinted catalyst was inactive because of the insufficient metal content. Notably, owing to the use of surface imprinting, catalytic efficiency of the Cu^{2+}-MIP was the highest among them.

Since it became known that, in general, chemical transformations occurred inside nanometre-sized microenvironments, the construction of MIPs with nanometre-sized active sites has become of significant importance to mimic activity of the natural enzymes. Accordingly, Zhang *et al.*[51] recently demonstrated that nanometre-sized active sites of an MIP catalyst were capable of functioning as regioselective nanoreactors for 1,3-dipolar cycloaddition of azides and alkynes. Furthermore, Li and Gong[52,53] reported on a series of MIP catalysts with substrate-selective nanometre-sized active sites (Scheme 11.6). The nanoreactor catalyst, made of a 4-nitrophenol-imprinted polymer and Ag nanoparticles, specifically bound 4-nitrophenol rather than its 4-nitrophenyl acetate and 2,6-dimethyl-4-nitrophenol analogs. Consequently, this catalyst efficiently catalyzed the 4-nitrophenol reduction. Nevertheless, a low rate constant increase was observed for its analogs. Importantly, the MIPs with nanometre-sized active sites exhibited substrate-selective catalysis, unlike simple Ag nanoreactors lacking binding sites for the substrate.

Moreover, Wulff and coworkers[54] prepared MIP macromolecules, each containing just one active site. They used a post-dilution method for the synthesis of more highly cross-linked macromolecular nanoparticles, thus aiding the molecular imprinting. That way the formed water-soluble macromolecular nanoparticles contained, on average, one active site per

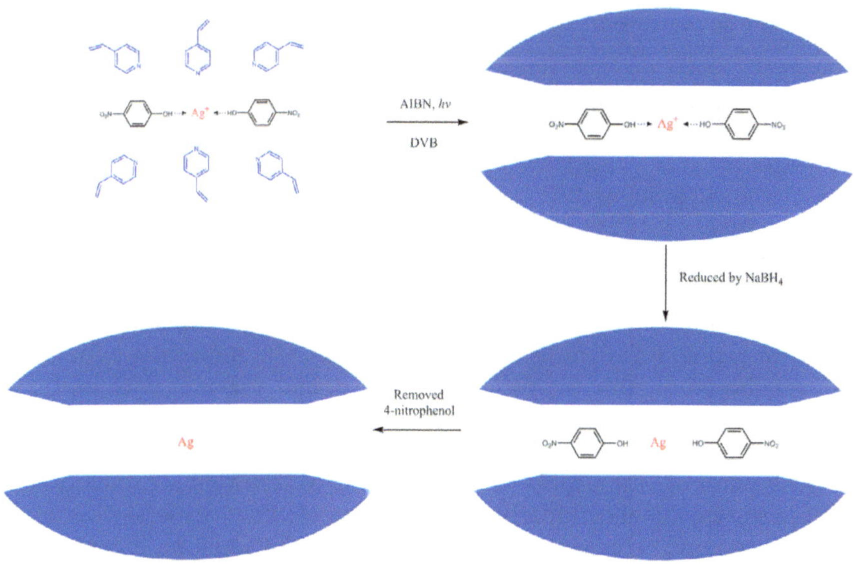

Scheme 11.6 Consecutive steps of the procedures of preparation of a MIP–Ag–NP nanoreactor. Reproduced from S. J. Li and S. Gong, A Substrate-Selective Nanoreactor Made of Molecularly Imprinted Polymer Containing Catalytic Silver Nanoparticles, *Advanced Functional Materials*, John Wiley and Sons, © 2009 Wiley-VCH Verlag GmbH &Co. KGaA, Weinheim.

each macromolecule, similar to the natural enzymes, but these nanoparticles were much more stable. These nanoparticles, forming nanogels, showed remarkable hydrolytic activity obeying a typical Michaelis–Menten kinetics.

11.3 Development of Biomimetic Supramolecular Molecularly Imprinted Catalysis

Natural enzymes represent three-dimensional dynamic supramolecular structures that are responsible for specific chemical transformations with high catalytic efficiency.[55] Previous studies strongly suggested that natural enzymes generally existed in ensembles of coupled conformational states.[56] Therefore, it is a delicate mimicry of natural enzymes to employ a supramolecular self-assembly strategy to construct artificial enzymes with three-dimensional dynamic structures.[7] Accordingly, supramolecular imprinting will significantly accelerate the development of MIPs. By using noncovalent interactions, self-organized templates can be rationally designed and prepared so as to construct the supramolecularly imprinted polymers (SIPs). During the preparation of the SIPs, a distinctive advantage is that the supramolecular template can be easily removed due to its dynamic nature. By means of a supramolecular strategy, therefore, it is obviously effective to generate a SIP structure and materials with the specific active microenvironment similar to that of the active centers of natural enzymes.

11.3.1 Biomimetic Supramolecular Imprinting of Catalysts by Self-assembly

In a biological system, molecularly imprinted self-assembling of functional building blocks leads to the formation of diverse supramolecular structures with a specific active microenvironment. The diverse structural morphologies including spherical micelles, vesicles, nanoparticles, and nanotubes, have been formed for the construction of supramolecular artificial enzymes.[12,57] Through rational designing of building blocks, the molecularly imprinted self-assembled nanomaterials can act as artificial enzymes with particular capacity to recognize the substrate and to perform the catalysis.

Recently, metal nanoparticle reactors attracted much attention because of their potential applications.[58] In contrast to traditional catalytic metal nanoparticles, metal nanoparticle reactors can offer active metal sites for different reactions including controllable catalysis.[59] Therefore, it is necessary to develop a new way for straightforward preparation of substrate-selective nanoreactors. As expected, the molecular imprinting appears to be a promising way to synthesize substrate-selective nanoreactors. During molecular imprinting, functional monomers and metal nanoparticles are, first, allowed to form a pre-organized supramolecular complex by self-assembly. Thus, the MIP substrate-selective reactors were prepared through polymerization in the presence of cross-linking monomers, after subsequent removal of the imprinted supramolecular template from the polymer matrix.

In biomimetic supramolecular chemistry, micelles represent a promising platform owing to their useful structure-related properties.[60] They have two typologically different domains and can be devised to serve as supramolecular artificial enzymes with the assistance of molecularly imprinted self-assembly. With this concept, Huang *et al.*[61] designed a smart micelle with a temperature-sensitive block copolymer bearing catalytic groups to simulate natural glutathione peroxidase. The micelle enzyme mimic revealed high catalytic efficiency and typical enzymatic kinetics. Very interestingly, the catalytic activity of the micelle catalyst could be changed by the temperature change because of the presence of temperature-sensitive poly(*N*-isopropylacrylamide) in the copolymer. Recently, Orth and Campos[62] reported on remarkable micellar catalysis for dephosphorylation of the bis(2,4-dinitrophenyl) phosphate diester and diethyl 2,4-dinitrophenyl phosphate triester with the polyvinylimidazole. Notably, polyvinylimidazole bound the reactive substrates on its domains through hydrophobic and electrostatic attractions, showing the polymer multifunctionality. Surprisingly, the supramolecularly imprinted micellar catalysts were capable of efficient catalyzing the dephosphorylation reactions with increments up to 10^7-fold. Likewise, the spherical vesicles formed by molecularly imprinted self-assembly have been used to build supramolecular artificial enzymes. For example, Liu and coworkers[63] recently prepared a vesicle-based glutathione peroxidase mimic made of the temperature-sensitive poly(*N*-isopropylacrylamide)-polyacrylamides diblock copolymer by a bending process. The molar ratio of functional building blocks is essential to produce a relatively stable vesicle of high catalytic activity. As expected, catalytic activity of the vesicle-based enzyme mimic was temperature dependent.

Beside spherical micelles and vesicles, tubular structures have also been used as scaffolds to prepare supramolecular artificial enzymes.[66,67] Recently, Liu and coworkers[68] designed and synthesized a class of glutathione peroxidase models by molecular imprinting of self-assemblies. For that, functionalities were non-covalently assembled on the surface of giant nanotubes (Scheme 11.7). The results demonstrated that the selenium- and guanidine-derivatized cyclodextrins behaved as both the catalytic groups and binding sites in the supramolecular artificial enzymes. The highly-ordered tubular structures were thermodynamically stable. Moreover, the resulting catalytic activity of the supramolecular artificial enzymes was high (140 U), being just 40-fold lower than that of natural glutathione peroxidase (5700 U).

11.3.2 Biomimetic Imprinting of Catalysts Using Microgel Matrices

With the development of soft materials, microgels have been considered as excellent matrices for the fabrication of biomimetic supramolecular catalysts by means of molecularly imprinted self-assembly technology. Catalytic

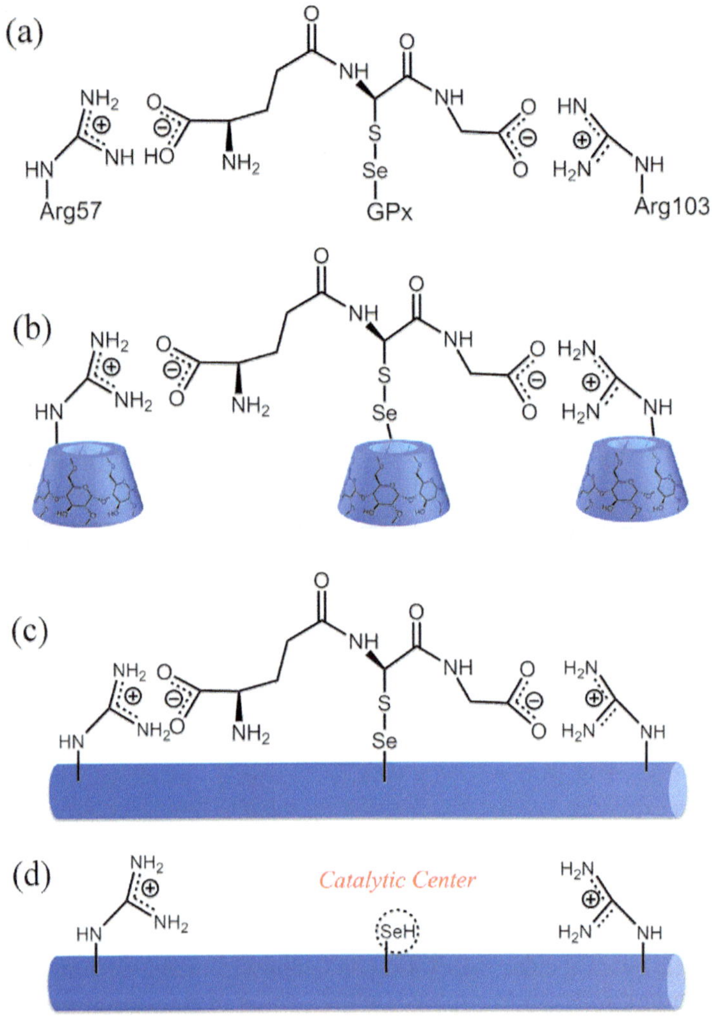

Scheme 11.7 Construction of supramolecular active centers on the surface of nanotubes by molecularly imprinted self-assembly. (a) The active center of glutathione peroxidase with a glutathione binding site.[64,65] (b) Supramolecularly imprinted complex of the glutathione substrate and guanidino-cyclodextrin. (c) Supramolecular stabilization of the active center on the surface of nanotubes by self-assembly. (d) The designer active center of an artificial enzyme with the glutathione recognizing site and the directed catalytic site. Reproduced from Y. Tang, L. Zhou, J. Li, Q. Luo, X. Huang, P. Wu, Y. Wang, J. Xu, J. Shen and J. Liu, Giant Nanotubes Loaded with Artificial Peroxidase Centers: Self-Assembly of Supramolecular Amphiphiles as a Tool to Functionalize Nanotubes, *Angew. Chem., Int. Ed.*,[68] John Wiley and Sons, © 2010 WILEY-VCH Verlag GmbH &Co. KGaA, Weinheim.

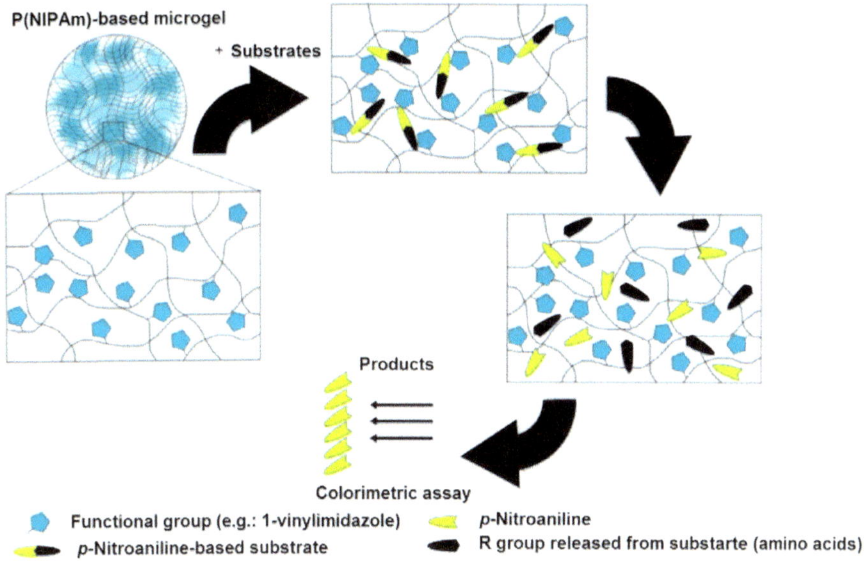

Scheme 11.8 Representation of a molecularly imprinted p(NIPAm)-based microgel catalyst capable of hydrolyzing *p*-nitroaniline-based substrate. Reprinted with permission from Y. M. Wong, Y. Hoshino, K. Sudesh, Y. Miura and K. Numata, *Biomacromolecules*, 2015, **16**, 411.[75] Copyright 2015 American Chemical Society.

efficiency and selectivity for some chemical transformations, such as carbonate hydrolysis,[69,70] Kemp eliminations,[71–73] and aldol condensations[29,74] *etc.*, of the imprinted microgel catalysts were high.

In order to mix effectively multiple functional groups onto the surface of polymeric microgels, Wong *et al.*[75] reported on an optimized poly(*N*-isopropylacrylamide) microgel bearing three different functionalities as a supramolecular catalyst used for acceleration of the hydrolysis of amide bonds under mild conditions (Scheme 11.8). They found that the pH adjustment was crucial to yield high catalytic efficiency of microgel catalysts because of a pH-sensitive balance between neutral and protonated forms of functional groups. Interestingly, the catalytic activity of microgel catalysts was enhanced when the distance between two functional groups was shortened. The substrate selectivity analysis and inhibition assay demonstrated that the molecularly imprinted microgel catalyst could act as an artificial amidase for efficient catalyzing of the hydrolysis of *p*-nitroaniline-based substrates.

Recently, Jorge *et al.*[76] synthesized imprinted gel catalysts by using rationally designed templates and active center Co-cyclen. The three-dimensional macromolecular matrix around the metal catalytic site slightly simulated the secondary coordination sphere of natural metalloenzymes. The coordination geometry and oxidation state of the metal catalytic center in the supramolecular complex were strongly affected by the chemical structure of the template, thus leading to a remarkable change in the catalytic features of the macromolecular gel matrix. Moreover, structural differences in the macromolecular matrices resulting from the molecularly imprinting technology with different complex templates were responsible for the substrate selectivity and the catalytic efficiency. To mimic the catalytic triad of natural protease enzymes, Karmalkar *et al.*[77] synthesized an imprinted hydrogel matrix by positioning imidazole as well as carboxyl and hydroxyl groups within the macromolecular matrix. Although the catalytic activity of the imprinted gel catalysts was lower than that of natural α-chymotrypsin, all of them could remarkably accelerated ester hydrolysis in comparison to the non-imprinted polymer.

11.4 Development of Biomimetic Molecularly Imprinted Polymers for Sensing

During the past years, molecular imprinting has largely been developed to become a powerful method for the preparation of substrate-binding polymers with high selectivity and sensitivity. The resultant MIP receptors have been widely utilized for chemical and biological sensing. That was because sensing was facile, providing a quick read-out analysis without the need of further experimental tests. Recently, great progress has been made in sensing applications for environmental monitoring, clinical diagnostics, detection of microorganisms and toxins, security and so on.[78,79] The MIP acceptors based on synthetic materials contained the desired recognition sites capable of detecting special targets with various advantages, including stability, potential reusability, resistance to biological decomposition, and multiple response signals. These advantages combined with the robustness and low cost, endowed biomimetic MIP receptors with important applicability for sensing.

Devising and fabricating MIP acceptors for sensing small-molecule compounds is a growing field in modern biotechnology, enabling detection of various toxic compounds in both the environment and foods. Sensing of MIP acceptors is capable of providing highly sensitive, selective, and reliable analysis regarding toxic molecules. For example, diphenols were detected by a catalytic free-standing MIP sensor system.[80] Generally, binding was identified using electrochemical, piezoelectric, and optical transducers. The electropolymerization of a mixture of phenylenediamine and glucose resulted in constructing an MIP chemosensor for conductometric determination of glucose.[81] Analytical methods for acetaminophen determination in pharmaceutical formulations are of paramount importance for quality control. Towards that, Tarley *et al.*[82] prepared a nanocomposite based on multi-walled carbon nanotubes grafted by molecularly imprinted poly(methacrylic acid–hemin) as a peroxidase-like catalyst for biomimetic sensing of acetaminophen. This MIP material was used as a biomimetic and catalytic unit of a chemosensor for electrochemical determination of acetaminophen by depositing it onto a glassy carbon electrode surface. The square wave voltammetry experiments showed that the limit of detection of this catalytic chemosensor for acetaminophen determination was low (1.1 μM L^{-1}). Moreover, 5-hydroxyindole-3-acetic acid, a urinary metabolite used as carcinoid tumor maker, was detected by an MIP enzyme mimic with peroxidase activity.[83] The product of catalytic oxidation of 5-hydroxyindole-3-acetic acid generated at the sensing phase was monitored by differential pulse voltammetry. The biomimetic chemosensor determined the 5-hydroxyindole-3-acetic acid with high sensitivity (0.72 nA μM^{-1}) and linear response over a concentration range covering a whole window of normal and pathologic urinary levels of 5-hydroxyindole-3-acetic acid (~0.2–9.6 ppm) with a limit of detection 0.27 ppm. On the basis of the substrate-binding feature of enzymes, Ye *et al.*[84] designed an enzyme-labeled molecularly imprinted polymer. This MIP was synthesized in the form of microspheres by precipitation polymerization. Its labeling with tobacco peroxidase allowed for colorimetric and chemiluminescence detection for the determination of herbicide 2,4-dichlorophenoxyacetic acid. Apparently, the competitive binding assay used appeared highly efficient. This MIP selectively bound the analyte in an aqueous solution by the chemiluminescence assay. Scheller *et al.*[85] demonstrated a successful example by combining an MIP with a biocatalyst in a catalytic biomimetic sensor. They applied both highly active horseradish peroxidase and the microperoxidase-11 to prepare an MIP by simultaneous electropolymerization and immobilization of the catalyst. The peroxide-dependent substrate was converted in a layer on top of the product-imprinted MIP deposited on the indicator electrode. This arrangement gave rise to the elimination of interfering signals for ascorbic acid and uric acid both in a phosphate buffer and in diluted serum samples. Furthermore, Lakshmi *et al.*[86] employed the approach of a transition state analogue to prepare an MIP with tyrosinase-like catalytic properties in the form of an electrochemical chemosensor for the determination of catechol and dopamine.

Although molecular imprinting procedures have successfully been applied to small-molecule templates in the area of artificial enzymes, extension of these procedures to macromolecular templates has been proved challenging. It is expected that synthetic materials exhibiting the ability of selective binding target macromolecular compounds would be advantageous over natural counterparts for a variety of applications. Recently, macromolecularly imprinted polymers, synthesized in the presence of macromolecule templates, are of particular importance because they open up the field for a whole new set of robust diagnostic tools.[87] There are a multitude of facile procedures already developed to form the 3-D binding sites for the entire macromolecules.[79,87] Sun *et al.*[88] prepared an MIP on the surface of the Au electrode by electrochemically mediated atom transfer radical polymerization with hemoglobin both as the catalyst and the template molecule. The

Au-MIP electrode was then used as the biosensor and successfully determined hemoglobin by differential pulse voltammetry. The linear dynamic concentration range of this biosensor was broad (1.0×10^{-10} to 1.0×10^{1} mg L^{-1}) and the limit of detection was low (7.8×10^{-11} mg L^{-1}) for hemoglobin. These analytical parameters were much better than those of the MIP-based hemoglobin sensors reported previously.[89,90]

11.5 Conclusions

The molecularly imprinted strategy provided a great potential to generate highly efficient catalysts with specific binding regions found on the surface of natural enzymes. The biomimetic molecularly imprinted catalysts are thus able to catalyze many chemical transformations with high efficiency and substrate selectivity under mild environmental conditions. By taking advantage of the cooperation between a catalytic site and a recognizing site, researchers made a remarkable progress in constructing biomimetic molecularly imprinted catalysts. For that, different scaffolds, such as macrocycles and acceptor molecules, polymers, supramolecular assemblies, nanometer-sized materials, catalytic antibodies, natural proteins, *etc.*, were used. Although some drawbacks, including instability, incompatibility, tenuous synthesis, low co-operativity, and low productivity, still occur in some artificial enzyme systems, biomimetic molecularly imprinted catalysts are able to overcome these difficulties promisingly. The preparation ease and the proper cooperation between catalysis and recognition allow for preparing tailor-made catalyst materials with desired properties and diverse catalytic functions. So far, the rate of a large number of chemical reactions, such as oxidation, reduction, elimination, and carbon–carbon bond formation, has significantly been increased by using molecularly imprinted catalysts. Owing to specific features of catalysis and molecular recognition originating from molecularly imprinted catalysts, these synthetic materials will not only be promising to apply in the catalytic industry, but also they will be widely used in sensing applications for environmental monitoring, clinical diagnostics, the detection of microorganisms and toxins, and security.

List of Abbreviations and Symbols

MIP Molecularly imprinted polymer
SIP Supramolecularly imprinted polymer
U Enzyme unit

Acknowledgements

The financial support from the Natural Science Foundation of China (Nos. 21574054, 21574056, 21234004, 21420102007, 91527302) and the Chang Jiang Scholars Program of China is acknowledged.

References

1. Z. D. Nagel and J. P. Klinman, *Chem. Rev.*, 2006, **106**, 3095.
2. B. P. English, W. Min, A. M. V. Oijen, K. T. Lee, G. B. Luo, H. Y. Sun, B. J. Cherayil, S. C. Kou and X. S. Xie, *Nat. Chem. Biol.*, 2006, **2**, 87.
3. D. D. Boehr, D. McElheny, H. J. Dyson and P. E. Wright, *Science*, 2006, **313**, 1638.
4. K. A. Henzler-Wildman, V. Thai, M. Lei, M. Ott, M. Wolf-Watz, T. Fenn, E. Pozharski, M. A. Wilson, G. A. Petsko, M. Karplus, C. G. Hübner and D. Kern, *Nature*, 2007, **450**, 838.
5. L. Pauling, *Nature*, 1948, **161**, 707.
6. Y. Xu, N. Yamamoto and K. D. Janda, *Bioorg. Med. Chem.*, 2004, **12**, 5247.
7. Z. Y. Dong, Q. Luo and J. Q. Liu, *Chem. Soc. Rev.*, 2012, **41**, 7890.
8. Y. Z. Yin, Z. Y. Dong, Q. Luo and J. Q. Liu, *Prog. Polym. Sci.*, 2012, **37**, 1476.
9. V. L. Schramm, *Annu. Rev. Biochem.*, 2011, **80**, 703.
10. A. J. Kirby and F. Hollfelder, *J. Am. Chem. Soc.*, 2010, **49**(10), 3637.
11. M. J. Wiester, P. A. Ulmann and C. A. Mirkin, *Angew. Chem., Int. Ed.*, 2011, **50**, 114.
12. Z. Y. Dong, Y. G. Wang, Y. Z. Yin and J. Q. Liu, *Curr. Opin. Colloid Interface Sci.*, 2011, **16**, 451.
13. P. J. Deuss, R. Heeten, W. Laan and P. C. J. Kamer, *Chem.–Eur. J.*, 2011, **17**, 4680.
14. Y. Murakami, J. Kikuchi, Y. Hisaeda and O. Hayashida, *Chem. Rev.*, 1996, **96**, 721.
15. J. Damen and D. C. Neckers, *J. Am. Chem. Soc.*, 1980, **102**(9), 3265.
16. K. J. Shea, E. A. Thompson, S. D. Pandey and P. S. Beauchamp, *J. Am. Chem. Soc.*, 1980, **102**, 3149.
17. G. Wulff and J. Vietmeier, *Macromol. Chem. Phys.*, 1989, **190**, 1717.
18. D. K. Robinson and K. Mosbach, *J. Chem. Soc., Chem. Commun.*, 1989, 969.
19. B. Sellergren and K. Shea, *Tetrahedron: Asymmetry*, 1994, **5**, 1403.
20. G. Wulff, T. Gross and R. Schonfeld, *Angew. Chem., Int. Ed.*, 1997, **36**, 1961.
21. R. Muller, L. I. Andersson and K. Mosbach, *Macromol. Rapid Commun.*, 1993, **14**, 637.
22. J. Beach and K. Shea, *J. Am. Chem. Soc.*, 1994, **116**, 379.
23. J. Matsui, I. A. Nicholls, I. Karube and K. Mosbach, *J. Org. Chem.*, 1996, **61**, 5414.
24. N. Kirsch, J. Hedin-Dahlström, H. Henschel, M. Whitcombe, S. Wikman and I. A. Nicholls, *J. Mol. Catal. B: Enzym.*, 2009, **58**, 110.
25. Z. Meng, T. Yamazaki and K. Sode, *Biotechnol. Lett.*, 2003, **25**, 1075.
26. A. N. Cammidge, N. J. Baines and R. K. Bellingham, *Chem. Commun.*, 2001, 2588.
27. G. Wulff, *Angew. Chem., Int. Ed.*, 1995, **34**, 1812.
28. S. Muratsugu and M. Tada, *Acc. Chem. Res.*, 2013, **46**, 300.
29. M. Resmini, *Anal. Bioanal. Chem.*, 2012, **402**, 3021.

30. G. Wulff and J. Q. Liu, *Acc. Chem. Res.*, 2012, **45**(2), 239.
31. D. K. Robinsona and K. Mosbach, *J. Chem. Soc., Chem. Commun.*, 1989, 969.
32. G. Wulff, T. Gross and R. Schoenfeld, *Angew. Chem., Int. Ed.*, 1997, **36**, 1962.
33. A. G. Strikovsky, D. Kasper, M. Gruen, B. S. Green, J. Hradil and G. Wulff, *J. Am. Chem. Soc.*, 2000, **122**, 6295.
34. G. Wulff, T. Gross and R. Schoenfeld, *Angew. Chem., Int. Ed.*, 1997, **36**, 1962.
35. J. Q. Liu and G. Wulff, *J. Am. Chem. Soc.*, 2004, **26**, 7452.
36. J. Q. Liu and G. Wulff, *Angew. Chem., Int. Ed.*, 2004, **43**, 1287.
37. J. Q. Liu and G. Wulff, *J. Am. Chem. Soc.*, 2008, **130**, 8044.
38. M. Emgenbroich and G. Wulff, *Chem.–Eur. J.*, 2003, **9**, 4106.
39. E. Yilmaz, K. Haupt and K. Mosbach, *Angew. Chem., Int. Ed.*, 2000, **39**, 2115.
40. A. Visnjevski, E. Yilmaz and O. Brüggemann, *Appl. Catal., A*, 2004, **260**, 169.
41. A. Volkmann and O. Brüggemann, *React. Funct. Polym.*, 2006, **66**, 1725.
42. D. Mathew, B. Thomas and K. S. Devaky, *J. Mol. Catal. A: Chem.*, 2016, **415**, 65.
43. B. Sellergren, *Molecularly Imprinted Polymers—man-made Mimics of Antibodies and Their Application in Analytical Chemistry*, Elsevier, Amsterdam, 2001.
44. K. Severin, *Curr. Opin. Chem. Biol.*, 2000, **4**, 710.
45. M. Yoshida, Y. Hatate, K. Uezu, M. Goto and S. Furusaki, *Colloids Surf., A*, 2000, **169**, 259.
46. E. Toorisaka, K. Uezu, M. Goto and S. Furusaki, *Biochem. Eng. J.*, 2003, **14**, 85.
47. M. Tada, Y. Akatsuka, Y. Yang, T. Sasaki, M. Kinoshita, K. Motokura and Y. Iwasawa, *Angew. Chem.,Int. Ed.*, 2008, **47**, 9252.
48. M. Tada, S. Muratsugu, M. Kinoshita, T. Sasaki and Y. Iwasawa, *J. Am. Chem. Soc.*, 2010, **132**, 713.
49. J. Czulak, A. Jakubiak-Marcinkowska and A. Trochimczuk, *Adv. Mater. Sci. Eng.*, 2013, **2013**, 9.
50. A. Jakubiak-Marcinkowska, M. Legan and J. Jezierska, *J. Polym. Res.*, 2013, **20**, 317.
51. H. Q. Zhang, T. Piacham, M. Drew, M. Patek, K. Mosbach and L. Ye, *J. Am. Chem. Soc.*, 2006, **128**, 4178.
52. S. J. Li and S. Q. Gong, *Adv. Funct. Mater.*, 2009, **19**, 2601.
53. J. Wang, M. Y. Zhu, X. J. Shen and S. J. Li, *Chem.–Eur. J.*, 2015, **21**, 7532.
54. G. Wulff, B. O. Chong and U. Kolb, *Angew. Chem., Int. Ed.*, 2006, **45**, 2955.
55. S. J. Benkovic and S. Hammes-Schiffer, *Science*, 2003, **301**, 1196.
56. C. Merlo, K. A. Dill and T. R. Weikl, *Proc. Natl. Acad. Sci. U. S. A.*, 2005, **102**, 10171.
57. L. Z. Gao, J. Zhuang, L. Nie, J. B. Zhang, Y. Zhang, N. Gu, T. H. Wang, J. Feng, D. L. Yang, S. Perrett and X. Y. Yan, *Nat. Nanotechnol.*, 2007, **2**, 577.

58. M. Graeser, E. Pippel, A. Greiner and J. H. Wendorff, *Macromolecules*, 2007, **40**, 6032.
59. Y. Lu, Y. Mei, M. Ballauff and M. Drechsler, *J. Phys. Chem. B*, 2006, **110**, 3930.
60. T. Dwars, E. Paetzold and G. Oehme, *Angew. Chem., Int. Ed.*, 2005, **44**, 7174.
61. X. Huang, Y. Z. Yin, X. Jiang, Y. Tang, J. Y. Xu, J. Q. Liu and J. C. Shen, *Macromol. Biosci.*, 2009, **9**, 1202.
62. E. S. Orth and R. B. Campos, *J. Braz. Chem. Soc.*, 2016, **27**, 285.
63. Y. Z. Yin, X. Huang, C. Y. Lv, L. Wang, S. J. Yu, Q. Luo, J. Y. Xu and J. Q. Liu, *Macromol. Biosci.*, 2010, **10**, 1505.
64. O. Epp, R. Ladenstein and A. Wendel, *Eur. J. Biochem.*, 1983, **133**, 51.
65. H. E. Ganther and R. J. Krauss, *Methods Enzymol.*, 1984, **107**, 593.
66. Q. X. Jin, L. Zhang, H. Cao, T. Y. Wang, X. F. Zhu, J. Jiang and M. H. Liu, *Langmuir*, 2011, **27**, 13847.
67. Y. Sugano, M. C. Vestergaard, M. Saito and E. Tamiya, *Chem. Commun.*, 2011, **47**, 7176.
68. Y. Tang, L. P. Zhou, J. X. Li, Q. Luo, X. Huang, P. Wu, Y. G. Wang, J. Y. Xu, J. C. Shen and J. Q. Liu, *Angew. Chem., Int. Ed.*, 2010, **49**, 3920.
69. S. C. Maddock, P. Pasetto and M. Resmini, *Chem. Commun.*, 2004, 536.
70. P. Pasetto, S. C. Maddock and M. Resmini, *Anal. Chim. Acta*, 2005, **542**, 66.
71. A. Servant, K. Haupt and M. Resmini, *Chem.–Eur. J.*, 2011, **17**, 11052.
72. A. Servant, S. Rogers, A. Zarbakhsh and M. Resmini, *New J. Chem.*, 2013, **37**, 4103.
73. P. Bonomi, A. Servant and M. Resmini, *J. Mol. Recognit.*, 2012, **25**, 352.
74. D. Carboni, K. Flavin, A. Servant, V. Gouverneur and M. Resmini, *Chem.–Eur. J.*, 2008, **14**, 7059.
75. Y. M. Wong, Y. Hoshino, K. Sudesh, Y. Miura and K. Numata, *Biomacromolecules*, 2015, **16**, 411.
76. A. R. Jorge, M. Chernobryva, S. E. J. Rigby, M. Watkinson and M. Resmini, *Chem.–Eur. J.*, 2016, **22**, 3764.
77. R. N. Karmalkar, M. G. Kulkarni and R. A. Mashelkar, *Macromolecules*, 1996, **29**, 1366.
78. R. C. Advincula, *Korean J. Chem. Eng.*, 2011, **28**(6), 1313.
79. M. J. Whitcombe, I. Chianella, L. Larcombe, S. A. Piletsky, J. Noble, R. Porter and A. Horgan, *Chem. Soc. Rev.*, 2011, **40**, 1547.
80. T. A. Sergeyeva, O. A. Slinchenko, L. A. Gorbach, V. F. Matyushov, O. O. Brovko, S. A. Piletsky, L. M. Sergeeva and G. V. Elska, *Anal. Chim. Acta*, 2010, **659**, 274.
81. Z. L. Cheng, E. K. Wang and X. R. Yang, *Biosens. Bioelectron.*, 2001, **16**, 179.
82. E. S. Moretti, J. F. Giarola, M. Kuceki, M. C. Prete, A. C. Pereira and C. R. T. Tarley, *RSC Adv.*, 2016, **6**, 28751.
83. D. Antuña-Jiménez, M. C. Blanco-López, A. J. Miranda-Ordieres and M. J. Lobo-Castañón, *Sens. Actuators, B*, 2015, **220**, 688.
84. I. Surugiu, L. Ye, E. Yilmaz, A. Dzgoev, B. Danielsson, K. Mosbach and K. Haupt, *Analyst*, 2000, **125**, 13.

85. A. Yarman and F. W. Scheller, *Angew. Chem.*, 2013, **125**, 11735.
86. D. Lakshmi, A. Bossi, M. J. Whitcombe, I. Chianella, S. A. Fowler, S. Subrahmanyam, E. V. Piletska and S. A. Piletsky, *Anal. Chem.*, 2009, **81**, 3576.
87. (a) D. R. Kryscio and N. A. Peppas, *Acta Biomater.*, 2012, **8**, 461; (b) A. Bossi, F. Bonini, A. P. F. Turner and S. A. Piletsky, *Biosens. Bioelectron.*, 2007, **22**, 1131.
88. Y. Sun, H. Y. Du, Y. T. Lan, W. J. Wang, Y. J. Liang, C. L. Feng and M. Yang, *Biosens. Bioelectron.*, 2016, **77**, 894.
89. S. Sun, L. Chen, H. Shi, Y. Li and X. He, *J. Electroanal. Chem.*, 2014, **734**, 18.
90. R. Zhang, S. Xu, J. Luo and X. Liu, *Microchim. Acta*, 2015, **182**, 175.

CHAPTER 12

Molecularly Imprinted Polymers: Providing Selectivity to Sample Preparation

ANTONIO MARTÍN-ESTEBAN

Departamento de Medio Ambiente, Instituto Nacional de Investigación y Tecnología Agraria y Alimentaria (INIA), Carretera de A Coruña km 7, E-28040 Madrid, Spain
*E-mail: amartin@inia.es

12.1 Introduction

During the past few decades, there has been a huge development of analytical instrumentation allowing eventually the determination of any compound in environmental, food, and biological samples. Typically, chromatographic techniques coupled to common detectors (UV, fluorescence) or, more recently, mass spectrometry (MS) or tandem MS are routinely used in analytical laboratories. However, direct injections of crude sample extracts are not recommended since it may invalidate the whole analysis, and thus a clean sample is generally convenient to improve separation and detection. Even when the selective detection provided by MS is used, matrix components can inhibit or enhance the analyte ionization, thus hampering accurate quantification. Therefore, sample preparation is a key step of the whole analytical process, being critical for unequivocal identification, confirmation and quantification of analytes.

Polymer Chemistry Series No. 28
Molecularly Imprinted Polymers for Analytical Chemistry Applications
Edited by Wlodzimierz Kutner and Piyush Sindhu Sharma

Published by the Royal Society of Chemistry, www.rsc.org

Sample preparation includes all those necessary steps for the isolation of analytes of interest from the selected sample by using an appropriate extracting phase. The main objectives of sample preparation are the removal of potential interferences, analyte preconcentration (especially in environmental water samples), converting (if needed) the analyte into a more suitable form for detection or separation, and providing a robust and reproducible method independent of variations in the sample matrix. More recently, new objectives have been set such as using smaller initial sample sizes, improvement of selectivity in extraction, facilitating the automation, and minimizing the amount of glassware and organic solvents to be used.[1] Organic compounds from liquid samples (*i.e.* environmental waters) are preferably extracted using solid phases by solid-phase extraction (SPE) or solid-phase microextraction (SPME) procedures, although for low volume samples, liquid–liquid extraction (LLE) can be also used. Extraction of target analytes from environmental or food solid samples is usually performed by mixing the sample with an appropriate extracting solution, subjecting that mixture to some process (agitation, microwave irradiation, *etc.*) to assist analytes transporting from the sample matrix to the extracting solution. For certain applications, matrix solid-phase dispersion (MSPD) can be also a good alternative. More recently, other techniques, such as stir-bar sorptive extraction (SBSE), micro solid-phase extraction (MSPE) or liquid-phase microextraction (LPME), among others, have also been incorporated to analytical laboratories.[2]

All the mentioned techniques suffer from a lack of selectivity making necessary an extensive optimization of the typical steps involved. However, even after careful optimization, some matrix components are co-eluted with target analytes making it difficult to reach detection limits according to the stringent regulations required nowadays. Some years ago, antibodies immobilized on an adequate support, called immunosorbents, were proposed as an alternative for use in SPE applications in order to overcome the aforementioned drawbacks associated with typical nonspecific sorbents.[3,4] Different immunosorbents have been employed for the determination of pesticides, mycotoxins, drugs and polyaromatic hydrocarbons, among others, showing an excellent degree of clean-up owing to the inherent selectivity of the antibodies used. However, the obtainment of antibodies is difficult, time-consuming, expensive, and in addition, it is difficult to guarantee *a priori* its success. Moreover, it is important to point out that after the antibodies have been obtained they have to be immobilized on an adequate support, which may result in poor antibody orientation or even complete denaturation.

More recently, as an alternative to immunosorbents, aptamers immobilised onto a solid support, *oligosorbents*, have been proposed.[5] Aptamers are identified for each target molecule within randomly synthesised nucleic acid libraries by an iterative *in vitro* process for selection and amplification. This process is called "systematic evolution of ligands by exponential enrichment" (SELEX). The SELEX procedure is characterised by the repetition of successive steps consisting of selection, amplification, and conditioning. In the first SELEX round, the library and the target analyte are incubated for binding. Unbound oligonucleotides are removed by washing and the target-bound oligonucleotides are eluted and subsequently amplified by PCR or reverse transcription PCR. This selected oligonucleotide pool is then used for the next selection round. After around 10–15 cycles, the enriched library is cloned and sequenced. The individual sequences are then analysed in order to identify the consensus motif, which corresponds to the minimal sequence required for target-specific binding providing aptamers with a high affinity against a specific region of the target nucleic acid molecule.[6–8] Unfortunately, the whole procedure can take from weeks to months to complete with no guarantee of resulting in an appropriated aptamer. Besides, aptamers have to be immobilised onto an appropriated support, which might affect its biospecific affinity. Thus, although some successful aptamer-based SPE methods have been proposed,[5] further research is necessary for decreasing the time and costs associated to synthesis of aptamers and oligosorbent preparation.

Molecularly imprinted polymers (MIPs) are synthetic, tailor-made, materials able to selectively bind an analyte target in preference to other closely related compounds.[9,10] These materials are obtained by polymerizing functional and cross-linking monomer molecules around a template molecule, leading to a highly cross-linked three-dimensional polymer network. The monomers are chosen considering their ability to interact with the binding sites of the template molecule. The non-covalent approach,[10] by far the most used for the preparation of MIPs, is based on the formation of relatively weak non-covalent interactions (*i.e.* hydrogen bonding, ionic interactions) between the template molecule and molecules of selected functional monomers before polymerization. The experimental procedure is rather simple and a wide variety of monomers able to interact with almost any kind of template are commercially available. However, it is not free from some drawbacks because template–monomer interactions are governed by an equilibrium process during the pre-polymerization step. Thus, in order to shift the equilibrium towards the formation of the template–monomer complex, excess monomer is used. Consequently, the excess of free monomers is randomly incorporated to the polymeric matrix leading to the formation of non-selective recognizing sites. Once polymerization has taken place, the template molecule is extracted and molecularly imprinted cavities bearing recognizing sites with their shape, size, and functionalities complementary to those of the target analyte are formed. The resulting imprinted polymers are stable, robust, and resistant to solutions of a wide range of pH, solvents and temperature. Therefore, MIPs emulate natural receptors but without the associated stability limitations. In addition, syntheses of MIPs are relatively cheap and easy, thus making MIPs a clear alternative to natural receptors.

The use of MIPs as selective sorbent materials allows for performing a customized sample treatment step prior to the final determination. Thus, their use in solid-phase extraction, so-called molecularly imprinted solid-phase extraction (MISPE), is by far the most advanced technical application of MIPs.[11–17] Besides, recent years have seen a growing interest in the combination of MIPs with other sample preparation techniques. Thus, in this chapter, the most recent MIP-based sample preparation techniques, the different approaches employed, as well as some selected applications, will be described.

12.2 Molecularly Imprinted Solid-phase Extraction

12.2.1 MISPE Modes

There are several configurations of MISPE, including batch SPE, where the MIP is left in contact with the sample in solution, and conventional off-line SPE, where the MIP is packed into cartridges and serves as in-line or on-line SPE. The former requires additional centrifugation and/or filtration steps, and thus it has been displaced by the conventional off-line SPE format. Nevertheless, it is important to highlight the work by Andersson *et al.*[18] for the determination of sameridine in human plasma samples. In this seminal work, an MIP prepared using a sameridine analogue as the template, was placed in contact for one hour with the semeridine analyte spiked human plasma. After the subsequent washing step, sameridine was eluted and further determined by gas chromatography leading to clean chromatographic traces thanks to the high selectivity provided by the MIP. In addition, it was demonstrated for the first time that the use of a close structural analogue of the analyte as the surrogate template prevents the problems associated with template bleeding, one of the typical drawbacks associated to the use of MIPs as cartridge and column packings in the SPE and chromatographic techniques, respectively.

12.2.1.1 Off-line Protocols

Off-line MISPE protocols do not differ from other SPE procedures. Typically, a small amount (15–500 mg) of an MIP is packed into polyethylene cartridges. Then, after the conditioning, loading and washing steps, analytes are eluted, ideally free of co-extractives, and the elution extract is further analysed by liquid chromatography, gas chromatography or capillary electrophoresis, as shown in Scheme 12.1. In general, sample is loaded onto the MIP cartridge in a low-polarity solvent, since in such a medium selective interactions are maximised, and after a washing step for the removal of compounds non-selectively bound to the polymeric matrix, analytes are eluted with a solvent able to disrupt the typical non-covalent interactions between the analyte and the imprinted polymer.

Aqueous samples can also be directly loaded onto MIP cartridges. In this manner, MIPs behave like a reverse-phase sorbent and thus both target analytes and matrix components are retained through non-selective interactions.

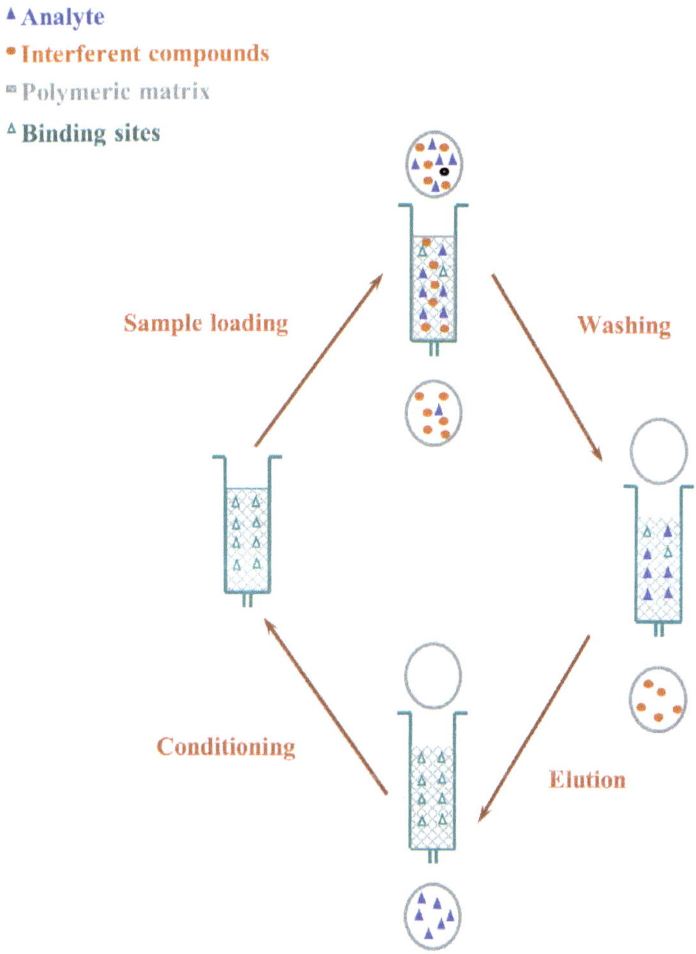

Scheme 12.1 Molecularly imprinted solid-phase extraction (MISPE) procedure. Reprinted from *Fresenius' Journal of Analytical Chemistry*, Molecularly imprinted polymers: new molecular recognition materials for selective solid-phase extraction of organic compounds, **370**(7), 2001, 795–802, A. Martin-Estban, © Springer-Verlag Berlin Heidelberg 2001, with permission of Springer.[11]

Then, the washing solvent has to be able to remove matrix components and, more important, of re-distributing non-selectively bound analytes to the selective imprints. Unfortunately, the success of this procedure is not always achieved. Therefore, efforts have been directed towards the synthesis of water-compatible MIPs by incorporating hydrophilic surface properties to the polymer in order to decrease non-selective hydrophobic interactions. This goal can be mainly achieved by using polar porogens,[19–21] hydrophilic comonomers, *e.g.* 2-hydroxyethyl methacrylate, acrylamide, or cross-linking monomers, *e.g.* pentaerythritoltriacrylate, methylenebis(acrylamide),[22–24] and/

or especially designed monomers capable of stoichiometrically interacting with the template binding sites.[25] These approaches have provided recognition of target analytes by MIPs in aqueous solutions to a certain extent. Therefore, further research in this field is expected in the coming years.

Originally, bulk polymerization, due to its simplicity, has been the preferred strategy to synthesize MIPs. Using this procedure, and after the unavoidable grinding and sieving steps, the particle size distribution was invariably heterogeneous with poor binding-site accessibility for the target analyte, thus limiting the MIP performance in certain applications. New polymerization strategies to obtain MIP beads with proper physical characteristics (*i.e.* size, porosity, pore volume, and surface area) have been developed in recent years. Among these new alternatives, multi-step swelling and polymerization,[26] precipitation polymerization,[27] and polymerization into the pores of silica beads, with the latter etched away with NH_4HF_2 after polymerization,[28] have been used for the preparation of MIPs for SPE of several analytes in biological, environmental, and food samples.

In order to improve mass transfer, some *in situ* polymerization strategies have also been proposed. At this regard, Du *et al.*[29] described the synthesis of a kinetin molecularly-imprinted monolith in a syringe by an *in-situ* polymerization technique. The synthesis procedure is rather simple and easily performed in any laboratory equipped with basic instrumentation. However, in order to guarantee a proper porous morphological structure in the MIP, thus providing flow-paths through the monolithic column, an appropriate amount of a polar solvent (dodecanol) is included in the mixture of solvents used as porogens. This solvent may disrupt the typical hydrogen bonding interactions between the template and monomers during polymerization, thus limiting its applicability to certain templates.

The *in-situ* synthesis of MIPs on the surface of microfiltration glass-fibre membranes in multi-well filter plates has been proposed as an alternative.[30–32] Briefly, the synthesis procedure is, as follows. First, membranes of a 24-well filter plate are washed with methanol and dried before use; then, 30–50 μL aliquots of a polymerization mixture are transferred onto the filter plate membranes under oxygen-free argon atmosphere. The plate is then covered with a UV-transparent cling film and placed under a UV lamp for plate irradiation for 3 h. After polymerization, the template is removed by successive washing, thus making the membranes ready for SPE experiments. The obtained MIP on the surface of the modified membranes forms a veil-like web between adjacent fibres, whereas inside the membrane, the polymer is partly deposited in clusters on the glass fibres. This polymer structure permits fast filtration on the composite membrane. The synthesis procedure has been further improved by using high-viscosity solvents (*i.e.* paraffin oil, room-temperature ionic liquids) as porogens. In this manner, with monomers-to-porogen ratios typical of bulk polymerization, it is possible to obtain regular shaped polymer micro (nano)spheres. The well-defined morphology and large surface area of the particles may increase the specific binding capacity of the membranes. Figure 12.1 shows, as an example, the scanning

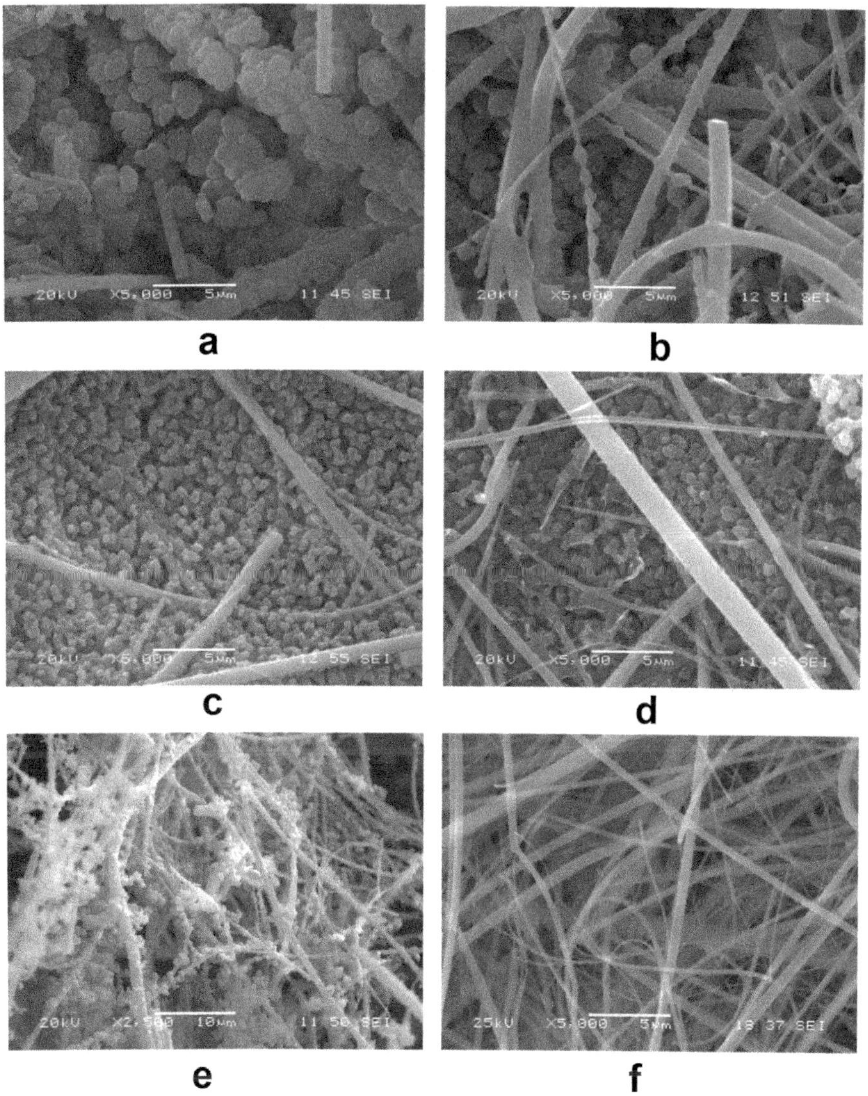

Figure 12.1 SEM micrographs of MIP composite membranes prepared with different polymerization solvents: (a) caprylonitrile/paraffin oil, (b) toluene/paraffin oil, (c) trihexyl(tetradecyl) phosphonium *tris*(pentafluoroethyl) trifluorophosphate (PH3 T FAP), (d) 1-butyl-3-methylimidazolium tetrafluoroborate, (e) cross-sectional view of a PH3 T FAP MIP, and (f) unmodified glass microfibre membrane. Reproduced from T. Renkecz, K. László and V. Horváth, *In situ* synthesis of molecularly imprinted nanoparticles in porous support membranes using high-viscosity polymerization solvents, *Journal of Molecular Recognition*,[32] John Wiley & Sons, © 2012 WILEY-VCH Verlag GmbH & Co. KGaA, Weinheim.

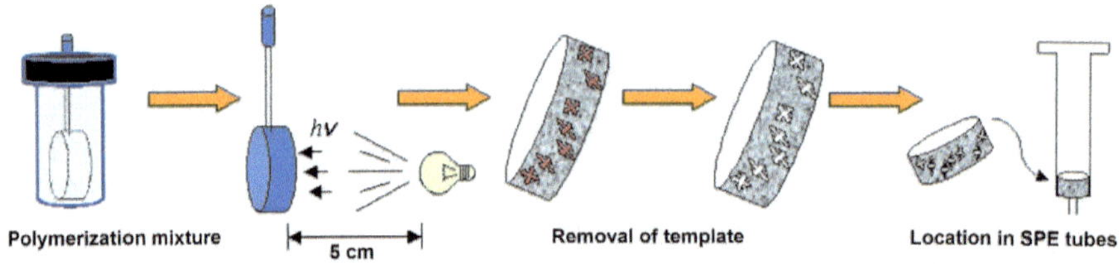

Scheme 12.2 Illustration of the synthesis procedure of imprinted frits. Reprinted from *Journal of Chromatography A*, **1218**(40), F. Barahona, E. Turiel and A. Martín-Esteban, Molecularly imprinted polymer grafted to porous polyethylene frits: A new selective solid-phase extraction format, 7065–7070, Copyright (2011) with permission from Elsevier.[33]

electron microscopy (SEM) micrographs of the composite membranes prepared using different high-viscosity solvents, where polymer microspheres entrapped within the glass fibre network are clearly present. This approach is a cost effective, one-step procedure for preparing MIP-composite membranes. It offers a viable alternative to those offered by existing MISPE cartridges. Besides, a much easier and faster synthesis method and high throughput analysis should be also pointed out as clear advantages. However, breakthrough volumes (<1 mL) are low, which excludes possible applications of these membranes to environmental samples. However, they suit pre-treatment of biological samples perfectly well.

A similar approach was recently proposed for the synthesis of MIPs grafted onto porous polyethylene (PE) frits.[33,34] The synthesis procedure, depicted in Scheme 12.2, is based upon the immersion of PE frits in a saturated solution of benzophenone (a photo-initiator capable of starting polymerization from the surface of the support material). Next, the frits, with the pores still wetted by this solution, were immersed in a vessel containing the polymerization mixture, and then the polymerization was initiated by UV-irradiation. After polymerization, the template was removed from the imprinted frits by Soxhlet extraction, and finally the modified frits were placed inside SPE glass tubes. Under optimum polymerization conditions, polymer aggregates are spread throughout the PE frit surface without significantly affecting its macroporous structure. The obtained imprinted frits were applied for clean-up of spiked extracts of thiabendazole (a post-harvest systemic fungicide) from citrus samples in the range of concentrations permitted according to the European legislation.[35]

12.2.1.2 On-line Protocols

In this format, a small precolumn packed with the MIP (typically ~50 mg) is placed in the loop of a six-port rotary injection valve. After loading the sample and washing out interfering compounds, the analytes are eluted by the mobile phase and then separated in the analytical column. Masqué *et al.*[36] described the first application of an on-line MISPE procedure coupled to HPLC for the selective extraction of 4-nitrophenol from a mixture of phenolic compounds in river water samples. However, in spite of the clear advantages of the on-line protocols, few papers have been published on-line MISPE studies. This low activity might be attributed to the usual lack of compatibility of the mobile phase required to perform the separation on the analytical column and the elution solvent necessary to desorb analytes from the MIP precolumn. This problem can be circumvented by directing the eluent from the MIP precolumn to the injection loop, and subsequently injecting on the chromatographic system,[37] or by mixing the organic elution solvent with an aqueous-rich solvent before reaching the analytical column.[38] However, these approaches increase the complexity of the on-line system requiring the incorporation of extra-instrumentation (*i.e.* pumps).

Besides, in on-line systems, MIPs must provide high permeability and fast mass transfer of analytes, which make necessary the synthesis of MIPs with improved properties. In this regard, a novel sample-preparation technique, termed dynamic liquid–liquid–solid microextraction (DLLSME), was developed and coupled on-line to HPLC for direct extraction, desorption, and determination of trace estrogens in complex samples.[39] The DLLSME consists of an aqueous donor phase, an organic medium phase, and molecularly imprinted polymer filaments (MIPFs) as the solid acceptor phase. The diagram of the whole device is shown in Scheme 12.3. The organic solvent with lesser density was directly added on top of the aqueous sample, and then the dynamic extraction was performed by circulating the organic solvent through the MIPFs inserted into a PEEK tube. This tube served as an

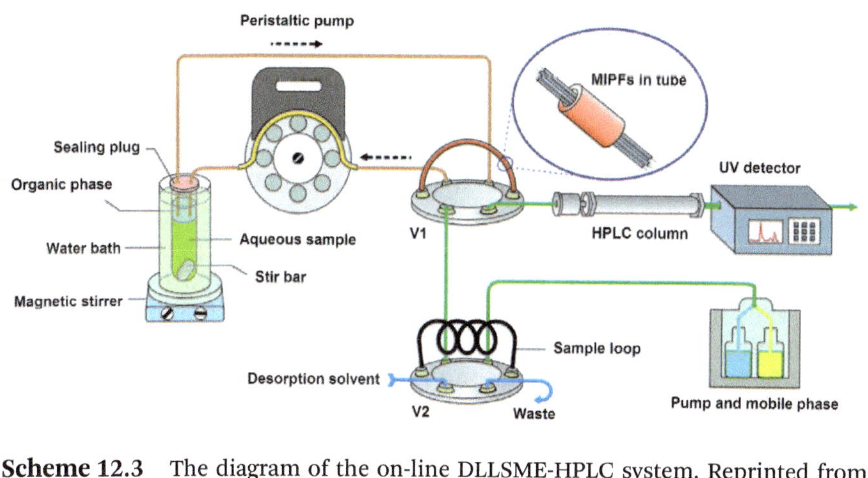

Scheme 12.3 The diagram of the on-line DLLSME-HPLC system. Reprinted from *Journal of Chromatography A*, **1241**, Q. Zhong, Y. Hu, Y. Hu and G. Li, Dynamic liquid-liquid-solid microextraction based on molecularly imprinted polymer filaments on-line coupling to high performance liquid chromatography for direct analysis of estrogens in complex samples, 13–20, Copyright (2012) with permission from Elsevier.[39]

extraction and desorption chamber. Afterwards, the extracted analytes on the MIPFs were on-line desorbed, and then introduced into the HPLC for analysis. According to the authors, permeability was greater, compared with that of monolithic capillary or column packed with finely dispersed sorbents. Besides, the small inner diameter of the tubing used could produce a high linear flow rate on the surface of the coatings, thus resulting in a fast transfer of target analytes between the solid and liquid phases. Evidently, the use of MIPFs presents clear advantages compared to the use of other on-line systems and, likely, new procedures based on its use will be further developed in the near future.

12.2.1.3 In-line Protocols

Extraction, enrichment, separation, and determination of target analytes can be achieved in one single step by the direct coupling of an MIP column in-line with the detection system thanks to the high selectivity provided by MIPs. MISPE with direct in-line UV-detection was first described by Sellergren[40] for the determination of pentamidine in urine. In that work, after loading 100 mL of diluted urine sample, the column was washed with 100 mL of a buffer solution (pH = 9), and then pentamidine eluted in ~1.5 ml of buffer of pH = 3. The selective elution performed allowed the successful detection of pentamidine free of co-extractives by UV at 270 nm. However, a high amount of matrix compounds were retained on MIP columns by hydrophobic interactions because a large volume of aqueous samples was loaded. Thus, it was necessary to use a large volume of washing solutions, which led to a quite long analysis time.

As mentioned above, in general, the recognition capabilities of MIPs are enhanced in organic media. In this regard, Turiel *et al.*[41] developed an analytical method for the determination of the fungicide thiabendazol (TBZ) in fruit sample extracts in acetonitrile. In this case, organic sample extracts (50 µl) were directly injected onto a TBZ-imprinted polymer column using a 100% CH_3OH mobile phase. Then, after 2.4 min, the mobile phase was switched to a methanol–acetic acid (80/20, *v*/*v*) solution in 0.1 min, keeping these conditions constant for 5 min before returning to the initial solution conditions. Following this strategy, the high selectivity of MIP column permitted the target analyte to be retained on the column while the interferences were rapidly eluted, thus making possible the detection and quantitation of TBZ in less than 15 min.

Peak broadening and peak-tailing phenomena are usually observed in the elution profiles of in-line MISPE procedures, leading to irreproducibility of peak-area measurements and long analysis time. These phenomena are attributed to a slow mass transfer and the existence of a heterogeneous distribution of molecularly imprinted cavities bearing recognizing sites, even if using imprinted beads. As an alternative, surface imprinting, which sets in place a thin film of an imprinted polymer grafted onto the surface of preformed beads, may solve some of these drawbacks, thereby conferring higher efficiency on the extraction process. In this regard, MIP microspheres with core–shell morphology and narrow particle-size distribution prepared by a two-step precipitation-polymerization procedure, have been proposed for in-line MISPE.[42,43] This synthetic procedure involves two steps: (i) polymer microspheres (the core particles) are prepared by precipitation polymerization of divinylbenzene-80 (DVB-80) in acetonitrile and (ii) the core particles are used as seeds in the synthesis of molecularly imprinted core–shell particles with well-developed pore structures, by copolymerization of DVB-80 with methacrylic acid (MAA) in the presence of the template in a mixed acetonitrile–toluene solvent porogen. Using these core–shell particles, the chromatographic performance was clearly improved compared to that offered by imprinted polymer microspheres, where the imprinting is not confined to the outer surfaces. This result demonstrates that the slow mass-transfer kinetics associated with conventional MIP stationary phases can be dramatically improved by surface-imprinting techniques.

MISPE with pulsed elution (MISPE-PE) represents an alternative, in which a small volume of an elution solvent instead of a steady solvent-switch, is used.[44] In this first work, theophylline in chloroform-diluted serum samples was injected onto a stainless steel column packed with theophylline-imprinted polymer beads using chloroform as the mobile phase. Subsequently, after interfering compounds were washed out from the column, 20 μL of methanol was injected, resulting in theophylline being eluted free of co-extractives and determined spectrophotometrically at 270 nm. MISPE-PE has been subsequently improved by the application of successive 20 μL pulses of different solvents, a procedure known as MISPE with differential pulsed elution (MISPE-DPE),[45] in order to wash out remaining interfering compounds and having the analyte fraction non-selectively retained. Both MISPE-PE and MISPE-DPE allow analyte enrichment through injection of large volumes of a sample and provide a high analytical sensitivity, thanks to the narrow band obtained by pulsed elution.

The in-line MISPE methods described above are focused on the determination of only one analyte due to the typical poor chromatographic performance associated with MIP columns. A notable exception to this statement was the work accomplished by Tamayo and Martín-Esteban[46] for the screening of phenylurea herbicides in vegetable samples. In this work, a linuron-imprinted polymer was prepared by filling the pores of preformed silica beads with the polymerisation mixture and subsequent dissolution of silica. Interestingly, target analytes were not only separated from matrix-interfering compounds but also in two different groups of phenylureas depending upon the presence or absence of a methoxy group in its chemical structure. Figure 12.2 shows the chromatograms obtained for blank and spiked (100 ng g^{-1} linuron and isoproturon) potato sample extracts. Both analytes were separated from each other and also from interferences, thus allowing the screening of these analytes at the concentration level required by recent legislation in less than 10 min.

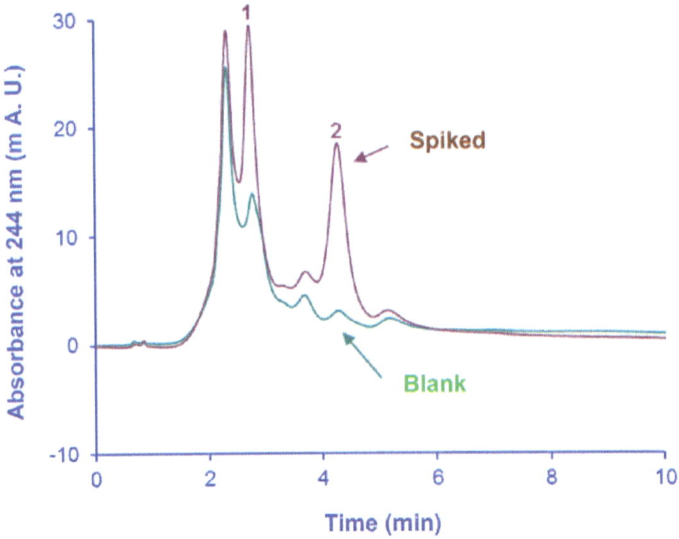

Figure 12.2 Chromatograms recorded after the injection of a blank and spiked potato sample extracts directly onto a linuron-imprinted column. Injection volume was 20 μL, mobile phase acetonitrile at a flow rate of 1 mL min^{-1}. Peak numbers: (1) linuron, (2) isoproturon. Reprinted from *Journal of Chromatography A*, **1098**(1–2), F. G. Tamayo and A. Martín-Esteban, Selective high performance liquid chromatography imprinted-stationary phases for the screening of phenylurea herbicides in vegetable samples, 116–122, Copyright (2005) with permission from Elsevier.[46]

12.2.1.4 Improved Batch Protocols

At present, the use of magnetic nanoparticles as the support for different stationary phases in batch SPE is one of the most active research area in sample preparation. These modified magnetic particles, after extraction, can be separated from the media by a simple magnet, thereby avoiding tedious filtration and/or centrifugation steps. The surface, and thus the properties, of magnetic particles can be modified by the incorporation of MIPs, providing selectivity to the extraction.[15,47] Several procedures have been described for the synthesis of magnetic MIPs. One of the simplest is outlined in Scheme 12.4 for the preparation of magnetic core–shell MIP particles.[48] Firstly, magnetite nanoparticles are encapsulated in silica through a typical sol–gel reaction using tetraethyl orthosilicate (TEOS), which result in hybrid Fe_3O_4@SiO_2 particles. The silica shells prevent oxidation and disturb magnetic dipolar interactions between magnetite nanoparticles, thus favouring their dispersion in liquid media. Moreover, silica shells are easily modifiable. Therefore, desirable functional groups can be grafted to the surface of the nanoparticles. Accordingly, vinyl groups, necessary for the subsequent surface imprinting, can be introduced on particles surface by reaction of Fe_3O_4@SiO_2 particles with 3-methacryloyloxypropyltrimethoxysilane (MPS).

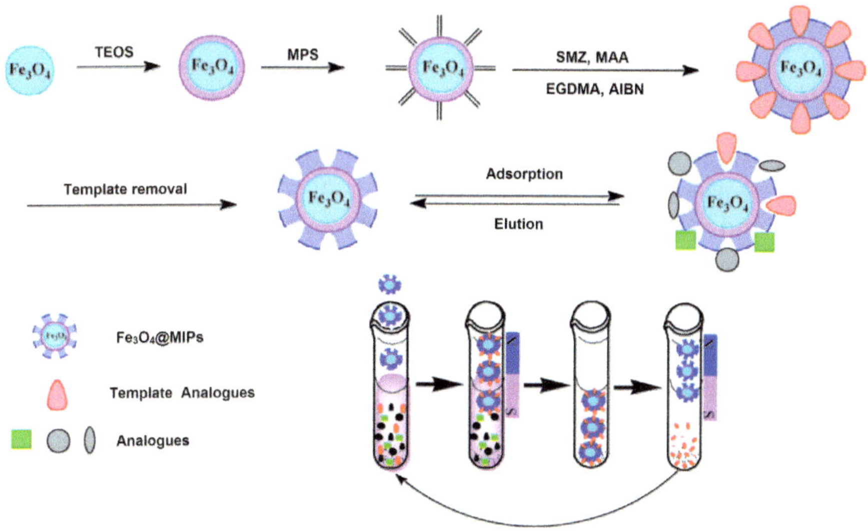

Scheme 12.4 Overview of synthesis of Fe_3O_4@MIPs nanoparticles and their application for extraction of sulfonamides with the help of an applied magnetic field. Reprinted from *Journal of Chromatography A*, **1245**, X. Kong, R. Gao, X. He, L. Chen, Y. Zhang, Synthesis and characterization of the core–shell magnetic molecularly imprinted polymers (Fe3O4@MIPs) adsorbents for effective extraction and determination of sulphonamides in the poultry feed, 8–16, Copyright (2012) with permission from Elsevier.[48]

Finally, the MIPs are formed on the surface of Fe_3O_4@SiO_2 by the copolymerization of vinyl end groups with MAA and ethylene glycol dimethacrylate (EGDMA) as the functional and cross-linking monomer, respectively, in the presence of sulfamethazine (SMZ) as the template. Once the magnetic MIP beads are prepared, they can be used in batch SPE procedures. These procedures have successfully been used for the extraction of triazines from soil, soybean, lettuce, and millet,[49] tetracycline antibiotics from egg and tissue,[50] sulfonamides in poultry feed[48] and microcystins in environmental waters,[51] among others.

Finally, a rather simple approach, so-called molecularly-imprinted micro-SPE (MI-MSPE) has been proposed. An MI-MSPE device comprises MIP particles enclosed within a porous polypropylene flat-sheet membrane envelope. Once the MI-MSPE device is prepared, it is conditioned and placed in the sample. The device tumbles freely within the sample solution during extraction for a certain period of time under stirring. After extraction, the device is manually removed with the help of tweezers and placed in a vial for elution of analytes. This approach has been successfully applied to the selective extraction of phenolic compounds from tap, river and sewage waters,[52] ochratoxin A from coffee, grape juice and urine,[53] and cocaine and its metabolites from human urine.[54]

12.2.2 Selected Applications

Since the first MISPE work by Sellergren[40] for the determination of pentamidine in urine samples appeared, many papers dealing with the use of MIPs in sample preparation have been published. In this section, some relevant MISPE studies on the selective extraction of analytes from biological, environmental and food samples will be described. For a more detailed and wide coverage, some recent reviews on selective solid-phase extraction of biological, environmental and food samples[12,55–57] using MIPs are recommended.

MISPE has been employed for the isolation and clean-up of a great variety of target analytes from biofluids (urine, serum, plasma) and tissue samples. Biofluids are generally diluted with an organic solvent to minimise the aqueous content of the sample, thus enhancing the recognition mechanism on the MIP, and to precipitate proteins. A high complexity of biological samples makes necessary a careful optimisation of the washing step in order to remove interfering compounds, which leads in some cases to the simultaneous loss of target analytes and thus low recovery. An alternative to improvement of the analytical performance of MISPE-based methods is the development of two-step SPE procedures. Firstly, a typical SPE sorbent is used to clean the sample before its application to the MIP sorbent. In this manner, the amount of impurities reaching the MIP is lower and target analytes are quantified easily. Besides, analytes are eluted from the first cartridge using an organic solvent, thus they reach the MIP cartridge in a suitable solvent for selective recognition. At this regard, Mullet *et al.*[58] developed an on-line two-step SPE-LC-MS procedure for the determination of verapamil and its metabolites in urine, plasma and cell cultures using a restricted access material (RAM) cartridge and a verapamil-imprinted polymer (Scheme 12.5). Briefly, the sample was loaded onto the RAM cartridge, in which analytes were retained by the alkyl hydrophobic bonded phase, whereas hydrophilic electroneutral diol groups bound to the external surface of the spherical particles protected the sorbent from contamination by proteins. Then, after this first clean-up, analytes were desorbed from the RAM precolumn using a mobile phase of pure acetonitrile, and then transferred to the MIP precolumn. Finally, the analytes retained by the MIP were desorbed and transferred to the analytical column for chromatographic separation. With this approach, verapamil and its metabolites were easily quantified by LC-MS in less than 35 min.

Environmental samples (mainly waters, sediments, and soils) have been subjected to MISPE for the extraction of pesticides, pharmaceuticals, and phenolic compounds, among others. The first off-line MISPE work describing an environmental application involved the extraction and enrichment of six chlorotriazines from groundwater and sediment samples using a terbutulazine-imprinted polymer.[59] An aliquot of 100 mL of a water sample was percolated and, after a clean-up step with dichloromethane, the matrix components were removed whereas target analytes were selectively retained

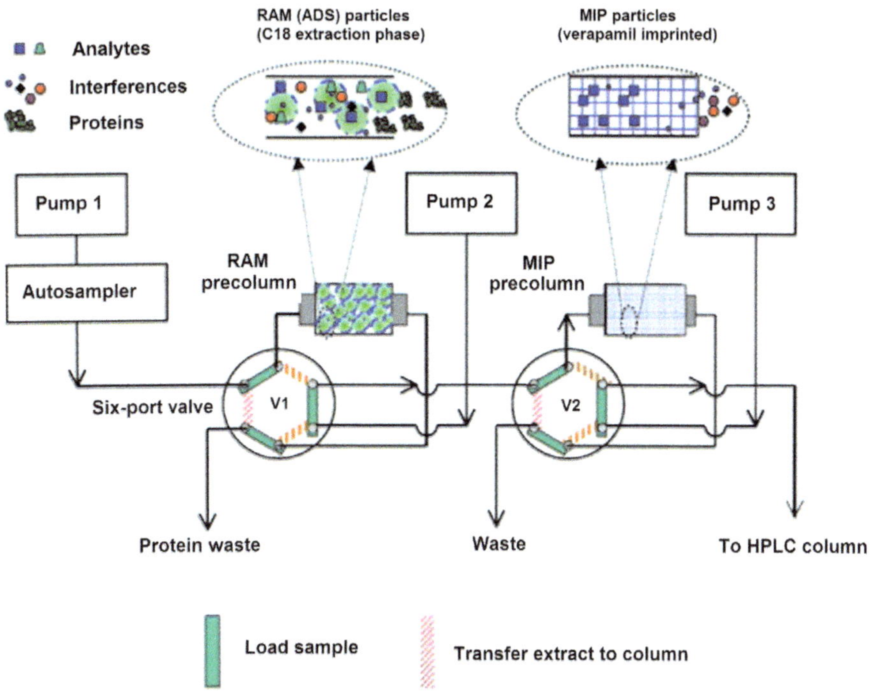

Scheme 12.5 Experimental configuration of the on-line two-step SPE-LC procedure for the determination of verapamil and its metabolites using a restricted access material (RAM) cartridge and a verapamil-imprinted polymer. Reprinted from *Journal of Chromatography B*, **801**(2), W. M. Mullett, M. Walles, K. Levsen, J. Borlak, J. Pawliszyn, Multidimensional on-line sample preparation of verapamil and its metabolites by a molecularly imprinted polymer coupled to liquid chromatography-mass spectrometry, 297–306, Copyright (2004) with permission from Elsevier.[58]

with recoveries higher than 80% (except for propazine, with a recovery of 53%). In spite of these results, subsequent works have concluded that small amounts of water on the MIP have a negative effect on the retention of analytes making it necessary to include an exhaustive drying step before the clean-up. Besides, matrix effects due to the presence of divalent cations in real water (such as calcium ions) have been observed. An ion-exchange interaction between these cations and the hydrogen of the carboxyl groups of the MIP occurs, thus preventing subsequent interactions (by hydrogen bonding) between the triazines and the MIP.[60] However, this matrix effect can easily be avoided by introducing an acidic washing step after the percolation of the water sample. Alternatively, as mentioned above, it is possible to perform a two-step SPE procedure by firstly loading water samples onto conventional sorbents (*i.e.* C_{18}, DVD) and then eluting the retained analytes with an organic solvent suitable for further loading on the MIP.[61,62] Although to a lesser extent, MISPE has also been successfully applied to the clean-up of soil sample extracts for the determination of triazines,[62,63] monosulfuron[64] and fluoroquinolones.[65]

Currently, the application of MISPE to food and beverage analysis represents only a small fraction of published papers. However, it increases faster than other MISPE areas. Therefore, high activity in this field is expected in the near future. Until now, MISPE applications for determination of pesticides, antibiotics and mycotoxins have been more widely studied. MISPE of triazines have been applied to the clean-up of fruits, vegetables and juices.[62,63,66–68] In general, high selectivity has been achieved but, in some cases, the complexity of the samples under study has led to the co-extraction of some matrix components. This problem has been circumvented by introducing a previous clean-up step using a non-imprinted polymer (NIP).[67] In this approach, the sample is first loaded onto the corresponding NIP. Therefore, analytes and matrix components are retained non-selectively. Then, target analytes (and a small amount of interfering compounds) are eluted during the washing steps whereas the majority of the matrix compounds remains bound to the NIP, thus achieving a first clean-up of the sample extracts. Subsequently, the obtained washing solution is loaded onto the corresponding MIP for further clean-up. Similar approaches have been followed for the determination of phenylurea herbicides.[69,70]

The application of MISPE to the extraction and clean-up of antibiotics from food samples have also demonstrated to be superior in analytical performance to traditional methods. For instance, MISPE has been applied to the determination of fluorquinolones in baby food.[71] In this work, after extraction of target analytes by ultrasound-assisted extraction, sample extracts in methanol were loaded onto a ciprofloxacin-imprinted polymer. Then, a simple washing step using 2 mL of methanol was performed before elution with 5 ml of a methanol–acetic acid (50 : 50, *v* : *v*) solution. This fraction was evaporated to dryness and dissolved in 0.5 ml of a methanol–water (25 : 75, *v* : *v*) solution for final determination with HPLC-UV. Figure 12.3 shows the chromatograms obtained for the spiked and unspiked baby food samples after conventional SPE using strong anion-exchange cartridges (Figure 12.3a and b) and MISPE (Figure 12.3c and d). Clearly, a clean baseline was obtained by MISPE, thus allowing the determination of target analytes at a trace concentration level. The detection limits reached were by one order of magnitude lower than those obtained by conventional SPE using strong anion-exchange cartridges.

Evidently, the main advantage attributed to MISPE is its inherent selectivity. Importantly, however, this selectivity leads to a considerable decrease of the analysis time and the cost. This statement was nicely demonstrated by Mohamed *et al.*[72] for the determination of chloramphenicol in milk (raw milk, skimmed milk, and milk powder). In this work, the advantages of MISPE, using a commercial MIP, were assessed by comparing the data generated from a traditional procedure. The MIP-based procedure did only require a previous centrifugation step whereas the traditional approach

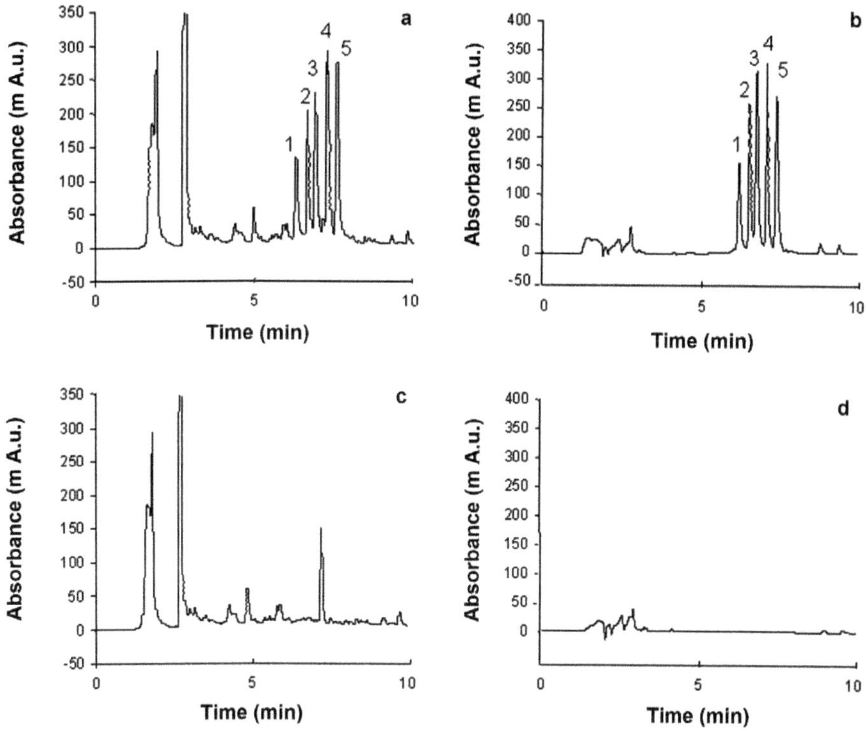

Figure 12.3 Chromatograms recorded after ultrasound-assisted extraction of baby-food samples and clean-up using strong anion exchange (a and c) and MIP cartridges (b and d). (a and b) Sample spiked at 1 μg g^{-1}; (c and d) unspiked sample. Peak assignment: 1 – enoxacin, 2 – norfloxacin, 3 – ciprofloxacin, 4 – danofloxacin, 5 – enrofloxacin. Reproduced from *Analytical and Bioanalytical Chemistry*, Selective sample preparation for the analysis of (fluoro)quinolones in baby food: molecularly imprinted polymers *versus* anion-exchange resins, **393**(3), 2008, 899–905, M. Díaz-Alvarez, © Springer-Verlag 2008, with permission from Springer.[71]

involved a protein precipitation step, an SPE clean-up and three liquid–liquid extraction steps. Apart from the expected cleaner chromatograms obtained following the MISPE procedure, sample throughput was considerably decreased (18 samples processed within 3 h compared to 8 h with the classical approach), thus clearly demonstrating the suitability of MISPE in food analysis.

Finally, template bleeding is considered one of the main drawbacks of MISPE in trace analysis. Apparently, template traces remain in the polymer network even after exhaustive washing by different methodologies, thus affecting the accurate determination of target analytes. The best manner of preventing this bleeding is using an analogue of the target analyte as template. Therefore, the bleeding of the template does not interfere in the quantification of the target analyte. As mentioned in Section 12.2.1 (above), this approach was originally proposed by Andersson *et al.*[18] for the selective extraction of sameridine on the MIP synthesized using a compound closely related to a target analyte as the template. Later, some other authors have followed this approach, allowing for the precise and accurate quantification of a great variety of analytes in different kinds of samples. However, the main drawback of template-analogue imprinted polymers is the resulting inferior molecular recognition ability compared to those prepared using the analyte as template. In this sense, a very interesting alternative is the use of a stable isotope labelled compound as template molecule (isotope imprinting).[73,74] In this way, the target analyte can be accurately detected by selective ion monitoring, although template bleeding might occur. Obviously, its applicability is limited to the availability of the corresponding isotope labelled compound analogue to the target analyte and the use of MS detection, although the potential of isotope imprinting to overcome the template bleeding problem is evident.

12.3 Molecularly Imprinted Solid-phase Microextraction

Solid-phase microextraction (SPME) has become one of the most used technique for sample preparation in analytical laboratories since its introduction by Arthur and Pawliszyn[75] in 1990. SPME is based on the partitioning of target analytes between the sample and a stationary phase, typically coating the surface of a fused silica fibre. After extraction, analytes can thermally be desorbed directly onto the injection port of a gas chromatograph or eluted with a suitable solvent for further analysis by chromatographic techniques. Its simplicity of operation, solvent less nature and the availability of commercial fibres have allowed SPME to become a tool routinely used. However, the variety of commercially available fibres is limited and just roughly covers the scale of polarity, thus leading to a lack of selectivity during the extraction. In this sense, the combination of molecular imprinting and SPME would ideally provide a powerful analytical tool with the characteristics of both these techniques (simplicity, flexibility, and selectivity).

The simplest way of combining these both techniques was proposed by Mullett *et al.*[76] in 2001, which consisted in packing a capillary with the MIP particles for in-tube SPME of propanolol in serum samples. However, this method requires using some extra instrumentation (pump, multiport valves), thus presenting the typical drawback associated with on-line MISPE protocols including the lack of compatibility between the solvent needed to desorb analytes from the MIP and the mobile phase used.

Recently, an MIP phase modified with a hydrophilic external layer (restricted access material), synthesized by *in situ* polymerization in a fused silica capillary, was used as the sorbent for in-tube SPME to determine parabens in breast milk samples by UHPLC-MS/MS.[77] The devised capillaries were coupled in a conventional HPLC microsyringe (250 μL). Scheme 12.6 illustrates

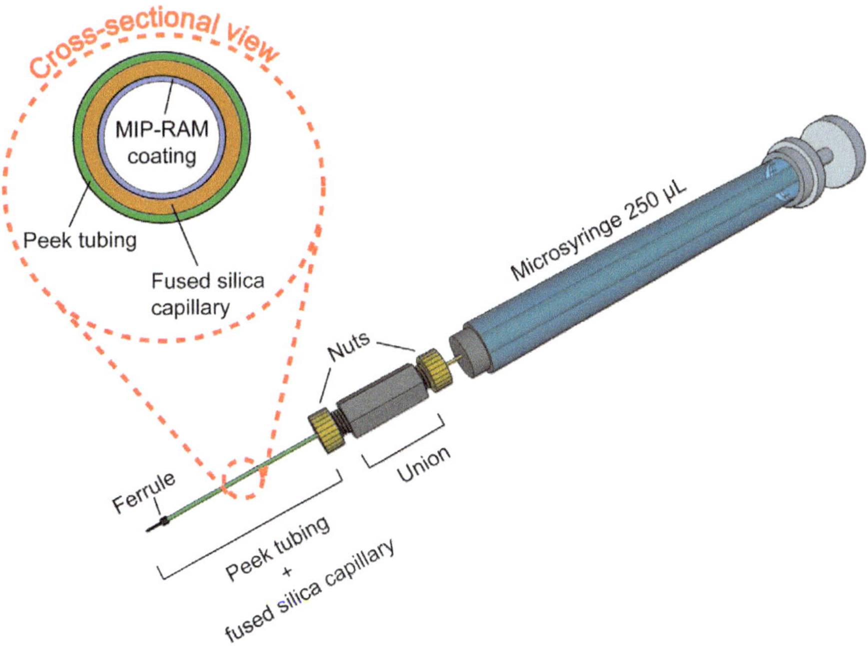

Scheme 12.6 An overview of an in-tube SPME device containing the fused silica capillary coated with the RAM-MIP sorbent. Reprinted from *Analytica Chimica Acta*, **932**, I. D. Souza, L. P. Melo, I. C. S. F. Jardim, J. C. S. Monteiro, A. M. S. Nakano, M. E. C. Quiroz, Selective molecularly imprinted polymer combined with restricted access material for in-tube SPME/UHPLC-MS/MS of parabens in breast milk samples, 49–59, Copyright (2016) with permission from Elsevier.[77]

the in-tube device containing the fused silica capillary coated with MIP-RAM. The milk samples are manually drawn through the MIP-RAM sorbent and ejected three times to pre-concentrate the analytes. Then, after the corresponding washing steps, elution was carried out by means of a single draw/eject cycle employing 100 μL of a methanol–ethanol (1 : 1, v:v) solution.

Although these approaches have successfully been employed and might open different ways of combining MIPs and SPME, MIP-based fibres, similar to the SPME fibres commercially available, would be accepted more easily in analytical laboratories. In this sense, MIP-coated fibres and MIP monoliths have been widely explored and their synthesis and performance will be described in this section.

12.3.1 MIP-coated Fibres

The preparation of the first MIP-coated silica fibre was proposed by Koster *et al.*[78] for the SPME of brombuterol from human urine. Silica fibres were activated by silylation and subsequently immersed in a polymerization solution composed by clenbuterol, MAA, EGDMA and AIBN in acetonitrile to carry out polymerization during 12 h at 4 °C under irradiation at 350 nm. According to authors, fibres with a polymeric film thickness of ~75 µm were obtained in a reproducible manner.

Key factors influencing both the morphology and the width of the obtained MIP coating were systematically studied in a series of papers dealing with the preparation of different MIP-coated silica fibres.[79–81] In these works, silica fibres were activated by silylation before immersion in the solution for polymerization (polymerization took place at 60 °C). Although the procedure and further pulling out of the fibre are tricky, the resulted coatings were homogenous and dense. Moreover, high reproducibility could be achieved by a strict control of polymerization conditions. The polymerization time is a key aspect of this procedure because it affects both coating thickness and fibre preparation reproducibility. In general, short polymerization time leads to low uniformity and thin coating, and longer time produces a too rigid polymer preventing the safe pulling out of the fibre. Accordingly, a polymerization time of 6 h was reported as optimum. Besides, a proper selection of the porogen is crucial. Apparently, the use of apolar solvents (*i.e.* toluene) lead to homogeneous MIP coating with thickness of 25.0 µm, whereas a thickness of 1.5 µm was achieved using acetone.[81] This problem can be circumvented by immersing the obtained fibres again in fresh polymerization solutions, thus resulting in repeated fibre coating by identical polymerization procedures until the desired thickness is reached. The SEM micrographics of a tetracycline MIP-coated fibre (Figure 12.4), not only show a homogeneous and dense morphological structure (Figure 12.4a) but also a highly cross-linked and porous structure (Figure 12.4b).[80]

The preparation of MIP-coated silica fibres has further been improved by surface reversible addition–fragmentation chain transfer polymerization (RAFT), which permits a higher control of the thickness of an MIP coating.[82] Briefly, silylated silica fibres were further functionalized with phenyl magnesium bromide (a RAFT agent) in the presence of carbon disulfide. Finally, the RAFT-agent-functionalized fibre was immersed in the corresponding polymerization solution for 48 h at 60 °C. Thanks to the controllable radical growing and chain propagation in the surface RAFT polymerization, MIP-coated fibres with thickness of 0.55 µm can be obtained in a robust and reproducible manner. The ultra-thin MIP coating guarantees sufficient analyte accessibility to selective imprinted cavities bearing recognition sites and the porous morphological structure of the MIP coating, thus increasing the diffusion rate of analytes in the coating.

The preparation of MIP-coated fibres by electropolymerization of suitable monomers (*i.e.* pyrrole) was recently explored.[83,84] Using this approach, the MIP was synthesized and deposited directly onto a support (*i.e.* platinum, or stainless steel) of any size and shape in a single step. The great advantage of this method is that the thickness of the polymer can be tuned and controlled by varying the polymerization conditions (*e.g.*, circulated current). Although this method is at a very early stage, the results obtained are rather promising and further development is expected in the near future.

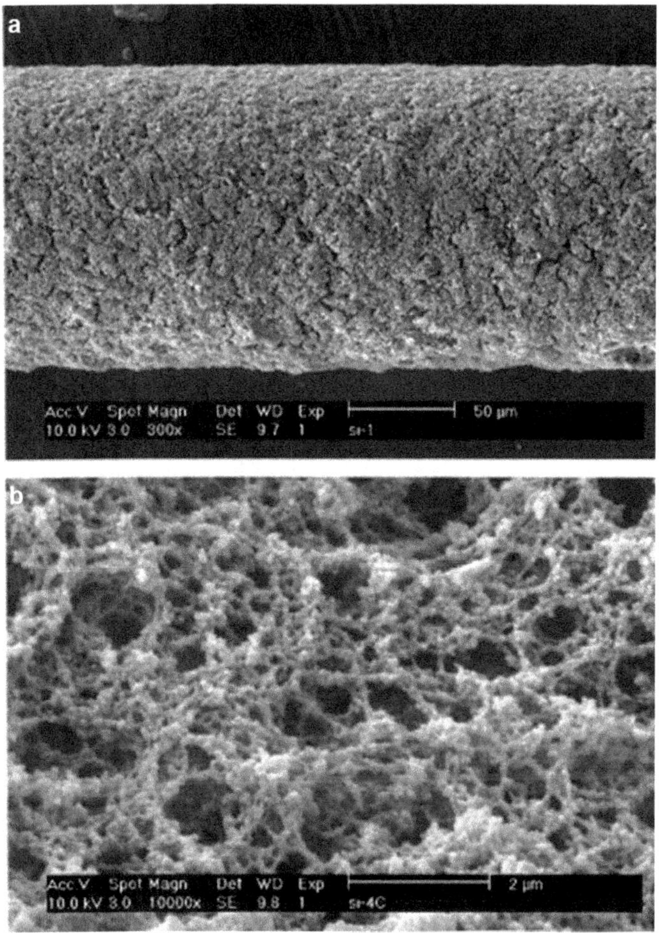

Figure 12.4 Scanning electron micrographs of tetracycline MIP coated fibre at (a) 300× and (b) 10 000× magnification. Preparation conditions of the MIP coated fibre. Solvent - acetone, functional monomer - acrylamide, cross-linking monomer - triethylolpropane trimethacrylate, initiator - azo(bis)-isobutyronitrile, polymerization time - 6 h, coating times - 10×. Reprinted from *Journal of Chromatography A*, **1188**(2), X. Hu, J. Pan, Y. Hu, Y. Huo and G. Li, Preparation and evaluation of solid-phase microextraction fiber based on molecularly imprinted polymers for trace analysis of tetracyclines in complicated samples, 97–107, Copyright (2008) with permission from Elsevier.[80]

12.3.2 MIP Fibres (Monoliths)

A completely different method to the direct preparation of MIP fibres (monoliths), using fused silica capillaries as moulds, has been proposed. This approach is sketched in Scheme 12.7.[85,86] In this method, silica capillaries are cut to approximately 30 cm long pieces and four windows of about 1.0 cm

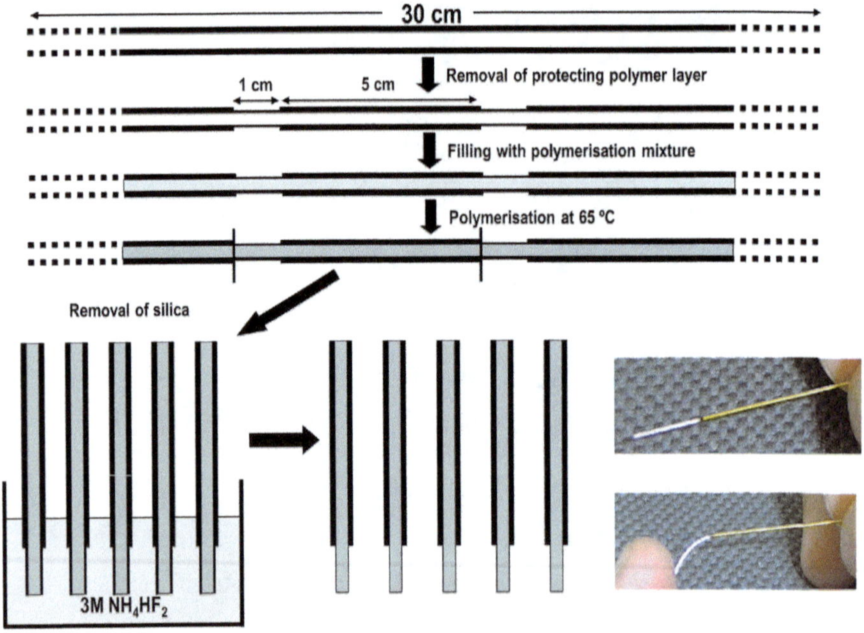

Scheme 12.7 Outline of preparation of molecularly imprinted fibres. Reprinted with permission from E. Turiel, J. L. Tadeo and A. Martín-Esteban, *Analytical Chemistry*, 2007, **79**, 3099.[85] Copyright (2007) American Chemical Society.

are prepared by burning the protecting polymer layer. Then, the capillary is filled with the polymerization mixture and both capillary ends are closed with two small pieces of rubber. The filled capillaries are introduced in an oven and polymerization takes place typically at temperatures higher than 60 °C for a certain period of time. Finally, capillaries are cut and immersed in an aqueous solution of NH_4HF_2 under agitation with silica walls being etched away. In this manner, MIP fibre monoliths of 1 cm length are obtained, with its thickness dependent on the inner diameter of the silica capillary used. Moreover, the obtained fibres were flexible and it was possible to bend them to a certain extent, preventing the easy breakage traditionally associated with the coated fused silica fibres. Similarly to MIP coated fibres, solvent, cross-linking monomer and polymerization time have a direct influence on the porosity, and thus on the final analytical performance, of the fibres prepared. As an illustrative example, Figure 12.5 shows the chromatogram obtained after SPME of parabens from soil sample extracts using an imprinted monolith prepared with *N*,*O*-bismethacryloyl ethanolamine, as a single functional monomer.[87] The high selectivity provided by the imprinted fibre allows for detection of target analytes at a very low concentration levels, which would be extremely difficult without performing any clean-up.

Desorption of target analytes from imprinted monolith fibres is mostly performed by immersing the fibre in a small volume of a solvent able to

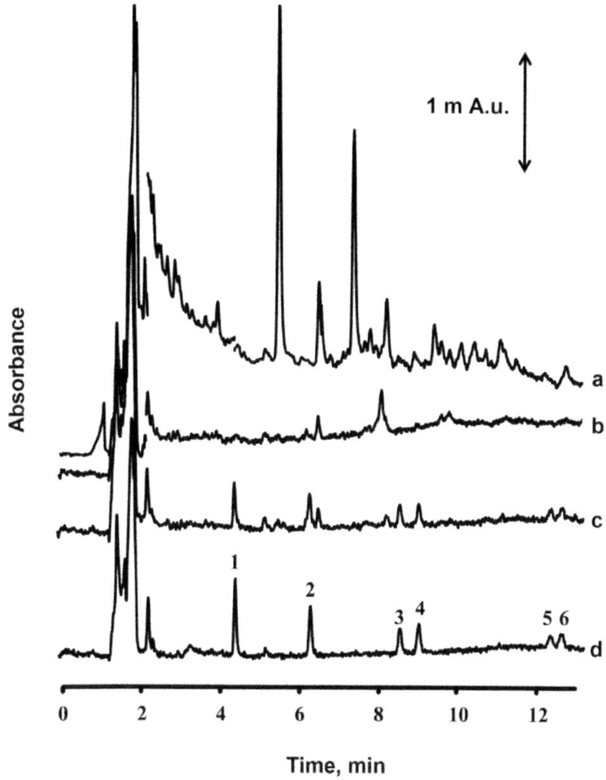

Figure 12.5 LC-UV chromatograms of extracts of solid samples recorded for (a) a non-spiked soil sample extract without clean-up, (b) a non-spiked soil sample extract after MI-SPME, (c) soil sample spiked with parabens at 10 ng g^{-1} after MI-SPME, and (d) standard solution of parabens at 60 μg L^{-1} after MI-SPME. Peak assignments: (1) mcthyl-paraben, (2) ethyl-paraben, (3) isopropyl-paraben, (4) propyl-paraben, (5) isobuthyl-paraben, and (6) buthyl-paraben. Reprinted from M. Díaz-Álvarez, S. P. Smith, D. A. Spivak and A. Martín-Esteban, Preparation of molecularly imprinted polymeric fibers using a single bifunctional monomer for the solid-phase microextraction of parabens for environmental solid samples, *Journal of Separation Science*,[87] John Wiley & Sons. © 2015 WILEY-VCH Verlag GmbH & Co. KGaA, Weinheim.

disrupt interactions of the analyte molecules with imprinted cavities. Implicitly, only a small fraction of the total volume used for the elution is injected to the chromatographic system. Thus, the on-line combination of SPME to HPLC is a useful alternative to improve detection limits. This approach has been followed by Li's group for the SPME-HPLC analysis of triazines in soybean, corn, lettuce and soils,[79] tetracyclines in chicken feed, chicken muscle and milk,[80] β-blockers in urine and plasma[81] and prometryn in soybean and corn[90] using MIP-coated fibres. After extraction and washing, fibres were immersed in the chamber of the specially designed SPME-HPLC six-way injection valve. Then, the target analytes were eluted statically, for a fixed time, with a suitable solvent or directly by the mobile phase used for the chromatographic separation of the analytes. Regardless the method used, clean chromatograms were obtained reaching detection limits at very low concentration levels.

Importantly, most of the reported imprinted fibres are thermally stable up to temperatures ranging from 250 to 300 °C, thus making them suitable for SPME-GC, which is the simplest, cheapest and more environmental friendly approach for desorbing target analytes. This method has been used for the determination of diacetylmorphine and analogous compounds in street heroin,[86] triazines in water, rice, onion and maize,[88,89] and polybrominated diphenyl ethers in municipal wastewaters[91] with high sensitivity and selectivity. Although the applications reported until now are rare, the thermal stability of imprinted fibres will surely help the development of MIP-based SPME procedures and their implementation in analytical laboratories in the near future.

12.4 Molecularly Imprinted Stir Bar Sorptive Extraction

Stir bar sorptive extraction (SBSE) is based on the partitioning of target analytes between a liquid sample and a stationary-phase coated stir bar. Until now, only polydimethylsiloxane (PDMS)-coated stir bars are commercially available, restricting the range of applications to the extraction of hydrophobic compounds (organochlorine and organophosphorous pesticides) due to the apolar character of PDMS. It is obvious that the preparation and use of MIP-coated stir bars would extend the applicability of SBSE in sample preparation. In this regard, Zhu *et al.*[92] devised stir bars coated with an MIP consisting of a film formed from a formic acid solution of nylon-6 polymer, imprinted with monocrotophos, to extract successfully and selectively several organophosphorus pesticides from dichloromethane solution for the analysis of environmental soil samples. In spite of the success of the obtained MIP-coated stir bars, this synthesis procedure has not been further explored.

More recently, the use of MIP-coated stir bars prepared by chemical bonding of the MIP to silylated stir bars has been proposed. This procedure is similar to that described above for the preparation of MIP-coated fibres for SPME. It has been applied for the determination of β_2-agonists,[93] triazole fungicides,[94] sulfa drugs,[95] thiabendazole,[96] and environmental estrogens[97] in different samples. The MIP-coated stir bar synthesis procedure seems to be a more generally applicable. It can easily be performed in any laboratory. Figure 12.6 shows the chromatograms, obtained after SBSE, using a dual-template MIP-coated stir bar, of several estrogens from the lake and river waters clearly demonstrating the suitability of the proposed imprinted stir bar for the selective extraction of environmental estrogens.

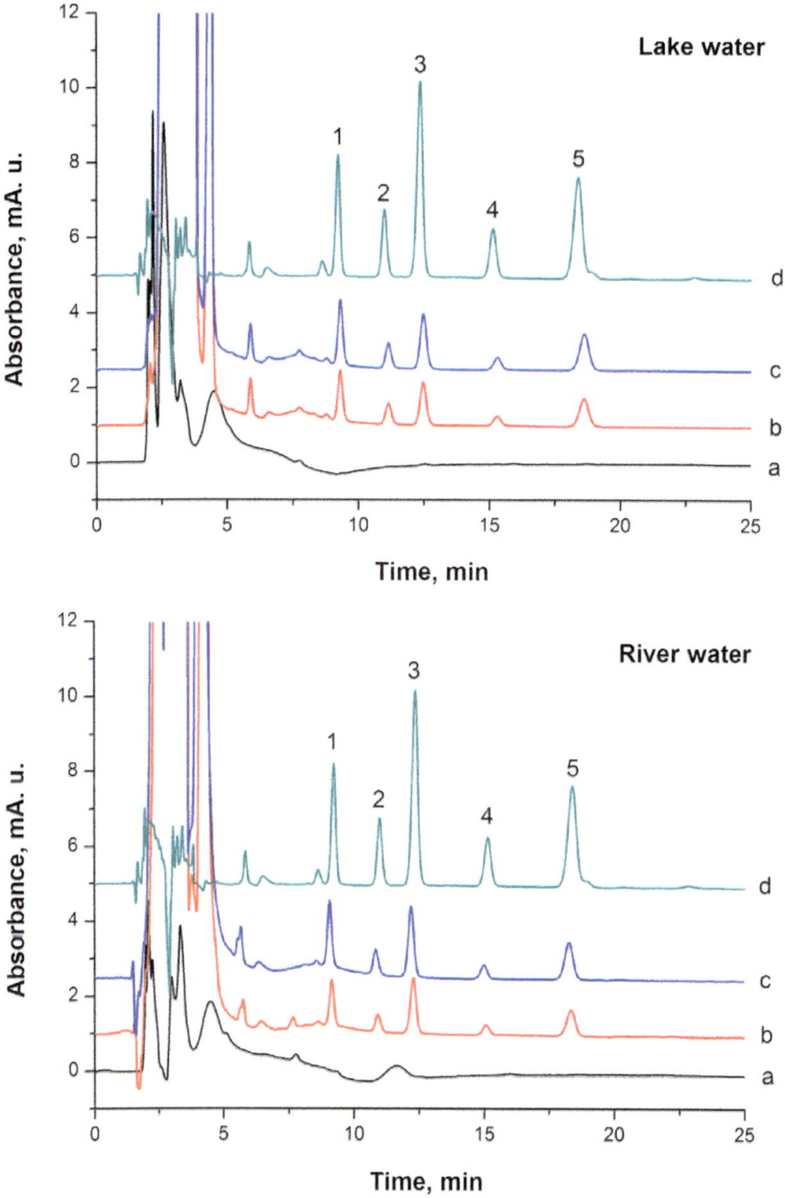

Figure 12.6 Chromatograms of 10 mg L^{-1} spiked water samples. (a) Direct injection of the spiked sample. (b) A spiked sample with the NIP-coated stir bar sorptive extraction. (c) A spiked sample with the MIP-coated stir bar sorptive extraction. (d) Standard solution of mixed 1 mg L^{-1} estrogens. Peak assignments: (1) bisphenol A, (2) estradiol, (3) bisphenol B, (4) estrone, and (5) diethylstilbestrol. Reprinted from the *Journal of Chromatography A*, **1358**, Z. Xu, Z. Yang and Z. Liu, Development of dual-templates molecularly imprinted stir bar sorptive extraction and its application for the analysis of environmental estrogens in water and plastic samples, 52–59, Copyright (2014) with permission from Elsevier.

12.5 Other Formats

12.5.1 Matrix Solid-phase Dispersion

Matrix solid-phase dispersion (MSPD) is based on a complete disruption of the sample (liquid, viscous liquid, semi-solid, or solid) by a solid sorbent, thus allowing sample components to disperse into it. Experimentally, the sample is placed in a glass mortar and blended with the sorbent until disruption and dispersion of the sample in the sorbent is complete. Then, the mixture is directly packed into an empty SPE cartridge and analytes are eluted after a proper washing step to remove interfering compounds. Direct use of MIPs as the dispersant sorbents is not straightforward because of a high water content of the samples typically extracted by MSPD, as selective recognition by MIPs is not favoured in aqueous samples. However, the use of water-compatible MIPs in MSPD has been demonstrated for the selective extraction and clean-up of fluoroquinolones from eggs and tissue,[20] with average recovery in the range of 72.2–114.1%, depending upon the analyte and the sample nature, and relative standard deviations of less than 7.0%. The MIP performance superior to that achieved using conventional sorbents (C_{18} and Florisil), where recovery was in the range of 24.9–84.6%, was clearly demonstrated. Recently, similar methods have been applied to the MSPD of steroids in goat milk.[98,99] The simplicity of operation of MSPD might appear attractive but further development in this area will depend upon obtaining MIPs with improved recognition capabilities in water-rich samples.

12.5.2 Combination of Liquid Membranes and MIPs

For the first time, Chimuka *et al.*[100] combined liquid membranes and MIPs for the extraction of 17β-estradiol from aqueous samples. In this work, a specially designed stainless steel extraction device, composed of two compartments separated by a membrane, was used. First, the membranes are equilibrated with a water-immiscible organic solvent, which will act as an acceptor solution. The lower compartment is filled with the aqueous sample, with a membrane as the separation interface. The upper compartment is filled with the organic acceptor phase and with a small amount of MIP particles (30–50 mg). Then, target analytes are liquid–liquid extracted, over a fixed extraction time (60–90 min), from the aqueous donor phase into the organic acceptor phase and from it to the imprinted cavities of the MIP. Finally, MIP particles are separated from the organic phase by filtration through a syringe filter, and then selectively bound compounds are released from the particles by passing a suitable elution solvent. Following this approach, enrichment factors ranging from 6 to 40 were achieved, depending upon the analyte and the sample, with an impressive selectivity, even with complex samples, *e.g.* wastewater.

In order to avoid filtration steps, a novel liquid–liquid–solid microextraction (LLSME) technique based on porous membrane-protected MIP-coated silica fibre was proposed.[101] A terbutylazine-imprinted polymer-coated silica fibre was protected with a length of porous polypropylene hollow fibre membrane, which was filled with a water-immiscible organic phase. Subsequently, the whole device was immersed into an aqueous sample for extraction of triazines used as model compounds. The target analytes were firstly extracted from the aqueous sample through the membrane, and then extracted onto the MIP fibre. This procedure was successfully applied to the determination of triazines in sludge water, watermelon, milk, and urine samples,[101] and thiabendazole for citrus sample extracts.[102] The main drawback of this approach is, however, the fragility of the fibres used, making it necessary to handle them cautiously in order to prevent their breakage. As an alternative, a similar device for the extraction of sulphonamides[103] and triazines[104] in surface environmental waters was recently proposed. In this case, a small amount (a few milligrams) of imprinted beads, were packed into the lumen of a porous polypropylene capillary in order to protect and separate them from aqueous media. Then, the packed fibre was immersed in toluene, filling the pores of the membrane, thus allowing target analytes to be extracted from aqueous samples to the organic solvent where MIPs performed selective recognition. Finally, target analytes were easily eluted with methanol from polypropylene capillary, and then further determined by liquid chromatography.

Recently, a novel molecularly imprinted polymer-coated hollow fibre (MIP-HF) for selectively extracting triazines directly from environmental waters has been proposed.[105] Scheme 12.8 shows the procedure for the preparation of MIP-HFs as well as the SEM images of the wall surface of the polypropylene hollow fibre used before and after polymerisation. Similarly to the above described devices, target analytes pass from the aqueous medium to the thin film of toluene retained in the pores of the membrane, according to a liquid–liquid extraction process ruled by the partition coefficients of the species involved; and then analytes diffuse from the supported liquid to the selective imprinted cavities of the MIP, where it is adsorbed by selective interactions. Importantly, these described approaches allow the selective recognition of target analytes in water samples, which open a new path to circumvent the frequent lack of selective recognition of MIPs in aqueous media.

12.6 Conclusions

In conclusion, MIPs are clearly compatible with current sample pretreatment procedures providing an unprecedented selectivity to the extraction process. MIPs are capable of cleaning up complex samples, thus leading to selective, sensitive, fast, and robust analytical methods without the need for using expensive detectors (*i.e.* MS detectors). Although the use of MIPs in SPE can be considered a robust alternative to non-selective sorbents, the development of new SPE formats, improving mass diffusion, as well as the use of new monomers favouring molecular recognition in aqueous media, is expected in the near future and thus will ease its implementation in analytical laboratories. Moreover, the synthesis of imprinted fibres for SPME and

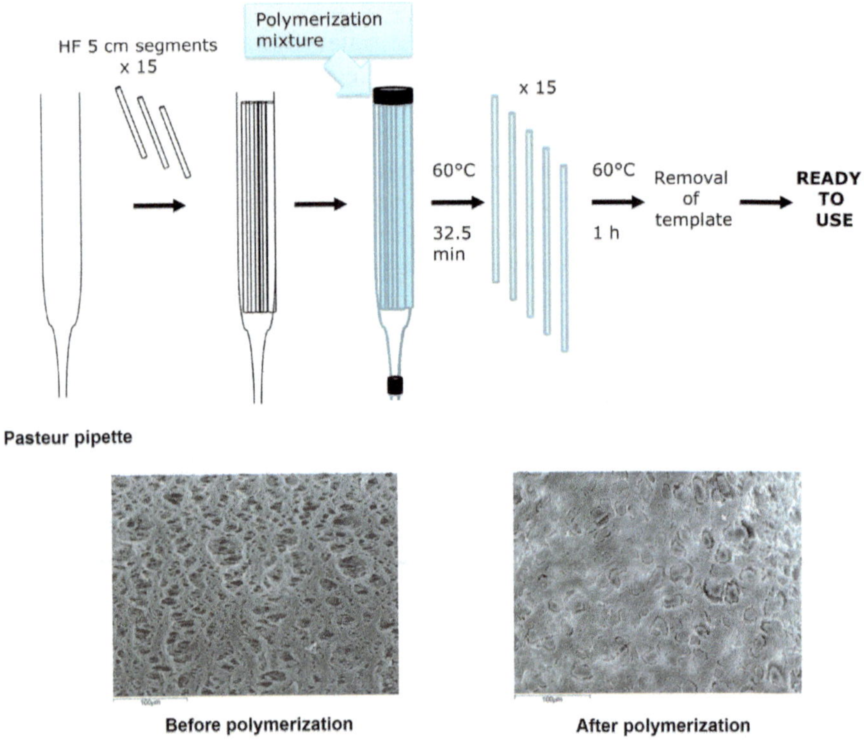

Scheme 12.8 Illustration of the polymerisation procedure for the preparation of MIP-HFs, and scanning electron micrographs of wall surface of original polypropylene hollow fibre and wall surface of MIP-HF at 300× magnification. Reprinted from *Journal of Chromatography A*, **1442**, F. Barahona, M. Díaz-Álvarez, E. Turiel and A. Martín-Estevan, Molecularly imprinted polymer-coated hollow fiber membrane for the microextraction of triazines directly from environmental waters, 12–18, Copyright (2016) with permission from Elsevier.[105]

the incorporation of MIPs to other extraction techniques demonstrate the adaptability of MIPs to almost any extraction technique. In this regard, there is plenty of room for improvement of their use in these techniques, their combination with others (*i.e.* liquid membranes) and the development of micro-MISPE devices. These improvements will bring new selective and simple analytical methods making MIPs one of the breakthrough technologies in sample preparation in the coming years.

List of Abbreviations

AIBN	2,2′-Azobisisobutyronitrile
DLLSME	Dynamic liquid–liquid–solid microextraction
DVB-80	Divinylbenzene-80
EGDMA	Ethylene glycol dimethacrylate

HF	Hollow fibre
LLE	Liquid–liquid extraction
LLSME	Liquid–liquid–solid microextraction
LPME	Liquid-phase microextraction
MAA	Methacrylic acid
MI-MSPE	Molecularly-imprinted microsolid-phase extraction
MIP	Molecularly imprinted polymer
MIPF	Molecularly imprinted polymer filament
MISPE	Molecularly imprinted solid-phase extraction
MISPE-DPE	Molecularly imprinted solid-phase extraction with differential pulsed elution
MISPE-PE	Molecularly imprinted solid-phase extraction with pulsed elution
MI-SPME	Molecularly-imprinted solid-phase microextraction
MPS	3-Methacryloyloxypropyltrimethoxysilane
MS	Mass spectrometry
MSPD	Matrix solid-phase dispersion
MSPE	Micro solid-phase extraction
NIP	Non-imprinted polymer
PCR	Polymerase chain reaction
PDMS	Polydimethylsiloxane
PE	Polyethylene
PH3 T FAP	Trihexyl(tetradecyl)phosphonium *tris*(pentafluoroethyl) trifluorophosphate
RAFT	Reversible addition–fragmentation chain transfer polymerization
RAM	Restricted access material
SBSE	Stir bar sorptive extraction
SELEX	Systematic evolution of ligands by exponential enrichment
SEM	Scanning electron microscopy
SMZ	Sulfamethazine
SPE	Solid-phase extraction
SPME	Solid-phase microextraction
TBZ	Thiabendazole
TEOS	Tetraethyl orthosilicate

References

1. R. Smith, *J. Chromatogr. A*, 2003, **1000**, 3.
2. J. Pawliszyn and H. L. Lord, *Handbook of Sample Preparation*, John Wiley & Sons, Inc., Hoboken, New Jersey, 2010.
3. M. Hennion and V. Pichon, *J. Chromatogr. A*, 2003, **1000**, 29.
4. A. Martín-Esteban, P. Fernandez, D. Stevenson and C. Camara, *Analyst*, 1997, **122**, 1113.
5. V. Pichon, F. Brothier and A. Combes, *Anal. Bioanal. Chem.*, 2015, **407**, 681.
6. R. Stoltenburg, C. Reinemann and B. Strehlitz, *Biomol. Eng.*, 2007, **24**, 381.
7. E. Peyrin, *J. Sep. Sci.*, 2009, **32**, 1531.

8. M. Mascini, I. Palchetti and S. Tombelli, *Angew. Chem., Int. Ed.*, 2012, **51**, 1316.
9. G. Wulff and A. Sarhan, *Angew. Chem.*, 1972, **84**, 364.
10. R. Arshady and K. Mosbach, *Makromol. Chem.*, 1981, **182**, 687.
11. A. Martín-Esteban, *Fresenius. J. Anal. Chem.*, 2001, **370**, 795.
12. E. Caro, R. Marce, F. Borrull, P. Cormack and D. Sherrington, *TrAC, Trends Anal. Chem.*, 2006, **25**, 143.
13. V. Pichon and K. Haupt, *J. Liq. Chromatogr. Relat. Technol.*, 2006, **29**, 989.
14. J. Haginaka, *J. Sep. Sci.*, 2009, **32**, 1548.
15. Y. Hu, J. Pan, K. Zhang, H. Lian and G. Li, *TrAC, Trends Anal. Chem.*, 2013, **43**, 37.
16. A. Martín-Esteban, *TrAC, Trends Anal. Chem.*, 2013, **45**, 169.
17. B. Bui and K. Haupt, *Anal. Bioanal. Chem.*, 2010, **398**, 2481.
18. L. Andersson, A. Paprica and T. Arvidsson, *Chromatographia*, 1997, **46**, 57.
19. B. Sellergren, J. Wieschemeyer, K. Boos and D. Seidel, *Chem. Mater.*, 1998, **10**, 4037.
20. H. Yan, F. Qiao and K. Row, *Anal. Chem.*, 2007, **79**, 8242.
21. H. Sun, F. Qiao, G. Liu and S. Liang, *Anal. Chim. Acta*, 2008, **625**, 154.
22. B. Dirion, Z. Cobb, E. Schillinger, L. Andersson and B. Sellergren, *J. Am. Chem. Soc.*, 2003, **125**, 15101.
23. E. Benito-Pena, S. Martins, G. Orellana and M. Moreno-Bondi, *Anal. Bioanal. Chem.*, 2009, **393**, 235.
24. T. Lv, H. Yan, J. Cao and S. Liang, *Anal. Chem.*, 2015, **87**, 11084.
25. J. Urraca, M. Moreno-Bondi, A. Hall and B. Sellergren, *Anal. Chem.*, 2007, **79**, 695.
26. J. Wang, P. Cormack, D. Sherrington and E. Khoshdel, *Angew. Chem., Int. Ed.*, 2003, **42**, 5336.
27. K. Hosoya, K. Yoshizako, Y. Shirasu, K. Kimata, T. Araki, N. Tanaka and J. Haginaka, *J. Chromatogr. A*, 1996, **728**, 139.
28. R. Abou-Hany, J. Urraca, A. Descalzo, L. Gomez-Arribas, M. Moreno-Bondi and G. Orellana, *J. Chromatogr. A*, 2015, **1425**, 231.
29. F. Du, G. Ruan, S. Liang, F. Xie and H. Liu, *Anal. Bioanal. Chem.*, 2012, **404**, 489.
30. G. Ceolin, F. Navarro-Villoslada, M. Moreno-Bondi, G. Horvai and V. Horvath, *J. Comb. Chem.*, 2009, **11**, 645.
31. T. Renkecz, G. Ceolin and V. Horvath, *Analyst*, 2011, **136**, 2175.
32. T. Renkecz, K. Laszlo and V. Horvath, *J. Mol. Recognit.*, 2012, **25**, 320.
33. F. Barahona, E. Turiel and A. Martín-Esteban, *J. Chromatogr. A*, 2011, **1218**, 7065.
34. M. Díaz-Álvarez and E. Turiel, *Mol. Imprinting*, 2015, **2**, 30.
35. European Community, *Council Directive 93/58/EEC*, OJ L211, 1993.
36. N. Masque, R. Marce, F. Borrull, P. Cormack and D. Sherrington, *Anal. Chem.*, 2000, **72**, 4122.
37. G. Theodoridis, C. Zacharis, P. Tzanavaras, D. Themelis and A. Economou, *J. Chromatogr. A*, 2004, **1030**, 69.
38. J. Ou, L. Hu, L. Hu, X. Li and H. Zou, *Talanta*, 2006, **69**, 1001.

39. Q. Zhong, Y. Hu, Y. Hu and G. Li, *J. Chromatogr. A*, 2012, **1241**, 13.
40. B. Sellergren, *Anal. Chem.*, 1994, **66**, 1578.
41. E. Turiel, J. L. Tadeo, P. A. G. Cormack and A. Martín-Esteban, *Analyst*, 2005, **130**, 1601.
42. F. Barahona, E. Turiel, P. A. G. Cormack and A. Martín-Esteban, *J. Polym. Sci., Part A: Polym. Chem.*, 2010, **48**, 1058.
43. F. Barahona, E. Turiel, P. A. G. Cormack and A. Martín-Esteban, *J. Sep. Sci.*, 2011, **34**, 217.
44. W. Mullett and E. Lai, *Anal. Chem.*, 1998, **70**, 3636.
45. W. Mullett and E. Lai, *Microchem. J.*, 1999, **61**, 143.
46. F. G. Tamayo and A. Martín-Esteban, *J. Chromatogr. A*, 2005, **1098**, 116.
47. L. Chen and B. Li, *Anal. Methods*, 2012, **4**, 2613.
48. X. Kong, R. Gao, X. He, L. Chen and Y. Zhang, *J. Chromatogr. A*, 2012, **1245**, 8.
49. Y. Zhang, R. Liu, Y. Hu and G. Li, *Anal. Chem.*, 2009, **81**, 967.
50. L. Chen, J. Liu, Q. Zeng, H. Wang, A. Yu, H. Zhang and L. Ding, *J. Chromatogr. A*, 2009, **1216**, 3710.
51. S. Pan, X. Chen, X. Li, M. Cai, H. Shen, Y. Zhao and M. Jin, *J. Chromatogr. A*, 2015, **1422**, 1.
52. Q. Feng, L. Zhao and J. Lin, *Anal. Chim. Acta*, 2009, **650**, 70.
53. T. Lee, B. Saad, W. Khayoon and B. Salleh, *Talanta*, 2012, **88**, 129.
54. J. Sanchez-Gonzalez, M. Tabernero, A. Bermejo, P. Bermejo-Barrera and A. Moreda-Pineiro, *Anal. Chim. Acta*, 2015, **898**, 50.
55. L. Andersson, M. Abdel-Rehim, L. Nicklasson, L. Schweitz and S. Nilsson, *Chromatographia*, 2002, **55**, S65.
56. V. Pichon and F. Chapuis-Hugon, *Anal. Chim. Acta*, 2008, **622**, 48.
57. C. Baggiani, L. Anfossi and C. Giovannoli, *Anal. Chim. Acta*, 2007, **591**, 29.
58. W. Mullett, M. Walles, K. Levsen, J. Borlak and J. Pawliszyn, *J. Chromatogr. B*, 2004, **801**, 297.
59. I. Ferrer, F. Lanza, A. Tolokan, V. Horvath, B. Sellergren, G. Horvai and D. Barcelo, *Anal. Chem.*, 2000, **72**, 3934.
60. F. Chapuis, V. Pichon, F. Lanza, S. Sellergren and M. Hennion, *J. Chromatogr. A*, 2003, **999**, 23.
61. T. Pap, V. Horvath, A. Tolokan, G. Horvai and B. Sellergren, *J. Chromatogr. A*, 2002, **973**, 1.
62. E. Turiel, A. Martín-Esteban, P. Fernandez, C. Perez-Conde and C. Camara, *Anal. Chem.*, 2001, **73**, 5133.
63. F. Chapuis, V. Pichon, F. Lanza, B. Sellergren and M. Hennion, *J. Chromatogr. B*, 2004, **804**, 93.
64. X. Dong, N. Wang, S. Wang, X. Zhang and Z. Fan, *J. Chromatogr. A*, 2004, **1057**, 13.
65. E. Turiel, A. Martín-Esteban and J. L. Tadeo, *J. Chromatogr. A*, 2007, **1172**, 97.
66. B. Bjarnason, L. Chimuka and O. Ramstrom, *Anal. Chem.*, 1999, **71**, 2152.

67. C. Cacho, E. Turiel, A. Martín-Esteban, C. Perez-Conde and C. Camara, *Anal. Bioanal. Chem.*, 2003, **376**, 491.
68. C. Cacho, E. Turiel, A. Martín-Esteban, D. Ayala and C. Perez-Conde, *J. Chromatogr. A*, 2006, **1114**, 255.
69. F. G. Tamayo, J. L. Casillas and A. Martín-Esteban, *Anal. Chim. Acta*, 2003, **482**, 165.
70. F. G. Tamayo, J. L. Casillas and A. Martín-Esteban, *Anal. Bioanal. Chem.*, 2005, **381**, 1234.
71. M. Diaz-Alvarez, E. Turiel and A. Martín-Esteban, *Anal. Bioanal. Chem.*, 2009, **393**, 899.
72. R. Mohamed, J. Richoz-Payot, E. Gremaud, P. Mottier, E. Yilmaz, J. Tabet and P. Guy, *Anal. Chem.*, 2007, **79**, 9557.
73. M. Kawaguchi, Y. Hayatsu, H. Nakata, Y. Ishii, R. Ito, K. Saito and H. Nakazawa, *Anal. Chim. Acta*, 2005, **539**, 83.
74. H. Sambe, K. Hoshina, K. Hosoya and J. Haginaka, *J. Chromatogr. A*, 2006, **1134**, 16.
75. C. Arthur and J. Pawliszyn, *Anal. Chem.*, 1990, **62**, 2145.
76. W. Mullett, P. Martin and J. Pawliszyn, *Anal. Chem.*, 2001, **73**, 2383.
77. I. D. Souza, L. P. Melo, I. C. S. F. Jardim, J. C. S. Monteiro, A. M. S. Nakano and M. E. C. Queiroz, *Anal. Chim. Acta*, 2016, **932**, 49.
78. E. Koster, C. Crescenzi, W. den Hoedt, K. Ensing and G. de Jong, *Anal. Chem.*, 2001, **73**, 3140.
79. X. Hu, Y. Hu and G. Li, *J. Chromatogr. A*, 2007, **1147**, 1.
80. X. Hu, J. Pan, Y. Hu, Y. Huo and G. Li, *J. Chromatogr. A*, 2008, **1188**, 97.
81. X. Hu, J. Pan, Y. Hu and G. Li, *J. Chromatogr. A*, 2009, **1216**, 190.
82. X. Hu, Y. Fan, Y. Zhang, G. Dai, Q. Cai, Y. Cao and C. Guo, *Anal. Chim. Acta*, 2012, **731**, 40.
83. E. Mazzotta, C. Malitesta, M. Diaz-Alvarez and A. Martín-Esteban, *Thin Solid Films*, 2012, **520**, 1938.
84. M. Diaz-Alvarez, E. Mazzotta, C. Malitesta and A. Martín-Esteban, *J. Mol. Recognit.*, 2014, **27**, 415.
85. E. Turiel, J. L. Tadeo and A. Martín-Esteban, *Anal. Chem.*, 2007, **79**, 3099.
86. D. Djozan and T. Baheri, *J. Chromatogr. A*, 2007, **1166**, 16.
87. M. Diaz-Alvarez, S. P. Smith, D. A. Spivak and A. Martín-Esteban, *J. Sep. Sci.*, 2016, **39**, 552.
88. D. Djozan and B. Ebrahimi, *Anal. Chim. Acta*, 2008, **616**, 152.
89. D. Djozan, M. Mahkam and B. Ebrahimi, *J. Chromatogr. A*, 2009, **1216**, 2211–2219.
90. X. Hu, Y. Hu and G. Li, *Anal. Lett.*, 2007, **40**, 645.
91. M. Li, N. Lei, C. Gong, Y. Yu, K. Lam, M. Lam, H. Yu and P. Lam, *Anal. Chim. Acta*, 2009, **633**, 197.
92. X. Zhu, J. Cai, J. Yang, Q. Su and Y. Gao, *J. Chromatogr. A*, 2006, **1131**, 37.
93. Z. Xu, Y. Hu, Y. Hu and G. Li, *J. Chromatogr. A*, 2010, **1217**, 3612.
94. Y. Hu, J. Li and G. Li, *J. Sep. Sci.*, 2011, **34**, 1190.
95. Z. Xu, C. Song, Y. Hu and G. Li, *Talanta*, 2011, **85**, 97.

96. E. Turiel and A. Martín-Esteban, *J. Sep. Sci.*, 2012, **35**, 2962.
97. Z. Xu, Z. Yang and Z. Liu, *J. Chromatogr. A*, 2014, **1358**, 52.
98. J. Ganan, A. Gallego-Pico, R. Garcinuno, P. Fernandez-Hernando, S. Morante, I. Sierra and J. Durand, *Anal. Bioanal. Chem.*, 2012, **403**, 3025.
99. J. Ganan, S. Morante-Zarcero, A. Gallego-Pico, R. Garcinuno, P. Fernandez-Hernando and I. Sierra, *Talanta*, 2014, **126**, 157.
100. O. Nemulenzi, B. Mhaka, E. Cukrowska, O. Ramstrom, H. Tutu and L. Chimuka, *J. Sep. Sci.*, 2009, **32**, 1941.
101. Y. L. Hu, Y. Y. Wang, Y. F. Hu and G. K. Li, *J. Chromatogr. A*, 2009, **1216**, 8304.
102. F. Barahona, E. Turiel and A. Martín-Esteban, *Anal. Chim. Acta*, 2011, **694**, 83.
103. M. Diaz-Alvarez, F. Barahona, E. Turiel and A. Martín-Esteban, *J. Chromatogr. A*, 2014, **1357**, 158.
104. E. Turiel, M. Diaz-Alvarez and A. Martín-Esteban, *J. Chromatogr. A*, 2016, **1432**, 1.
105. F. Barahona, M. Diaz-Alvarez, E. Turiel and A. Martín-Esteban, *J. Chromatogr. A*, 2016, **1442**, 12.

CHAPTER 13

Electrosynthesized Molecularly Imprinted Polymers for Chemosensing: Fundamentals and Applications

E. MAZZOTTA[a,b], S. RELLA[a], A. TURCO[a] AND C. MALITESTA*[a]

[a]Laboratorio di Chimica Analitica, Dipartimento di Scienze e Tecnologie Biologiche ed Ambientali (Di.S.Te.B.A.), Università del Salento, via Monteroni, 73100 Lecce, Italy; [b]IFAC-CNR, Via Madonna del Piano, 10, 50019 Sesto Fiorentino (FI), Italy
*E-mail: cosimino.malitesta@unisalento.it

13.1 Introduction

Chemical sensors are analytical devices consisting of two fundamental parts: a recognition element residing in an active layer which is integrated with a transducer (Scheme 13.1).

The role of the first one is to selectively interact with the analyte producing a chemical signal (*e.g.* a product of a selective reaction between the analyte and the recognition element or simply the bonding between them, *etc.*), which is translated by the transducer into an electric signal. The occurrence and the strength of this signal are related to the presence and to the concentration of the analyte, respectively.

Polymer Chemistry Series No. 28
Molecularly Imprinted Polymers for Analytical Chemistry Applications
Edited by Wlodzimierz Kutner and Piyush Sindhu Sharma

Published by the Royal Society of Chemistry, www.rsc.org

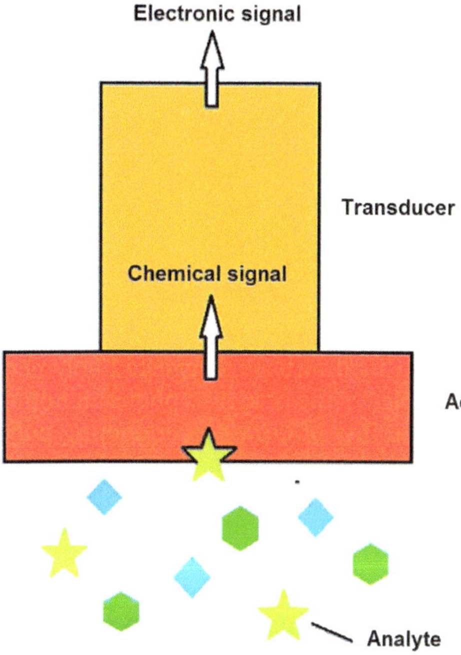

Scheme 13.1 Simplified representation of a generic chemical sensor. Reproduced from ref. 1 with permission from The Royal Society of Chemistry.

Chemosensing is an alternative approach to analysis of complex samples in respect to chromatographic techniques. In the latter, components of the sample are, in principle, physically separated and hence singly determined by a universal detector. Generally, this strategy requires complex instrumentation and expert analysts, while chemical information is incorporated in the low-cost device in chemosensing, which can be applied also by analysts with limited expertise.

Classification of sensors is based on nature of their recognition elements[2] and transducers employed.[3] So, for example, sensors with biologically active layers are named biosensors and electrochemical sensors are those where an electrochemical process is the transduction process.

While glass electrodes could be classified as a chemical sensor for H^+, the beginning of the history of chemical sensors is generally considered the development of a glucose biosensor based on glucose oxidase enzyme by Clark and Lyons.[4]

Replacement of biological components by artificial recognition elements to give biomimetic sensors has been more recently considered as an important advance. Among them, molecularly imprinted polymers (MIPs) have assumed a prominent position. In fact, they are highly stable, able to work in harsh environments, cheap and free of ethical problems related to biological components. In addition, they can be prepared in principle for any analyte.

Introduction to molecularly imprinting polymerization is reported in several previous chapters. This chapter is aimed at a particular way to perform the preparation of MIPs, *i.e.* the electrochemical method, which can be easily applied by those with no experience in polymer chemistry. The choice of non-covalent imprinting, by far the most applied, has the same advantage.

Performing polymerization by an electrochemical way was first reported in 1949.[5] On the other hand, this possibility could be suggested by the (unwanted) growth of passive films on electrode surfaces as a consequence of electrolysis of organic substances, a phenomenon reported since long time. The general observation that working at low concentrations of the organic electroactive species avoids this drawback, suggested polymerization as the chemical reaction responsible for the observed phenomenon.

The discovery of the possibility of growing conducting polymers by electrosynthesis has represented a milestone in materials chemistry research. In general, the synthesis of conducting polymers has been so important to merit in 2000 a Nobel Prize for their first proposers Heeger, McDiarmid and Shirakawa.

Electrochemical polymerization is a process in which a solution containing one monomer (or more monomers for copolymers) is subjected to electrode oxidation or reduction producing as a result a polymeric material, as an insoluble phase or as electrode (thick or thin) film or altogether. Radical polymerization represents an important class among different electrochemical polymerization types. In this respect, a possibility is the case where few radicals are produced by the electrode reaction and initiate the polymerization by a chain process similar to chemical polymerization.

Another possibility is stoichiometric polymerization, in which the amount of charge spent in the process is equivalent to monomer moles included in the resulting polymer, *i.e.* electrode oxidation or reduction of each monomer molecule is necessary for obtaining the material. An example of this mechanism is represented by the electrochemical polymerization of pyrrole[6] (Scheme 13.2).

The possibility of obtaining a film on the electrode surface depends on different factors. The stability of the cation resulting from oxidation of analyte can be of paramount importance. A very reactive cation generally reacts rapidly with solvent and other nucleophiles in the vicinity of electrode surface instead of participating in oligomerization, while very stable cation can diffuse away far from the electrode surface, thus limiting the possibility of filming.[7]

An interesting feature of stoichiometric electrochemical polymerization is represented by the possibility to finely control the amount of deposited polymer by circulating charge on electrodes of different shape and hence properties like the thickness of the film and, in MIP application, the number of active sites in the polymer. In addition, films produced in this way are quite adherent to the electrode surface and rigid, which is beneficial in terms of integration with transducer and in mechanical stability of active sites.

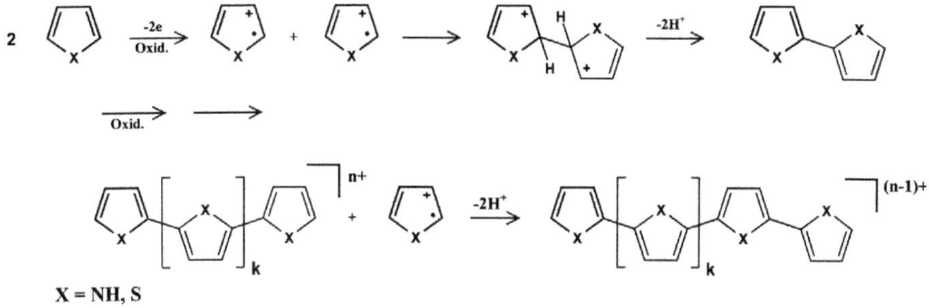

Scheme 13.2 Classical formation mechanism of conducting polymers. Reprinted with permission from J. Heinze, B. A. Frontana-Uribe and S. Ludwigs, *Chem. Rev.*, 2010, **110**, 4724.[6] Copyright (2010) American Chemical Society.

After a few attempts of apparent coupling of molecular imprinting with electrochemical polymerization,[8,9] in 1999 two groups independently proposed deliberate application of electrochemical synthesis to molecular imprinting.[10,11] MIP film growth under potentiodynamic conditions and a quartz crystal microbalance (QCM) response of an MIP sensor for glucose are reported in Figure 13.1.

Parameters to be controlled in development of electrosynthesis of an MIP are in general the same ones as those in the chemical counterpart, *i.e.* the monomer efficiently interacting with the analyte, the monomer-to-analyte ratio, (porogen) solvent, *etc.* Preliminary theoretical or experimental investigations on those parameters are a good practice in the electrosynthesized MIP (eMIP) field. In this respect, spectroscopic studies of the mononer–analyte complex[12,13] can throw light on the system to be polymerized. In contrast to chemical polymerization, the use of a cross-linking monomer is not necessary with eMIPs, even if some attempts to its use has been reported.[13–16] In addition, attention should be paid to avoid a plausible range of syntheses in which the analyte is electroactive for two main reasons: products of analyte electrochemistry could alter the polymerization process and depletion of analyte molecules at the electrode surface or in the near region could decrease the yield of imprinting.

After preparing an eMIP, verification of imprinting and selectivity of material are mandatory. Both these goals are generally pursued by performing analyte binding experiments on template-extracted eMIPs and electrosynthesized non-imprinted polymers NIPs, *i.e.* polymers prepared under the same conditions of eMIPs apart from the presence of the analyte (template), by the analyte and possible interferences (*e.g.* structural analogues). However, it must be kept in mind that selectivity depends not only on the properties of eMIPs, but also on those of the transducers. In fact, substances present in the sample, even if totally unrelated to the analyte, if they are able to get transducer surface and to give a signal, represent a problem with the accuracy of the device. A classic example in the chemical

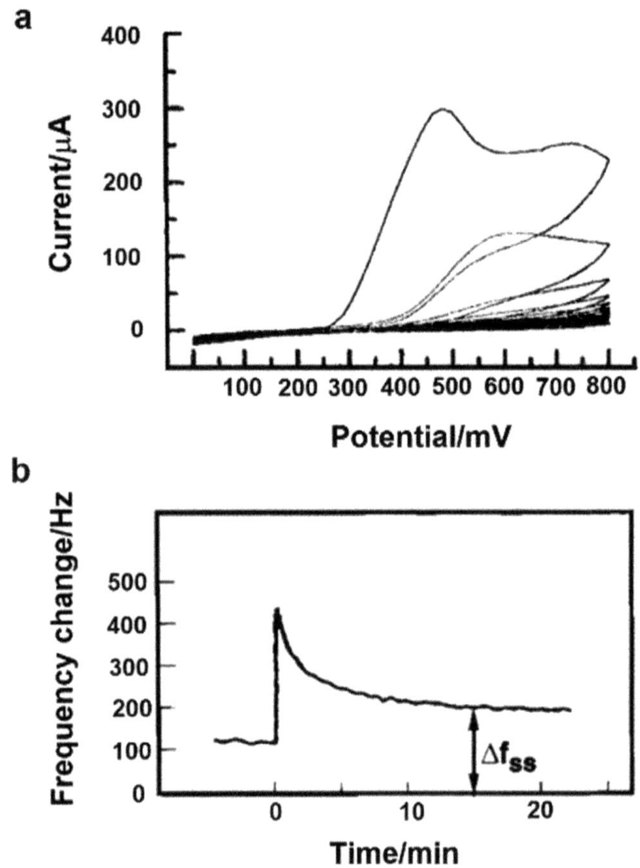

Figure 13.1 (a) Multi-scan cyclic voltammetry curve for *o*-phenylenediamine (*o*-PD) 5 mM in acetate buffer (pH = 5.2) on a Pt electrode of a QCM. Scan rate, 50 mV s^{-1}. Number of scans, 20. Glucose concentration, 20 mM. (b) QCM response of an imprinted poly(*o*-phenylenediamine) (PPD) based sensor to glucose injection. The curve is relevant to the evolution of frequency upon addition of glucose up to 20 mM following the steady value for glucose10 mM. Δf_{ss} represents the steady-state total value. Reprinted with permission from C. Malitesta, I. Losito and P. G. Zambonin, *Anal. Chem.*, 1999, **71**, 1366.[10] Copyright (1999) American Chemical Society.

sensor field is represented by electroactive species (like ascorbic acid) in glucose sensor operation.[17]

The chemical characterization of materials (comprising eMIPs loaded with the analyte) generally parallels the above work to increase confidence in interpretation of imprinting studies. Finally, the usual determination of analytical figures of merit, such as sensitivity, LOD, working stability and shelf life, and recovery in application to real/spiked samples, complete the work.

Three main different types of electrochemical sensors based on eMIPs have been reported in the literature.

1. Sensors based on transduction process based on direct electroactivity of analyte.
2. Sensors based on a change in the electrochemical signal produced by an electroactive probe as a consequence of the occupation of active sites by the analyte. In fact, the binding of analyte molecules in active sites has been shown to increase the signal of the probe. This phenomenon has been interpreted by a so-called "gate effect".[18,19] In practice, shrinking of the MIP caused by non-covalent reoccupation of binding sites increases the size of the micro- and macro-pores and hence probe diffusivity through eMIP layer. Actually, a different behaviour has been also reported (see, *e.g.* Karimian *et al.*[20] and Li *et al.*[21]), *i.e.* a decrease of probe signal with increasing concentrations. In this case, occupation of binding sites by the analyte has been considered to decrease the pores available for probe diffusion. Probably both behaviours can be observed depending on concentration range, *i.e.* a gate effect at low concentrations and a blocking effect at high concentrations, as supported by a work[22] showing both phenomena, each prevailing in two different concentrations regimes.
3. Sensors based on an electroactive probe included in the eMIP structure, whose electrochemical signal changes as a consequence of its reaction/interaction with the analyte.[23]

For all the above chemosensors, two detection schemes are presented in literature, *i.e.*

- direct measurement in sample solutions;
- measurement in a proper test solution after a pre-concentration step in sample solution.

The two schemes have obviously different performances in term of response time and sensitivity.

Application of eMIPs in chemosensing has been reviewed quite recently.[24,25] However, the field continues to grow at a high rate so that the timely consideration of the work published in the last five years reported in this chapter can be helpful for those interested in entering this still exciting field of technology. An overview of typical monomers used in the electrosynthesis of MIPs for sensor applications is reported, including a section dedicated to the latest achievements in the application of nanomaterials to the development of eMIP chemosensors.

13.2 Pyrrole

Since the first reports on electrochemical imprinting, pyrrole has been one of the most used monomers[24] in this research field, due to some beneficial aspects of the corresponding polymer, polypyrrole (PPy), such as high conductivity, simple doping–dedoping process, and the possibility of being electrosynthesized in aqueous solution with high stability of the resulting polymer;[26] all features making PPy very attractive as a molecular imprinted polymer, especially for electrochemical sensing.

Although several studies on the imprinting of PPy have been performed in a time-span of about 10 years after the first reported works,[24] it has still attracted a significant research interest during the last few years. This is clearly demonstrated by the high variety of target molecules used as templates with the aim of combining PPy properties with high selectivity of molecular imprinting in developing novel, highly selective and robust chemosensing devices. Among these, several drugs (*e.g.* glicazide,[27] halofuginone,[28] phenothiazine[29]), antibiotics (*e.g.* sulfanilamide,[30] dapsone,[31] sulfadimethoxine[32]), analytes of interest in food analysis [*e.g.* chlorogenic acid,[33] bisphenol A (BPA),[34] nicotinamide[20]] and in environmental monitoring (*e.g.* chlorinated pesticides[35] and explosives[36]) have been successfully used as templates in the imprinting of PPy over the last few years, still demonstrating the wide applicability of MIP based chemosensors in important fields such as environment, health and diagnostic.

The trend observed during the past decade in the imprinting technology passing from small molecular weight to macromolecular targets,[37] can be observed also in electrochemical imprinting, with the introduction of several strategies aiming at overcoming the barriers of protein imprinting, such as their conformational changes, large size, and large number of potential interaction sites on their surface responsible for possible nonspecific absorption. Pyrrole was one of the first proposed monomers to this aim in a work describing a very preliminary attempt in the imprinting of electrosynthesized PPy for bovine leukemia virus glycoprotein.[38] Since then, other strategies for PPy macromolecular imprinting have been explored, mainly based on the surface imprinting approach.[39,40] In one case,[40] it was successfully achieved with a straightforward strategy consisting of the formation of the pre-polymerization complex by simply dissolving the template in a monomer solution. Among the other parameters influencing the developed electrochemical sensor for the detection of the target protein (*i.e.* bovine hemoglobin, BHb), the authors demonstrated that the thickness of the imprinted polymer, related to the number of scan cycles in electropolymerization, was crucial because an increase of cycles number led to a thicker film with less accessible imprinted sites. Although an estimation of the resulting polymer thickness was not reported, an optimum of 6 cycles of polymerization was selected to form a very thin polymer film with a thickness comparable to the size of the protein template.

The immobilization of the target protein through a self-assembled layer offered advantages over electropolymerization from a protein–monomer mixture, generating more uniformly accessible binding sites.[37] This was proved also in the case of PPy imprinting for BHb by Li *et al.*,[39] who covalently bound the template protein to aldehyde group-functionalized silica microspheres deposited on an Au electrode, before electropolymerization. With this approach, a three-dimensional macroporous imprinted PPy was obtained after etching silica and extraction of BHb (Scheme 13.3), with all the imprinted cavities situated on the surface of the inside-wall surfaces of

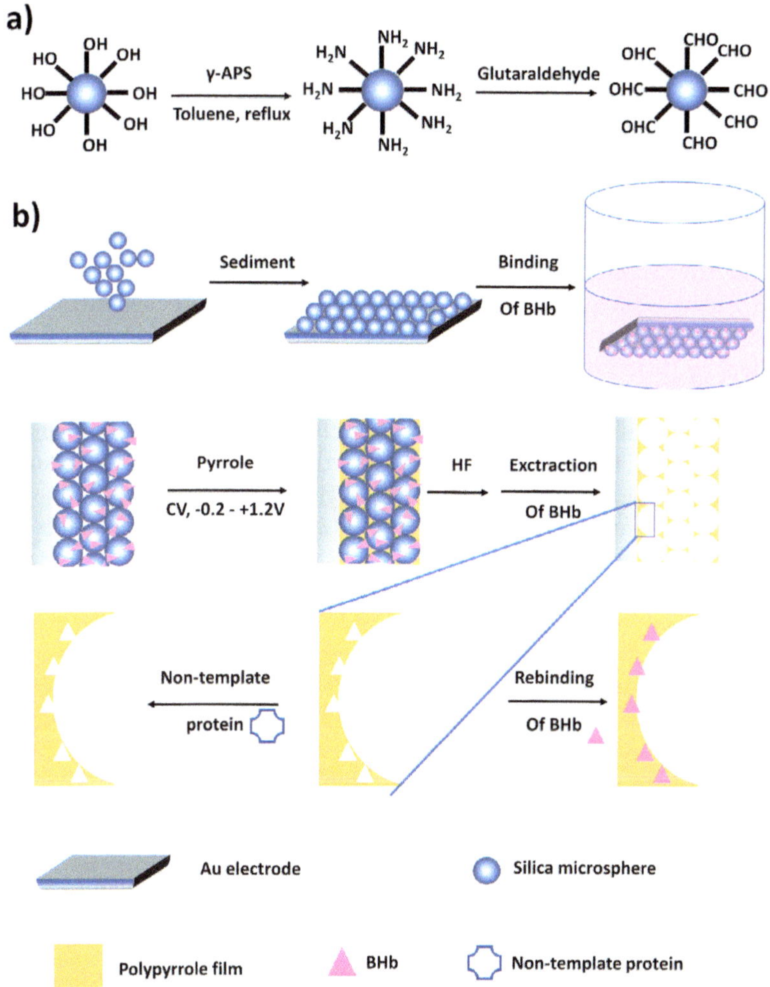

Scheme 13.3 Simplified representation of (a) the preparation of aldehyde group modified silica microspheres, and (b) the fabrication procedure of the PPy-MIPs on Au electrode. γ-APS – 3-aminopropyltriethoxysilane, BHb – bovine hemoglobin, CV – cyclic voltammetry. Adapted from ref. 39 with permission from The Royal Society of Chemistry.

the macroporous polymer. The as-achieved surface imprinting was responsible for reaching the imprinting factor of 20 and an increase in sensitivity to BHb from 0.127 μA $(g\ L^{-1})^{-1}$ to 0.136 μA $(g\ L^{-1})^{-1}$ with respect to results reported in Kan *et al.*[40]

Moreover, a quite similar approach was proposed in the fabrication of an electrochemical sensor for dopamine based on imprinted PPy,[41] exploiting preliminary modification of the electrode surface with silica microspheres and subsequent electropolymerization from a pyrrole/dopamine mixture.

Also in this case, the etching of silica and the extraction of the template at the end of polymerization aimed at forming an ordered macroporous MIP, with a significant increase of the surface area and, consequently, of easily accessible binding sites, resulting in an enhancement of dopamine peak current in differential pulse voltammetry (DPV) measurements in comparison with what was observed on the imprinted PPy film.

Another interesting strategy for dopamine detection by an imprinted electrochemical sensor has recently been proposed using a pyrrole derivative as the monomer, namely pyrrole-phenylboronic acid (py-PBA),[42] *ad hoc* synthesized with the aim of conferring double recognition capabilities to the resulting imprinted polymer, combining the imprinting effect with boronic acid functionalities ability to covalently interact with *cis*-diol groups of dopamine. On the other hand, pyrrole moieties enable electropolymerization, thus ensuring a simple and rapid anchoring of the sensing film to the electrode surface. It was demonstrated that the prepared chemosensor exhibited a much higher imprinting factor than that obtained when using pyrrole or PBA alone as the functional monomer (18.4, 5.1, and 2.1, respectively).

The above-mentioned examples clearly show that both electroactive and non-electroactive compounds have successfully been used as templates in the electrochemical imprinting of PPy, mainly coupled to direct electrochemical transduction in the first case,[27–33,35,36,41,42] while in the second case exploiting potentiometric[33] or indirect electrochemical detection based on the change in the signal of a redox probe, such as hexacyanoferrate(II).[20,39,40] An alternative to direct electrochemical detection was proposed by Zhang *et al.*[34] in the design of an imprinted PPy-based chemosensor for BPA. The aim was to prevent the formation of a nonconductive polymer film from the electrochemical oxidation of BPA at a relatively high potential that can passivate the electrode, thus hindering further BPA redox process. The authors developed a ruthenium-mediated photoelectrochemical sensor for BPA determination using a PPy-based MIP as the recognition element, an SnO_2 nanoparticle as the semiconductor material, and a blue 473 nm LED as the excitation light source (Scheme 13.4). BPA was photoelectrochemically oxidized with Ru^{3+}, thus leading to a photocurrent enhancement that was dependent on the BPA concentration. Along with satisfactory chemosensor performance, the remarkable aspect of the proposed sensing platform is its potential applicability to determination of organic compounds that can be photoelectrochemically oxidized by Ru^{3+}.

Importantly, flexibility of electrochemical imprinting of PPy in terms of a suitable template molecule relies on some properties of the MIP itself. The choice of pyrrole as the monomer, in some cases supported by computational studies[29,43] typically evidencing monomer–template interaction by hydrogen bonds, offers advantages over other electropolymerizable monomers because conductivity of the resulting MIP can be easily modified by changing polymerization conditions, which, in turn, can determine the formation of a highly or poorly conductive polymer, the latter resulting from PPy overoxidation. PPy can be overoxidized by applying high potentials (*e.g.*

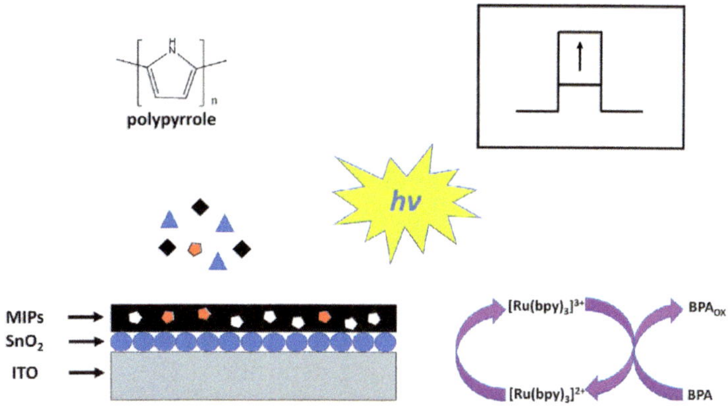

Scheme 13.4 The diagram of MIPs-based photoelectrochemical sensor for BPA detection. ITO – indium tin oxide. Adapted from *Analytica Chimica Acta*, **887**, B. Zhang, L. Lu, F. Huang and Z. Lin, $[Ru(bpy)_3]^{2+}$-mediated photoelectrochemical detection of bisphenol A on a molecularly imprinted polypyrrole modified SnO_2 electrode, 59–66, Copyright (2015) with permission from Elsevier.

1.0 V *vs.* Ag/AgCl), on an already synthesized PPy or during PPy deposition, in buffer or in an alkaline solution in the presence of oxygen, thus inducing the formation of additional functionalities (*e.g.* carbonyl and carboxyl groups), along with the loss of conductivity,[44] The use of such insulating PPy is doubly advantageous in the imprinting because it allows for preparing thin films in the nanometre range (particularly useful in protein imprinting, as mentioned above) and in detecting non-electroactive molecules by an indirect redox-mediated electrochemical sensing scheme. Noticeably, the ionic permselectivity of overoxidized PPy is beneficial in the latter case, being responsible for the repulsion of the anionic probe, thus limiting the electrochemical response to the probe diffusion through imprinted cavities to the electrode surface. Alternatively, when overoxidation of imprinted PPy is coupled with direct detection of electroactive species, it is responsible for template removal from the imprinted matrix. In this approach, proposed for the first time in 1999 for the imprinting of glutamic acid,[45] template entrapment/removal and subsequent analyte binding are based on acid–base equilibria of the template. Recently, we exploited this strategy in the fabrication of an electrochemical sensor for SDM.[32] The efficacy of different overoxidation conditions in template removal was demonstrated by the X-ray photoelectron spectroscopy (XPS) analysis highlighting a relationship among chemical modifications due to overoxidation, the amount of removed template and MIP binding properties, evaluated by amperometric measurements. Moreover, the role of the cationic electrolyte used in electropolymerization significantly influenced MIP electrochemical behavior, along with its morphology, evidencing a correlation between film homogeneity and stability/reproducibility of the amperometric responses.

13.3 *o*-Phenylenediamine

A homopolymer prepared from *o*-phenylendiamine (*o*-PD), poly(*o*-phenylenediamine) (PPD), was one of the first successful examples of electrosynthesized MIPs.[10] After that, a high number of papers reporting its applications in the electrochemical imprinting appeared in the literature, typically involving electrochemical and piezoelectric detection of the template.[25,46] This trend has continued to grow during the last years, presumably supported by the easy electrochemical preparation of PPD in aqueous solutions (typically performed potentiodynamically in acetate buffer at pH = 5.2) and by the non-conducting nature of the resulting polymer, with a thickness of a few nanometres due to the self-limiting deposition. These interesting features make PPD an attractive candidate for the electrochemical imprinting of proteins[47,48] and of other not electroactive targets such as benzophenone,[49] 4-aminobutyric acid,[50] glucose and fructose,[51] furosemide[52] and isocarbophos.[53] In these cases, analyte binding is typically monitored by voltammetric evaluating the modification of hexacyanoferrate(II)/hexacyanoferrate(III) redox signals as a consequence of MIP interaction with the template analyte. Based on the same principle, electrochemical impedance spectroscopy (EIS) has also been proposed as an efficient tool for studying the interface properties of electrodes coated with PPD-based MIPs.[47,53] The EIS spectrum provides information on the charge-transfer resistance, which is related to the charge-transfer kinetics of the redox probe at the electrode surface and thus it can be used for gaining information about MIP deposition and on releasing/binding steps.

Such a simple scheme of PPD-based MIP preparation and characterization has been applied successfully in the first example of electrochemical sensor for human cardiac troponin based on an MIP[47] prepared by electropolymerization. Although the chemosensor performance in determining troponin in a human blood serum sample was affected by unspecific protein adsorption, it exhibited a wide linear concentration range (0.009–0.8 ng mL^{-1}) to troponin in buffer solutions as well as a low detection limit (9 pg mL^{-1}), high sensitivity, and short response time.

Imprinting of PPD for protein has successfully also been performed by Tretjakov *et al.*[48] who proposed the integration of imprinting with surface acoustic wave (SAW) sensing for label-free detection of immunoglobuline G (IgG). Before PPD electropolymerization, IgG was anchored to the gold surface of a SAW chip through a cleavable cross-linker so as to be removed at the end of polymer deposition by bonding cleavage and subsequent washing. The template removal by this approach was allowed by the low PPD film thickness, which was carefully evaluated by monitoring the amount of circulated charge during electropolymerization. Precise control of MIP thickness, estimated by ellipsometry, was crucial for ensuring surface imprinting, thus facilitating protein template removal and subsequent protein analyte binding. The developed sensing platform represented one of the first examples of MIP integration with SAW devices, which, therefore, allowed observing the real-time binding event that could be potentially suitable for the cost effective fabrication of MIP-based multiplexed chemosensors. The same authors demonstrated its successful application also to a PPD-based MIP mediated sensing of an antibiotic, sulfamethizole.[54]

Along with the optimization of MIP film thickness, other strategies have been explored for improving performances of chemosensors integrating PPD-based MIPs, aiming at increasing detectable signals and/or enhancing the imprinting. Among these, the use of composite materials mainly resulting from the integration of imprinted PPD with nanomaterials is widely reported (for composite nanomaterial/eMIPs, see Section 13.7).

Hybrid imprinted PPD has been prepared in recent works[55,56] from the same research group, describing the use of silver dendrites (AgDs) electrochemically deposited on the film of an MIP of the PPD copolymer for the fabrication of a voltammetric chemosensor for minoxidil[55] and mesalamine.[56] AgDs, consisting of a long central trunk with nano-scale branches, as revealed by SEM analysis, play an important role in improving chemosensor performance, due to their large specific surface area, high conductivity, and useful electrical properties. Nonetheless, the use of AgDs, supported also by their easy preparation and high stability in water, was not demonstrated to effectively amplify the electrochemical responses.

Another strategy for improving PPD electrochemical imprinting consists of performing co-polymerization with other monomers, in some cases rationally selected by a computational approach[55,56] with the aim of introducing more functionalities potentially interacting with the template. Examples of (*o*-PD)-based copolymers appeared in literature during the last years include PPD co-polymerization with gallic acid and aminobenzoic acid.[53,55,56] and with hydroquinone.[57] In an interesting work,[15] PPD co-polymerization with amine terminated *N*-isopropylacrylamide (NIPAAm) has been explored for synthesizing a temperature-controlled on/off-switchable MIP for folic acid. The choice relies on the thermosensitivity of poly(NIPAAm), which switches from a swollen hydrophilic form under the lower critical solution temperature (LCST, ranging from 22 to 34 °C) to a shrunk hydrophobic form above the LCST. On this basis, the effect of temperature on template removal and analyte binding was studied, evidencing, respectively, an optimum value of 50 °C and 22 °C.

Although, as demonstrated by the above-reported examples and by reviews on the topic,[24] the great majority of MIPs-PPD have been successfully coupled with electrochemical transduction, their integration with other sensing platforms has also been recently explored.[21,58]

A photoelectrochemical sensor for lindane based on imprinted PPD and TiO_2 nanotubes has been developed by Wang *et al.*[58] In the proposed sensing scheme (Scheme 13.5), the imprinted PPD acts not only as the recognition element, but also participates in the enhancement of photocurrents due to its fast electron-transfer capability and strong light absorption in the visible region. Under visible light irradiation, the MIP-PPD film can generate the photoelectric HOMO–LUMO transition, thus delivering the excited electrons

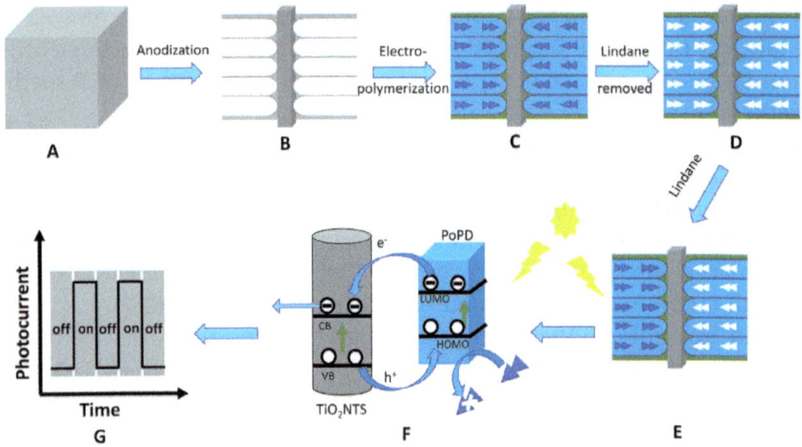

Scheme 13.5 Illustration for fabrication (a, b, c, d, and e) and detection mechanism (f and g) of the photoelectrochemical sensor for lindane based on imprinted PPD. CB – conduction band, VB – valence band, HOMO – highest occupied molecular orbital, LUMO – lowest unoccupied molecular orbital, NT – nitroaromatic compound, PPD – poly(*o*-phenylenediamine). Adapted from the *Journal of Inorganic and Organometallic Polymers and Materials*, Photoelectrochemical Sensor Based on Molecularly Imprinted Polymer-coated TiO2 Nanotubes for Lindane Specific Recognition and Detection, **23**, 2013, 703–711, P. Wang,[58] © Springer Science + Business Media New York 2013, with permission of Springer.

to the conduction band of TiO_2 nanotubes. Simultaneously, a positive charged hole (h^+) of MIP-PPD is consumed in the lindane oxidation process, thus promoting amplification of the photocurrent response, which is proportional to lindane concentration.

A novel transduction mechanism was proposed by Li *et al.*[21] in the assembly of an electrochemiluminescence (ECL) chemosensor for the clopyralid herbicide, based on the gate effect, analogous to what was used for electrochemical detection of non-electroactive analytes by MIP films. Also in this case, template elution and analyte binding acted as a gate to control the diffusion of a probe (an electrochemiluminescent system) through the PPD imprinted cavities and its reaction on the electrode surface (Scheme 13.6). Measurements were performed using the luminol–H_2O_2 system so that changes of its ECL intensity as a consequence of template removal, and then analyte binding could be monitored, showing a linear dependence on the clopyralid concentration in the range of 1 nM–0.8 μM.

13.4 Thiophene

Since 2005, thiophene and thiophene derivatives have been widely used in electrochemical imprinting[24] and still attract great research interest in the field nowadays.

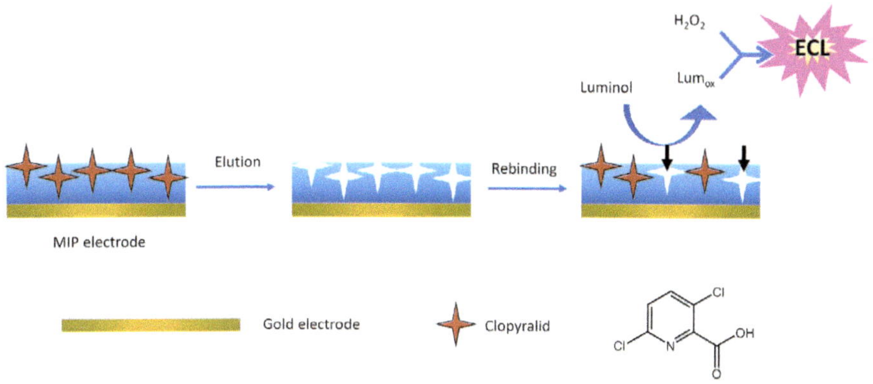

Scheme 13.6 Illustration of operation of the ECL-MIP sensor based on the gate controlled effect. Adapted from *Journal of Solid State Electrochemistry*, Clopyralid detection by using a molecularly imprinted electrochemical luminescence sensor based on the "gate controlled" effect, **18**(7), 2014, 1815–1822, X. Li,[21] © Springer-Verlag Berlin Heidelberg 2014, with permission of Springer.

Recently, Chen *et al.*[59] proposed the use of 3-thiophenemalonic acid (3-TMA) as the functional monomer in the assembly of a MIP-based chemosensor for the determination of terbuthylazine (TBA). Among other parameters, the authors focused on the monomer/template ratio in the polymerization mixture, testing different ratios (2 : 1, 3 : 1, 4 : 1 and 5 : 1) and evidencing that the increase of the functional monomer concentration was responsible for the formation of a higher amount of imprinted cavities, as revealed by DPV measurements.

Thiophene and different thiophene derivatives have been used, both in homo- and in co-polymerization, for electrochemical imprinting of atrazine,[60–62] namely poly(3,4-ethylenedioxythiophene-*co*-thiophene acetic acid),[60,62] 3-thiophene acetic acid (3-TAA),[61] 3-thiophenemalonic acid (TMA),[61] and thiophene methanol (TMeOH).[61]

Lattach *et al.*[60] reported on the development, characterization and optimization of an electrochemical sensor based on poly(3,4-ethylenedioxythiophene-*co*-thiophene acetic acid) molecularly imprinted conducting polymers. The aim of the work was to understand the influence of the chemical functionalities and the structural properties of the sensing films on the recognition of atrazine. In particular, the authors studied the recognition properties of the MIP sensing film as a function of atrazine concentration by square wave voltammetry (SWV) measurements. They observed that SWV peak current varied for atrazine concentrations exceeding 10^{-8} mol L^{-1}, with a current signal linear decrease with the increase of the atrazine concentration in the range of 10^{-8}–10^{-4} mol L^{-1} as a consequence of variation in the electroactivity of conducting polymers in the presence of increasing amounts of target molecules. The same imprinting protocol was applied by the same authors to the development of an electrochemical surface acoustic wave sensor (ESAW) for atrazine.[62]

In another work,[61] the same authors experimentally and theoretically studied the influence of the nature of the chemical functionalities of a polythiophene-based MIP on its ability to bind and detect atrazine. Differently functionalized thiophene-based monomers (thiophene, EDOT, 3-TAA, TMA, TMeOH) were used in order to form pre-polymerization complexes with atrazine in the acetonitrile solvent. They demonstrated, thanks to electrochemical measurements, the clear effect of the nature of the functional monomer on the amount of atrazine entrapped during electropolymerization and on the sensitivity of the MIP films toward the template molecule, determined on the basis of circulated charge in voltammetric measurements on MIP and NIP films in the solution in the absence and presence of atrazine in increasing concentrations, respectively. Both values decreased in the following order of functional monomers: TMA, TAA, TMeOH, EDOT, and thiophene. Due to a high correlation between intensity of the measured electrochemical signals and the concentrations of the prepolymerization complexes, the authors attributed these experimental results to the strength of the functional monomer–ATZ interaction.

Dabrowski *et al.*[63] have recently developed a piezomicrogravimetric chemosensor based on an MIP prepared from 2,2-bithiophene-5-boronic acid for determination of D-arabitol. The monomer was adopted with the aim of allowing the formation of weak reversible covalent bonds between vicinal hydroxyl groups of arabitol and boronic acid substituents of the bithiophene. The work represented the first application of such a monomer in the imprinting procedure and the first example of an artificial recognition-based chemosensor for arabitol. Characterization of MIP and NIP films was performed by XPS to investigate the chemical composition and to confirm the efficiency of the imprinting strategy, monitoring the entrapment and subsequent extraction of the template, through the analysis of the boron and sulfur binding energy regions.

The use of electrochemical quartz crystal microbalance (EQCM) to simultaneously measure mass and current variations in an electrochemical process to investigate thiophene-based performance of MIPs was demonstrated by Vergara *et al.*[64] to detect a chemical nerve agent analog, pinacolyl methylphosphonate (PMP). In this work, the terthiophene monomer with a carboxyl functional group (2-(2,5-di(thiophen-2-yl)thiophen-3-yl)acetic acid) was used. The advantage of using the terthiophene carboxylic acid monomer was that the carboxylic acid functionality could interact with the template to form a noncovalent linkage *via* H-bonding, while the terthiophene moiety can polymerize or cross-link, thus resulting in an intramolecular or intermolecular reaction with another terthiophene unit. Furthermore, as evidenced by the authors, the terthiophene monomer is most likely to form a more ordered and homogeneous polymeric film known to have a relatively lower oxidation potential than that of the monothiophene and bithiophene counterparts.

Coupling thiophene-based MIPs with piezoelectric microgrametry (PM) at QCM detection was also proposed for highly sensitive and selective determination of melphalan (mel)[65] and several explosive nitroaromatic compounds (NTs)[66] including 2,4,6-trinitrophenol (TNP), 2,4,6-trinitrotoluene (TNT), 1,3,5-trinitrobenzene (TNB), and 2,4-dinitrotoluene (DNT), by using 3-TAA and bis(2,2-bithienyl)-(4-aminophenyl)methane as the functional monomer, respectively. While the former was already recognized as a monomer suitable for imprinting due to the presence of the carboxyl group not involved in polymerization and available for interaction with the analyte or template, the latter was synthesized *ad hoc* with the aim of getting several benefits in imprinting of NTs. The bis(2,2-bithienyl)methane moiety played a fluorophore role for the fluorescence titration, decreased the polarity of the monomer, thus making it soluble in an organic solvent of low polarity, and enabled electropolymerization. Besides, the phenylamine group was an appreciable π–π recognizing site of NT, as demonstrated by the fluorescence titration, computational molecular modeling, and the CV anodic half-peak potential shift of the phenylamine moiety electro-oxidation. The recognition of NT analytes was in fact ascribed to the π–π interactions between benzene ring of the NT molecule and the phenylamine group of the functional monomer used.[66]

In another work,[67] the application of imprinted thiophene derivatives has been explored for carnosine sensing, due to its relevance in diagnosis of Alzheimer and Parkinson's diseases. The authors selected 4-bis(2,2′-bithien-5-yl)methylbenzoic acid and bis(2,2-bithienyl)-benzo[18-crown-6]methane as functional monomers to prepare MIP films, due to the availability of a variety of recognition sites. In particular, both monomers were composed of two parts. One part is an electroactive bis(2,2-bithiophene) moiety capable of electropolymerization while the other is a recognition site capable of interacting with the available binding sites of the carnosine template. Impedimetric capacity (IC) and PM measurements were performed under flow-injection analysis (FIA) conditions for carnosine sensing with an appreciable detectability of 20 μM (in IC measurements). Due to multiple modes of carnosine binding to MIP recognition sites, the IC chemosensor was more selective to carnosine than to its common interferences including anserine, carcinine, and histidine. Advantageously, the imprinting factor, determined by PM, was high, equaling 14.9.

A very recent application of a synthetic polymer receptor in the clinical field has been described by Sharma *et al.*[14] to detect neopterin, a cancer biomarker. In this work, the authors used two functional monomers (bis-bithiophene derivatized with cytosine and bithiophene derivatized with boronic acid) considering the presence of two distinct moieties of neopterin template molecule, and a cross-linking monomer having the function of preserving the structure of the molecular cavity. In particular, the cytosine was used for effective complexing the pterin moiety by Watson–Crick type hydrogen bonding, whereas the boronic acid was used for covalent bonding of the vicinal diol moiety of neopterin. Measurements of open-circuit potential (OCP) under optimized FIA conditions determined neopterin detection in the concentration range of 0.15–2.5 mM, with the limit of detection of 22 μM with good anti-interference properties against 6-biopterin and other pterin analogs of neopterin.

13.5 Phenol

Within the panorama of phenol-based MIPs, an interesting approach has recently been used by Li *et al.*[68] for doxycycline determination through the use of an MIP prepared by electropolymerization of pyrogallol doped with alizarin red (AR), coupled with ECL detection. In the proposed sensing scheme, the excited intermediate of polypyrogallol (PPG*) was generated by the auto-oxidation of the polypyrogallol MIP. The slow process of auto-oxidation could be accelerated using hydrogen peroxide. Meanwhile, AR was oxidized on the electrode to generate oxidized derivatives of AR, which reacted with PPG* to form excited AR*, responsible for chemiluminescence emission. The ECL increased after doxycycline analyte binding by MIP-PPG due to the energy transfer from AR* to doxycycline, and subsequent formation of excited doxycycline. As a result, the ECL intensity was greatly enhanced, being linearly dependent on the doxycycline concentration. The devised chemosensor was characterized by high reproducibility and stability, with the response not significantly changing after 10 days and decreasing by ~10% after one month.

Since the beginning of research activity into electrochemical imprinting in poly(phenol)s,[24] phenols bearing the amino and/or thiol group have shown to be suitable materials for designing electrosynthesized MIPs because of the possibility of creating functionalized imprinted cavities, thus providing more selective interactions with the template molecule.

A preliminary self-assembly of the functional monomer (*i.e.* *o*-aminothiophenol, *o*-AT) on the surface of the Au electrode was performed in the assembly of an MIP-based electrochemical sensor for ractopamine (RAC).[69] The formation of Au–S bonds between the monomer and the electrode determined the formation of a compact and stable *o*-AT monolayer, responsible for the improvement of the stability of the chemosensor response, indirectly evaluated with hexacyanoferrate(II) as the electrochemical probe.

Moreover, *o*-aminophenol (*o*-AP) has been used for fabrication of MIP-based chemosensors for determination of several analytes including norepinephrine,[70] dicyandiamide (DCD),[71] and tris(2,3-dibromopropyl) isocyanurate (TBC),[72] a persistent organic pollutant. Characteristic of poly(*o*-aminophenol) is the presence of an electron-donating –OH group that represents a potential coordinating site.

In the imprinting of DCD,[71] the authors observed the formation of dense and uniform poly(*o*-AP) membranes with an open structure through *o*-AP electropolymerization in acidic solution. Chemosensor sensitivity and detection limit were improved by optimizing some fundamental parameters including electropolymerization cycle numbers, template–monomer molar ratio, extraction time, and time of chemosensor exposure to the analyte. The current response was evaluated indirectly by using hexacyanoferrate(II)/hexacyanoferrate(III) as the electroactive probe, and satisfactory recovery results were obtained in the DCD determination in milk samples.

Indirect electrochemical detection was used also for TBC sensing, because of its electroinactivity.[72] Interestingly, the authors compared their results with those using an aniline-based MIP for the same target molecule, prepared with the same experimental procedure. This comparison aimed at demonstrating that the presence of amino and hydroxyl groups simultaneously led to an improvement of the detection limit and sensitivity of the MIP chemosensor thanks to the presence of hydrogen bonds, which facilitates selective recognition of the analyte.

13.6 Electrosynthesized MIPs Based on Other Functional Monomers

In addition to the above-reported commonly used functional monomers, several other monomers have been employed in electrochemical imprinting and applied to chemosensing in different fields, including the medical,[73,74] biological,[16,75,76] and environmental[77] fields.

In designing of an electrochemical sensor for L-phenylalanine, Chen *et al.*[16] proposed the use of a novel chiral hybrid MIP, resulting from hybridizing insulating poly(methacrylic acid) with conducting polycarbazole by a covalent bond. To this aim, a pendant polymer precursor, poly[2-(*N*-carbazolyl)ethylmethacrylate-*co*-meth-acrylic acid]s (PCEMMAs) was prepared by electrochemical cross-linking and subsequent electropolymerization (Scheme 13.7). In this way, the electroactive carbazole groups in PCEMMAs were used to cross-link the pendant polymer (PCEMMAs), which could stabilize the imprinted cavities without using an additional cross-linking monomer. On the other hand, the poly(methacrylic acid) in PCEMMAs acted also as the functional monomer to interact with the template, and its flexibility might balance the rigidity of polycarbazole to improve the imprinting efficiency of the MIP. Moreover, the conductive polycarbazole in the hybrid MIP can provide an efficient matrix for signal transmission between the electrode and the sensitive layer, and simultaneously give better adhesion for the sensitive layer to contact with the electrode.

Another novel monomer successfully used in electrochemical imprinting is scopoletin, recently employed by Dechtrirat *et al.*[75] in protein surface imprinting. For chemosensor fabrication, a thiolated oligoethyleneglycol (OEG)/sugar conjugate was first immobilized on the transducer surface. Afterwards, concanavalin A (ConA), a carbohydrate binding protein, was immobilized on the modified surface, and then an ultrathin polyscopoletin film was deposited by electropolymerization on top (Scheme 13.8), with a thickness comparable to that of the protein. The system was tested through measurement with a QCM and selectivity with respect the target protein analogues was high.

Another approach to protein sensing was proposed by Wang and Wei[76] who prepared a novel surface plasmon resonance (SPR) chip integrated with an MIP for determination of bovine serum album (BSA). For the first time, they used 3-aminophenylboronic acid (3-APBA) as the conductive material for this sensing, directly deposited on the SPR sensor chips by electropolymerization. The nanoscale cavities formed by removing BSA templating molecules played a great role in BSA recognition, resulting in higher BSA adsorption on the MIP film in comparison to that on the NIP film. The effects of pH and

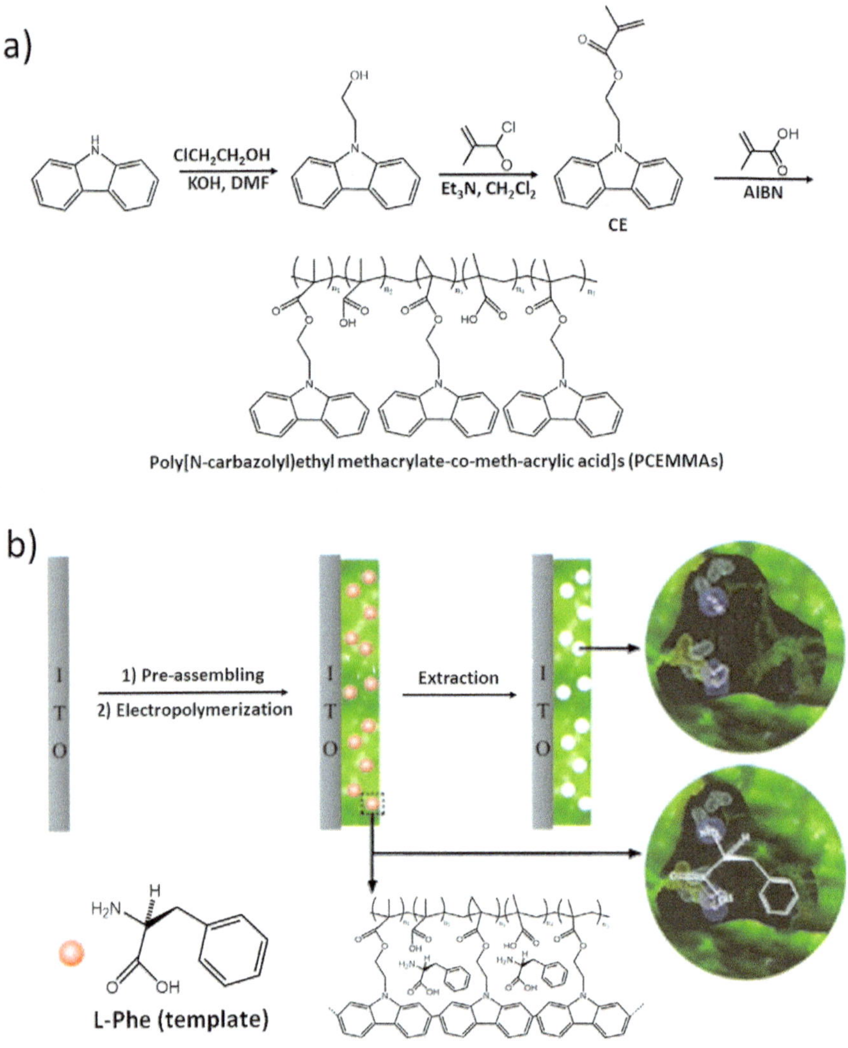

Scheme 13.7 (a) Synthesis route of PCEMMAs (b) Schematic illustration of the MIP sensor fabrication by electrochemically cross-linking PCEMMAs. DMF – *N*,*N*-dimethylformamide, Et_3N – triethylamine, AIBN – azobis(isobutyronitrile), CE – 2-(*N*-carbazolyl)ethyl methacrylate, L-Phe – L-phenylalanine. Adapted from *Analytica Chimica Acta*, **754**, Y. Chen, L. Chen, R. Bi, L. Xu and Y. Liu, A potentiometric chiral sensor for L-Phenylalanine based on crosslinked polymethylacrylic acid-polycarbazole hybrid molecularly imprinted polymer, 83–90,[16] Copyright (2012) with permission from Elsevier.

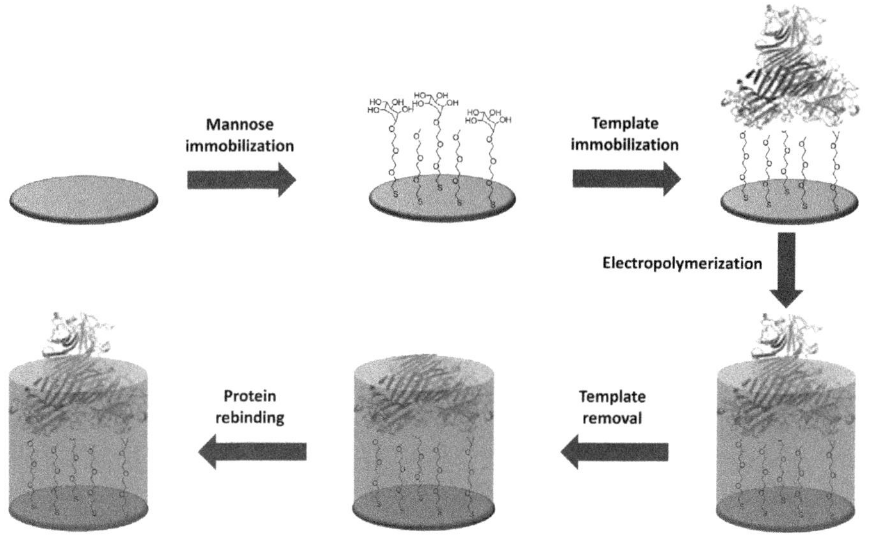

Scheme 13.8 Illustration of the template oriented surface imprinting for preparation. Adapted from *Advanced Functional Materials*, Hybrid material for protein sensing based on electrosynthesized MIP on a mannose terminated self-assembled monolayer, D. Dechtrirat, N. Gajovic-Eichelmann, F. F. Bier and F. W. Scheller,[75] John Wiley and Sons, © 2013 WILEY-VCH Verlag GmbH & Co. KGaA, Weinheim.

ionic strength on BSA adsorption were studied within the pH range between the pI value of poly(3-APBA) and the BSA protein without adding NaCl, thus resulting in the best adsorption conditions.

N-Isopropylacryamide has been also tested as a functional monomer with the aim to synthesize a thermo-responsive MIP allowing combination of the electrocatalytic activity with the response to temperature stimulus.[73] The authors investigated the temperature influence on the DPV response to the target analyte, theophylline. To this aim, the hexacyanoferrate(II) redox probe was used to investigate properties of the prepared chemosensor because of the electroinactivity of theophylline. Experimental results revealed that at 20 °C the MIP film was swollen, thus promoting probe permeation with subsequent peak current increase, whereas at 40 °C the peak current dramatically decreased because of the film shrinkage.

13.7 Composite System Electrosynthesized MIP–Nanomaterials

In recent years, we witness an increasing interest in the devising MIP chemosensors with shorter analysis time and higher sensitivity. This requires the improvement of MIP binding kinetics, a higher number of imprinted sites and complete template removal. Integration of nanomaterials into sensing film can solve these issues by increasing the chemosensor surface area and favoring electronic communication to the imprinted cavities.

Metallic nanoparticles, and gold nanoparticles (AuNPs) in particular, have attracted research interest in electrochemical sensing, thus enabling the development of ultrasensitive assays. AuNPs can dramatically increase the intensity of the electrochemical signal recorded after the analyte binding to MIP in two ways. On the one hand they can be used as highly conductive materials to make electronic communication faster and more intense, and on the other they can increase roughness of the materials, thus increasing the number of imprinted sites. In the last years, different approaches have been used in order to entrap AuNPs on the electrode surface before MIP deposition, including electrochemical deposition, physical entrapment during MIP electropolymerization or in a sol–gel networks, and surface assembly on the electrode surface.

In the case of electrochemical deposition, AuNPs were synthesized and deposited on the electrode surface in a single electrochemical step, thus resulting in an AuNP homogeneously covered electrode, on which a surface MIP can be deposited by electropolymerization starting from a solution containing the template and monomers, followed by template elution from the as fabricated nanocomposite film.[78,79] In some cases, after electrochemical deposition of AuNPs, certain monomers, such as *p*-aminothiophenol, in the presence of the target analyte were self-assembled on these AuNPs. Then, self-assembled monomers were electropolymerized and the template eluted. With this approach, it was possible to obtain a thinner MIP film, in which molecule diffusion was easier and the imprinted cavities were more easily accessible, consequently increasing the binding kinetics of the analyte molecules to the MIP film.[80]

Another approach, successfully applied to formation of the MIP/AuNPs nanocomposite, was used previously by Willner *et al.*[81] and more recently by Sun *et al.*[82,83] and Jaffrezic-Renault *et al.*[84–86] These authors exploited the strong Au–S bond between *p*-aminothiophenol (*p*-ATP) and AuNPs in solution in order to prepare functionalized nanoparticles. Then the *p*-ATP functionalized AuNPs were electropolymerized in the presence of the template thus producing a microporous metal organic framework (MOF), in which AuNPs were present inside the MIP (Scheme 13.9). Then, as usual, the template was removed by washing with selected solvents. A further advancement on this way has been recently proposed by Willner *et al.*,[87] who exploited a double functionalization of AuNPs with both *p*-ATP and dithiothreitol before electropolymerization of MIP. The prepared surface was used for ultrasensitive heavy metals detection by SPR.

In all the approaches the strong interaction between the MIP and AuNPs can favor electronic communication between them during the sensing step, thus decreasing LOD and widening the linear concentration range compared to the film prepared in the absence of AuNPs.

In an attempt to increase the performance of the chemosensor for dopamine, Jiang *et al.*[88] proposed a mixed approach, in which AuNPs functionalized with *p*-ATP were assembled on the gold electrode surface in order to increase the specific surface area. After that, they performed electropolymerization

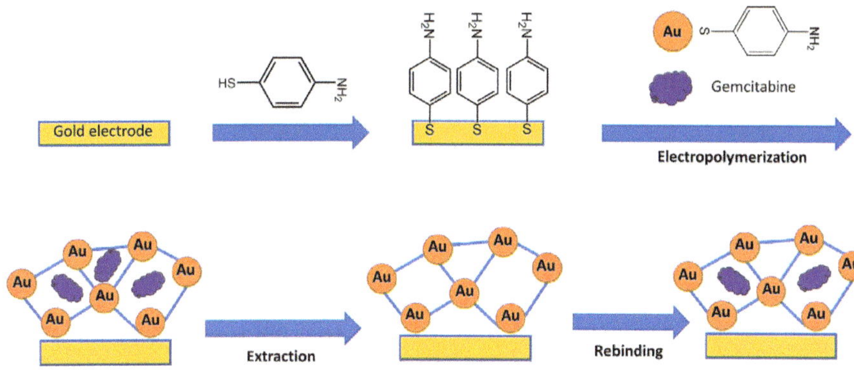

Scheme 13.9 Illustration of the fabrication of the MIP sensor for gemcitabine based on microporous metal organic framework. Adapted from *Talanta*, **138**, A. Florea, Z. Guo, C. Cristea, F. Bessueille, F. Vocanson, F. Goutaland, S. Dzyadevych, R. Săndulescu and N. Jaffrezic-Renault, Anticancer drug detection using a highly sensitive molecularly imprinted electrochemical sensor based on an electropolymerized microporous metal organic framework, 71–76,[85] Copyright (2015) with permission from Elsevier.

in the presence of the dopamine template and, again, AuNPs functionalized with *p*-ATP favored dopamine diffusion inside the MIP film and electronic communication after template binding through the formation of a microporous MOF. With this approach, the authors could prepare an AuNPs/MIP film with improved conductivity and electrocatalytical activity, thus permitting satisfactory detection of trace dopamine in human serum.[88]

Ensafi *et al.*[89,90] proposed, for the first time, a simple approach to improve the electron conductions from a sol–gel MIP to an electrode by entrapping a conductive polymer, prepared by electropolymerization, such as PPy and AuNPs within the sol–gel network. They prepared a mixture where pyrrole, template and $HAuCl_4$ were dispersed in the sol. Then, the MIP film was deposited by potentiodynamic electropolymerization, and then by applying an opposite (negative) potential the AuNPs were deposited in a single step. Then, the template was removed by washing. The authors prepared reproducible chemosensors with high performance that could be ascribed to the fact that, with this procedure, it was possible to prepare a porous imprinted composite film with many selective imprinted cavities. In particular, they prepared MIP chemosensors for caffeine and lorazepam. In both cases, they fabricated an electrochemical sensor with LOD in the subnanomolar range.

AuNPs were also used by Ensafi *et al.*[91,92] to enhance electron transfer in an electrochemical sensor. In this sensor, a DNA-MIP made of DNA imprinted with Sudan II served as the recognition element. After the synthesis of DNA-MIP, they entrapped this element in a film of PPD and AuNPs electropolymerized under potentiodynamic conditions using a solution of the *o*-phenylendiamine monomer and $HAuCl_4$. The composite deposited on the electrode surface was able to detect Sudan II satisfactorily in real samples.

In the last few years, other kinds of nanoparticles have been used in combination with MIPs to increase the chemosensor response. For example, Sun *et al.*[93] electrochemically deposited metallic Ni–Cu nanostructured composites, and then Cu was oxidized at an opposite potential in order to remove it from the film. In this way, the authors produced a Ni nanoporous structure. Then, *o*-PD was electropolymerized on this structure in the presence of the metronidazole template, which was subsequently removed by washing in order to produce an MIP (Scheme 13.10). The highly controlled electrochemical approach ensured the stability and reproducibility of the chemosensor, and the presence of porous Ni nanostructure favored increasing the surface area and accessibility of imprinted cavities, thus widening the linear dynamic concentration range and lowering the LOD, compared to those of the MIP deposited on a non-porous Ni electrode.

Other metallic nanostructures have also been used as scaffolds to prepare 3-D MIP arrays. For example, Wu *et al.*[94] electrodeposited Zn nanorods on the ITO substrate forming a Zn 3-D scaffold on the surface of which a thin PPy

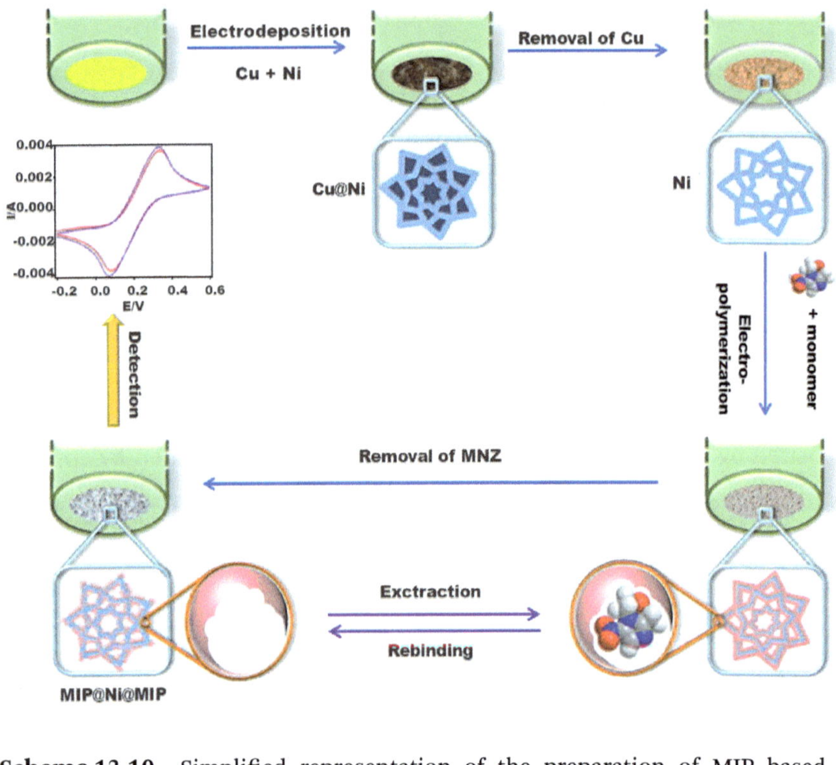

Scheme 13.10 Simplified representation of the preparation of MIP based on porous Ni nanostructures. MNZ – metronidazole. Adapted with permission from Y. Li, Y. Liu, Y. Yang, F. Yu, J. Liu, H. Song, J. Liu, H. Tang, B. Ye and Z. Sun, *ACS Appl. Mater. Interfaces*, 2015, 7, 15474.[93] Copyright (2015) American Chemical Society.

film was deposited by electropolymerization in the presence of epinephrine as the template. After extraction of this template from the resulting MIP film, a 3-D MIP array was obtained (Scheme 13.11). The authors highlighted that both the 3-D structure and the fact that imprinted cavities are located only on the surface of the MIP composite gave rise to high accessibility to these cavities for the epinephrine analyte, thus providing higher sensitivity and selectivity toward epinephrine.

In the last few years, interest in carbon based materials, such as carbon nanotubes (CNTs) or graphene, and their use in the development of electrochemical sensors has grown exponentially. Basically, these materials are carbon allotropes like diamond, graphite, and fullerenes, but with unique interesting properties, such as their outstandingly high rate of electron transfer, large specific surface area, thermal and mechanical properties, high chemical stability, and high biocompatibility derived from their electronic structure and nanodimensionality. A great problem in handling these materials arises from their insolubility in all the organic solvents and in water, thus limiting the manipulation and processability of the material. Several techniques are used to disperse these carbon-based materials, such as sonication and ultrasonication which are useful for breaking π–π and van der Waals

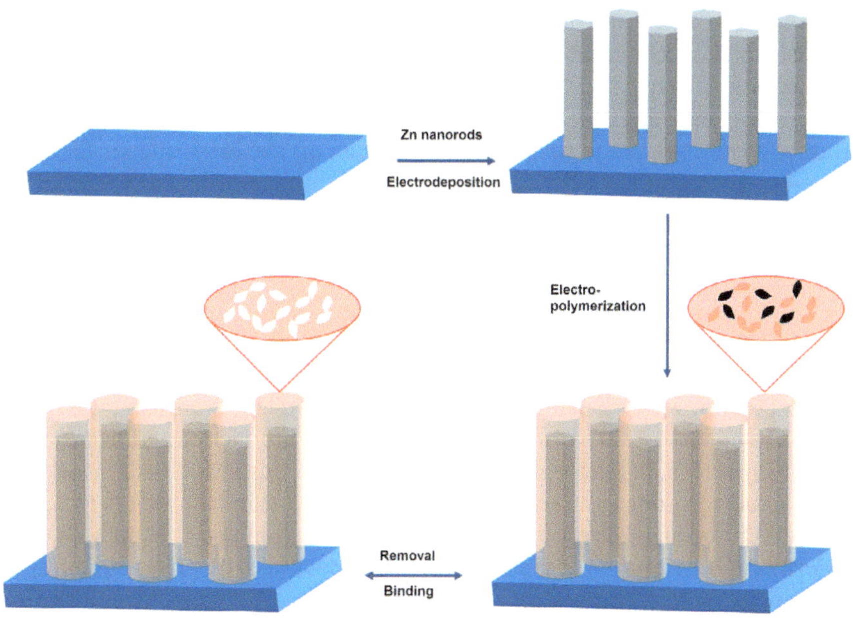

Scheme 13.11 Illustration of the preparation of 3-D MIP arrays for electrochemical sensor. Reprinted from *Sensors and Actuators B: Chemical*, **222**, H. Li, H. Wang, W. Li, X. Fang, X. Guo, W. Zhou, X. Cao, D. Kou, Z. Zhou and S. Wu, A novel electrochemical sensor for epinephrine based on three dimensional molecularly imprinted polymer arrays, 1127–1133,[94] Copyright (2016) with permission from Elsevier.

interactions. However, usually after a relatively short time, the material starts to aggregate and precipitate again. In order to avoid these problems, different functionalization methods have been developed including both covalent and non-covalent approaches. These aspects are also critical in application of these materials in chemosensors in general and in MIP chemosensors in particular. In the latter, soluble and homogeneous materials are needed in order to obtain reproducible chemosensor surfaces and, consequently, reproducible chemosensor performance. The most common approach to solve the problem consists of chemical functionalization of the nanomaterials. Wang *et al.*[95] non-covalently functionalized CNTs. First, they dispersed CNTs using the sodium dodecyl sulfate surfactant by ultrasonic agitation. Then, they cast an aliquot of the obtained suspension on the electrode surface and left it to dry. A film of the MIP for brucine was then deposited by electropolymerization on the as prepared electrode surface in the presence of *o*-PD and the brucine template. The CNT presence on the electrode increased electrochemically active surface area and electron transfer, thus widening the linear concentration range and lowering the limit of detection. This kind of approach, involving initial deposition by casting of functionalized carbon nanotubes suspension, and then MIP deposition by electropolymerization on as prepared surface is the most commonly used approach for the preparation of this MIP/CNTs nanocomposites. Moreover, different kind of CNTs functionalization have been used to this aim, in which covalent functionalization with –COOH on the CNTs surface, prepared by oxidation in an acidic solutions is most commonly used.[96–98] In some cases, CNTs were also dispersed in dimethylformamide (DMF) without any further functionalization in order to produce a homogeneous suspension to be cast on the electrode surface.[99,100] However, the stability of the CNT film prepared that way is really low. In the electropolymerization step of MIP deposition on the film of CNTs, different polymers have been recently used including *o*-PD[95,97,98] and pyrrole[99,100] but, in theory, it is possible to use equally well all other polymers prepared by electropolymerization thanks to the excellent electronic properties of these materials that favor redox processes on their surface.

Deposition techniques, such as drop casting, could often result in a thick film that can slow down diffusion of the analyte and electronic transfer inside MIP/CNTs composite. To obtain more homogeneous, reproducible and thinner films of CNTs, Prasad *et al.*[96] used –COOH functionalized CNTs, and then spin coated in order to homogeneously deposit a thin film of CNTs before MIP deposition by electropolymerization, thus enhancing the chemosensor performance. As a proof of that, the authors examined the effect of spin time and velocity of spin coating on performance of the MIP based chemosensor. At one extreme a thicker film inhibited chemosensor response, but at the other an ultra-thin coating caused a decrease in the number of molecular cavities after MIP deposition by electropolymerization thus lowering the chemosensor performance.

A fast, scalable and easy method of deposition of CNTs before deposition of MIPs by electropolymerization is electrodeposition of CNTs. Kan *et al.*[101,102] produced an electrochemical sensor for dopamine by a two-step procedure. First, –COOH functionalized CNTs were electrodeposited from a solution onto a glassy carbon electrode (GCE) surface by applying a potential of +1.70 V *vs.* SCE for 400 s. This way, it was possible to prepare rapidly and reproducibly an electrode surface functionalized with CNTs. The as prepared electrode was immersed in a solution of pyrrole containing dopamine as template, and then electropolymerization was started for MIP deposition. The authors obtained satisfactory results for dopamine recognition and the as proposed totally electrochemical approach resulted in a reproducible electrode surface in a short time.

Similar approaches, used to prepare CNTs/MIP nanocomposites, were used also with graphene. In this case, the most common approach consists of graphene chemical oxidation in order to obtain graphene oxide (GO) as a more easily handled material. The next steps typically consist of casting a suspension of GO on the electrode surface, and then starting MIP deposition by electropolymerization on the as prepared surface.[103] This simple approach has been recently used in the assembly of GO/(imprinted PPD) for determination of imidacloprid[104] and moxifloxacin.[105] In the latter case, after the deposition of a graphene suspension on the electrode, an MIP was prepared by co-electropolymerization of *o*-PD and lysine, thus simultaneously taking advantage of the presence of graphene responsible for current enhancement and of co-polymerization. The additional presence of L-lysine moieties in MIP was in fact expected to improve electrostatic interaction with the template, thus increasing the number of imprinted cavities. Among strategies explored for enhancing the imprinting, the deposition of sulfonated graphene on the Au electrode was also proposed by Wu *et al.*[106] in the assembly of a PPD-based MIP electrochemical sensor for dopamine. This choice was related not only to high stability and electrical conductivity of graphene modified with $-SO_3^-$ groups, but also to the availability of a large number of negatively charged functionalities, thus improving MIP interaction with the cationic form of dopamine. In another work,[107] a bilayer consisting of the imprinted PPD, deposited by electropolymerization on a composite material integrating PPy, GO and cobalt(II) phthalocyanine (PPy-GO-CoPc), preliminarily deposited on GCE, was prepared for exploiting the synergistic effect and electro-catalytic activity of the PPy-GO-CoPC toward redox activity of the target analyte, quinoxaline-2-carboxylic acid (QCA). Properties of the composite material were complementary to those of individual components with a synergistic effect in catalyzing redox processes of QCA, which was further enhanced in the presence of an imprinted PPD film, also responsible for the observed chemosensor selectivity.

In some cases, both graphene and CNTs were also used to improve performance of an MIP electrochemical sensor. Huang *et al.*[108] devised an MIP-graphene-CNT electrochemical sensor for triptamine. A multi-step (electro) chemical procedure was used to this aim. That is, first, a sulfonated graphene (GS) in the presence of pyrrole deposited on the GCE by electropolymerization. Next, carboxylated CNTs were non-covalently functionalized with hyaluronic acid (HA) in order to solubilize them. A drop of HA-CNTs was then deposited on the PPy-GS functionalized GCE. Finally, poly(aminobenzoic acid) imprinted with tryptamine was deposited by electropolymerization on the GS/PPy/HA-CNTs surface. Both the loading of GS and CNTs affected the chemosensor response. That is, an increase in the GS loading resulted in the current increase until an optimum loading was found, at which the current response was the largest. The HA-CNTs revealed a similar behavior. In this case, at a higher concentration with respect to that considered optimal, CNTs start to aggregate, thus hindering diffusion of analyte molecules during sensing.

In recent years, researchers combined carbon nanostructured materials with metallic nanoparticles for MIP preparation to obtain hybrid organic/inorganic nanocomposites capable of widening the linear concentration range and lowering the LOD of MIP chemosensors. To this end, the most used approach consists of, first, physical deposition, by drop casting, of carbon nanomaterials functionalized with metallic nanoparticles on the electrode surface, and then subsequent MIP deposition by electropolymerization on their surface. A metallic nanoparticle can be bound to CNT or graphene before[109–111] or after[112–115] the deposition of the carbon nanomaterial on the electrode surface. In the latter case, electropolymerization is usually used in order to deposit nanoparticles on a functionalized electrode surface. In a few cases, metallic nanoparticles are not directly attached to the carbon nanomaterial but entrapped in the MIP matrices during electropolymerization, thus forming a microporous MOF on the deposited CNTs or graphene film.[116–118]

A different approach for simultaneous deposition of graphene and metallic nanoparticles was proposed by Zhou *et al.*[119] who proposed a one-step fabrication of an MIP membrane by magnetic-field directed self-assembly. They immersed a magnetic GCE in a suspension of Fe_3O_4 nanoparticles functionalized with polyaniline (Fe_3O_4@PANI) bound by π–π stacking to graphene. That way, the graphene/Fe_3O_4@PANI composite could be adsorbed on the electrode surface by exploiting magnetic interactions in the solution, where the pyrrole functional monomer and the glutathione template were also present, for MIP preparation. After magnetically driven adsorption of graphene/Fe_3O_4@PANI on the electrode surface, an MIP film was deposited by potentiostatic electropolymerization in the same solution (Scheme 13.12). The glutathione template was then removed from the resulting MIP matrix. Parameters of MIP film preparation, such as the amount of graphene and Fe_3O_4 magnetic nanoparticles as well as magnetic field strength and deposition time, were optimized. Apparently, the presence of both graphene and Fe_3O_4 magnetic nanoparticles were important not only to control deposition but also to improve the chemosensor performance due to uniform anisotropic structures and suitable thickness. The approach proposed has been developed as an effective strategy to control assembly of a novel nanocomposite, for which a traditional self-assembly could not be used.

13.8 Conclusions

Electrochemical polymerization represents a fruitful way to prepare MIPs. It emerged later than the chemical way, but it soon attracted the interest of several researchers. This is because it can be applied also by those with no

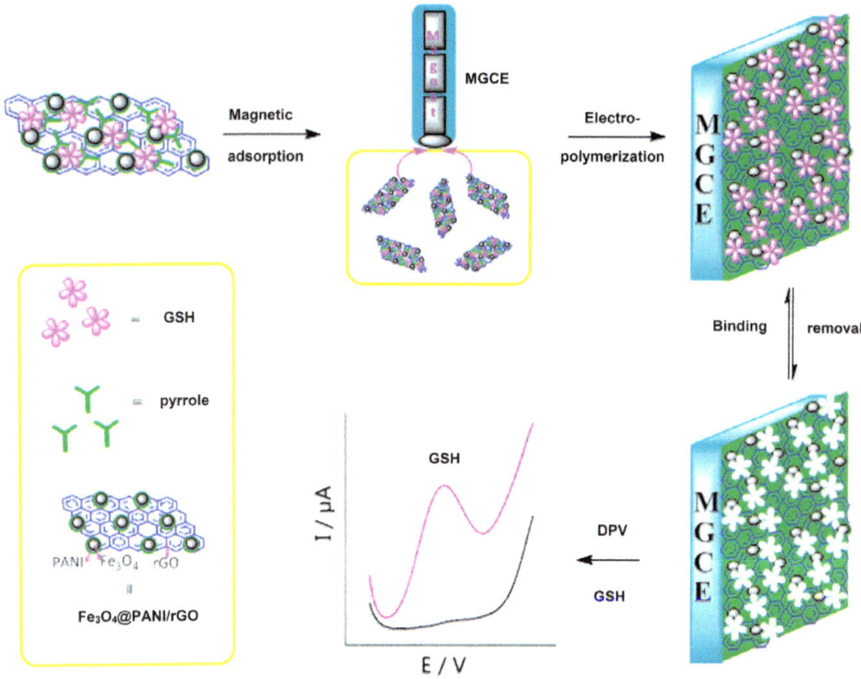

Scheme 13.12 Simplified representation of the preparation of glutathione MIP on graphene/Fe_3O_4@PANI electrode. Reproduced from *Analytica Chimica Acta*, **886**, W. Zhu, G. Jiang, L. Xu, B. Li, Q. Cai, H. Jiang and X. Zhou, Facile and controllable one-step fabrication of molecularly imprinted polymer membrane by magnetic field directed self-assembly for electrochemical sensing of glutathione, 37–47,[119] Copyright (2015) with permission from Elsevier.

expertise in polymer chemistry. In addition, the possibility to finely control the amount of MIP deposited and to easily produce thin MIP films integrated with transducers has promoted its application in chemosensing.

Since then, intense work has been performed in the field, by employing several monomers and several templates/analytes. This procedure has synergistically combined knowledge of both the chemical synthesis of MIPs and the field of electrosynthesized polymers, in particular conducting polymers,[6] even if the great wealth of information in electropolymerization does not seem extensively applied. A great role in devising eMIPs has been characterization of materials. A lacking aspect seems investigation of detailed mechanisms by which MIPs grow, based on the corresponding information available for electrosynthesized polymers. This kind of investigation could be of great help in predicting both population and selectivity of molecularly imprinted cavities in the polymer depending on experimental conditions.

Frontiers of this field are certainly represented by the imprinting of macromolecules (*e.g.* proteins) presented in another chapter of this book, inclusion of catalytic centers in the eMIPs and the extension of the electrochemical method to the preparation of nano-eMIPs. For sure, there is a large room for the original and novel research work of newcomers to the field ready to face this challenge.

List of Abbreviations and Symbols

3-APBA	3-Aminophenylboronic acid
3-TAA	3-Thiophene acetic acid
3-TMA	3-Thiophenemalonic acid
AgD	Silver dendrite
AIBN	Azobis(isobutyronitril)
γ-APS	3-Aminopropyltriethoxysilane
AR	Alizarin red
ATZ	Atrazine
AuNP	Gold nanoparticle
BHb	Bovine hemoglobin
BPA	Bisphenol A
BSA	Bovine serum album
CB	Conduction band
CE	2-(*N*-Carbazolyl)ethyl methacylate
CNT	Carbon nanotube
ConA	Concanavalin A
CoPc	Cobalt(II) phtalocyanine
CV	Cyclic voltammetry
DCD	Dicyandiamine
DMF	*N*,*N*-Dimethylformamide
DNT	2,4-Dinitrotoluene
DPV	Differential pulse voltammetry
ECL	Electrochemiluminescence
EDOT	3,4-Ethylenedioxythiophene
EIS	Electrochemical impedance spectroscopy
eMIP	Molecularly imprinted polymer prepared by electropolymerization
EQCM	Electrochemistry quartz microbalance
Et_3N	Triethylamine
ESAW	Electrochemical surface acoustic wave sensor
FIA	Flow injection analysis
Δf_{ss}	Steady-state frequency change
GCE	Glassy carbon electrode
GO	Graphene oxide
GS	Sulfonated graphene
HA	Hyaluronic acid
HOMO	Highest occupied molecular orbital
IC	Impedimetric capacity
IgG	Immunoglobuline G
ITO	Indium tin oxide
LCST	Lower critical solution temperature

LOD	Limit of detection
L-phe	L-Phenylaline
LUMO	Lowest unoccupied molecular orbital
MIP	Molecularly imprinted polymers
MNZ	Metronidazole
MOF	Metal organic framework
NIP	Non imprinted polymer
NIPAAm	*N*-Isopropylacrylamide
NT	Nitroaromatic compound
o-AT	*o*-Aminothiophenol
OEG	Thiolated oligoethyleneglycol
o-PD	*o*-Phenylenediamine
PANI	Polyaniline
p-ATP	*p*-Aminothiophenol
PCEMMA	Poly[2-(*N*-carbazolyl)ethyl methacrylate-*co*-methacrylic acid]
PM	Piezoelectric microgravimetry
PMP	Pinacolyl methylphosphonate
PPD	Poly(*o*-phenylenediamine)
PPG	Polypyrogallol
PPy	Polypyrrole
py-PBA	Pyrrole-phenylboronic acid
QCA	Quinoxaline-2-carboxylic acid
QCM	Quartz crystal microbalance
RAC	Ractopamine
SAW	Surface acoustic wave
SDM	Sulfadimethoxine
SEM	Scanning electron microscopy
SPR	Surface plasmon resonance
SWV	Square wave voltammetry
TBA	Terbuthylazine
TBC	Tris(2,3-dibromopropyl) isocyanurate
TMA	3-Thiophenemalonic acid
TMeOH	Thiophene methanol
TNB	1,3,5-Trinitrobenzene
TNP	2,4,6-Trinitrophenol
TNT	2,4,6-Trinitrotoluene
VB	Valence band
XPS	X-ray photoelectron spectroscopy

Acknowledgements

The support from Italian Ministry of University and Research (MIUR) Futuro in Ricerca (FIR) programme under Grant no. RBFR122KL1(SENS4BIO) is acknowledged. Apulia Region is acknowledged for support by Future in Research programme under Grant no. 9EC1495 (Ultrasensitive sensor for food analysis) and no. YFN6JP8 (Gold nanoparticle-enzyme assembly for environmental biosensors), programme cofinanced by ERDF and Cohesion

Fund 2007-13. Caripuglia is acknowledged for support by financing the project "Materiali innovativi porosi nanocompositi per la rimozione e il recupero di composti fenolici da acque di vegetazione olearie". Authors are grateful to Dr Nicole Jaffrezic and co-workers for original Figure 2 of ref. 85, presented as Scheme 13.9 and to Dr Xuemin Zhou and co-workers for the original graphical abstract of ref. 119, presented as Scheme 13.12 in the present chapter.

References

1. E. Mazzotta, S. Rella, A. Turco and C. Malitesta, *RSC Adv.*, 2015, **5**, 83164.
2. D. R. Thevenot, K. Tóth, R. A. Durst and G. S. Wilson, *Pure Appl. Chem.*, 1999, **71**, 2333.
3. A. Hulanicki, S. Glab and F. Ingman, *Pure Appl. Chem.*, 1991, **63**, 1247.
4. L. Clark and C. Lyons, *Ann. N. Y. Acad. Sci.*, 1962, **102**, 29.
5. E. Dineen, T. C. Schwan and C. L. Wilson, *J. Electrochem. Soc.*, 1949, **96**, 226.
6. J. Heinze, B. A. Frontana-Uribe and S. Ludwigs, *Chem. Rev.*, 2010, **110**, 4724.
7. T. A. Skotheim, *Handbook of Conducting Polymers*, M. Dekker, 1986.
8. R. S. Hutchins and L. G. Bachas, *Anal. Chem.*, 1995, **67**, 1654.
9. L. D. Spurlock, A. Jaramillo, A. Praserthdam, J. Lewis and A. Brajter-Toth, *Anal. Chim. Acta*, 1996, **336**, 37.
10. C. Malitesta, I. Losito and P. G. Zambonin, *Anal. Chem.*, 1999, **71**, 1366.
11. T. L. Panasyuk, V. M. Mirsky, S. A. Piletsky and O. S. Wolfbeis, *Anal. Chem.*, 1999, **71**, 4609.
12. L. Longo, G. Vasapollo, A. Scardino, R. A. Picca and C. Malitesta, *J. Porphyrins Phthalocyanines*, 2006, **10**, 1061.
13. C. Malitesta, M. R. Guascito, E. Mazzotta and R. A. Picca, *Thin Solid Films*, 2010, **518**, 3705.
14. P. S. Sharma, A. Wojnarowicz, M. Sosnowska, T. Benincori, K. Noworyta, F. D'Souza and W. Kutner, *Biosens. Bioelectron.*, 2016, **77**, 565.
15. N. Karimian, M. H. A. Zavar, M. Chamsaz, A. P. F. Turner and A. Tiwari, *Electrochem. Commun.*, 2013, **36**, 92.
16. Y. Chen, L. Chen, R. Bi, L. Xu and Y. Liu, *Anal. Chim. Acta*, 2012, **754**, 83.
17. C. Malitesta, F. Palmisano, L. Torsi and P. G. Zambonin, *Anal. Chem.*, 1990, **62**, 2735.
18. Y. Yoshimi, R. Ohdaira, C. Iiyama and K. Sakai, *Sens. Actuators, B*, 2001, **73**, 49.
19. S. A. Piletsky, E. V. Piletskaya, K. Yano, A. Kugimiya, A. V. Elgersma, R. Levi, U. Kahlow, T. Takeuchi, I. Karube, T. I. Panasyuk and A. V. El'skaya, *Anal. Lett.*, 1996, **29**, 157.
20. N. Karimian, M. H. Arbab Zavar, M. Chamsaz, N. Ashraf, A. P. F. Turner and A. Tiwari, *RSC Adv.*, 2015, **5**, 35089.
21. X. Li, J. Li, W. Yin and L. Zhang, *J. Solid State Electrochem.*, 2014, **18**, 1815.
22. Y. Yoshimi, K. Sato, M. Ohshima and E. Piletska, *Analyst*, 2013, **138**, 5121.

23. E. Mazzotta and C. Malitesta, *Sens. Actuators, B*, 2010, **148**, 186.
24. C. Malitesta, E. Mazzotta, R. A. Picca, A. Poma, I. Chianella and S. A. Piletsky, *Anal. Bioanal. Chem.*, 2012, **402**, 1827.
25. P. S. Sharma, A. Pietrzyk-Le, F. D'Souza and W. Kutner, *Anal. Bioanal. Chem.*, 2012, **402**, 3177.
26. A. Turco, E. Mazzotta, C. Di Franco, M. V. Santacroce, G. Scamarcio, A. G. Monteduro, E. Primiceri and C. Malitesta, *J. Solid State Electrochem.*, 2016, 1.
27. H. Hrichi, M. R. Louhaichi, L. Monser and N. Adhoum, *Sens. Actuators, B*, 2014, **204**, 42.
28. A.-E. Radi and I. Abd-Elaziz, *Anal. Methods*, 2015, **7**, 8152.
29. A. Nezhadali, Z. Rouki and M. Nezhadali, *Talanta*, 2015, **144**, 456.
30. K. K. Tadi, R. V. Motghare and V. Ganesh, *Electroanalysis*, 2014, **26**, 2328.
31. A. Afkhami, F. Gomar and T. Madrakian, *J. Electrochem. Soc.*, 2015, **162**, B109.
32. A. Turco, S. Corvaglia and E. Mazzotta, *Biosens. Bioelectron.*, 2015, **63**, 240.
33. K. Koirala, F. B. Sevilla III and J. H. Santos, *Sens. Actuators, B*, 2016, **222**, 391.
34. B. Zhang, L. Lu, F. Huang and Z. Lin, *Anal. Chim. Acta*, 2015, **887**, 59.
35. Z. O. Uygun and Y. Dilgin, *Sens. Actuators, B*, 2013, **188**, 78.
36. S. Mamo and J. Gonzalez-Rodriguez, *Sensors*, 2014, **14**, 23269.
37. J. Erdőssy, V. Horváth, A. Yarman, F. W. Scheller and R. E. Gyurcsányi, *TrAC, Trends Anal. Chem.*, 2016, **79**, 179.
38. A. Ramanaviciene and A. Ramanavicius, *Biosens. Bioelectron.*, 2004, **20**, 1076.
39. L. Li, L. Yang, Z. Xing, X. Lu and X. Kan, *Analyst*, 2013, **138**, 6962.
40. X. Kan, Z. Xing, A. Zhu, Z. Zhao, G. Xu, C. Li and H. Zhou, *Sens. Actuators, B*, 2012, **168**, 395.
41. X. Kan, C. Li, H. Zhou, A. Zhu, Z. Xing, Z. Zhao and G. Xu, *Am. J. Biomed. Sci.*, 2012, **4**, 184.
42. M. Zhong, Y. Teng, S. Pang, L. Yan and X. Kan, *Biosens. Bioelectron.*, 2015, **64**, 212.
43. A. Nezhadali and R. Shadmehri, *Sens. Actuators, B*, 2013, **177**, 871.
44. C. Malitesta, I. Losito, L. Sabbatini and P. G. Zambonin, *J. Electron Spectrosc. Relat. Phenom.*, 1995, **76**, 629.
45. B. Deore, Z. Chen and T. Nakaoka, *Anal. Sci.*, 1999, **15**, 827.
46. C. Malitesta, E. Mazzotta, R. A. Picca, A. Poma, I. Chianella and S. A. Piletsky, *Anal. Bioanal. Chem.*, 2012, **402**, 1827.
47. N. Karimian, M. Vagin, M. H. A. Zavar, M. Chamsaz, A. P. F. Turner and A. Tiwari, *Biosens. Bioelectron.*, 2013, **50**, 492.
48. A. Tretjakov, V. Syritski, J. Reut, R. Boroznjak and A. Öpik, *Anal. Chim. Acta*, 2016, **902**, 182.
49. H. Li, H. Guan, H. Dai, Y. Tong, X. Zhao, W. Qi, S. Majeed and G. Xu, *Talanta*, 2012, **99**, 811.
50. X. Zheng, R. Lin, X. Zhou, L. Zhang and W. Lin, *Anal. Methods*, 2012, **4**, 482.

51. Q. Wang, L. L. Paim, X. Zhang, S. Wang and N. R. Stradiotto, *Electroanalysis*, 2014, **26**, 1612.
52. K. Kor and K. Zarei, *Talanta*, 2016, **146**, 181.
53. X. Yan, J. Deng, J. Xu, H. Li, L. Wang, D. Chen and J. Xie, *Sens. Actuators, B*, 2012, **171**, 1087.
54. A. G. Ayankojo, A. Tretjakov, J. Reut, R. Boroznjak, A. Öpik, J. Rappich, A. Furchner, K. Hinrichs and V. Syritski, *Anal. Chem.*, 2016, **88**, 1476.
55. N. Karimian, M. B. Gholivand and F. Taherkhani, *J. Electroanal. Chem.*, 2015, **740**, 45.
56. M. Torkashvand, M. B. Gholivand and F. Taherkhani, *Mater. Sci. Eng., C*, 2015, **55**, 209.
57. F. Zhao, X. Li, W. Xu, W. Zhang and X. Ying, *Anal. Lett.*, 2014, **47**, 1712.
58. P. Wang, L. Ge, M. Li, W. Li, L. Li, Y. Wang and J. Yu, *J. Inorg. Organomet. Polym. Mater.*, 2013, **23**, 703.
59. J. Chen, G. Xu, L. Bai, J. Tao, Y. Liu and Y. Zhang, *Int. J. Electrochem. Sci.*, 2012, **7**, 9812.
60. Y. Lattach, F. Garnier and S. Remita, *ChemPhysChem*, 2012, **13**, 281.
61. Y. Lattach, P. Archirel and S. Remita, *J. Phys. Chem. B*, 2012, **116**, 1467.
62. Y. Lattach, N. Fourati, C. Zerrouki, J.-M. Fougnion, F. Garnier, C. Pernelle and S. Remita, *Electrochim. Acta*, 2012, **73**, 36.
63. M. Dabrowski, P. S. Sharma, Z. Iskierko, K. Noworyta, M. Cieplak, W. Lisowski, S. Oborska, A. Kuhn and W. Kutner, *Biosens. Bioelectron.*, 2016, **79**, 627.
64. A. V. Vergara, R. B. Pernites, S. Pascua, C. A. Binag and R. C. Advincula, *J. Polym. Sci., Part A: Polym. Chem.*, 2012, **50**, 675.
65. A. Kumar Singh and M. Singh, *Biosens. Bioelectron.*, 2015, **74**, 711.
66. T.-P. Huynh, A. Pietrzyk-Le, K. C. C. Bikram, K. R. Noworyta, J. W. Sobczak, P. S. Sharma, F. D'Souza and W. Kutner, *Biosens. Bioelectron.*, 2013, **41**, 634.
67. A. Wojnarowicz, P. S. Sharma, M. Sosnowska, W. Lisowski, T.-P. Huynh, M. Pszona, P. Borowicz, F. D'Souza and W. Kutner, *J. Mater. Chem. B*, 2016, **4**, 1156.
68. S. Li, J. Li, Q. Lin and X. Wei, *Analyst*, 2015, **140**, 4702.
69. L.-J. Kong, M.-F. Pan, G.-Z. Fang, K. Qian and S. Wang, *Anal. Bioanal. Chem.*, 2012, **404**, 1653.
70. Rosy, H. Chasta and R. N. Goyal, *Talanta*, 2014, **125**, 167.
71. H. Wang, Y. Liu, S. Wei, S. Yao and S. Gong, *Int. J. Electrochem. Sci.*, 2015, **10**, 8834.
72. X. Ma, J. Liu, D. Wu, L. Wang, Z. Zhang and S. Xiang, *New J. Chem.*, 2016, **40**, 1649.
73. Z. Zhao, Y. Teng, G. Xu, T. Zhang and X. Kan, *Anal. Lett.*, 2013, **46**, 2180.
74. S. Des Azevedo, D. Lakshmi, I. Chianella, M. J. Whitcombe, K. Karim, P. K. Ivanova-Mitseva, S. Subrahmanyam and S. A. Piletsky, *Ind. Eng. Chem. Res.*, 2013, **52**, 13917.
75. D. Dechtrirat, N. Gajovic-Eichelmann, F. F. Bier and F. W. Scheller, *Adv. Funct. Mater.*, 2014, **24**, 2233.

76. Y. Wang and T.-X. Wei, *Chin. Chem. Lett.*, 2013, **24**, 813.
77. P. Lenain, S. De Saeger, B. Mattiasson and M. Hedström, *Biosens. Bioelectron.*, 2015, **69**, 34.
78. H. Li, Z. Wang, B. Wu, X. Liu, Z. Xue and X. Lu, *Electrochim. Acta*, 2012, **62**, 319.
79. X. Zhang, Y. Peng, J. Bai, B. Ning, S. Sun, X. Hong, Y. Liu, Y. Liu and Z. Gao, *Sens. Actuators, B*, 2014, **200**, 69.
80. X. Sun, L. Zhang, H. Zhang, H. Qian, Y. Zhang, L. Tang and Z. Li, *J. Agric. Food Chem.*, 2014, **62**, 4552.
81. M. Riskin, R. Tel-Vered, T. Bourenko, E. Granot and I. Willner, *J. Am. Chem. Soc.*, 2008, **130**, 9726.
82. Q. Wang, J. Ji, D. Jiang, Y. Wang, Y. Zhang and X. Sun, *Anal. Methods*, 2014, **6**, 6452.
83. J. Sun, J. Ji, Y. Wang, Y. Zhao, Y. Zhang and X. Sun, *Anal. Bioanal. Chem.*, 2015, **407**, 1951.
84. Z. Guo, A. Florea, C. Cristea, F. Bessueille, F. Vocanson, F. Goutaland, A. Zhang, R. Săndulescu, F. Lagarde and N. Jaffrezic-Renault, *Sens. Actuators, B*, 2015, **207**, 960.
85. A. Florea, Z. Guo, C. Cristea, F. Bessueille, F. Vocanson, F. Goutaland, S. Dzyadevych, R. Săndulescu and N. Jaffrezic-Renault, *Talanta*, 2015, **138**, 71.
86. M. Bougrini, A. Florea, C. Cristea, R. Sandulescu, F. Vocanson, A. Errachid, B. Bouchikhi, N. El and N. Jaffrezic-renault, *Food Control*, 2016, **59**, 424.
87. Y. Ben-Amram, R. Tel-Vered, M. Riskin, Z.-G. Wang and I. Willner, *Chem. Sci.*, 2012, **3**, 162.
88. C. Xue, Q. Han, Y. Wang, J. Wu, T. Wen, R. Wang, J. Hong, X. Zhou and H. Jiang, *Biosens. Bioelectron.*, 2013, **49**, 199.
89. B. Rezaei, M. Khalili Boroujeni and A. A. Ensafi, *Biosens. Bioelectron.*, 2014, **60**, 77.
90. B. Rezaei, M. K. Boroujeni and A. A. Ensafi, *Electrochim. Acta*, 2014, **123**, 332.
91. B. Rezaei, M. K. Boroujeni and A. A. Ensafi, *Biosens. Bioelectron.*, 2015, **66**, 490.
92. B. Rezaei, M. K. Boroujeni and A. A. Ensafi, *Sens. Actuators, B*, 2016, **222**, 849.
93. Y. Li, Y. Liu, Y. Yang, F. Yu, J. Liu, H. Song, J. Liu, H. Tang, B. Ye and Z. Sun, *ACS Appl. Mater. Interfaces*, 2015, **7**, 15474.
94. H.-H. Li, H. Wang, W. Li, X. Fang, X. Guo, W.-H. Zhou, X. Cao, D. Kou, Z. Zhou and S. Wu, *Sens. Actuators, B*, 2016, **222**, 1127.
95. P. Liu, X. Zhang, W. Xu, C. Guo and S. Wang, *Sens. Actuators, B*, 2012, **163**, 84.
96. B. B. Prasad, I. Pandey, A. Srivastava, D. Kumar and M. P. Tiwari, *Sens. Actuators, B*, 2013, **176**, 863.
97. H. Shekarchizadeh, A. A. Ensafi and M. Kadivar, *Mater. Sci. Eng. C*, 2013, **33**, 3553.

98. Y. Peng, Z. Wu and Z. Liu, *Anal. Methods*, 2014, **6**, 5673.
99. J. G. Pacheco, M. Castro, S. Machado, M. F. Barroso, H. P. A. Nouws and C. Delerue-Matos, *Sens. Actuators, B*, 2015, **215**, 107.
100. H. da Silva, J. Pacheco, J. Silva, S. Viswanathan and C. Delerue-Matos, *Sens. Actuators, B*, 2015, **219**, 301.
101. X. Kan, H. Zhou, C. Li, A. Zhu, Z. Xing and Z. Zhao, *Electrochim. Acta*, 2012, **63**, 69.
102. H. Zhou, G. Xu, A. Zhu, Z. Zhao, C. Ren, L. Nie and X. Kan, *RSC Adv.*, 2012, **2**, 7803.
103. H. da Silva, J. G. Pacheco, J. MCS Magalhães, S. Viswanathan and C. Delerue-Matos, *Biosens. Bioelectron.*, 2014, **52**, 56.
104. L. Kong, X. Jiang, Y. Zeng, T. Zhou and G. Shi, *Sens. Actuators, B*, 2013, **185**, 424.
105. Z. Jiang, G. Li and M. Zhang, *RSC Adv.*, 2016, **6**, 32915.
106. D. Wu, H. Li, X. Xue, H. Fan, Q. Xin and Q. Wei, *Anal. Methods*, 2013, **5**, 1469.
107. Y. Yang, G. Fang, X. Wang, M. Pan, H. Qian, H. Liu and S. Wang, *Anal. Chim. Acta*, 2014, **806**, 136.
108. X. Xing, S. Liu, J. Yu, W. Lian and J. Huang, *Biosens. Bioelectron.*, 2012, **31**, 277.
109. B. Ertan, T. Eren, İ. Ermiş, H. Saral, N. Atar and M. L. Yola, *J. Colloid Interface Sci.*, 2016, **470**, 14.
110. X. Wang, J. Dong, H. Ming and S. Ai, *Analyst*, 2013, **138**, 1219.
111. F. Wang, L. Zhu and J. Zhang, *Sens. Actuators, B*, 2014, **192**, 642.
112. M. Cui, J. Huang, Y. Wang, Y. Wu and X. Luo, *Biosens. Bioelectron.*, 2015, **68**, 563.
113. J. Ji, Z. Zhou, X. Zhao, J. Sun and X. Sun, *Biosens. Bioelectron.*, 2015, **66**, 590.
114. X. Meng, W. Guo, X. Qin, Y. Liu, X. Zhu, M. Pei and L. Wang, *RSC Adv.*, 2014, **4**, 38649.
115. B. Rezaei, O. Rahmanian and A. A. Ensafi, *Sens. Actuators, B*, 2014, **196**, 539.
116. M. B. Gholivand and M. Torkashvand, *Mater. Sci. Eng. C*, 2016, **59**, 594.
117. W. Lian, S. Liu, J. Yu, X. Xing, J. Li, M. Cui and J. Huang, *Biosens. Bioelectron.*, 2012, **38**, 163.
118. L. Kong, M. Pan, G. Fang, X. He, Y. Xia and S. Wang, *RSC Adv.*, 2015, **5**, 11498.
119. W. Zhu, G. Jiang, L. Xu, B. Li, Q. Cai, H. Jiang and X. Zhou, *Anal. Chim. Acta*, 2015, **886**, 37.

CHAPTER 14

Molecularly Imprinted Polymer Sensor Arrays

PING LI, WILLIAM J. RICHARDSON, DI SONG AND KEN D. SHIMIZU*

University of South Carolina, Department of Chemistry and Biochemistry, Columbia, SC 29208, USA
*E-mail: shimizu@mailbox.sc.edu

14.1 Introduction

14.1.1 Advantages of MIP Sensors

Molecularly imprinted polymers (MIPs) provide rapid access to molecular recognition elements with rationally tailorable selectivity making them well-suited for sensing applications.[1–4] Molecular imprinting is typically a simple one-pot process and thus does not require time consuming multistep syntheses or fabrication.[5–7] The synthetic efficiency and directed selectivity of imprinting arises from self-assembly in the pre-polymerization solution of functional monomers and a template molecule. The polymerization then fixes the orientation and distances between the recognition groups of the functional monomers into the rigid polymer matrix. Removal of the templates creates binding cavities with affinity and selectivity for the template and structural analogues. This approach allows for the rapid preparation of polymeric recognition elements with selectivity for many different types of analytes including pharmaceuticals, environmental contaminants, volatile

Polymer Chemistry Series No. 28
Molecularly Imprinted Polymers for Analytical Chemistry Applications
Edited by Wlodzimierz Kutner and Piyush Sindhu Sharma

Published by the Royal Society of Chemistry, www.rsc.org

organic compounds, metal ions, proteins, and even microorganisms like viruses and bacteria.[4] Molecular imprinting also yields recognition elements that reveal excellent mechanical and chemical stability, which further ameliorates their sensing utility.

14.1.2 Challenges for MIP Sensors

The primary factor that has hindered the adoption of MIPs in commercial sensors is their low selectivity and high cross-reactivity. This is due to the generally low fidelity with the molecular imprinting process, which leads to a broad distribution of binding sites of varying structure, affinity and selectivity.[8–10] Most depictions of the imprinting process show every monomer and template forming a templated recognizing site (Scheme 14.1A). However, in most cases, imprinting proceeds with low fidelity, resulting in the majority of functional monomers and/or templates not participating in this process (Scheme 14.1B).[8] Thus, a high percentage of the functional monomers form non-templated background binding sites, lowering the overall affinities and selectivity of MIPs.

Due to the problems associated with binding site heterogeneity, much of the research on molecular imprinting has focused on developing methods to improve the fidelity of the imprinting process. However, solutions to these problems usually require significantly more synthetic effort or involve more complex and time-consuming processes. Thus, addressing the problem of recognizing site heterogeneity often comes at the cost of forfeiting the primary advantage of MIPs, which is their rapid and efficient synthesis.

14.1.3 MIP Arrays as a Solution

An alternative solution to improve the selectivity of MIP chemosensors is to group multiple MIPs together into a sensor array (Scheme 14.2A).[2,11] Each analyte is tested against the MIP array resulting in a unique response pattern that can be used to differentiate and classify the analytes (Scheme 14.2B).[12–14] MIPs appear to be ideally suited to serve as recognition elements in the chemosensor array format. First, chemosensor arrays do not require individual sensing elements with high selectivity. In fact, there have been many

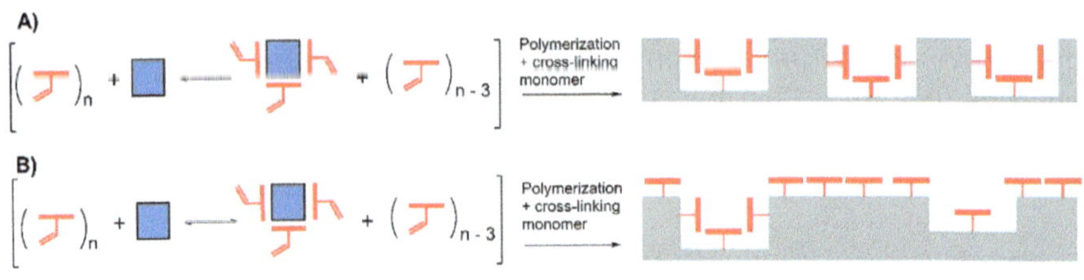

Scheme 14.1 Illustration of the idealized (**A**) homogeneous and (**B**) more typical heterogeneous distributions of binding sites in MIPs.

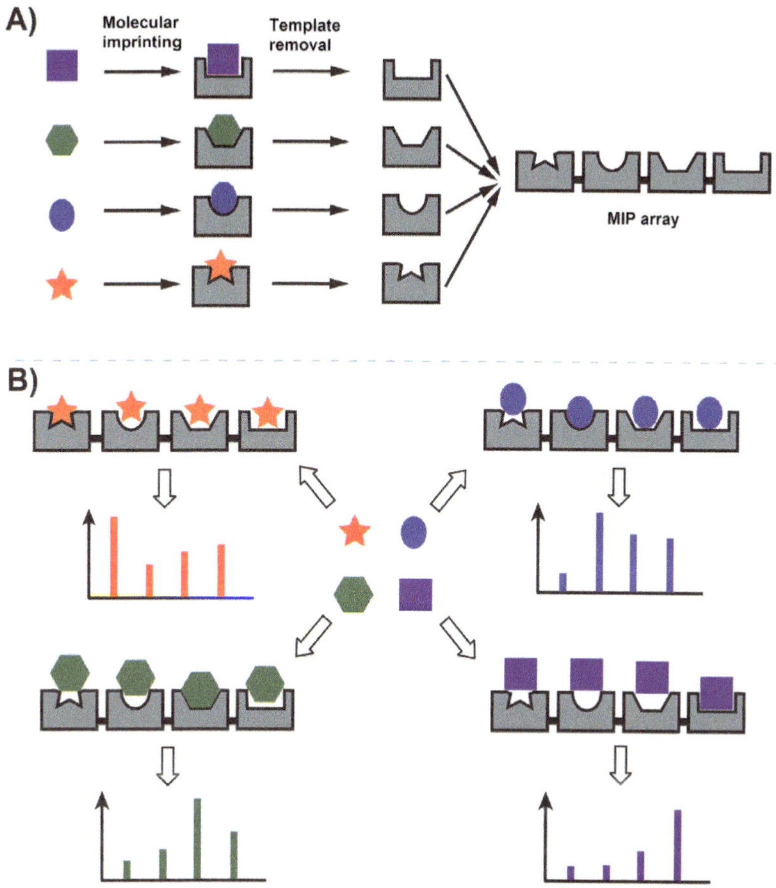

Scheme 14.2 (**A**) Schematic representation of the preparation of an MIP sensor array *via* the molecular imprinting process to yield a series of polymer sensors with unique selectivity patterns and (**B**) the unique response patterns of four different analytes against an MIP sensor array.

successful examples of chemosensor arrays using non-specific recognition materials, such as polymer films or blends.[15]

Second, the molecular imprinting process solves one of the standing challenges in the chemosensor array field. Synthesizing and collecting a sufficient number of sensing elements with different selectivity is very difficult. Individually tailored recognition elements with high selectivity, such as synthetic organic receptors,[16–20] peptides, and nucleic acid,[21] or antibodies[22–24] require extensive effort *via* organic synthesis or biochemical engineering. Alternatively, low-cost materials, such as simple polymer films, can be assembled into arrays. However, they cannot be readily tailored and have low or no selectivity. MIPs appear to combine the attractive benefits of the two general types of recognition elements. MIPs can specially be tailored to recognize a particular analyte or class of analytes. At the same time, MIPs can be produced quickly and inexpensively *via* molecular imprinting.

14.1.4 Early Examples of MIP Sensor Arrays

The use of molecular imprinting to fabricate elements of a chemosensor array can be traced back to the early 2000s. In 2003, Mirsky and co-workers reported the first example of a molecularly imprinted chemosensor array.[25] An unconventional molecular imprinting process was used to tailor the selectivity of a self-assembled monolayer (SAM). Purine and pyrimidine thiols created molecule-shaped defects in the dodecanethiol SAM surface (Figure 14.1A), which served as recognizing sites for similarly shaped aromatic analytes. Recognizing events were monitored *via* the capacitance change in the gold substrate. SAMs imprinted with five different purine and pyrimidine thiols were assembled into a capacitance chemosensor array and tested against six non-thiol analytes. Multivariate analysis of the similarities and differences in the response patterns confirmed that the imprinted chemosensor array was able to clearly distinguish the six analytes (Figure 14.1B). A more detailed interpretation of multivariate analysis is provided in Section 14.3.5. This example not only highlighted the potential of imprinted materials in sensor arrays but also demonstrated the potential of these arrays to detect and classify non-template analytes, thus greatly broadening the utility of this approach.

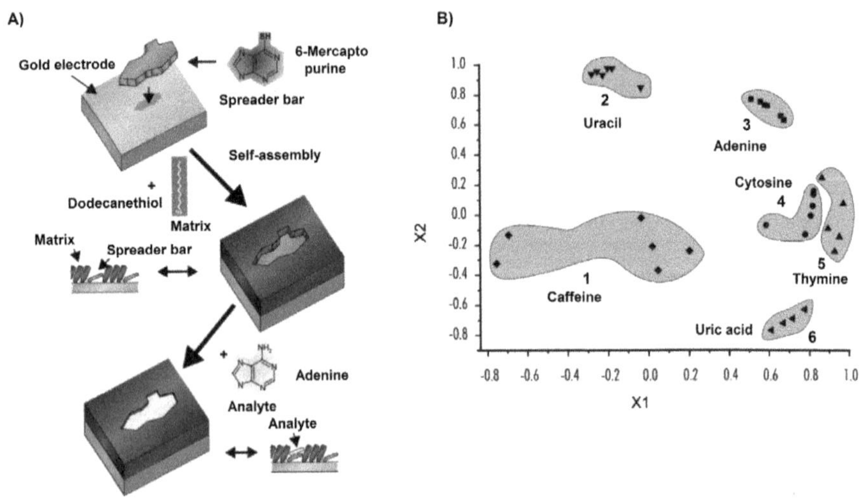

Figure 14.1 (**A**) Schematic representation of the preparation of the molecularly imprinted self-assembly monolayer (SAM) of dodecanethiol on the gold substrate using 6-mercaptopurine as template. (**B**) Two-dimensional principal component analysis (PCA) plot of the electronic responses of six analytes on the capacitance sensors with imprinted SAM coating. Reproduced from ref. 25 with permission from The Royal Society of Chemistry.

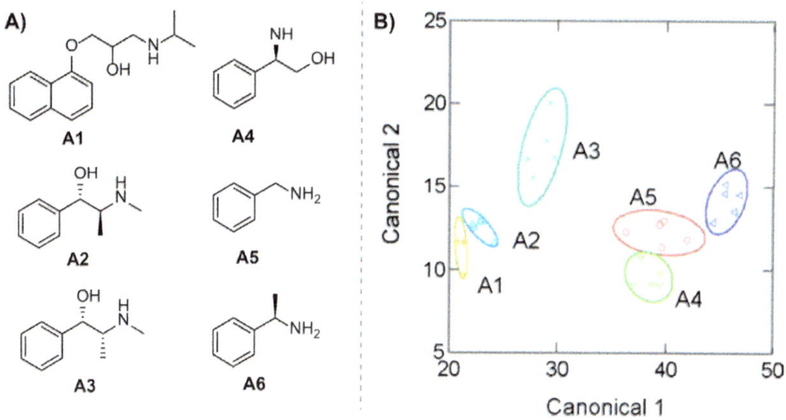

Figure 14.2 (**A**) Structural formulas of the six amine analytes (A1–A6). (**B**) Two-dimensional linear discriminant analysis (LDA) plot of the spectral responses of analytes A1–A6 against the MIP array during the analyte binding. Reproduced from ref. 26 with permission from The Royal Society of Chemistry.

The next year,[26] we reported the first example of an imprinted polymer array for sensing applications. The MIPs were rapidly prepared *via* the free-radical polymerization of commercially available functional monomer, methacrylic acid (MAA), and a cross-linking monomer, ethylene glycol dimethacrylate (EGDMA), in the presence of seven different small-molecule amine templates (Figure 14.2A).[26] Despite the low fidelity of the individual MIPs, the resulting array consisting of the seven MIPs and one non-imprinted polymer (NIP) was able to differentiate seven biogenic and pharmaceutically active amines with 94% accuracy (Figure 14.2B). Particularly impressive was the ability of the MIP array to differentiate structurally similar analytes, such as diastereomers (ephedrine *versus* pseudoephedrine) and amines that differed just by one methyl substituent (benzyl amine *versus* α-methyl benzylamine). One limitation of this early example was the necessity for the analytes to be UV active. This was addressed in our later work by employing a dye-displacement strategy.[27] The colorimetric dye-displacement MIP assay was able to monitor both UV-active and inactive substrates.

In 2007, Takeuchi and co-workers extended the imprinted polymer array approach to biomacromolecule profiling.[28] Three protein imprinted polymers (PIPs) were prepared using acrylic acid (AA) and glycosyloxylethyl methacrylate as functional monomers, methylene bis(acrylamide) (MBAA) as the cross-linking monomer, and three different protein templates. The individual PIPs showed moderate selectivity (Figure 14.3A). However, the array made up of three PIPs and one NIP was able to differentiate the three templated proteins: cytochrome *C* (Cyt *C*), ribonuclease A (Rib), and α-lactalbumin (Lac) as well as two additional non-template proteins: albumin (Alb) and myoglobin (Myo) (Figure 14.3B). This example demonstrated that the

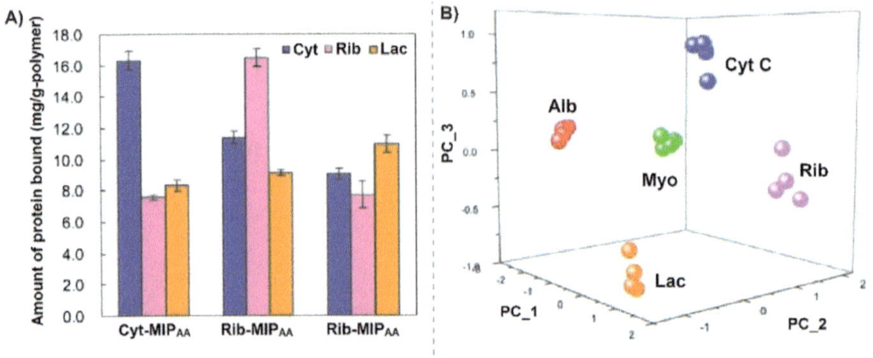

Figure 14.3 (**A**) Absorption capacity of the three protein-imprinted polymers [(Cyt-*C*)-MIP$_{AA}$, Rib-MIP$_{AA}$, and Lac-MIP$_{AA}$] toward the template analytes. (**B**) The three-dimensional principal component analysis (PCA) plot of the absorption responses of three protein analytes (four replicates) against the three-channel MIP array. Alb – albumin, Cyt *C* – cytochrome *C*, Lac – α-lactose, Myo – myoglobin, Rib – ribonuclease A, MIP$_{AA}$ – molecularly imprinted polymer prepared using acrylic acid. Reproduced from ref. 28 with permission from The Royal Society of Chemistry.

MIP-based sensing strategy is not restricted to small molecules and is also applicable to larger biomolecules. This example also reiterated the ability of imprinted arrays to distinguish analytes beyond those used as templates.

14.1.5 Outline of the Chapter Goals

Early examples of MIP arrays in sensing applications were summarized in our 2010 review of the topic.[11] In the years since, an increasing number of researchers have been attracted to this topic. In the present chapter, we provide an update to MIP-based chemosensor array research since 2010 and provide a guide to future researchers that are interested in developing imprinted polymer arrays for dedicated sensing applications. First, a comprehensive table was compiled summarizing all examples of MIP chemosensors and assays in the literature that used an array format. Next, the design characteristics and trends in successful MIP arrays will be discussed. These include: monomer and matrix, polymer morphology, signal transducers, and pattern recognition analysis methods. Finally, illustrative examples will be discussed in more detail to highlight novel approaches and techniques.

14.2 Survey of Imprinted Chemosensor Arrays and Assays in the Literature

Table 14.1 is a compilation of reports in the literature, in which the molecular or protein imprinting process was used to prepare the individual elements for a chemosensor array or assay. While the total number of examples

Table 14.1 Literature survey of MIP sensor arrays.[a]

	Analyte and template	Polymer matrix	Morphology	Signal platform	Statistical analysis	Ref.
1	Templates/analytes: 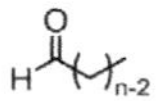 n = 6, 8 and 9	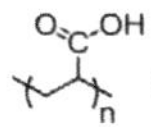 PAA	Thin film coating	QCM	PCA and SVM	29
2	Templates/analytes: cytochrome *C*, ribonuclease A, lysozyme, and myoglobin	Functional monomer: 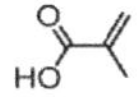 MAA 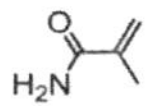 AM; Cross-linking monomer: 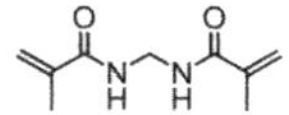 MBAA	Surface imprinting (2-step)	Competitive binding assay	N/A	30
3	Templates: five azo dyes: amaranth, tartrazine, sunset yellow, orange II, chrysoidin and para red; Analytes: five azo dyes and rhodamine B	Functional monomer: MAA; Cross-linking monomer: 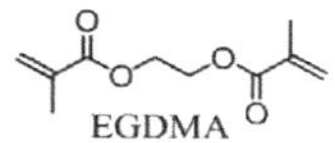 EGDMA	Ground powder	Binding (UV-vis)	CDA	31
4	Templates: *n*-propanoic acid, *n*-hexanoic acid, and *n*-octanoic acid Analytes: *n*-hexanal, *n*-heptanal, and *n*-nonanal	PAA	Thin film coating	QCM	PCA and SVM	32
5	Templates/analytes: bovine hemoglobin, equine myoglobin, and cytochrome *C*	Functional monomer: AM; Cross-linking monomer: MBAA	Solution deposition	CV	PCA	33
6	Templates: 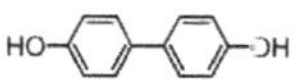 BIP 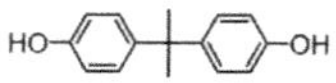 BPA 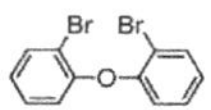 DBDE 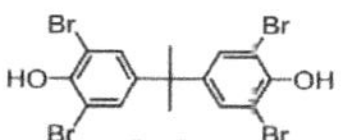 TBBPA 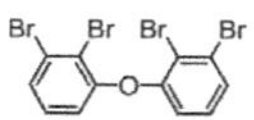 TBDE Analytes: BPA, BIP, and TBBPA	Functional monomer: 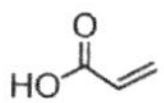 AA; Cross-linking monomer: EGDMA	Surface imprinting (sacrificial mold)	Light refraction	PCA	34

(*continued*)

Table 14.1 (continued)

	Analyte and template	Polymer matrix		Morphology	Signal platform	Statistical analysis	Ref.
7	Templates/analytes: 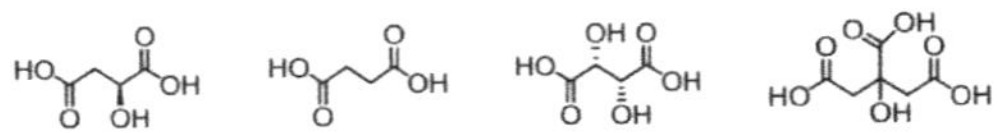	Precursor: $Ti(O\textit{i}Pr)_4$ HCl (cat.), and formamide		Ground powder	Dye-displacement assay	PCA	35
8	Templates: 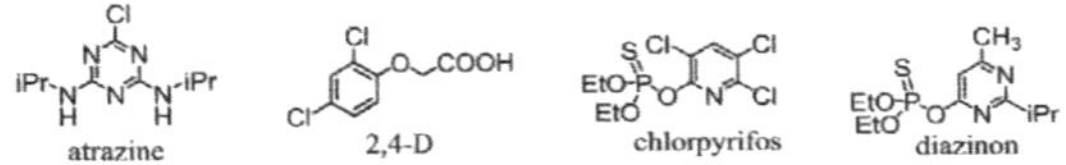 Analytes: atrazine, 2,4-D, chlorpyrifos, diazinon, 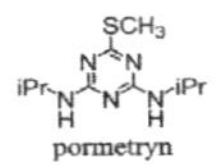 and 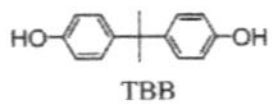	Functional monomer: MAA	Cross-linking monomer: EGDMA	Surface imprinting (sacrificial template)	Light refraction	PCA	36
9	Templates: 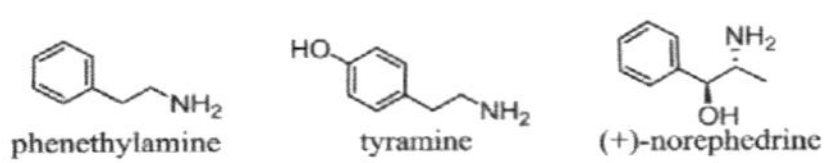 Analytes: Phenethylamine, tyramine, (+)-norephedrine, 4-methylbenzyl amine, and L-tyrosine methyl ester	Functional monomer: MAA	Cross-linking monomer: EGDMA	Ground powder	Binding analysis (UV-vis)	PCA	37
10	Templates/analytes: 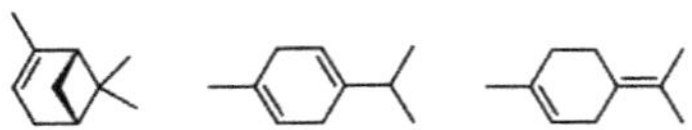	Functional monomer: MAA	Cross-linking monomer: EGDMA	Deposition	Capacitance sensor and QCM	N/A	38
11	Templates/analytes: 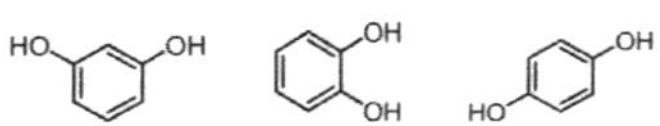	Functional monomer: MAA	Cross-linking monomer: EGDMA	Surface imprinting (grafting)	Chemiluminescence assay	N/A	39

12	Templates: 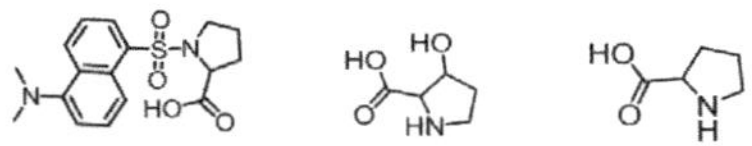 *N*-dansyl proline hydroxyproline proline Analytes: *N*-dansylated Pro, Lys, Hyp, Leu, Phe, Trp, Ile, Glu, Gly, Ser, Cys, and Asp	Functional monomer: MAA	Cross-linking monomer: EDGMA	Micro-spheres	Chemi-lumi-nescence assay	N/A	40
13	Templates: D-fructose and D-xylose Analytes: D-fructose; D-xylose, D-arabinose, D-glucose, D-galactose, D-mannose, L-sorbose, D-ribose, L-rhamnose, and sucrose	Precursor: 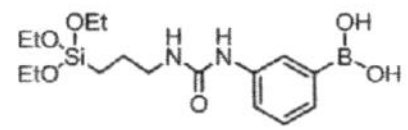 $(EtO)_4Si$ and $NH_3{\cdot}H_2O$ (cat.)		Ground powder	Dye-displace-ment assay	PCA	41
14	Templates: Zn^{2+} and Cd^{2+} Analytes: Zn^{2+}, Cd^{2+}, Mg^{2+}, Ca^{2+}, and Al^{3+}	Precursor: 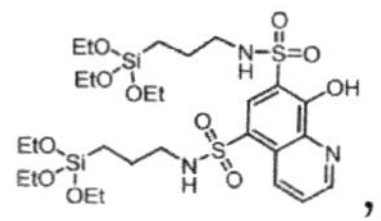 , $(EtO)_4Si$ and NaOH (cat.)		Ground powder	MIP fluo-rescence turn-on	PCA	42
15	Templates: Cytochrome *C*, ribonuclease A, and a-lactalbumin Analytes: Cytochrome *C*, ribonuclease A, a-lactal-bumin, albumin, and myoglobin	Functional monomer: AA, 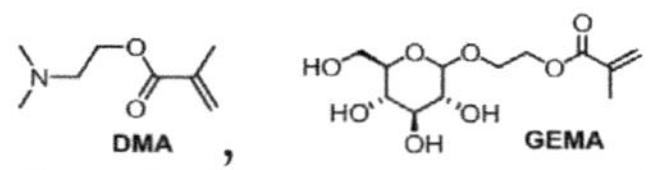 Cross-linking monomer: MBAA;		Ground powder	Rebinding (HPLC)	PCA	28

(*continued*)

Table 14.1 (*continued*)

	Analyte and template	Polymer matrix		Morphology	Signal platform	Statistical analysis	Ref.
16	Templates/analytes: 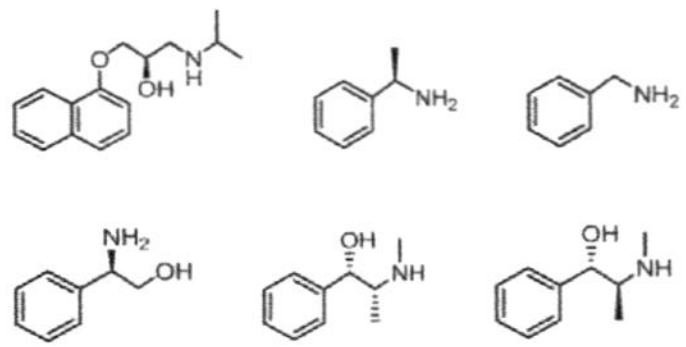	Functional monomer: MAA	Cross-linking monomer: EGDMA	Ground powder	Rebinding (UV-vis) & dye-displacement assay	LDA	26 and 27
17	Templates/analytes: Water, ethyl acetate, butanol, propanol, and limonene	Functional monomer: AA, styrene; 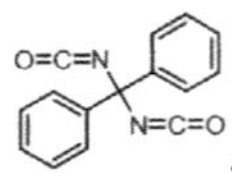 , 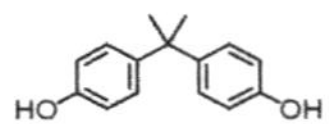	Cross-linking monomer: polyvinyl alcohol, divinylbenzene; 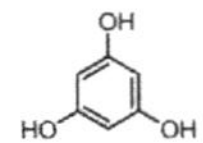	Thin film coating	QCM	N/A	43
18	Templates: 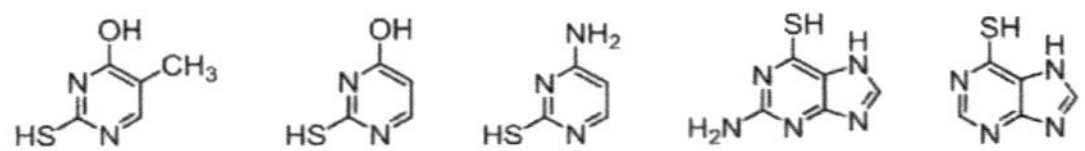 Analytes: 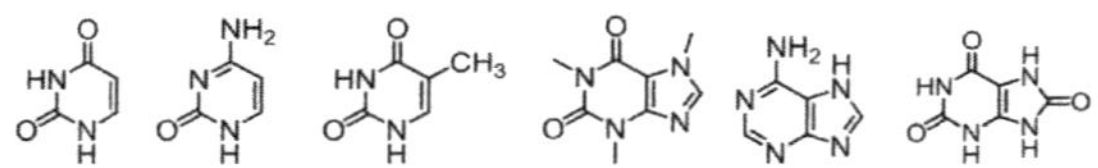	Dodecanethiol on gold substrate		Self-assembled monolayer	QCM	PCA	25

[a] 2,4-D – 2,4-dichlorophenoxyacetic acid, AA – acrylic acid, AM – acylamide, BIP – 4,4′-diphenol, BPA – bisphenol A, CDA – categorical data analysis, CV – cyclic voltammetry, DBDE – 2,2′-dibromodiphenylether, DMA – dimethylaminoethylmethacrylate, DVB – divinylbenzene, EGDMA – ethyleneglycoldimethacrylate, GEMA – glycosyloxyethylmethacrylate, HPLC – high-performance liquid chromatography, LDA – linear discriminant analysis, MAA – methacrylic acid, MBAA – methylenebisacylamide, PAA – poly(acrylic acid), QCM – quartz crystal microbalance, SVM – support vector machine, TBDE – 2,2′3,3′-tetrabromophenylether.

is still relatively modest, they span a range of polymer types, polymer morphologies, analytes, sensing platforms, and applications, demonstrating the viability and versatility of this approach. In general, the choices of analytes, polymers, and methods of analyses for these MIP chemosensor and array assays follow the trends in the larger molecular imprinting community.[6,7] For example, the most common polymer matrix for the MIP chemosensor arrays is MAA and EGDMA co-polymer, which is also the most popular MIP polymer matrix in MIP studies. Similarly, the majority of analytes in Table 14.1 contain polar functionalities, such as amines, carbonyls, and phenols that form strong directional hydrogen bonds with common functional monomers, such as MAA, acrylic acid (AA) and acrylamide (AM). The ability of the template to bind the functional monomers is critical for efficient molecular imprinting. Finally, the majority of the examples are assays and not true chemosensors as MIPs do not contain any intrinsic signaling mechanisms. In these assays, the recognizing events were indirectly monitored by measuring the disappearance of an unbound analyte or the release of dyes into solution rather than from a direct signal from the polymer matrix.

14.3 Design Choices

Analysis of the characteristics of each MIP array in Table 14.1 assisted in identifying successful imprinting matrices, analytes and templates, polymer formats, sensing platforms, and pattern recognition analyses. Each of these topics will be discussed below in the context of the examples in Table 14.1.

14.3.1 Templates and Analytes

Selection of analytes for an MIP chemosensor array project is determined by the particular application as well as compatibility with molecular imprinting. The most common analytes and templates for imprinted polymer arrays are small molecules that contain multiple polar functional groups. Polar functional groups on the templates include amine, alcohol, ketones/aldehyde, ether, and carboxyl groups. The presence of multiple polar functional binding groups on the templates facilitates the interactions with multiple functional monomers, which helps improve the probability that the templated binding sites would be lined with multiple recognition groups. Moreover, smaller molecules are easier to imprint as they are more easily removed from the polymer matrix after the imprinting.

Other types of successful templates include metal ions[42] and proteins,[28,30,33] each presenting unique challenges. Metal ions are very similar in terms of shape as they are spherical cationic species, which only differ subtly in their size, coordination strengths, and coordination geometries. Proteins are much larger templates, which often become physically entrapped during polymerization and, thus, their imprinting protocols typically require more flexible polymer matrices with lower degrees of crosslinking. In addition, the solubility of proteins is usually restricted to aqueous solvent systems, which are not compatible with common MIP polymerization conditions. Thus, water-soluble initiators, functional and cross-linking monomers are required.

For sensing applications, it is common to use a structural analog or epitope of the analyte as a template. The use of these analogs as templates addresses the problem of slow release of the templates from the MIP matrix, which is particularly problematic for sensing applications.[44] Some template molecules are tightly bound or physically entrapped during the imprinting. Therefore, they are not removed during subsequent washing step of the imprinting process and will slowly leach out of the polymer over time.[45] Second, the use of a structural analog as the template can improve the imprinting efficiency especially when the analytes lack polar functionalities. Finally, the ability of an imprinted chemosensor array to identify and distinguish analytes beyond those used as templates greatly extends the utility of the imprinting methodology. This enables MIP chemosensor arrays to distinguish a broader range of analytes including analytes that were unknown or unavailable when MIPs were prepared.

14.3.2 Polymer Matrix

The functional monomers and polymer matrices are the most important factors in developing effective MIP arrays because of their profound impact on the affinity and selectivity of the MIPs. The functional monomer choice is primarily dictated by the nature of a template and targeted analyte. The functional monomer needs to form reversible covalent or non-covalent bonds with the template. By far the most common and successful functional monomers used in the arrays are the commercially available vinyl carboxylic acids, such as AA and MAA. One underappreciated reason for the success of carboxylic acid monomers is their ability to suppress the formation of non-specific background binding sites by forming hydrogen bonded dimers.[46]

Other functional monomers that can form strong directional non-covalent, covalent, and coordination interactions with the template, such as hydrogen bonds, electrostatic interactions, and metal–ligand interactions, have also been successfully employed. These included monomers bearing basic recognition groups, *e.g.*, 2-dimethylaminoethyl methacrylate (DMA),[28] hydrogen bonding functionality, *e.g.*, AM,[30,33] diol recognition groups, *e.g.*, boronic acids,[41] and metal coordinating groups, *e.g.*, quinoline.[42]

The major component of MIP matrices is the cross-linking monomer. The primary role of this monomer is to form a rigid matrix that can hold the functional monomers at complementary distances and orientations to the template. Again, those most commonly used are commercially available vinyl cross-linking monomer, *e.g.*, EGDMA,[26,27,31,34,36–40] methylenebisacrylamide (MBAA),[28,30,33] and divinylbenzene (DVB),[43] which polymerize by free-radical polymerization (FRP). Moreover, there were examples of polycondensation cross-linking monomers including polyisocyanates,[43] tetraalkoxyl silanes,[41,42] and titanium tetraalkoxide.[35] In some cases, the cross-linking monomers can also bear recognition groups that can form hydrogen bonds, *e.g.*, MBAA and titania (hydrolyzed from titanium tetraalkoxide), or aromatic stacking functionality such as that in the DVB and bisphenol A (BPA).

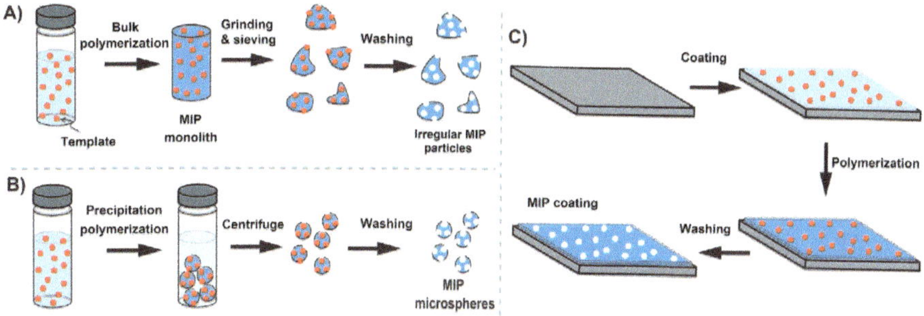

Scheme 14.3 Examples of polymer morphologies used in MIP chemosensor and assay elements: (**A**) irregular MIP particles, (**B**) MIP microspheres, and (**C**) MIP thin film coating.

14.3.3 Polymer Morphologies

The choice of polymer morphology is crucial for the integration of MIPs with the sensing platform (Scheme 14.3). The majority of MIPs in Table 14.1 are heterogeneous particles prepared by grinding MIP monoliths. This is the easiest MIP morphology to prepare. Bulk molecular imprinting procedures typically yield macroporous monoliths that have high interior surface areas. The irregular size and shape of the ground particles can lead to heterogeneous binding kinetics which can be improved by preparing uniform microspherical particles using emulsion or slow precipitation polymerization. However, the conditions to prepare microspherical MIPs are often detrimental to the imprinting process and, thus, the imprinted microspheres often have weaker or fewer imprints than the corresponding bulk monolithic MIPs.

For most sensing platforms, irregular ground particles are not ideally suited. In most cases, such as quartz crystal microbalance (QCM) or electrochemical sensors, a uniform MIP thin film is preferable. The preparation of imprinted thin films, however, is technically challenging. MIPs are highly cross-linked matrices with high rigidity and low flexibility. Thus, the preparation of thin films using standard imprinting recipes leads to brittle films containing cracks and defects due to the inability to alleviate the mechanical stresses formed during the polymerization process. Strategies to avoid these issues include the use of lower crosslinking percentages, self-assembled layers, grafting, sol/gel processes,[47,48] and electrochemical polymerization.[6,7]

14.3.4 Sensing Platforms

One of the most challenging aspects of developing MIP chemosensors is integrating MIPs into a signaling platform. MIPs are mainly recognition platforms, which do not possess any innate signaling properties. The key is to have efficient coupling of the binding events in an MIP to a signal transduction mechanism, such as calorimetry, piezoelectric microgravimetry at QCM, electrochemistry, or optical spectroscopy. Each of these sensing platforms will be discussed below.

14.3.4.1 Binding Assays

The majority of examples in Table 14.1 are assays as opposed to true chemosensors, sidestepping the problem of integration with a sensing platform. The binding events in assays are indirectly monitored by tracking the disappearance of unbound analyte in solution or the displacement of preloaded dyes from the binding sites. Assays that monitor the unbound analyte concentration measure the decrease in unbound analyte in solution before and after equilibration with the MIP. While easy to implement, analyte-monitoring methods have limited applicability as they require the analytes to be spectroscopically active. Competitive binding assays address this limitation by monitoring the displacement of a spectroscopically active analog (UV-vis, radiolabeled, or fluorescent) from the binding sites.[13] First, the dyes are loaded into the MIP binding sites. The binding of the analyte to the MIP displaces the dyes, releasing them into solution and generating a spectroscopic response. The dye-displacement method is advantageous as it broadens the spectrum of potential analytes. The disadvantage of competitive binding assays is a general decrease in selectivity as the preloaded dyes or labeled competitors are not precise structural analogs of the individual analytes.

14.3.4.2 Summary of Other Signaling Mechanisms

Other signaling platforms that have been successfully integrated with MIP arrays include mass-sensors, electrochemical including capacitance chemosensors, optical or emissive chemosensors, and chemiluminescent chemosensors.[6,7] The most popular are piezomicrogravimetric mass sensors using QCM. The changes in mass are measured from shifts in the resonance frequency of a thin piezoelectric substrate. MIPs are easily integrated with QCM chips as they can be either adhered to the piezoelectric substrate as small particles or as thin films. In addition, a QCM can monitor binding of volatile analytes in the gas phase or in solution. Another common sensing platform involves electrochemical including capacitance chemosensors for sensing either electroactive or electroinactive analytes with electrodes coated with MIP films. Finally, colorimetric or emissive MIP chemosensors have been developed by integrating MIPs into an optical signaling platform. While colorimetric or emissive MIP chemosensors can be more difficult to design and develop, they have the advantage of simple and low cost read-out.

14.3.5 Data Processing for Chemosensor Arrays

14.3.5.1 Introduction to Chemosensor Array Signal Processing

The volume and complexity of the data obtained from chemosensor arrays requires multivariate analysis to systematically identify similarities and differences in the response patterns. These pattern recognition protocols involve two unique steps, *i.e.*, the compilation of response patterns for a set of known standards and matching the response patterns of unknown samples to the patterns of the known standards. The known and unknown response patterns are a collection of signals measured for a single analyte against each element in an MIP array. Ideally, the response pattern for an analyte can be collected from a single solution of the analyte against a physical array of all of the sensing elements. However, since most examples were assays, the response patterns were more commonly collected by compiling the data from separate measurements of the analytes against each polymer or sensing element in the array. Note that multiple measurements can be made for each element in an array. These could be measurements of signal intensity at different wavelengths or time points or spectroscopic parameters. Thus, an array with five MIP polymers could yield a response pattern of 5, 10, or even 100 measurements for a single analyte.

14.3.5.2 Principal Component Analysis (PCA)

Principal component analysis (PCA) is the most common and simple method for processing sensor array data. PCA is available in many mathematical and statistical software packages including: Excel plug-ins, Mathematica, MatLab, and R. PCA is also one of the easier analysis methods to understand and, thus, is often utilized to provide a general overview of pattern recognition algorithms. PCA systematically identifies similarities and differences in the respective patterns. In PCA, the unique or orthogonal variations in the patterns are extracted and quantified. These orthogonal variations are called principal components (PC). For example, PC1 is a measure of the average signal intensity for all the data points in each pattern. PC2 is the variation in the patterns after the variation due to PC1 has been removed by normalizing the intensity of each pattern to its average intensity. PC3 is the variation in the patterns after the variations quantified by PC1 and PC2 have been removed. The total number of PCs is equal to the number of data points in a single pattern, which can be quite large for some data sets. However, only the first 3 PCs are typically analyzed because this is usually sufficient to characterize the majority of variation in the patterns. For instance, in Figure 14.4 below, the combination of PC1 and PC2 describe 93.5% of the variance in the patterns. The percentage of variation characterized by each PC is the value in parentheses on the axis labels. If a PC's percentage of variation is small (*i.e.* <5% or <1%), then the differences being measured could be smaller than the error of the measurements. In the case of Figure 14.4, PC3 may contain enough variation to be meaningful, because only 6.5% of the variance is left undescribed by PC1 and PC2. Negative or positive PC values can correlate with physical properties or parameters but not in a predicable fashion. The PC values simply describe whether a pattern is higher or lower than the average of a particular measure of variation.

The major advantage of pattern recognition methods, such as PCA, is that they can reduce the complexity of the data and provide systematic and quantitative measures of the differences and similarity between patterns. The typical output for a PCA is a two- or three-dimensional plot of the PCs. The most common is a plot of PC1 *versus* PC2 as these PCs encompass the

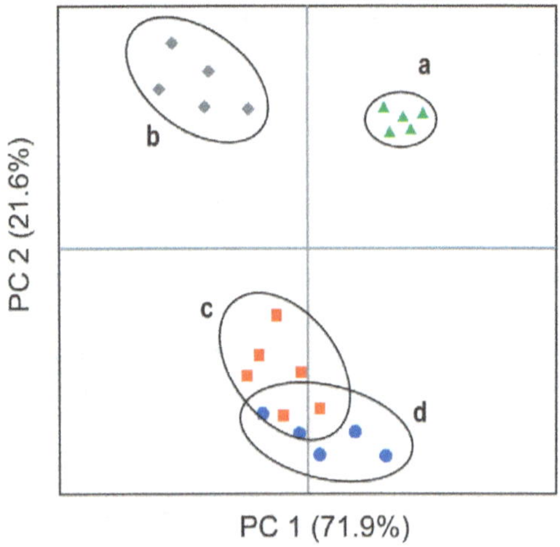

Figure 14.4 An illustrative two-dimensional PCA plot highlighting the clustering of data points of four type (a, b, c, and d) of response patterns with five replicates each.

majority of variation in the patterns. In the PC plots, each point represents one pattern that was measured for one sample of an analyte. Response patterns that are similar will be close to one another in the PC plots. Very different response patterns will be further apart as seen for samples a and b in Figure 14.4. Ideally, multiple measurements of the same analyte will yield similar response patterns, which will cluster together in the PC plots. To achieve accurate differentiation, the individual data points in the PCA plots for each analyte should be well separated and not overlap. In Figure 14.4, samples c and d are clearly not well separated, and distinguishing between the two will prove problematic based on PC1 and PC2. The PCA software will often provide a 95% confidence ellipse to aid the researchers in determining if the clusters for each analyte patterns overlap.

14.3.5.3 Linear Discriminant Analysis (LDA)

The second most common pattern recognition algorithm for analyzing sensor array data is linear discriminant analysis (LDA). This analysis is very similar to PCA with two key advantages. First, the similarities and differences in the patterns are weighted or supervised by the identity of the samples. LDA tries to find similarities in patterns for the same analytes and tries to find differences in patterns for different analytes. Thus, the discriminant plots for LDA will be very similar to the PC plots but with tighter grouping of patterns for similar analytes and larger separation for different analytes. The second advantage of LDA is the ability to classify unknown samples. While this is possible using PCA, it is much easier in most software packages' implementations of LDA.

There are two general strategies for using sensor arrays to classify samples. Ideally, the library of response patterns will be large enough that it can be randomly divided into two groups: 'known' and 'unknown'. To assess the accuracy of the array, the identities of the known and unknown sample will actually be known to the user. In the known group, it is important to have multiple response patterns for every analyte to create a stable library of patterns to characterize the unknown samples. Thus, the LDA analysis is run on the known sample patterns, and the resulting discriminants are used to classify the unknown patterns. The classification accuracy can be tested by comparing the LDA classification of the unknown samples *versus* the actual identity of the samples. Once the accuracy of the sensor array has been established, the array can be used to classify truly unknown samples. If only a small library of patterns is available (*i.e.* 5 replicate patterns for each analyte), then a jackknife or 'leave-one-out' LDA analysis can be performed. The software will perform a series of LDA analyses in which all the patterns except for one will be used to build the library of known patterns and classify the one 'unknown' pattern. This is repeated until all of the patterns have been treated as an 'unknown' pattern. The jackknife classification accuracy is almost always higher than the previous method as the jackknife analysis has a larger fraction of the patterns in the known library.

14.3.5.4 Other Methods

In addition to PCA and LDA, a variety of other multivariate analyses can be applied to sensor array analysis. One other notable pattern recognition method is support vector machines (SVMs). Like LDA, SVM is a supervised pattern recognition algorithm. However, SVM separates data based on non-linear boundaries. These boundaries are produced using a kernel function that translates the data into a new space capable of linear separation by way of two hyperplanes. The parameterization of this method is more complex and difficult than LDA, but if used correctly, the increase in classification accuracy can be substantial. Hayashi and co-workers used SVM along with PCA in their MIP arrays for classifying aldehydes[29,32] and acids.[49] Significant overlap in the PCA plots suggested that discrimination could not be achieved using linear boundaries. However, a kernel based on binary class datasets was used to train the SVM, which achieved an average recognition accuracy of 74% for the mixtures and single analytes.[29]

Like SVM, artificial neural network (ANN) is another non-linear classification method that is often effective when other linear methods fail. The down-side of this technique is that it is complex to parameterize and it requires a nuanced understanding of the method by the user.

14.4 Literature Studies

Selected examples from literature were chosen to demonstrate the efficacy of the approach and to highlight key design choices when developing an MIP chemosensor array. In particular, the MIP composition and morphology along with signaling mechanism, analyte identity, and pattern recognition analysis will be discussed.

14.4.1 Example 1: Imprinted Photonic Polymer Array for Detection of Polybrominated Flame Retardants

Li and coworkers[34] developed a molecularly imprinted photonic polymer (MIPP) sensor array for the detection of polybrominated diphenyl ethers (PBDs). PBDs are a class of commonly used flame retardants in consumer products and electronics, which can have adverse health and environmental impacts. One challenge in imprinting PBDs is the absence of functionality that can form strong non-covalent interactions to common functional monomers. To address this problem, the authors used three phenolic analogues of PBDs that can form hydrogen bonds and thus could be effectively imprinted (Figure 14.5A).

Another unique aspect of this system was the photonic signaling mechanism based on an opalescent 3-D macroporous matrix (Figure 14.5B). The authors assembled three separate uniform MIPPs[50] which had a narrow absorption band in the optical spectra due to Bragg diffraction. The wavelengths of the diffraction band were very sensitive to the microsphere spacing (Figure 14.5C). Therefore, the binding of analytes to the MIPPs could be monitored very accurately as they swelled the imprinted matrix, thus changing the array dimensions. This approach has the advantage that neither the analytes nor the polymer matrix needs to be optically active or labeled to yield an optical signal. A colloidal crystal array (CCA)[51] of SiO_2 microspheres (Figure 14.5D left) was used as a sacrificial framework to construct MIP films with uniformly sized and spaced spherical cavities (Figure 14.5D right). The SiO_2 CCA was soaked with functional and cross-linking monomers, and a phenol template (BPA, BIP, or TBBPA). Thus, the MIP was formed in the gaps between the CAA microspheres. After polymerization, the SiO_2 mold was etched away using hydrofluoric acid to uncover the ordered macroporous MIPP matrix.

Analysis of the selectivity patterns for the different MIPPs demonstrated their ability to serve as highly sensitive colorimetric MIP sensors. Each MIPP showed different spectral responses to the three phenolic templates and the two non-phenolic PBD analytes (TBDE and DBDE). Although high levels of cross-reactivity were observed, three-dimensional PCA plots revealed that the chemosensor array could accurately distinguish the five analytes even at very low concentrations (10^{-8} to 10^{-12} M) (Figure 14.5E). Leave-one-out cross-validation measured a 100% classification accuracy for the five analytes. This example highlights the advantages of integrating MIPs into well-defined micro- or nano-structured materials,[52] whose collective properties can amplify existing properties and generate new properties.[53]

14.4.2 Example 2: Protein Imprinted Hydrogel Array for Electrochemical Protein Profiling

Reddy and co-workers developed a protein imprinted polymer (PIP)-based electrochemical sensor array for protein profiling.[33] This imprinted polymer chemosensor array is comprised of four PIPs templated with bovine

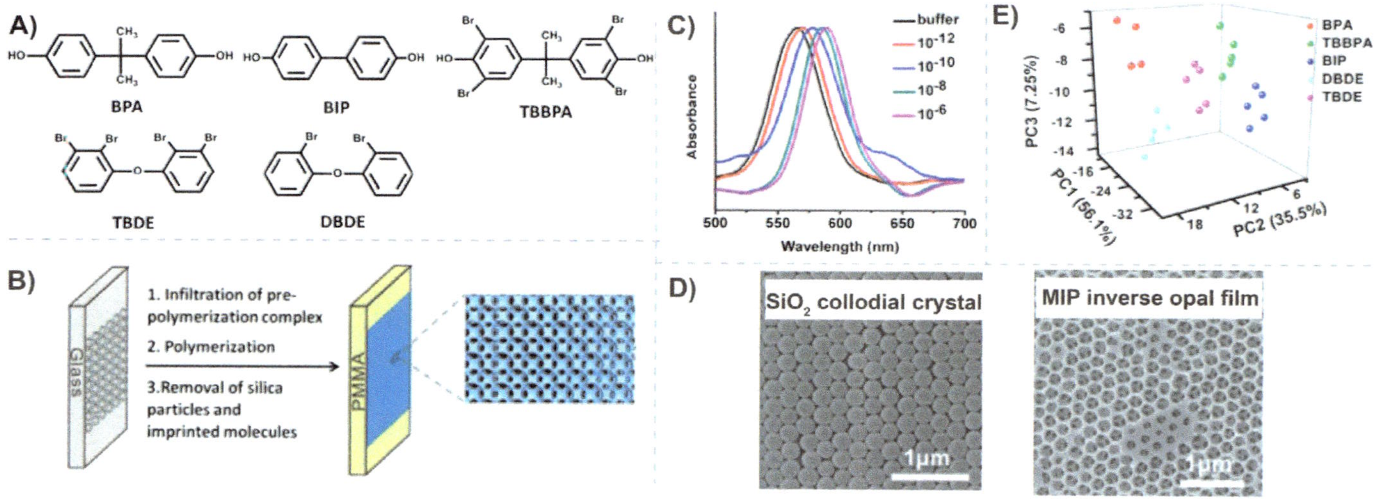

Figure 14.5 **(A)** Structural formulas of five analytes: bisphenol A (BPA), 4,4′-diphenol (BIP), tetrabromobisphenol (TBBPA), 2,2′,3,3′-tetrabromophenylether (TBDE), and 2,2′-dibromodiphenylether (DBDE). **(B)** Schematic illustration of the preparation of molecularly imprinted photonic polymers (MIPPs). **(C)** Optical response of the BIP-imprinted inverse opal photonic film for different DBDE concentrations. **(D)** SEM images of the SiO_2 colloidal crystal template and the resultant MIPP film. **(E)** The three-dimensional PCA plot for the identification of five analytes at the concentration of 10^{-10} M. PMMA – poly(methyl methacrylate). Reproduced from ref. 34 with permission from The Royal Society of Chemistry.

hemoglobin (BHb), equine myoglobin (EMb), cytochrome *C* (Cyt *C*), and bovine serum albumin (BSA). The PIP compositions were prepared using acrylamide as the functional monomer, *N*,*N*′-methylenebisacrylamide as the cross-linking monomer, and ammonium persulfate as the radical initiator. The obtained MIP hydrogels were deposited on a working glassy carbon electrode (GCE) and fixed with a covering polycarbonate membrane. The PIP layer significantly improved the cyclic voltammetric discrimination of the protein analytes. The voltammetric responses of PIP-coated electrodes against four template protein analytes were studied using hierarchical cluster analysis (HCA) and PCA.

This work highlights the challenges of imprinting protein analytes as well as the advantages of integrating imprinted arrays with electrochemical sensors. Because of their fragile nature, large size, and limited solubility in common organic solvents, the imprinting of proteins is challenging. The PIPs were prepared in phosphate buffered water at room temperature to maintain the folded structures of the protein templates. Ammonium persulfate was used as an FRP initiator. It does not require elevated temperatures or UV irradiation like more common FRP procedures using azobisisobutyronitrile (AIBN) or benzoyl peroxide initiators. To accommodate the larger size of the protein templates, the cross-linking levels (10% *w*/*w*) were much lower than those in traditional imprinting protocols, which are usually 50–80% (*w*/*w*) cross-linked. These modifications yielded PIP hydrogels with satisfactory imprinting effects (Figure 14.6A).

The protein binding to the PIP hydrogels was monitored by cyclic voltammetry. GCEs with a deposited layer of PIP hydrogels showed subtle differences in their cyclic voltammograms for the four proteins (Cyt *C*, BSA, EMb,

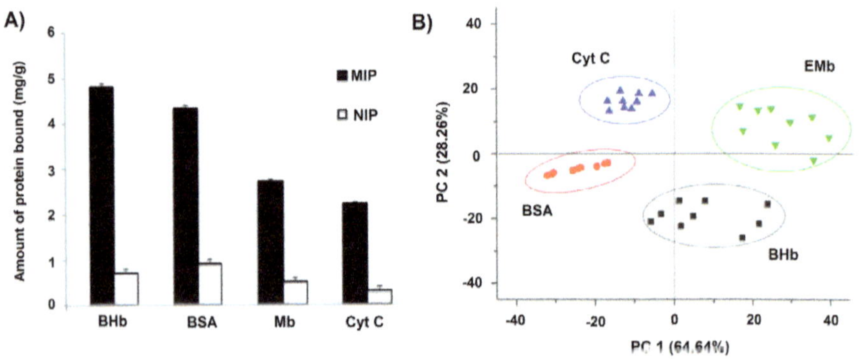

Figure 14.6 (**A**) The amount of protein bound for four MIP and one NIP hydrogels against the template protein analytes. (**B**) The two-dimensional PCA plot of the voltammetric responses of the four proteins on the MIP-coated GC electrode. BHb – bovine hemoglobin, BSA – bovine serum albumin, Emb – equine myoglobin, Cyt *C* – cytochrome *C*. Reprinted from *Sensors and Actuators B: Chemical*, **204**, L. Bueno, H. F. El-Sharif, M. O. Salles, R. D. Boehm, R. J. Narayan, T. R. L. C. Paixão and S. M. Reddy, MIP-based electrochemical protein profiling, 88–95, Copyright 2014 with permission from Elsevier.[33]

and BHb). The PCA analysis of patterns made up of key parameters (E_p, I_p, $E_{1/2}$, $I_{at\ -0.8V}$, $\Delta I_{decay\ peak\ current\ at\ -0.8V}$) for each cyclic voltammogram was able to accurately classify the four protein analytes. The data points corresponding to each analyte were well separated in the PCA plot (Figure 14.6B).

14.4.3 Example 3: MIP Array for Discrimination of Water-soluble Azo Dyes

Qiu and co-workers[31] developed a seven-channel MIP chemosensor array for the classification of water-soluble azo-dyes using a very similar method and analysis as that described in our initial 2004 report.[26] This array is comprised of one NIP (P0) and six MIPs (P1–P6) that were prepared using azo dyes, namely, amaranth (B1), tartrazine (B2), sunset yellow (B3), orange II (B4), chrysoidin (B5), and para red (B6) (Figure 14.7A). The MIPs and NIP were prepared using MAA as the functional monomer and EGDMA as the cross-linking monomer under bulk free-radical polymerization conditions. The seven-channel (P0–P6) array was used to discriminate seven dyes, *i.e.*, six template dyes (B1–B6) and another non-template rhodamine B (D1). Because of the lack of signaling elements in the MAA-EGDMA matrix, the binding events were monitored indirectly by measuring the change in the UV-vis absorption at 480 nm for the unbound dye in solution before and after equilibration with the individual polymers (Figure 14.7B). The response patterns for the 7 dyes were measured five times against the 7-channel MIP array yielding a 7 × 5 × 7 data set that was analyzed using multivariate analysis.

This work highlights the ability of the array format to transform low selectivity MIPs into an effective and accurate chemosensor or assay. The low selectivity and high cross-reactivity of the individual MIPs can be seen in the response patterns for individual dyes (Figure 14.7B). While each dye shows the strongest response to its corresponding imprinted polymer, all the dyes also show similar levels of binding to the other polymers (including the non-imprinted polymer). However, each dye produces a unique binding pattern against the 7-channel array. This was quantified using categorical data analysis (CDA) which is a multivariate analysis similar to PCA and LDA. The CDA plot of the first and second determinants showed high levels of discrimination (Figure 14.7C). The data points for the different dyes were well separated and replicate patterns for the same types of dyes were clustered tightly together.

14.4.4 Example 4: An Imprinted Titania Array for Discrimination of Small Organic Acids

Jiang and co-workers developed an imprinted titania array for discrimination of four low-molecular-weight carboxylic acids: L-malic acid (MA), succinic acid (SA), L-tartaric acid (TA), and citric acid (CA) (Figure 14.8A).[35] These acids are commonly found in fruits, and their concentrations and ratios can be used to classify fruit products, such as juices and wines. Two imprinted

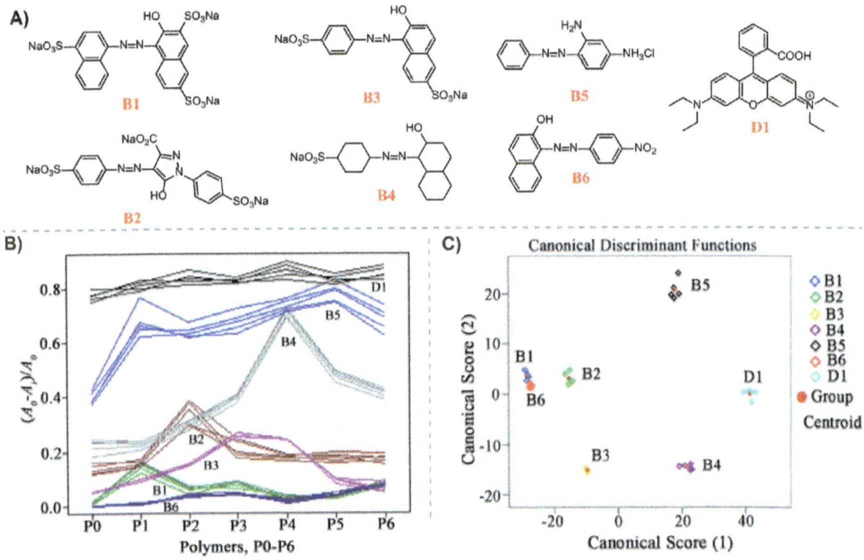

Figure 14.7 **(A)** Structural formulas of the six water-soluble azo dyes (B1–B6) and rhodamine B (D1). **(B)** The plot of the UV-vis spectral response at 480 nm of the seven analytes against the six MIPs (P1–P6) and one NIP (P0) during the rebinding experiments (five replicates). **(C)** The two-dimensional categorical data analysis (CDA) plot of the seven analytes against the seven-channel MIP chemosensor array (five replicates). Reproduced with permission from H. Qiu, Selective recognition and discrimination of water-soluble azo dyes by a seven-channel molecularly imprinted polymer sensor array, *Journal of Separation Science*, John Wiley and Sons, © 2014 Wiley-VCH Verlag GmbH & Co. KGaA.[31]

titania matrices were prepared *via* a sol–gel process. HCl, $Ti(O\textit{i}Pr)_4$, and formamide were reacted in the presence of MA or CA templates. A non-imprinted matrix was prepared under similar conditions but without the template to comprise a three-channel array. Recognizing was monitored by dye-displacement of rhodamine-B from the binding sites (Figure 14.8B). Unique absorption constants were observed for the four acid analytes (CA, MA, SA, and TA). The imprinted titania matrices non-linearly responded in the concentration ranges that these acids are present in fruit juices (Figure 14.8C). The three-channel chemosensor array was then used to categorize thirteen freshly prepared fruit juices with satisfactory performance.

This example highlights the ability to molecularly imprint inorganic matrices.[48] In this example, the imprinted titania was prepared by the sol–gel process in the presence of the templates. The CA- and MA-imprinted titania matrices showed enhanced affinity towards their corresponding acid templates in comparison with the non-imprinted titania. PCA of the spectral responses for thirteen juice samples showed that the three-channel imprinted array could effectively distinguish juice samples prepared from different fruits (Figure 14.8D). Impressively, this excellent performance

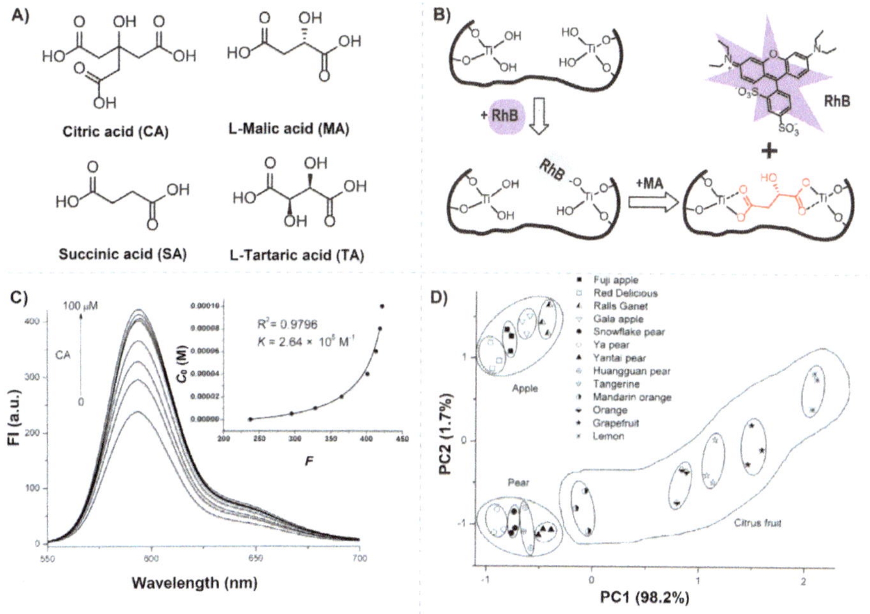

Figure 14.8 (**A**) Structural formulas of the four carboxylic acid analytes. (**B**) Schematic representation of the fluorescent dye-displacement method for monitoring the binding of acid analytes to the imprinted titania matrix. (**C**) Spectral changes in the dye-displacement study of CA-imprinted titania for different CA concentrations. Inset shows the non-linear fit between CA concentration, C_0, and fluorescence intensity, F. (**D**) The two-dimensional PCA plot of the spectral responses (three replicates) of thirteen fruit juices against the three-channel titania array (CA-, MA-imprinted, and non-imprinted). Reprinted from *Food Chemistry*, **165**, J. Tan, R. Li and Z.-T. Jiang, Discrimination of fresh fruit juices by a fluorescent sensor array for carboxylic acids based on molecularly imprinted titania, 35–41, Copyright 2014 with permission from Elsevier.[35]

required only two imprinted and one non-imprinted matrices. The high PC1 value (98.2%) can be traced to the fact that a single variable, the citric acid (CA) concentration, was the primary differentiation factor in the juice samples. This was later confirmed by the HPLC analysis of the concentrations of the four acids in different juice samples.

14.4.5 Example 5: MIP-coated QCM Chemosensor Array for Detection of Low-molecular-weight Aldehydes

Hayashi and co-workers[29] developed an MIP film-coated QCM chemosensor array for the detection of three volatile aliphatic aldehydes, *viz.*, *n*-hexanal, *n*-heptanal, and *n*-nonanal. These low-molecular-weight aldehydes are found in human body odor and, therefore, their profiling and quantitative detection

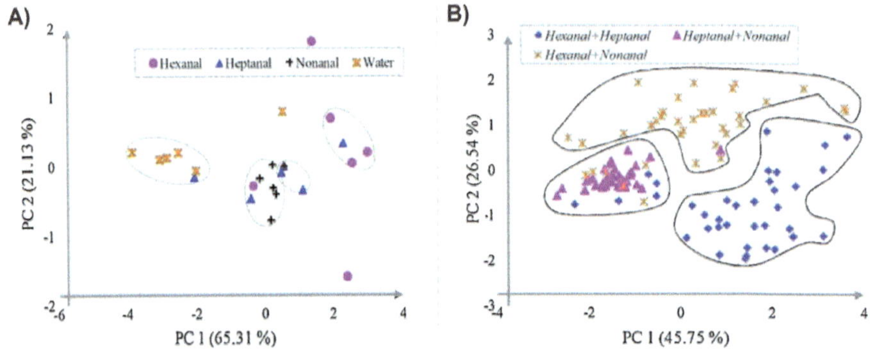

Figure 14.9 **(A)** The two-dimensional PCA plot of the frequency responses of three aldehyde analytes and water against the four coated QCM chemosensors (4 analytes × 6 concentrations × 4 sensors). **(B)** The two-dimensional PCA plot of the frequency responses of three binary aldehyde mixtures against the four coated QCM chemosensors (3 analytes × 36 concentrations × 4 sensors). Reprinted from *Sensors and Actuators B: Chemical*, **206**, S. K. Jha and K. Hayashi, Polyacrylic acid polymer and aldehydes template molecule based MIPs coated QCM sensors for detection of pattern aldehydes in body odor, 471–487, Copyright 2015 with permission from Elsevier.[29]

could lead to valuable applications in medical, biometric, and forensic sciences. Three MIP thin films were spin-coated on the surface of quartz resonators using an ethanol solution of poly(acrylic acid) (PAA) and three aldehyde templates. The binding of volatile analytes to these polymers coating on the QCM chemosensors was measured *via* shifts in resonance frequency.

This example demonstrates the ability of MIP arrays to monitor and sense chemical vapors.[54] The MIP coatings on QCM resonators yielded selective chemosensors, which could quickly detect chemical vapors without interference from moisture. Moreover, this example highlighted the power of MIP chemosensor arrays to monitor and distinguish multi-component mixtures. The four-channel QCM chemosensor array was used to detect the three template aldehydes as well as binary and ternary mixtures of the aldehydes. Statistical analyses of the frequency response patterns, using PCA and SVM methods, revealed that the MIP-coated QCM chemosensor array could discriminate the three aldehyde analytes (Figure 14.9A) as well as their binary mixtures (Figure 14.9B). Three binary mixtures of aldehydes with varying concentration ratios were successfully distinguished with a maximum recognition accuracy of 89%.

14.5 Conclusion

The chemosensor array format is a powerful tool that can help unleash the full potential of MIPs in analytical applications. Molecular imprinting is a particularly efficient and low-cost method for preparing a series of tailor-made recognition elements. Moreover, the chemosensor array format addresses one of the major impediments in MIP chemosensor devising, which is their poor overall selectivity and high cross-reactivity. The individual elements in a chemosensor array do not need to have to reveal high levels of selectivity. They only need to have different selectivity, which is readily achievable *via* molecular imprinting.

The success of the MIP array format has been demonstrated by the increasing number of successful reports in the literature over the past two decades, covering a range of analytes from small organic compounds, such as chiral amines, carboxylic acids, volatile aldehydes, to proteins. MIP chemosensor arrays can also incorporate imprinted materials beyond the traditional monolithic polymers, such as MAA- and AA-based hybrid particles. Molecularly imprinted hydrogels, nanoparticles, thin film coatings, colloidal crystal arrays, and porous organic or hybrid matrices have all been successfully adapted to the array format. The diversity of MIP morphologies is critical to the development of new MIP sensing platforms, such as colorimetric, fluorimetric and chemiluminescence assays, as well as the adaptations of mature sensing platforms, such as acoustic wave-based and electrochemical sensors. A noteworthy trend is the development of molecularly imprinted organic or inorganic microstructured matrices. These hybrid materials can improve the binding performance of MIPs due their high surface areas and endow MIPs with unique electrical and optical signal transduction mechanisms. Finally, small improvements in individual MIP selectivity often translate into dramatic collective performance improvements of the array. With these advancements, it is not unrealistic to expect MIP-based chemosensor arrays to challenge conventional analytical methods in terms of cost, limit of detection, and accuracy.

List of Abbreviations and Symbols

AA	Acrylic acid
AIBN	Azobisisobutyronitrile
Alb	Albumin
AM	Acrylamide
ANN	Artificial neural network
BHb	Bovine hemoglobin
BIP	4,4′-Diphenol
BPA	Bisphenol
BSA	Bovine serum albumin
CA	Citric acid
CCA	Colloidal crystal array
CDA	Categorical data analysis
CV	Cyclic voltammetry
Cyt *C*	Cytochrome *C*
DBDE	2,2′-Dibromodiphenylether
DMA	2-Dimethylaminoethyl methacrylate
DVB	Divinylbenzene
EGDMA	Ethylene glycol dimethacrylate

EMb	Equine myoglobin
$E_{1/2}$	Half-peak potential (in cyclic voltammetry)
E_p	Peak potential (in cyclic voltammetry)
F	Fluorescence intensity
FRP	Free-radical polymerization
GCE	Glassy carbon electrode
GEMA	Glycosyloxyethylmethacrylate
HCA	Hierarchical cluster analysis
HPLC	High performance liquid chromatography
I_p	Peak current (in cyclic voltammetry)
Lac	α-Lactalbumin
LDA	Linear discriminant analysis
MA	L-Malic acid
MAA	Methacrylic acid
MBAA	Methylene *bis*(acrylamide)
MIP	Molecularly imprinted polymer
MIPP	Molecularly imprinted photonic polymer
Myo	Myoglobin
NIP	Non-imprinted polymer
PAA	Poly(acrylic acid)
PBD	Polybrominated diphenyl ether
PC	Principal components
PCA	Principal component analysis
PIP	Protein imprinted polymers
PMMA	Poly(methyl methacylate)
QCM	Quartz crystal microbalance
Rib	Ribonuclease A
SA	Succinic acid
SAM	Self-assembled monolayer
SVM	Support vector machine
TA	L-Tartaric acid
TBBPA	Tetrabromobisphenol
TBDE	2,2′,3,3′-Tetrabromophenylether

Acknowledgement

The authors acknowledge support from NSF grant OISE 1103904.

References

1. K. Haupt and K. Mosbach, *Chem. Rev.*, 2000, **100**, 2495.
2. T. Huynh and W. Kutner, *Biosens. Bioelectron.*, 2015, **74**, 856.
3. J. Wackerlig and P. A. Lieberzeit, *Sens. Actuators, B*, 2015, **207**, 144.
4. L. Uzun and A. P. F. Turner, *Biosens. Bioelectron.*, 2016, **76**, 131.
5. G. Wulff, *Angew. Chem., Int. Ed. Engl.*, 1995, **34**, 1812.

6. L. Chen, S. Xu and J. Li, *Chem. Soc. Rev.*, 2011, **40**, 2922.
7. L. Chen, X. Wang, W. Lu, X. Wu and J. Li, *Chem. Soc. Rev.*, 2016, **45**, 2137.
8. R. J. Umpleby, S. C. Baxter, A. M. Rampey, G. T. Rushton, Y. Z. Chen and K. D. Shimizu, *J. Chromatogr. B*, 2004, **804**, 141.
9. Y. Zhang, D. Song, L. M. Lanni and K. D. Shimizu, *Macromolecules*, 2010, **43**, 6284.
10. M. Schauperl and D. W. Lewis, *J. Phys. Chem. B*, 2015, **119**, 563.
11. K. D. Shimizu and C. J. Stephenson, *Curr. Opin. Chem. Biol.*, 2010, **14**, 743.
12. K. J. Albert, N. S. Lewis, C. L. Schauer, G. A. Sotzing, S. E. Stitzel, T. P. Vaid and D. R. Walt, *Chem. Rev.*, 2000, **100**, 2595.
13. K. L. Diehl and E. V. Anslyn, *Chem. Soc. Rev.*, 2013, **42**, 8596.
14. J. R. Askim, M. Mahmoudi and K. S. Suslick, *Chem. Soc. Rev.*, 2013, **42**, 8649.
15. K. J. Johnson and S. L. Rose-Pehrsson, *Annu. Rev. Anal. Chem.*, 2015, **8**, 287.
16. M. W. Peczuh and A. D. Hamilton, *Chem. Rev.*, 2000, **100**, 2479.
17. M. H. Lee, J. S. Kim and J. L. Sessler, *Chem. Soc. Rev.*, 2015, **44**, 4185.
18. A. T. Wright and E. V. Anslyn, *Chem. Soc. Rev.*, 2006, **35**, 14.
19. P. A. Gale and C. Caltagirone, *Chem. Soc. Rev.*, 2015, **44**, 4212.
20. R. Kubota and I. Hamachi, *Chem. Soc. Rev.*, 2015, **44**, 4454.
21. W. Zhou, P.-J. J. Huang, J. Ding and J. Liu, *Analyst*, 2014, **139**, 2627.
22. R. M. Lequin, *Clin. Chem.*, 2005, **51**, 2415.
23. P. B. Luppa, L. J. Sokoll and D. W. Chan, *Clin. Chim. Acta*, 2001, **314**, 1.
24. C. D. Moore, O. Z. Ajala and H. Zhu, *Curr. Opin. Chem. Biol.*, 2016, **30**, 21.
25. T. Hirsch, H. Kettenberger, O. S. Wolfbeis and V. M. Mirsky, *Chem. Commun.*, 2003, 432.
26. N. T. Greene, S. L. Morgan and K. D. Shimizu, *Chem. Commun.*, 2004, 1172.
27. N. T. Greene and K. D. Shimizu, *J. Am. Chem. Soc.*, 2005, **127**, 5695.
28. T. Takeuchi, D. Goto and H. Shinmori, *Analyst*, 2007, **132**, 101.
29. S. K. Jha and K. Hayashi, *Sens. Actuators, B*, 2015, **206**, 471.
30. T. Kuwata, A. Uchida, E. Takano, Y. Kitayama and T. Takeuchi, *Anal. Chem.*, 2015, **87**, 11784.
31. Z. Long, Y. Lu, M. Zhang and H. Qiu, *J. Sep. Sci.*, 2014, **37**, 2764.
32. S. K. Jha and K. Hayashi, *Talanta*, 2015, **134**, 105.
33. L. Bueno, H. F. El-Sharif, M. O. Salles, R. D. Boehm, R. J. Narayan, T. R. L. C. Paixao and S. M. Reddy, *Sens. Actuators, B*, 2014, **204**, 88.
34. D. Xu, W. Zhu, C. Wang, T. Tian, J. Li, Y. Lan, G. Zhang, D. Zhang and G. Li, *Chem. Commun.*, 2014, **50**, 14133.
35. J. Tan, R. Li and Z.-T. Jiang, *Food Chem.*, 2014, **165**, 35.
36. D. Xu, W. Zhu, C. Wang, T. Tian, J. Cui, J. Li, H. Wang and G. Li, *Chem.–Eur. J.*, 2014, **20**, 16620.
37. J.-D. Lee and J.-I. Hong, *Tetrahedron Lett.*, 2013, **54**, 2890.
38. H. F. Hawari, N. M. Samsudin, A. Y. M. Shakaff, Y. Wahab, U. Hashim, A. Zakaria, S. A. Ghani and M. N. Ahmad, *Sens. Actuators, B*, 2013, **187**, 434.

39. H. Qiu, C. Luo, M. Sun, F. Lu, L. Fan and X. Li, *Anal. Chim. Acta*, 2012, **744**, 75.
40. P. Chang, Z. Zhang and C. Yang, *Anal. Chim. Acta*, 2010, **666**, 70.
41. J. Tan, H.-F. Wang and X.-P. Yan, *Anal. Chem.*, 2009, **81**, 5273.
42. J. Tan, H.-F. Wang and X.-P. Yan, *Biosens. Bioelectron.*, 2009, **24**, 3316.
43. F. L. Dickert, P. A. Lieberzeit, P. Achatz, C. Palfinger, M. Fassnauer, E. Schmid, W. Werther and G. Horner, *Analyst*, 2004, **129**, 432.
44. R. A. Lorenzo, A. M. Carro, C. Alvarez-Lorenzo and A. Concheiro, *Int. J. Mol. Sci.*, 2011, **12**, 4327.
45. A. Ellwanger, C. Berggren, S. Bayoudh, C. Crecenzi, L. Karlsson, P. K. Owens, K. Ensing, P. Cormack, D. Sherrington and B. Sellergren, *Analyst*, 2001, **126**, 784.
46. T. Zhang, F. Liu, W. Chen, J. Wang and K. Li, *Anal. Chim. Acta*, 2001, **450**, 53.
47. X. Ding and P. A. Heiden, *Macromol. Mater. Eng.*, 2014, **299**, 268.
48. M. E. Diaz-Garcia and R. B. Laino, *Microchim. Acta*, 2005, **149**, 19.
49. S. K. Jha, C. J. Liu and K. Hayashi, *Sens. Actuators, B*, 2014, **204**, 74.
50. X. Hu, Q. An, G. Li, S. Tao and J. Liu, *Angew. Chem., Int. Ed.*, 2006, **45**, 8145.
51. S.-H. Kim, S. Y. Lee, S.-M. Yang and G.-R. Yi, *NPG Asia Mater.*, 2011, **3**, 25.
52. S. Tokonami, H. Shiigi and T. Nagaoka, *Anal. Chim. Acta*, 2009, **641**, 7.
53. A. K. Yetisen, I. Naydenova, F. da Cruz Vasconcellos, J. Blyth and C. R. Lower, *Chem. Rev.*, 2014, **114**, 10654.
54. A. Afzal, N. Iqbal, A. Mujahid and R. Schirhagl, *Anal. Chim. Acta*, 2013, **787**, 36.

Subject Index